AF333952

Microarrays

Second Edition

METHODS IN MOLECULAR BIOLOGY™

John M. Walker, SERIES EDITOR

METHODS IN MOLECULAR BIOLOGY™

Microarrays

Volume 1: Synthesis Methods

SECOND EDITION

Edited by

Jang B. Rampal

Beckman Coulter, Inc.
Brea, CA

HUMANA PRESS ✳ TOTOWA, NEW JERSEY

Dedication

To my parents and gurus.

Preface

To meet the emerging needs of genomics, proteomics, and the other omics, microarrays have become unique and important tools for high-throughput analysis of biomolecules. Microarray technology provides a highly sensitive and precise technique for obtaining information from biological samples. It can simultaneously handle a large number of analytes that may be processed rapidly. Scientists are applying microarray technology to understand gene expression, to analyze mutations and single-nucleotide polymorphisms, to sequence genes, and to study antibody–antigen interactions, aptamers, carbohydrates, and cell functions, among many other research subjects.

The objective of *Microarrays* is to enable the researcher to design and fabricate arrays and binding studies with biological analytes. An additional goal is to provide the reader with a broader description of microarray technology tools and their potential applications. In this edition, *Microarrays* is divided in two parts: *Volume 1* deals with methods for preparation of microarrays, and *Volume 2* with applications and data analysis. Various methods and applications of microarrays are described and accompanied by exemplary protocols. *Volume 2* also covers topics related to bioinformatics, an important aspect of microarray technologies because of the enormous amount of data coming out of microarray experiments. Together, the two volumes provide useful information to the novice and expert alike.

From this point onward, I will discuss the contents of *Volume 1: Synthesis Methods*. For readers just entering the array technology field, as well as those who are well versed, the history of microarray technology from its conception is covered in the first chapter. Surface activation chemistries and various types of matrices involved in the synthesis of microarrays are summarized in Chapters 2 and 3. As the major objective of this volume is to provide detailed synthesis methods for constructing microarrays, so the emphasis of the remaining chapters is on methods and protocols. I tried to include various types of protocols. Some may look very similar, but in fact each protocol has a unique utility based on the research problem or individual interests. Chapter 4 details array optimization processes based on numerous factors, for example, the printing quality, spot morphology, and quantification of hybridized target. Chapter 5 presents array-based comparative genomic hybridization (array CGH) and includes procedures for making bacterial artificial chromosome

DNA arrays. Chapter 6 describes the 60-mer oligonucleotide probes immobilized on coated glass slides to study the effect of target concentration, retention, signal linearity, and properties of fluophores in quantitative gene expression measurements. The array production method using premodified DNA can be directly applied to construction of oligo or cDNA arrays. Such arrays can be used for detection of chromosomal abnormalities in complex genomes. Chapter 7 highlights the use of unique bifunctional reagents, NTMTA and NTPAC for building glass and plastic biochips. Chapter 8 explains the use of sensitive reagents for the determination of the functional group density in the microarray system by spectrophotometric methods. Chapter 9 illustrates the synthesis of high-density arrays using a digital microarray synthesis platform. The use of long optimized oligonucleotide probes (150-mers) for high and specific signal intensity for the measurement of gene expression is described in Chapter 10. In addition to sequence and probe length, the importance of other parameters, such as the surface of the glass slide, linkers/spacers, and the conditions for hybridization are also highlighted in Chapter 10. Chapter 11 deals with *in situ* synthesized oligoarrays using the Southern Array Maker (SAM) synthesizer and standard phosphoramidite chemistry. The array probes, including cystic fibrosis, were synthesized onto the flat surface of aminated polypropylene. The printing and use of the synthetic oligonucleotide probes for the detection of multiplex ligation-dependent probe amplification products is explained in Chapter 12. The synthesis and the use of grafted pyrrole oligonucleotide probes are demonstrated in Chapter 13. Chapter 14 deals with the optimization of hybridization conditions for *in situ* synthesized oligoarrays on plastic. Chapters 15 and 16 are devoted for the synthesis of peptide arrays. Creation of protein microarrays in microplate is described in Chapter 17. Arrays of the captured monoclonal antibodies corresponding to specific interleukins are printed down onto the bottom of the wells. A Biomek® 2000 workstation equipped with a high-density replicating tool is used for printing the low-density arrays. For higher density arrays, a microarrayer system (BioDot, Inc.) is employed. Printing the protein arrays onto specially polymer-coated glass slide while maintaining the activity and structure of the protein is described in Chapter 18. Chapter 19 focuses on the printing of cell microarrays for the functional exploration of genomes. Chapter 20 is related to suspension arrays. The oligo-coated microparticles are hybridized with the target molecule. The protocol for quantification of oligohybridization complex is analyzed by a europium (111) detection system. Glyco-bead array for calculating the sugar-binding lectins is described in Chapter 21. As we all are aware that array technology is moving forward from micro to nano, Chapter 22 describes this emerging technology.

The chapter highlights the utility of nanoarrays, particularly the analysis of nanoarrays by using label-free nucleic acids and proteins and others.

I believe this volume, *Synthesis Methods*, will provide valuable information to scientists at all levels, from the novice to those intimately familiar with array technology. I would like to thank all the contributing authors for providing manuscripts. My thanks are also due to colleagues for their help in completing this work. I thank John Walker for editorial guidance and the staff of Humana Press for making it possible to include large body of available microarray technologies in this volume. Finally, my thanks to my family, especially to my sweet wife Sushma Rampal, for providing all sorts of incentives to complete this project successfully.

Jang B. Rampal

Contents

CONTENTS OF THE COMPANION VOLUME

Volume 2: Applications and Data Analysis

Contributors

HAFID ABAIBOU • *Apibio, Grenoble, France*

BOGDAN V. ANTOHE • *MicroFab Technologies, Inc., Plano, TX*

IAN R. BERRY • *Regional DNA Lab, Regional Genetics Service, St. James's University Hospital, Leeds, UK*

WEI-WEN CAI • *Department of Molecular and Human Genetics, Houston, TX*

DAVID CASTEL • *CEA, DSV, DRR, Functional Genomics Department, Evry, France*

TOM S. CHAN • *Beckman Coulter, Inc., Fullerton, CA*

KAILASH CHAND GUPTA • *Nucleic Acids Research Laboratory, Institute of Genomics and Integrative Biology, Delhi University Campus, Delhi, India*

HUA CHEN • *NASA Ames Research Center, Moffett Field, CA*

HUANG-TSU CHEN • *Full Moon BioSystems, Inc., Sunnyvale, CA*

JYOTI CHOITHANI • *Nucleic Acids Research Laboratory, Institute of Genomics and Integrative Biology, Delhi University Campus, Delhi, India*

CHENG-CHUNG CHOU • *Department of Life Science and Institute of Molecular Biology, National Chung Cheng University, Chia-Yi, Taiwan, Republic of China*

CLAUS B. V. CHRISTENSEN • *Department of Micro and Nanotechnology, Technical University of Denmark, Kongens Lyngby, Denmark*

PHILIPPE CLEUZIAT • *Apibio, Grenoble, France*

PETER J. COASSIN • *Aurora BioSciences Discovery, San Diego, CA*

PATRICK W. COOLEY • *MicroFab Technologies, Inc., Plano, TX*

MICHAEL C. CRESS • *Beckman Coulter, Inc., Fullerton, CA*

MARIE-ANNE DEBILY • *CEA, DSV, DRR, Functional Genomics Department, Evry, France*

CAROL A. DELANEY • *Regional DNA Lab, Regional Genetics Service, St. James's University Hospital, Leeds, UK*

MARTIN DUFVA • *Department of Micro and Nanotechnology, Technical University of Denmark, Kongens Lyngby, Denmark*

A. K. GAIGALAS • *Biotechnology Division, National Institute of Standards and Technology, Gaithersburg, MD*

INCA GHOSH • *New England Biolabs, Beverly, MA*

XAVIER GIDROL • *CEA, DSV, DRR, Functional Genomics Department, Evry, France*

PING GONG • *Department of Chemical Engineering, Polytechnic University, Brooklyn, NY*

DAVID W. GRAINGER • *Department of Pharmaceutics and Pharmaceutical Chemistry, University of Utah, Salt Lake City, UT*

CHARLES H. GREEF • *Accelr8 Technology Corporation, Denver, CO*

ALEXANDRE HO-PUN-CHEUNG • *Centre de Recherche en Cancérologie, Centre Régional de Lutte contre le Cancer Val d'Aurelle-Paul Lamarque, Parc Euromédecine, Montpellier Cedex, France*

SEIICHIRO ITO • *Life Science Division, Hitachi Software Engineering Co. Ltd., Kanagawa, Japan*

KAISA KETOMÄKI • *Department of Chemistry, University of Turku, Turku, Finland*

SAMVEL KOCHINYAN • *New England Biolabs, Beverly, MA*

PRADEEP KUMAR • *Nucleic Acids Research Laboratory, Institute of Genomics and Integrative Biology, Delhi University Campus, Delhi, India*

JUN LI • *NASA Ames Research Center, Moffett Field, CA*

MICHAEL J. LOCHHEAD • *Accelr8 Technology Corporation, Denver, CO*

HARRI LÖNNBERG • *Department of Chemistry, University of Turku, Turku, Finland*

EVELYNE LOPEZ-CRAPEZ • *Centre de Recherche en Cancérologie, Centre Régional de Lutte contre le Cancer Val d'Aurelle-Paul Lamarque, Montpellier Cedex, France*

SHWETA MAHAJAN • *Nucleic Acids Research Laboratory, Institute of Genomics and Integrative Biology, Delhi University Campus, Delhi, India*

ROBERT S. MATSON • *Beckman Coulter, Inc., Fullerton, CA*

MARK R. McCORMICK • *NimbleGen Systems Inc., Madison, WI*

J. NOBLE • *Biotechnology Group, Quality of Life Division, National Physical Laboratory, Middlesex, United Kingdom*

RAYMOND C. MILTON • *Beckman Coulter, Inc., Fullerton, CA*

KONAN PECK • *Institute of Biomedical Sciences, Academia Sinica, Taipei, Taiwan, Republic of China*

AMANDINE PITAVAL • *CEA, DSV, DRR, Functional Genomics Department, Evry, France*

JANG B. RAMPAL • *Beckman Coulter, Inc., Brea, CA*

GOPAL REMBHOTKAR • *Nucleic Acids Research Laboratory, Institute of Genomics and Integrative Biology, Delhi University Campus, Delhi, India*

TODD A. RICHMOND • *NimbleGen Systems Inc., Madison, WI*

M. SALIT • *Biotechnology Division, National Institute of Standards and Technology, Gaithersburg, MD*

M. B. SATTERFIELD • *Biotechnology Division, National Institute of Standards and Technology, Gaithersburg, MD*

HARTMUT SELIGER • *Arbeitsgruppe Chemische Funktionen in Biosystemen, Universitat Ulm, Ulm, Germany*

REBECCA R. SELZER • *NimbleGen Systems Inc., Madison, WI*

YAXIAN SHI • *Full Moon BioSystems, Inc., Sunnyvale, CA*

LUO SUN • *New England Biolabs, Beverly, MA*

GRAHAM R. TAYLOR • *Regional DNA Lab, Regional Genetics Service, St. James's University Hospital, Leeds, UK*

BHASHYAM VAIJAYANTHI • *Department of Chemistry, Gargi College, University of Delhi, Siri Fort Road, Delhi, India*

LILI WANG • *Biotechnology Division, National Institute of Standards and Technology, Gaithersburg, MD*

MING-QUN XU • *New England Biolabs, Beverly, MA*

KAZUO YAMAMOTO • *Frontier Science Bioscience, University of Tokyo, Kashiwa Chiba, Japan*

FUMIKO YASUKAWA • *Life Science Division, Hitachi Software Engineering Co. Ltd., Kanagawa, Japan*

SHANSHAN ZHANG • *Full Moon BioSystems, Inc., Sunnyvale, CA*

YAPING ZONG • *Full Moon BioSystems, Inc., Sunnyvale, CA*

YUNFEI ZONG • *Full Moon BioSystems, Inc., Sunnyvale, CA*

1

Introduction

Array Technology—An Overview

Hartmut Seliger

Summary

Microarray technology has its roots in high-throughput parallel synthesis of biomacromolecules, combined with combinatorial science. In principle, the preparation of arrays can be performed either by *in situ* synthesis of biomacromolecules on solid substrates or by spotting of *ex situ* synthesized biomacromolecules onto the substrate surface. The application of microarrays includes spatial addressing with target (macro) molecules and screening for interactions between immobilized probe and target. The screening is simplified by the microarray format, which features a known structure of every immobilized library element. The area of nucleic acid arrays is best developed, because such arrays are allowed to follow the biosynthetic pathway from genes to proteins, and because nucleic acid hybridization is a most straightforward screening tool. Applications to genomics, transcriptomics, proteomics, and glycomics are currently in the foreground of interest; in this postgenomic phase they are allowed to gain new insights into the molecular basis of cellular processes and the development of disease.

Key Words: Applications; array technology; cell and tissue arrays; combinatorial science; DNA arrays; overview; roots; potential and problems; preparation; protein arrays; saccharide arrays.

1. Introduction

Around the turn of the century molecular biology has moved toward new frontiers. The sequencing of the human genome allows to describe a complete picture of the genetic heritage of humans and the same is true for many other organisms. This, in turn, allows medicine to become, at least partly, a molecular science. In this primary state of development, the sequencing efforts yield a host of data. The interpretation of these data will require a long time and a huge effort. As it turns out, the cells, tissues, or even whole organs are an ever more complex network of biological interactions. How long will it take—or will it ever be possible—to

From: *Methods in Molecular Biology, vol. 381: Microarrays: Second Edition: Volume 1: Synthesis Methods*
Edited by: J. B. Rampal © Humana Press Inc., Totowa, NJ

understand more than the main mechanisms, which allows the organism to respond to the diverse internal and external signals and stimuli *(1)*?

When science takes a leap forward, this is generally also connected with the development of new technologies. Molecular medicine owes much of its impetus to combinatorial science. In the area of applied combinatorial science, arrays are perhaps the most popular technology. In the early years, when combinatorial chemistry was promoted by peptide chemists, the array technology was rapidly adopted by nucleic acid chemists, who developed arraying of nucleic acid sequences as a tool for genomic research. Today, the interest turns to more distant goals: protein arrays are in focus, arrays of whole cells or of tissue samples have started to be investigated.

With each generation of new arrays one can gain insight into more and more complex biological processes and their behavior in the "normal" and "pathological" state, which means into the development of disease. Arrays of oligonucleotides are now almost in routine use for the detection of mutations or other aberrations of the genome. From libraries of cDNA, one could learn to read out and compare the transcriptome, which in turn, is related to the proteome. More information about occurrence and function of proteins is further furnished by protein libraries. The step from the genome and the proteome to the "phenome" might take a longer time. Now, scientists are also investigating on the glycome and other "-omes" might lie ahead.

Of course, array techniques are not confined to molecular biology and molecular medicine. The development of new drugs has been accelerated by orders of magnitude through libraries of organic substances *(2)* and materials science also profits from these new developments *(3)*.

1.1. Definitions

At this point it might be necessary to introduce some definitions: arrays are subspecies of combinatorial libraries. In general, a combinatorial library of substances is an arrangement of molecules of closely related and systematically varied structure. Thus, an oligonucleotide or peptide library might contain all sequences of a given length, or a defined subset of sequences. In another case, for materials' research it might be interesting to look at a macromolecular scaffold substituted with groups systematically varied according to their electronic nature and many other variations are conceivable.

The different species, which sum up to a library, can be called "elements." Thus, a library of nuclcotide triplets would contain $4^3 = 64$ elements, a complete library of tripeptides made up of 20 genetically programmed amino acids would count $20^3 = 8000$ elements. The rationale behind constructing and using libraries is the possibility of subjecting all elements or a subset of individual elements at the same time, to the same kind of reaction or interaction with an analyte molecule. The information gained in this manner might lead to the elucidation of the analyte's sequence or structure, or in other cases, may identify lead structures for pharmaceutical development, to mention just two out of many possible applications.

There are different formats for combinatorial libraries. The simplest would be a solution containing a random number of differently structured molecules, for example, mixed sequences resulting from a random oligonucleotide *(4,5)* or peptide synthesis. Such a procedure might be valuable in certain cases, for example, in the generation of "aptamers;" however, the different elements are prepared as a mixture. The elements cannot be separated, i.e., spatially defined, and, thus, cannot be individually addressed.

Libraries, in which the individual elements are bound to particles (beads), are easily prepared by the ingenious "split-couple-combine" (or "split-and-mix") procedure *(6–8)*. In such libraries, the different elements are spatially separated. Each bead contains one individual sequence. However, the locus of each sequence is not defined, as the oligonucleotide or peptide sequence bound to a given bead is, *a priori*, not known. "Deconvolution" procedures are necessary, before individual sequences/beads can be addressed.

Simplicity of application is the striking feature of arrays. Arrays are libraries, in which the different elements are bound to the surface of a "substrate," for example, a semiconductor "chip," or a plastic plate, or film. Location and structure of each element bound to the substrate is predefined. Therefore, each element can be individually addressed *(9–12)*. Arrays can be prepared by spotting and immobilization of the ready-made elements, for example, oligonucleotide or peptide sequences. This mode might require an enormous preparative effort accompanying the preparation of the "chips." The other alternative is the *"in situ"* synthesis of the desired sequences on the surface of the substrates. This, in turn, requires specific surface chemistry and a good deal of engineering *(13,14)*. Despite these difficulties arraying techniques are available, which allow to combine a very large number of elements on a limited space (up to some $10^5/cm^2$) *(15,16)*. They have the potential of easy handling, for instance, in diagnostic tests. However, the higher the "density" of spots per unit square, the more complicated is the detection and evaluation of individual signals *(17)*.

Because there appears to be some confusion in the literature, the following nomenclature is adhered: the given macromolecule (sequence), which is used to analyze an unknown macromolecule and which is, in most cases, the immobilized interaction partner, will be called the "probe" or "capture probe." The unknown interaction partner whose structure or sequence is interrogated, will be called the "target" or "analyte."

1.2. Arrays Are the Result of Interdisciplinary Efforts

Evidently, despite this seeming simplicity of application, a number of problems are connected with the preparation and application of arrays; the development of array techniques is, in turn, the fruit of an interdisciplinary effort. First of all, the design of an array requires knowledge about the reactions or interactions, which are to be performed with the different library elements. Making the arrays further involves the solution of specific preparative problems. Given the example of oligonucleotide or peptide sequences, such arrays could not be prepared without the development of massively parallel automated synthesis of such oligomers of high purity, either as accompanying technology, or as on-chip fabrication. The knowledge regarding surface science, design of highly reactive surfaces, and study of their reaction with peptide or nucleotide sequences is necessary. At this point, the chemist and biochemist must join efforts with engineers to position elements exactly and reliably on given spots and to deliver automatically either nucleotide or peptide sequences or reagents for their *in situ* synthesis. The urge for miniaturization requires the exact positioning of ever more minute quantities of often highly labile substances.

The application of array technology to molecular biology or medicine essentially requires a new type of analytical science. Crucial is the detection of reactions or interactions on an extremely small scale, essentially down to the single molecule respective the single event of interaction *(18)*. To this end, ultrasensitive detection methods have to be developed. Most of them are based on labeling techniques *(19)*, but more recently also on electrochemical methods *(20)*. The application of such methods to the screening, particularly of high-density arrays, also requires engineering and instrumentation.

As stated before, the result of such screening procedures is a host of data and their validation is the first task followed by their interpretation. Here, combinatorial science has to collaborate with information technology and biomathematics *(17)*. The final step is the interpretation, which is mainly in the hands of medical science, or pharmacology, or whatever users of the information.

1.3. Roots

Having described the interdisciplinary efforts to develop arrays, it might be adequate to retrace some of the roots of this technology. Based on the author's expertise, the discussion is confined mainly to the example of nucleic acids and their interactions, with an occasional flashlight on peptides.

Two background developments has to be mentioned, when reviewing the beginning of those lines of work, which later converged to array technology:

1. The availability of synthetic nucleic acid fragments and their immobilization to solid supports.
2. The development of nucleic acid hybridization to the point, where it could be exploited for diagnostic purposes.

Synthetic nucleic acid fragments (the term "oligonucleotide" is used for chain lengths up to 100 bases and beyond) were, in the beginning, rather exotic substances. Despite the fact that suitable protection schemes were available at an early date, the preparation of even short oligonucleotides in solution was quite laborious. The total synthesis of two genes in the laboratory of Khorana *(21)* was an undertaking that had the dimensions of "big science." Although these gene syntheses were going on and were viewed with admiration, a new and much simpler methodology was being developed. Merrifield *(22)* and Letsinger *(23,24)* developed—in parallel and independent from each other—the polymer supported synthesis of peptides and oligonucleotides. Whereas, the "solid-phase" technique was an immediate success in the peptide field, polymer-supported oligonucleotide synthesis took many years to develop into routine, and finally to the automated methods that are used today.

The hybridization properties of nucleic acids have become evident, as soon as the double-stranded nature of nucleic acids was known. Yet, the application of support-bound oligonucleotides took a longer while and was closely connected to the availability of synthetic nucleic acid fragments. In 1972, Astell and Smith first described the immobilization of an oligonucleotide conventionally prepared in solution to cellulose and the use of these columns for chromatographic separation of the complementary strand, which was subsequently released by denaturation *(25)*. Wallace et al. *(26)* were the first to show that synthetic oligonucleotides could be used as discriminating hybridization probes. In their studies they used sets of probes of mixed sequence *(27)*, which corresponds to solution-phase libraries (*see* **Subheading 1.1.**). The notable feature of these probes was that for the first time they allowed to distinguish between full match and a single mismatch.

During this time, the synthesis of oligonucleotides was still performed by the nucleoside phosphotriester method, which, within the following years, was almost completely replaced by the use of nucleoside phosphite synthons *(28)*, especially the phosphoramidite chemistry *(29)*. This new chemistry and in particular its adaptation to mechanized synthesis, was undoubtedly the breakthrough regarding the application of synthetic nucleic acid fragments *(30)*. With a rapidly growing demand for synthetic oligonucleotides, mainly for use as probes and gene fragments, improvements to make the existing methodology faster and more efficient were necessary. The need was felt to prepare more than one oligonucleotide in a single run, which was, for example, realized manually by using a battery of syringes *(31,32)*. Yet, the performance of many parallel operations was still tedious and with the risk of error. Thus, the idea of using segmental supports for multiplexed oligonucleotide synthesis *(33)* was a definite step forward. This technique was adapted to semiautomated *(34,35)* synthesis and somewhat later, a prototype of a machine for fully automated parallel preparations of multiple oligonucleotide fragments was described *(36)*.

Today the engineering of multiplexed oligonucleotide synthesis has been improved to a point, where we talk of "high-throughput" synthesis. Every year new machines are being constructed, which allow the parallel preparation of ever increasing numbers of oligonucleotides. In a recent report 1536-well microplates were used to synthesize, in parallel, a corresponding number of oligonucleotides in about 10 h *(37)*, and more record-breaking news is probably in store. The result of these efforts is a dramatic decrease of cost for standard oligonucleotide sequences. Therefore, procedures requiring large numbers of prefabricated probes or primers, for example, spotting of arrays or cDNA preparation, can now be handled with much less expenditure.

1.4. Combinatorial Chemistry: The Beginning

Combinatorial chemistry results from multiplexed synthesis with the aim of screening individual elements for a given activity or interaction. Unlike the synthesis on segmental supports described before, the products of multiplexed synthesis now are supposed to remain bound to a support particle or to a support surface for the purpose of interaction and screening. The two lines have the same roots, but combinatorial chemistry deviated in the 1980s, when sets of peptides were synthesized by a multiplex technique with the aim of establishing structure–activity relationships *(38,39)*. The seminal idea probably was born in the research group of Geysen *(40)*, who prepared sets of peptides in small quantities, but large numbers on the heads of polyacrylate pins. The pins were arranged to a kind of brush; the pinheads could be dipped into multiwell plates, where each well would contain the individual reagents for chain elongation.

A parallel development was initiated by Furka *(6–8)*: the so-called "split-couple-combine" ("split-and-mix") method. The essence of this method, which was also originally developed in the peptide field, is that before every coupling cycle the amount of polymer support is equally divided into a given number of batches. To each batch is coupled a different amino acid. After finishing this cycle the different batches are mixed and then divided again into as many portions as amino acids are to be added. This coupling in split batches, combining and splitting again for the next coupling is continued until the combinatorial synthesis is completed. The feature of this method is that, in the end, each bead contains a single individual sequence, which could be used, for instance, for studying epitopes with affinity toward monoclonal antibodies *(41)*. However, the drawback of this method is that the sequence finally attached to the bead is, *a priori*, not known. This involves an identification procedure off single beads, a problem, that was first solved in the case of the peptide libraries *(8,41)*, and only much later for the case of oligonucleotide libraries *(42,43)*. Generally, it might be because of this "deconvolution" problem that the interest in beaded libraries has by far not matched the interest in arrays, to which we will now come back in the brief survey.

2. Array Technology Picks Up Speed: Masks or No Masks

In the beginning, there was some irritation among scientists that they should use wild mixtures of millions of compounds for the identification of lead structures instead of pure and well-characterized substances. But it was quickly recognized that "irrational drug design" *(44)* would speed up the identification of drug leads by orders of magnitude. Thus, since the early 1990s a fast development ensued, resulting in a nearly exponential growth of the number of array-based publications *(47)* (in the meantime probably >10,000) and a comparable number of patents *(45)*. In view of this enormous amount of publications and a number of recent books *(13,14,46–53)*, this introductory chapter does not attempt a complete coverage of the information available, not even of parts of array technology. Rather, an attempt is made to look at the background and beginnings of array technology, whereas, the second part that describes the recent progress and future projections is to a large extent, a "review-of-reviews" with links to important and helpful summaries and compilations.

In the first moment, if one had to concentrate first on DNA arrays and, here, on technology-oriented publications, it is difficult, to find a way through the numerous intertwined lines of development. A clear distinction is between *in situ* synthesis of oligonucleotides on substrate surfaces and the deposition ("spotting") of ready-made sequences, here again, between chemosynthetic oligonucleotides and enzymatically transcribed DNA (cDNA).

For *in situ* synthesis it is perhaps possible to distinguish three main tracks: (1) the use of masks with microfluidic devices; (2) addressing through light-directed deprotection; (3) maskless technologies, and also crossover developments occurring between (1), (2), and (3). The main steps of *in situ* synthesis of arrays are described in **Fig. 1A**.

2.1. In Situ *Synthesis Involving Masks*

The seminal papers in this area are a result of the work of Southern and his colleagues *(54–59)*. Arraying was done by masks, which create a system of microfluidic channels on the surface of a glass substrate, through which the reagents are delivered from a DNA synthesizer. By dislocation of the masks a "grid" of different oligonucleotides is prepared, which, by suitable chemistry, remain bound to the glass surface. Another research group used polypropylene substrates, which could be functionalized on the surface by plasma discharge in the presence of ammonia *(60)*. A 64-channel automated delivery system for

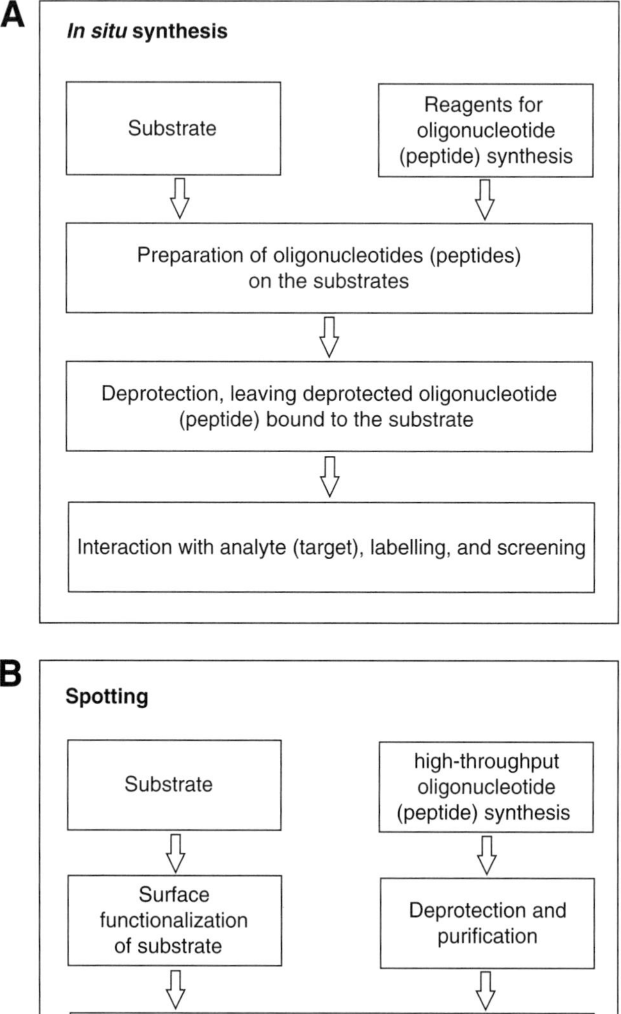

Fig. 1. Two routes to the preparation of oligonucleotide arrays.

reagents could be rotated 90° to produce a checkerboard array of 64 × 64 sequences *(61)*. Full mechanization was achieved using an "endless" polypropylene tape, functionalized by "wet chemical" oxidation + reduction, which is automatically positioned above a meander-shaped "reaction chamber," thus creating a "bar-coded" one-dimensional array of frame-shifted oligonucleotide sequences *(62,63)*.

Although the techniques described so far all use conventional phosphoramidite chemistry for oligonucleotide chain lengthening, a parallel route of development was initiated by the work of Fodor at the Affymax Research Institute. In his seminal publication *(64)*, Fodor described the use of photolabile protecting groups for the ends of the chains grown on the substrate surface. This chain lengthening method is based on a photolithographic process. Spatial addressing of given spots on the substrate is ensured by photolithographic masks. After photodeprotection all other steps of the synthesis cycle are performed simultaneously for all elements, but chain lengthening will occur only at the spots previously deprotected. This combination of solid-phase synthesis with photolithography, which was demonstrated for peptide preparations *(64)*, but rapidly adapted to oligonucleotide synthesis, is particularly featured by the possibility of generating extremely high density arrays *(65)* approx 300,000 spots cm^2 and beyond *(66)*. Crucial for the success of this method is the facile and quantitative removal of the 5′-protection, therefore, the search for new photolabile protecting groups has been greatly stimulated *(67–69)*. Despite these efforts, the yield of the light-directed deprotection is not as close to quantitative as that of the usual acid deprotection. Therefore, modifications have been developed, in which conventional acid deprotection is gated by photolithography using semiconductor photoresistent masks *(70,71)*. In an alternative method called micro wet-printing, structured membranes with integrated delivery system are pressurized onto the chip surface. This allows conventional phosphoramidite chemistry to be performed on given spots of 2 μm diameter with a coupling efficiency of >98% *(72)*.

Recently, new interest arises in arrays prepared by *in situ* synthesis of oligonucleotide analogs. Especially peptide nucleic acid (PNA) *(73)* or locked nucleic acid (LNA) *(74)* sequences are interesting as capture probes, as they show tighter and more specific binding toward complementary nucleic acids than DNA itself.

2.2. Maskless Arraying Techniques

Although the techniques using masks are now routine procedures and have reached a high degree of sophistication, they have not been found fully satisfactory. For independent addressing, masks essentially have to be restructured after each synthesis cycle. This makes the preparation of custom-synthesized arrays quite expensive. Therefore, techniques avoiding the use of masks have been promoted during recent years. Systems of computerized micromirrors have been used to direct the light for deprotection at given dots on the array, an innovation that makes the preparation of high density arrays faster and more flexible *(75,76)*. Coupling of this technique with the use of a sensitizer has considerably improved the yields of photochemical deprotection *(77)*. To avoid the undesired diffusion of reagents, particularly acid, the surface has been microstructured, thus creating small wells for each individual spot *(71)*. However, this procedure reduces the resolution, leading to a spot size around 150 μm.

2.3. Jetting or Dispensing Devices

Generally, if devices for jetting or dispensing chemicals are used for *in situ* synthesis, they have to cope with three problems, namely the evaporation of solvents, access of air and moisture, and the aggressive nature of some of the reagents used during the phosphoramidite cycle. Piezoelectric dispensing devices *(78)* solve these questions by fitting a closed "reaction chamber" over the spot to be reacted. For protection against air and moisture, while dispensing phosphoramidites or H-phosphonates, protogene has integrated the liquid reservoir, fluid system, and piezo jetting elements into closed ceramic substrate housing up to 64-jetting units *(79)*. Such precautions are not necessary, if acid deprotection is the only step handled by a dispensing unit. Combion handles detritylation by using $ZnBr_2/CH_2Br_2$/isopropanol in glass capillaries surrounded by piezo elements *(80)*. In the case of a shuttling device, the deprotection step is done by dispensing acid to predetermined spots on the array by a bubble-jet using diamond technology to avoid corrosion. For the other steps of the phosphoramidite cycle the total array is relocated over a microfluidic device, through which all reagents of the phosphoramidite cycle are delivered *(81)*. An "inverse" masking strategy deposits a cover layer of polymer over all elements, which are not included in the following chain elongation, then proceeds normally with the synthesis cycle and then washes the polymer cover away, before starting a new chain elongation *(82)*.

An important criterion, when using dispensing or jetting devices, is the drop size respective the quantity of liquid ejected with each droplet. Although the ejection of 10–100 pL is standard in many approaches *(83)*, it is possible to reduce the ejected volume to <2 pL with recent techniques *(81)*.

Quasi an "inversion" of Geysen's old strategy *(40)* is the "microcontact printing" technique, in which reagents are delivered to spots on the surface containing the growing oligonucleotide chains by silicon rubber stamps *(84–86)*.

2.4. Spotting is Different

The techniques of producing arrays by spotting of synthetic oligonucleotides or enzymatically prepared nucleic acids (cDNA) are different from *in situ* synthesis on substrate surfaces. Here, the problems of oligonucleotide synthesis are quasi "outsourced." Arrays are prepared from ready-made oligonucleotides or nucleic acids. The questions that have to be solved from the standpoint of chemistry, relate to the choice of a proper functionalized substrate and the immobilization of the (likewise functionalized) oligonucleotide chains to the activated substrate surface.

2.4.1. Noncovalent Linkage of Nucleic Acids to Surfaces

The possibility of immobilizing nucleic acids to membranes was already exploited in the early years of nucleic acids chemistry by Gillespie, who immobilized DNA noncovalently to a nitrocellulose surface *(87)*. In many cases, it is sufficient for immobilization, if the polyanionic nucleic acid chain is simply bound to a positively charged surface through electrostatic interaction *(88,89)*. Another "classical" way of noncovalent binding is through biotin–streptavidin interaction, which was introduced by Wilchek >30 yr ago *(90)*. Noncovalent adsorption to surfaces is also the rationale behind the different blotting procedures, which transfer nucleic acids from the surface of one substrate to another *(91–93)*.

2.4.2. Covalent Linkage of Nucleic Acids to Surfaces

For routine diagnostics and in view of multiple reuse a noncovalent attachment of the probe is usually not sufficient. Therefore, a multitude of methods for covalent anchoring of oligonucleotides to substrate surfaces have been developed. An overview of the different steps involved in spotting and covalent immobilization is given in **Fig. 1B**.

2.5. Substrates and Immobilization Techniques

Any spotting procedure starts with the choice of an appropriate substrate. For simplicity, glass (for instance in the form of microscopic slides) is by far the preferred substrate, but a variety of other polymers can optionally be used *(94)*. The substrate surface mostly contains hydroxyl, amino, or carboxyl groups as primary functionalities. By a secondary reaction these are converted into highly activated functional groups, such as isothiocyanate, N-hydroxysuccinimide, or tresyl chloride residues. Generally, this question of functionalization leaves a wide field for the imagination of chemists *(94)*. As a general rule, such highly activated species are more stable than in solution, when they are "isolated," i.e., prevented from diffusion, through immobilization to the surface of a polymer. For facile attachment, the preformed oligonucleotide chains are also functionalized, generally on one of their termini. The functional groups chosen for this purpose are mostly hydroxyl, amino, or thiol groups, i.e., groups different in their reactivity from the functionality of the nucleobases. An activation of the oligonucleotide termini is generally avoided, because there would be a risk of backbiting into the oligonucleotide chain. Analogous considerations can be forwarded for the spotting of peptide chains. Because the signal intensity can be increased by increasing the surface loading with oligonucleotide (or peptide) probes, attempts have been made to transform the planar into a structured surface (the term "three-dimensional surfaces" has been coined for this approach). This could be done by applying a porous coating to the glass plates, either with cationic or hydrophobic polymers *(95)*, or with dendrimers *(96)*, or with an ultrathin layer of star-shaped polyethylene glycol *(97)*. A 100-fold increase of the "density" of functional groups on the substrate surface has been reported *(95)*. An often used coating is also the application of a thin layer of sputtered gold onto the glass surface, which allows the attachment of thiol substituted oligonucleotides *(98–100)*.

An alternative to the direct immobilization of libraries of oligonucleotides or peptides to nonstructured or structured substrate surfaces is to cover the surface with a layer of oligonucleotide loaded microbeads *(101,102)*. Films of colloidal silica were applied onto flat glass surfaces *(103)*. Extensive studies on such a "three-dimensional" surface showed, that a silica particle layer of 0.5 μm thickness resulted in a 20-fold increase of hybridization signal, although about 20% of the surface area are inaccessible to probe synthesis and >30% are inaccessible to hybridization with target.

It is a small step forward from structured substrates to the introduction of microchannels, which contain the sites for oligonucleotide growth or immobilization *(104,105)*. Delivery of solutions through these microchannels speeds up washing and facilitates hybridization *(106)*. Additionally, enzymes can be immobilized inside the channels to process reactants, thus, progressing from microarray to lab-on-a-chip *(107)*.

A potential disadvantage of immobilizing oligonucleotides to substrates by chemical reactions is the necessity for removal of reactants and byproducts. This is avoided, if photochemical techniques are employed for the immobilization *(108,109)*. If this photoreaction is selective enough, it provides for a clean and convenient attachment to the substrate.

2.6. Spotting of cDNA and Other Long Nucleic Acid Fragments

Monitoring mRNA synthesis starts with the reverse transcription of mRNA into cDNA using synthetic oligonucleotide primers. In view of the small quantity of mRNA available, the cDNA synthesis is usually followed by PCR amplification. In PCR, 5′-amino-modified primers are used, which directly allow spotting of the PCR products onto activated microarrays *(110)*. Because amino-modified primers are more expensive than unmodified primers (generating unmodified cDNA), the spotting of such unmodified cDNA was investigated by several groups. Additionally, it was found necessary to increase the stability of the bound cDNA. For this purpose, a postimmobilization ultraviolet (UV) crosslinking can be applied *(111)*. Alternatively, the substitution with an aromatic amine allowed diazotization of the substrate surface. The diazonium groups could then be reacted with the nucleobases of unlabeled target DNA (the authors assume, that binding can occur to A, C, and G). This was then followed by hybridization with labeled probe *(112)*.

Judging from the literature and from the experience *(113)*, the mode of attachment of a DNA sequence to the substrate is not of utmost importance to the hybridization behavior. Generally, *in situ* synthesized probes are bound to the substrate through their 3′-end (or through the 5′-end; synthesis in reverse direction has been reported in **refs.** *114* and *115*). When spotting shorter oligonucleotides, terminal attachment is generally assumed, and has been investigated through the formation of self-assembled monolayers on gold surfaces *(98)*. In the case of longer DNA chains, especially cDNA, it is more likely, that a multipoint attachment to the substrate occurs, particularly if UV crosslinking is used for tighter binding. A comparison of hybridization behavior between 50 and 70 base oligonucleotides spotted and UV crosslinked onto aminated glass slides vs deposited PCR amplicons of cDNA did not show any significant difference in sensitivity *(116)*. A detection limit of about 10 mRNA copies per cell was reported *(117)*. It is hard to imagine, that in such a case a double-helix of the B-type could be formed on hybridization. The same question arises, for example, on spotting of DNA on nitrocellulose membranes, which is also a multipoint attachment. There is little mention in the literature regarding the structure that may be formed on double-strand formation. Lemeshko *(118)* postulates a partially unwound and significantly extended double strand, where the loss of base stacking is compensated by electrostatic interactions between DNA and substrate matrix.

3. Nucleic Acid Hybridization: Detecting the Difference

Most applications of nucleic acid arrays are based on hybridization techniques. In contrast to DNA- and RNA-hybridization in solution, which has been intensively studied, hybridization processes on surfaces are more complex and their assessment presents more difficulties. Here are some of the complications:

1. One of the interacting partners is immobilized.
2. The quantities of interacting partners are often extremely small.
3. The surface density of the immobilized probe cannot be increased indefinitely.

Furthermore, in the case of *in situ* synthesis the immobilized oligonucleotide partner is inevitably a mixture containing, in addition to probe sequence, a large number of truncated and failure sequences (for calculation of failure sequences *see* **ref.** *119*). Model studies done by Landegren *(120)* stress the necessity to prepare immobilized probes with a yield of chain elongation as high as possible.

In the case of long DNA fragments spotted either on nitrocellulose or on activated glass slides, the hybridization kinetics was found to be dependent on the concentration of the target solution, but also on the length of the arrayed DNA. The authors explain this result by the fact, that longer DNA needs less attachment sites and, thus, generates longer sequence lengths for hybridization. The nature of the substrate or the kind of surface was not found to be of importance *(121)*. In contrast, Belosludtsev *(122)* showed that conditions under which the substrate surface is slightly positively charged, results in very fast hybridization kinetics.

Another critical parameter, as previously stated, is obviously the surface density of the immobilized probe. Using comparable target concentrations, the hybridization kinetics were investigated corresponding to probe densities between 2×10^{12} and 12×10^{12} molecules of DNA/cm^2 (either single- or double strand). All samples exhibited similar hybridization kinetics; however, in the experiment done with 12×10^{12} molecules/cm^2 the hybridization efficiency reached only 10% of the efficiency found with the lower density sample (a plateau being attained already at 0.2 h) *(123)*. The authors explain this result by electrostatic repulsion between the immobilized sequences, which each require a certain space; additional thermodynamic considerations are forwarded in a subsequent article *(124)*. As a consequence, microstructured substrates were recommended for better mixing of the hybridization partners *(106)*.

Prerequisite to all hybridization-based investigations are efficient and highly sensitive detection methods. The efficiency of detection systems, in turn, is connected with different parameters, namely sensitivity, mismatch discrimination, spatial resolution, reproducibility, signal-to-noise ratio, and cost. Most detection methods rely on labeling of the target nucleic acid. After stringent hybridization and washing the presence of label on certain spots of the array indicates the formation of perfectly matched double strand. The conditions must be chosen so that duplexes with a single mismatch be excluded.

The detection technology depends largely on the type of label chosen. Fluorescent substituents are most widely used as labels. These require optical detection systems, most often confocal scanning devices. Recent technical advances are, for example, reported in **ref.** *125*. An overview of fluorescence detection methods is given in **refs.** *126* and *127*. In order to increase the signal intensity dye-loaded beads *(128)*, gold nanoparticles *(129)*, dendrimers *(130,131)* and a number of other labeling alternatives can be used. Quantitative fluorescence calibration for the analysis of microarrays has also been described *(132)*. Labeling of the target with radioisotopes provides for highly sensitive detection *(15,133–135)*; however, radiochemical methods are hazardous for routine clinical applications. Optical detection methods, which avoid the extra step of target labeling are, for instance, fluorescent staining *(136)*, surface plasmon resonance imaging *(137)*, or mass spectrometry *(138)*. Heller and colleagues have drawn attention to the fact that in the usual "passive" microarray approach all spots of capture probes are simultaneously exposed to the same hybridization conditions. The stringency of mismatch discrimination might suffer, if length, G–C content, or potential secondary structures of the probes are inadequate. Therefore, the use of an electric field is recommended to direct nucleic acid hybridization on microchips *(139)*.

Much attention has recently been drawn to the electrochemical detection of hybridization. Often it is still a label, for example, a redox-active substituent, whose change in redox potential on hybridization is measured *(140,141)*. A more elegant way would be to detect minute changes, for example, in impedance or capacitance, which occur on hybridization of unlabeled probe and target *(142,143)*. With enhanced specificity and increased sophistication of electrochemical-biosensor technology, future development might even be directed toward avoiding PCR on monitoring gene expression by cDNA spotting *(144)*.

A major concern in the application of DNA microarrays is the reproducibility and reliability of the hybridization results. Sterrenburg et al. *(145)* recommend the cohybridization of a common reference on all arrays containing spotted cDNA PCR products; a similar procedure was proposed by Hu *(146)*. A system for automatic quantification of hybridization signals from cDNA arrays has been described *(147)*. The validation of analysis results is closely connected to data handling and processing (*see* **Subheading 6.**). Chuaqui *(148)* proposes a dual strategy: in silico comparison with previous information on the one hand, and comparison with other experimental methods on the other hand (*see also* **ref. *149***).

4. Applications of DNA Chip Technology: Business Opportunities Drive Science

There might not be a reason for writing a review, or even editing a book, on array technology, if there was not such an enormous potential of applications. DNA array technology has, since long, outgrown the laboratory and entered the realm of business (40 million US $/yr is a current estimate with a 10-fold growth predicted up to the end of this decade *[150]*). Concomitantly, most of the new developments since the late 1980s have been submitted for patents; some of the resulting patent battles have been spotlighted by Gabig *(101)*. It is hardly possible in this introductory chapter to do more than highlight some of the most interesting applications, which are anyway the topic of following sections of this book. A huge volume of literature has accumulated over the last years. Most of the reviews and many articles cited in this chapter contain aspects of application in addition to novel procedures. Two recent surveys of applications, one in an article by Heller *(151)*, another in a review by McGall *(66)* are mentioned.

Array technology can be used to assess essentially the complete pathway of biological information processing. DNA arrays can interrogate the composition of genomes and find sequence aberrations (sequencing by hybridization, detection of mutations, and single nucleotide polymorphisms). By far the most effort is currently spent on the analysis of gene expression. Profiling is usually done in a differential way, i.e., by comparing the expression level of a disease-related gene to the level of expression in the healthy state. This is mostly done by two-color fluorescent labeling. An excellent introduction to microarray profiling and its potential for the study of disease was given earlier by Khan *(152)*. In the meantime, a wide range of diseases, as well as problems in medical research and related areas, is being studied by array technology. Just to quote a few very recent publications: research is being done in the field of autoimmunity *(153)*, parasitology *(154)*, toxicology *(155)*, occupational health *(156)*, microbial forensics *(157)*, and military medical science *(158)*. A wide and most interesting field of current applications is cancer research. Studies on the differential expression of cancer-related genes have yielded a much more diversified picture of cancer and might lead to the development of a defined and perhaps more individual therapy *(159)*. Printing out the "molecular portrait of cancer" *(160)* and taking advantage of progress in tumor immunology *(161)* might lead to new routes of cancer treatment *(162)*. Also in the spotlight of recent reviews is melanoma research *(163)* or genetic disorders *(164–166)*.

5. Microarrays Beyond DNA

5.1. Protein Microarrays

Although DNA arrays are extremely helpful in detecting the status of the transcriptome in a diseased against healthy state, this does not necessarily represent the status of the proteome, i.e., the ensemble of proteins resulting from the expression of a group of different genes. This is because mRNA and protein levels are often not correlated and many proteins are posttranslationally modified *(167)*. Therefore, considerable effort is now concentrated on constructing arrays, which might serve to analyze the protein status in a given sample *(168)*. The first question in the case of protein analysis concerns the choice of capturing probes. Often, specific antibodies or antibody fragments are arrayed, but also nonpeptidic "aptamers" or specific ligands can be used for immobilization *(169)*. Methods of detection and quantification include, for example, fluorescence tags *(170)* or mass-spectrometric analysis *(171)*, often parallel to detection methods for DNA arrays, for example, in the application of microfluidic devices. Peptides or proteins can likewise be used as immobilized ligands. In addition to the conventional methods of *in situ* synthesis *(172)* or spotting *(173)* specific techniques, such as electrospraying *(174)* or mass-spectrometric deposition *(175)* have been developed. "Reverse-phase protein microarrays" are often obtained by spotting representative protein fractions from cells or tissues. By interaction with disease-specific markers information is obtained, which can be used in a prognostic sense or in order to design an individual therapy *(176)*.

5.2. Saccharide Microarrays

Saccharide microarrays are the subject of very recent research activities. Many proteins, especially mammalian proteins, are substituted with oligosaccharide chains. Likewise, cell surfaces are "decorated" with glycans, mainly in the form of glycoconjugates (glycoproteins or glycolipids). These glycoconjugates have important functions within the cell: they are recognition markers for cell-to-cell communication, they promote cell adhesion, development, differentiation, and immune response *(177,178)*, and they are also involved in disease processes, such as inflammation, tumor metastasis, or bacterial infections. Even though nature uses only a limited set of sugars and their aminated analogs as monomers, the fact that these can be linked together in many different ways, can be differently substituted, and the possibility of anomerization at several chiral centers, create, in the sum, nearly unlimited possibilities of structural variation. Thus, the information content in relatively short oligoglycans might by far exceed the possibility of information storage in oligonucleotides of similar length *(179)*. Such considerations have opened the scene for a newly emerging science: glycomics. The task of glycomics is the investigation on the specific "inventory" of glycans, glycoproteins, glycolipids, and other glycoconjugates of a given organism, as well as the elucidation of genes and gene products responsible for their metabolism *(179)*.

What is now called glycomics, has long been a rather neglected field of science. Similar to nucleic acids and proteins (*see* **Subheading 1.3.**), technological advances have now paved the way *(180)*. On the one hand, modern analytical methods, like two-dimensional nuclear magnetic resonance or mass-spectrometric techniques, facilitate the elucidation of glycan structures, although purification and structural analysis still remain a formidable task. On the other hand, the construction of an automated carbohydrate synthesizer *(181)* has made it possible to prepare saccharides of given structure with a reasonable amount of time and effort *(182)*. New enzymatic methods of saccharide synthesis *(183)* are an additional preparative tool.

This, in turn, has been the starting point for combinatorial chemistry in the field of carbohydrates and glycans. Combinatorial carbohydrate research can help to elucidate all kinds of interactions of carbohydrates with carbohydrates, nucleic acids, proteins, and other biomolecules *(184)*. For these studies, a variety of specifically fluorescence-labeled saccharides are now available, in addition to the established "toolbox" of other labeled biomolecules. In addition to fluorescent labels quantum-dot conjugates, affinity and magnetic particle tagging, as well as surface plasmon resonance have been used to probe for carbohydrate interactions *(184)*.

From this basis the preparation of carbohydrate microarrays has made rapid progress during the last years. In contrast to nucleic acid and protein arrays, *in situ* synthesis is here a rather exceptional case *(185)*, but might become interesting with further development of solid-phase preparation methods. In addition to chemical methods, enzymatic transformations of carbohydrates bound to the surface of microarrays have been reported *(186)*. On the other hand, numerous methods for the immobilization of ready-made oligosaccharides and their conjugates have been developed and high-density microarrays have been constructed (for recent review *see* **ref. *177***). Among the currently available methods, noncovalent linkage of unmodified carbohydrates to, for instance, nitrocellulose-coated glass slides is the simplest *(187–189)*, but has also the disadvantage, that desorption can occur on repeated use. Tighter binding occurs with neoglycolipids (lipid-bound oligosaccharides) *(177)*, which give rise to the assumption, that this adsorption is mediated by hydrophobic interactions. However, the trend goes definitely toward covalent immobilization of carbohydrates. In order to effect stable and region-specific immobilization of glycosides (mostly at the reducing end), chemical modification is often necessary for both the substrate and the carbohydrate. A variety of substitution and immobilization methods have been developed using, for example, the attachment of maleimide-linked carbohydrates to sulfhydryl-coated glass slides *(190)*, of aminophenyl-substituted glycosides to slides coated with cyanuric chloride *(191)*, or of glycosides modified with azide groups with alkynylated surfaces *(192)*. Other alternatives are Diels–Alder *(193)* or Staudinger *(194)* ligation. Michael addition of glycosides modified with maleimide groups to thiol-derivatized glass slides also provides for stable attachment *(195)*.

Among applications of carbohydrate microarrays protein recognition is the most prominent topic. Several studies have proven the specificity of oligosaccharide microarrays for binding carbohydrate-recognizing proteins, such as antibodies *(196)* or lectins *(197)*. Lectin-carbohydrate binding affinities could be quantitatively assessed by determining IC50 values for soluble carbohydrates against the carbohydrate arrays *(198)*. By immobilizing a series of closely related structural determinants of glycans decorating the viral-surface envelope of HIV, the binding capacity of a series of gp120 binding proteins was investigated *(199)*. In view of antibiotic resistance acquired by bacteria through transfer of plasmid DNA or overexpression of endogenous enzymes modifying aminoglycoside antibiotics, the high-throughput search for new glycosides, that bind strongly to therapeutic targets, but show little affinity to resistance- and toxicity-causing proteins, is of great importance *(200–203)*.

Recent work by Seeberger has taken carbohydrate arrays one step further toward biomolecular diagnostics and therapy. Starting from the idea, that microarrays present carbohydrate ligands in a manner that mimics interactions at cell–cell interfaces, such arrays could be used to study carbohydrate-binding specificities of bacteria. Because only picomols of ligand are required for such studies and microarrays can harbor a large number of different structures, such investigations allow high-throughput screening for antiadhesion therapeutics *(204)*.

Some of the problems encountered in the preparation and use of carbohydrate microarrays have been summarized by Mellet et al. *(205)*. First of all, they recommend covalent linkage of oligosaccharides to the chip surface, because noncovalent attachment is not very stable in the case of shorter saccharides. If possible, this should be coupled with the formation of selfassembled monolayers (*see* **ref. *193***). This covalent immobilization, however, must (1) be highly efficient and (2) be done in such a way that the carbohydrate-containing macromolecules retain their recognition properties. The latter requirement is not trivial in view of the fact, that the immobilized ligands are generally very small compared with the proteins they are supposed to bind *(206)*.

Despite these potential restraints carbohydrate array technology is on a promising way. There is potential, for instance, for rapid diagnosis of infectious diseases, for detection of migrating tumor cells, for profiling of carbohydrate-binding proteins, and for identification of novel inhibitors of glycan–protein interactions *(205)*. Thus, carbohydrate arrays in the future might significantly contribute to biomolecular diagnostics and therapy.

5.3. Cell and Tissue Microarrays

Very recently, cell and tissue microarrays have come into the focus of research. Using very careful methods of deposition, small numbers of cells, or even single cells, can be spotted as library elements. This can be done by structuring the substrate surface *(207)*, by depositing cells in nanocraters *(208)* or in microchambers *(209)*, in order to prevent cross-contamination. A small number of cells (eventually a single cell *[209]*) could be immobilized within these nanostructured arrays. Standard microcontact printing methods were employed for spotting the cell suspension. Joos et al. *(210)* described another method of immobilization in which poly-L-lysine-coated glass slides was used as substrates for ionostatic fixation of the cells. After depositing the cells a hydrophobic pen was used to frame around the spots. With an average of 18.3 cells per spot 400 colonies could be arrayed. In all these cases the cells remained in a viable state after spotting. Self-assembled monolayers of cysteine-terminated oligopeptides on gold-coated substrates were used to immobilize *Escherichia coli* cells. These cells showed decrease in viability on addition of phenol, exemplifying the potential use of cell chips to detect environmental pollutants *(211)*. Some applications: in **ref. *209***, the authors describe the stimulation of deposited B cells with antigen, along with detection of their status before and after stimulation. Furthermore, the cells could be retrieved from the microchambers for analysis of antibody-specific DNA. The studies described in **ref.** *210* use antibodies to proteins displayed on the cell surface to monitor, in high-throughput format, specific binding to different cell types applying appropriate fluorescent-labeled detection molecules *(210)*. The monitoring of RNA interference-mediated loss of function was the topic of another publication *(212)*.

A considerable amount of research was triggered by a seminal paper by Ziauddin *(213)*. Their method of arraying differs from the earlier-described methods, in that they deposit cells not directly, but, in first instance, spot drops of cDNA-expressing plasmid, suspended in aqueous gelatin solution. To this is added a lipid-transfection agent, and finally the whole is overlayered with a cluster of cells. By "reverse transfection" the cells pick up the plasmid and thus, start to actively overexpress a defined gene product *(214)*. In this way, such "indirect" arraying of cells constitutes, essentially, a new route to proteomics *(215)*. Sabatini et al. *(216)* have demonstrated two potential applications of this technique: (1) screening for gene products involved in biological processes of pharmaceutical relevance, and (2) the high-throughput screening for drug leads directly in the cellular environment. Other groups have used this technique to analyze the binding affinities of membrane-displayed antibodies *(217)* and to quantitatively assess the efficiency of plasmid DNA and siRNA transfection *(218)*.

Tissue microarrays can be used for histochemical or immunohistochemical staining or for *in situ* hybridization *(219)*. In this aspect they complement DNA or protein chips in diagnostics and prognostics *(220–222)*. Several 10 or 100 specimens of tissue are arranged onto slides, constructed from paraffin-embedded or frozen tissue *(223)*. It is envisaged, that tissue microarrays might bring about a revolution in pathology and quickly become an integral part of clinical routine. However, there is a warning that, because of the small diameter, spotted samples might not be representative for the tissue as a whole *(220,223)*. This can especially be the case in heterogeneous tissues, such as tumors. In this case multiple samples must be sampled and screened, guided by experience.

6. Managing the "Data Deluge"

Being primarily a means of high-throughput diagnostics, array technologies create, first of all and above all, an enormous amount of data known as "Data Deluge" *(224)*. Especially for the processing of data from gene expression monitoring, a number of algorithms have been proposed. In addition to forwarding such algorithms for high-density oligonucleotide arrays, Zhou *(225)* provides a critical review of the literature up to 2003. Links to information processing and management are contained in the review by Khan *(152)*. Gene expression profiling coupled to DNA computing has been proposed *(226–228)*. Baldi *(229)* developed a Bayesian framework for the analysis of microarray expression data. More on the practical side are algorithms for filtering noise from data sets

(230), for increasing the reproducibility *(231)* or for selecting probes with optimized uniform hybridization properties *(232)*.

7. Array Technology: To Where Will it Proceed?

Array technology and combinatorial science in general, will certainly be one of the key technologies of this century. No wonder, that a certain amount of euphoria is expressed in many publications and reviews (e.g., **refs.** *233* and *234*). On the other hand, microarrays are not yet a common tool in the hands of non-specialists, especially in molecular biology laboratories *(12)* and in clinical routine *(152)*. So, there is still a demand for progress and improvements in many directions. New developments are also owing in the field of technology. This concerns, in particular, the quality control of the chips and the reliability and reproducibility of the results *(235)*. Utmost care should be taken in the validation of the array data, especially if these are the basis for clinical diagnostics or prognostics *(236)*. Furthermore algorithms will be necessary for this validation. Above all, progress must be made in the interpretation of the data collected from the application of arrays. New sequencing techniques promise to analyze whole genomes in a matter of months *(151,237)*. But much more time and effort is needed, in order to perfect expression profiling and to correlate expression profiles to the healthy or diseased status of the organism. A wide field of technologies and applications, awaiting further development, lies beyond arraying DNA. For instance, micro-RNA, a new genetic element, which obviously plays a role in silencing of mRNA translation and in the control of cell and tissue differentiation, is just coming into the focus of research and development *(151)*. Proteome analysis is on the agenda. Already the view goes beyond cDNA and protein arrays to arrays of tissues and cells. Novel techniques not only allow gene expression profiling in an array of spots containing small numbers each of deposited cells, or even single cells as elements *(238)*. In the longer run viable cells spotted on appropriate grids may be used as biological "labs-on-chips." From all these research efforts, much more has to be learned about the factors and parameters, which, in many ways, control gene expression. The ensemble of gene products, in turn, regulates the behavior of cells, and the crosstalk of cells the biological properties of tissues. Thus, chip technology may provide an "array of insights" *(239)* into the complex mechanisms active within biological systems *(240,241)*.

Beyond the gain of knowledge and insights for science, what is the benefit for society, where are the warnings? Medicine expects as a major benefit the development of more and more individualized therapeutical approaches *(242)*. This is expected to reduce side effects of current treatments or medication and to cut down the soaring cost of healthcare. On the other hand, ethical questions arise and will have to be solved by the world community. For instance: how can individuals escape stigmatization, if molecular diagnostics reveal the predisposition to a genetic malfunction or a severe disease *(243)*? Clearly, there is, in this controversial situation, a responsibility of science and medicine, but above all there is a demand for further research and development, since, as described by Neumaier *(243)*, "the furtherment of analytical/technical skills and diagnostic/medical knowledge, and quality management, are without doubt, a major integral element of ethical behavior or activity in the field of genetic diagnostics."

8. Some Helpful Links

Here are some links, which the reader can find in the more recent literature. Companies are listed with their activities, location and websites, status 2005, in a review by Gershon *(244)*. A list of producers of microarrays, biochips and lab-on-chips, status 2002, is given by Heller *(151)*. A list of different equipment for the construction of arrays (status 1999) can be found in an article by Bowtell *(245)*. Gwynne also gives a list of resources for microarray analysis (1999) *(246)*. Scanners for microarray analysis are compiled in *(247)*. Useful websites on applications of arrays are contained in a new review by Chaudhury *(17)* (2005) and also in two previous reviews *(152,234)*. Gerhold et al. *(234)* also list chip formats marketed by different companies. Wheeler and colleagues *(248)* have compiled a list of data resources, which can be used in addition to GenBank (2000). A link to the European Resource Center (RZPD) of the German Human Genome Project is given in an article by Lehrach *(249)*.

References

1. Cramer, F. (1979) Fundamental complexity. A concept in biological sciences and beyond. *Interdiscip. Sci. Rev.* **4,** 132–139.
2. DeWitt, S. H., Kiely, J. S., Stankovic, C. J., Schroeder, M. C., Cody, D. M. R., and Pavia, M. R. (1993) Diversomers: an approach to nonpeptide, nonoligomeric chemical diversity. *Proc. Natl. Acad. Sci. USA* **90,** 6909–6913.
3. Hsieh-Wilson, L. C., Xiang, X. -D., and Schultz, P. G. (1996) Lessons from the immune system: from catalysis to materials science. *Acc. Chem. Res.* **29,** 164–170.
4. Gillam, S., Waterman, K., and Smith, M. (1975) The base-pairing specificity of cellulose-pdT$_9$. *Nucleic Acids Res.* **2,** 625–634.
5. Dodgson, J. B. and Wells, R. D. (1977) Synthesis and thermal melting behavior of oligomer-polymer complexes containing defined lengths of mismatched dA-dG and dG-dG nucleotides. *Biochemistry* **16,** 2367–2374.
6. Furka, A., Sebestyen, F., Asgedom, M., and Dibo, G. (1988) More peptides by less labour, Abstr. *10th International Symposium of the Medical Chemist*, Budapest, pp. 288.
7. Furka, A., Sebestyen, F., Asgedom, M., and Dibo, G. (1988) Cornucopia of peptides by synthesis, in *Highlights of Modern Biochemistry,* Proceedings of the 14th International Congress of Biochemistry, VSP. *Utrecht* **5,** pp. 47.

8. Furka, A., Sebestyen, F., Asgedom, M., and Dibo, G. (1991) General method for rapid synthesis of multicomponent peptide mixtures. *Int. J. Pep. Protein Res.* **37,** 487–493.
9. Southern, E., Mir, K., and Shchepinov, M. (1999) Molecular interactions on microarrays. *Nat. Genet.* **21,** 5–9.
10. Gerhold, D., Rushmore, T., and Caskey, C. T. (1999) DNA chip: promising toys have become powerful tools. *Trends Biochem. Sci.* **24,** 168–173.
11. Ekins, R. and Chu, F. W. (1999) Microarrays: their origins and applications. *Trends Biotechnol.* **17,** 217–218.
12. Lockhart, D. J. and Winzeler, E. A. (2000) Genomics, gene expression and DNA arrays. *Nature* **405,** 827–836.
13. Schena, M. (ed.) (2002) *Microarray Analysis.* Wiley & Sons, New York.
14. Schena, M. and Davies, R. W. (2002) Genes, genomes and chips, in *DNA Microarrays: A Practical Approach,* (Schena, M., ed.), Oxford University Press, New York, pp. 1–16.
15. Lipshutz, R. J., Fodor, S. P. A., Gingeras, T. R., and Lockhart, D. J. (1999) High density synthetic oligonucleotide arrays. *Nat. Genet. Suppl.* **21,** 20–24.
16. McGall, G. H. and Christians, F. C. (2002) High-density genechip oligonucleotide probe arrays. *Adv. Biochem. Eng. Biotechnol.* **77,** 21–42.
17. Chaudhuri, J. D. (2005) Genes arrayed out for you: the amazing world of microarrays. *Med. Sci. Monit.* **11,** RA52–RA62.
18. Földes-Papp, Z., Kinjo, M., Tamura, M., Birch-Hirschfeld, E., Demel, U., and Tilz, G. P. (2005) An ultrasensitive way to circumvent PCR-based allele distinction: direct probing of genomic DNA by solution phase hybridization down to femtomolar allele concentrations and less using two-color fluorescence cross-correlation spectroscopy. *Exp. Mol. Pathol.* **78,** 177–189.
19. Epstein, J. R., Brian, I., and Walt, D. R. (2002) Fluorescence-based nucleic acid detection and microarrays. *Anal. Chim. Acta* **469,** 3–33.
20. Drummond, T. G., Hill, M. G., and Barton, J. K. (2003) Electrochemical DNA sensors. *Nat. Biotechnol.* **21,** 1192–1199.
21. Khorana, H. G. (1979) Total synthesis of a gene. *Science* **203,** 614–625.
22. Merrifield, R. B. (1963) Solid-phase peptide synthesis. I. The synthesis of a tetrapeptide. *J. Am. Chem. Soc.* **85,** 2149–2154.
23. Letsinger, R. L. and Kornet, M. J. (1963) Popcorn polymer as a support in multistep synthesis. *J. Am. Chem. Soc.* **85,** 3045–3046.
24. Letsinger, R. L. and Mahadevan, V. (1965) Oligonucleotide synthesis on a polymer support. *J. Am. Chem. Soc.* **87,** 3526–3527.
25. Astell, C. R. and Smith, M. (1972) Synthesis and properties of oligonucleotide-cellulose columns. *Biochemistry* **11,** 4114–4120.
26. Suggs, S. V., Wallace, R. B., Hirose, T., Kawashima, E. H., and Itakura, K. (1981) Use of synthetic oligonucleotides as hybridization probes: isolation of cloned cDNA sequences for human beta 2-microglobulin. *Proc. Natl. Acad. Sci. USA* **78,** 6613–6617.
27. Wallace, R. B., Johnson, M. J., Hirose, T., Miyake, T., Kawashima, E. H., and Itakura, K. (1981) The use of synthetic oligonucleotides as hybridization probes. II.

Hybridization of oligonucleotides of mixed sequence to rabbit beta-globin DNA. *Nucl. Acids Res.* **9,** 879–894.

28. Letsinger, R. L. and Lunsford, W. B. (1976) Synthesis of thymidine oligonucleotides by phosphite triester intermediates. *J. Am. Chem. Soc.* **98,** 3656–3661.
29. Caruthers, M. H. (1985) Gene synthesis machines: DNA chemistry and its uses. *Science* **230,** 281–285.
30. Itakura, K., Rossi, J. J., and Wallace, R. B. (1984) Synthesis and the use of synthetic oligonucleotides. *Annu. Rev. Biochem.* **53,** 323–356.
31. Tanaka, T. and Letsinger, R. L. (1982) Syringe method for stepwise chemical synthesis of oligonucleotides. *Nucleic Acids Res.* **10,** 3249–3260.
32. Seliger, H., Scalfi, C., and Eisenbeiss, F. (1983) An improved syringe technique for the preparation of oligonucleotides of defined sequence. *Tetrahedron Lett.* **24,** 4963–4966.
33. Frank, R., Heikens, W., Heisterberg-Moutsis, G., and Blöcker, H. (1983) A new general approach for the simultaneous chemical synthesis of large numbers of oligonucleotides: segmental solid supports. *Nucleic Acids Res.* **11,** 4365–4377.
34. Seliger, H., Herold, J., Kotschi, U., Lyons, J., Schmidt, G., and Eisenbeiss, F. (1987) Semimechanized simultaneous synthesis of multiple oligonucleotide fragments. *Nucleosides Nucleotides* **6,** 137–146.
35. Bannwarth, W. and Jaiza, P. (1986) A system for the simultaneous chemical synthesis of different DNA fragments on solid support. *DNA* **5,** 413–419.
36. Seliger, H., Kotschi, U., Lyons, J., and Singrün, B. (1989) Automatenunterstützte Simultansynthese von DNA-Fragmenten und ihre biomedizinische Anwendung. *BioEngineering* **5,** 144–147.
37. Cheng, J. -Y., Chen, H. -H., Kao, Y. -S., and Peck, K. (2002) High throughput parallel synthesis of oligonucleotides with 1536 channel synthesizer. *Nucleic Acids Res.* **30,** E93.
38. Hudson, D. (1999) Matrix assisted synthetic transformations: a mosaic of diverse contributions. I. The pattern emerges. *J. Comb. Chem.* **1,** 333–360.
39. Hudson, D. (1999) Matrix assisted synthetic transformations: a mosaic of diverse contributions. II. The pattern is completed. *J. Comb. Chem.* **1,** 403–457.
40. Geysen, H. M., Meloen, R. H., and Barteling, S. J. (1984) Use of peptide synthesis to probe viral antigens for epitopes to a resolution of a single amino acid. *Proc. Natl. Acad. Sci. USA* **81,** 3998–4002.
41. Lam, K. S., Salmon, S. E., Hersh, E. M., Hruby, V. J., Kazmierski, W. M., and Knapp, R. J. (1991) A new type of synthetic peptide library for identifying ligand-binding activity. *Nature* **354,** 82–84.
42. Rotte, B., Hinz, M., Bader, R., Astriab, A., Markiewicz, W. T., and Seliger, H. (1996) Synthetic oligonucleotide combinatorial libraries with single bead sequence identification. *Collect. Czech. Chem. Commun.* **61,** S311–S314.
43. Seliger, H., Bader, R., Hinz, M., et al. (2000) Polymer-supported nucleic acid fragments. Tools for biotechnology and biomedical research. *Reactive Functional Polymers* **43,** 325–339.

44. Brenner, S. (1991) Symposium on natural and artificial processes, Göttingen, cited from: von Kiedrowski, G. *Angew. Chem.* **103,** 831–840.
45. Recent patents in microarrays (2001) *Nat. Biotechnol.* **19,** 385.
46. Schena, M. (ed.) (2000) *Microarray Biochip & Technology.* BioTechniques/Eaton Publishing, Natick, MA.
47. Rampal, J. B. (ed.) (2001) *DNA Arrays.* Humana, Totowa, NJ.
48. Bowtell, D. and Sambrook, J. (eds.) (2002) *DNA Microarrays: A Molecular Cloning Manual.* Cold Spring Harbor Laboratory Press, Cold Spring Harbor, NY.
49. Hardimann, G. (2003) *Microarrays Methods and Applications: Nuts & Bolts.* DNA Press, Eagleville, PA.
50. Causton, H. C., Quackenbush, J., and Brazma, A. (2003) *Microarray Gene Expression Data Analysis: A Beginner's Guide.* Blackwell, Oxford.
51. Blalock, E. M. (ed.) (2003) *A Beginner's Guide to Microarray.* Kluwer Academic, Norwell, MA.
52. Simon, R. M., Edward, L., Korn, E. L., et al. (2004) *Design and Analysis of DNA Microarray Investigations.* Springer, New York.
53. Schena, M. (2004) *Protein Microarrays.* Jones & Bartless, Sudbury, MA.
54. Maskos, U. and Southern, E. M. (1992) Parallel analysis of oligodeoxyribonucleotide (oligonucleotide) interactions. I. Analysis of factors influencing oligonucleotide duplex formation. *Nucleic Acids Res.* **20,** 1675–1678.
55. Maskos, U. and Southern, E. M. (1992) Oligonucleotide hybridizations on glass supports: a novel linker for oligonucleotide synthesis and hybridization properties of oligonucleotides synthesised in situ. *Nucleic Acids Res.* **20,** 1679–1684.
56. Southern, E. M., Maskos, U., and Elder, J. K. (1992) Analyzing and comparing nucleic acid sequences by hybridization to arrays of oligonucleotides: evaluation using experimental models. *Genomics* **13,** 1008–1017.
57. Maskos, U. and Southern, E. M. (1993) A novel method for the analysis of multiple sequence variants by hybridisation to oligonucleotides. *Nucleic Acids Res.* **21,** 2267–2268.
58. Maskos, U. and Southern, E. M. (1993) A novel method for the parallel analysis of multiple mutations in multiple samples. *Nucleic Acids Res.* **21,** 2269–2270.
59. Southern, E. M., Case-Green, S. C., Elder, J. K., et al. (1994) Arrays of complementary oligonucleotides for analysing the hybridisation behaviour of nucleic acids. *Nucleic Acids Res.* **22,** 1368–1373.
60. Wehnert, M. S., Matson, R. S., Rampal, J. B., Coassin, P. J., and Caskey, C. T. (1994) A rapid scanning strip for tri- and dinucleotide short tandem repeats. *Nucleic Acids Res.* **22,** 1701–1704.
61. Matson, R. S., Rampal, J., Pentoney, S. L. Jr., Anderson, P. D., and Coassin, P. (1995) Biopolymer synthesis on polypropylene supports: oligonucleotide arrays. *Anal. Biochem.* **224,** 110–116.
62. Seliger, H., Bader, R., Birch-Hirschfeld, E., et al. (1995) Surface reactive polymers for special applications in nucleic acid synthesis. *Reactive Functional Polymers* **26,** 119–126.

63. Bader, R., Betz, O., Brugger, H., et al. (1998) An automated method to create a one dimensional array of chemically synthesized oligonucleotides, in *Microreaction Technology,* (Ehrfeld, H., ed.), Springer, Berlin, pp. 120–123.

64. Fodor, S. P. A., Read, J. L., Pirrung, C., Stryer, L., Lu, A. T., and Solas, D. (1991) Light-directed spatially addressable parallel chemical synthesis. *Science* **251,** 767–773.

65. Lipshutz, R. J., Fodor, S. P., Gingeras, T. R., and Lockhart, D. J. (1999) High density synthetic oligonucleotide arrays. *Nat. Genet.* **21,** 20–24.

66. McGall, G. H. and Christians, F. C. (2002) High-density genechip oligonucleotide probe arrays. *Adv. Biochem. Eng. Biotechnol.* **77,** 21–42.

67. Pirrung, M. C. and Bradley, J. -C. (1995) Comparison of methods for photochemical phosphoramidite-based DNA synthesis. *J. Org. Chem.* **60,** 6270–6276.

68. Beier, M. and Hoheisel, J. D. (1999) New developments in light-controlled synthesis of DNA-arrays. *Nucleosides Nucleotides* **18,** 1301–1304.

69. Walbert, S., Pfleiderer, W., and Steiner, U. E. (2001) Photolabile protecting groups for nucleosides: mechanistic studies of the 2-(2-nitrophenyl)ethyl group. *Helv. Chim. Acta* **84,** 1601–1611.

70. Beecher, J. E., McGall, G. H., and Goldberg, M. J. (1997) Chemically amplified photolithography for the fabrication of high density oligonucleotide arrays. *Polymeric Mater. Sci. Eng.* **76,** 597–598.

71. Gao, X., LeProust, E., Zhang, H., et al. (2001) A flexible light-directed DNA chip synthesis gated by deprotection using solution photogenerated acids. *Nucleic Acids Res.* **29,** 4744–4750.

72. Ermantraut, E., Schulz, T., Tuchscheerer, J., et al. (1998) Fluorescence array biosensor—biochemistry, in *Micro Total Analysis Systems 1998: Proceedings of the Utas '98 Workshop, Banff, 1998* (Harrison, D. J. and van den Berg, A., eds.), Kluwer Academic Publisher, Dordrecht, pp. 217–221.

73. Weiler, J., Gausepohl, H., Hauser, N., Jensen, O. N., and Hoheisel, J. D. (1997) Hybridisation-based DNA screening on peptide nucleic acid (PNA) oligonuclotide arrays. *Nucleic Acids Res.* **25,** 2792–2799.

74. Smith, C. (2005) Genomics: getting down to details. *Nature* **435,** 991–994.

75. Singh-Gasson, S., Green, R. D., Yue, Y., et al. (1999) Maskless fabrication of light-directed oligonucleotide microarrays using a digital micromirror array. *Nat. Biotechnol.* **17,** 974–978.

76. Nuwaysir, E. F., Huang, W., Albert, T. J., et al. (2002) Gene expression analysis using oligonucleotide arrays produced by maskless photolithography. *Genome Res.* **12,** 1749–1755.

77. Wöll, D., Walbert, S., Stengele, K. -P., et al. (2002) *More efficient photolithographic synthesis of DNA-chips by photosensitization,* Poster at EuroBiochips, Berlin. 2002, and at International Round Table: Nucleosides, Nucleotides Nucleic Acids, Leuven, 2002.

78. Blanchard, A. P. (1998) Synthetic DNA arrays, in *Genetic Engineering* **20** (Setlow, J. K., ed.) Plenum Press, New York, pp. 111–123.

79. Butler, J. H., Cronin, M., Anderson, K. M., et al. (2001) In situ synthesis of oligonucleotide arrays by using surface tension. *J. Am. Chem. Soc.* **123,** 8887–8894.

80. Theriault, T. P., Winder, S. C., and Gamble, R. C. (1999) Application of ink-jet printing technology to the manufacture of molecular arrays, in *DNA Microarrays: A Practical Approach,* (Schena, M., ed.), Oxford University Press, New York, pp. 101–119.

81. Adamschik, M., Hinz, M., Maier, C., et al. (2001) Diamond micro system for biochemistry, *Diamond Related Mater.* **10,** 722–730.

82. Vinet, F., Hoang, A., Mittler, F., and Rosilio, C. (2000) A new strategy for *in situ* synthesis of 3(10) oligonucleotide arrays for DNA chip technology, in *Proceedings, Congress on Microelectronics, Microsystems and Nanotechnology, (MMN 2000)* (Nassiopoulou, A. G. and Ziaumi, X., eds.) World Scientific, Athens, Greece, pp. 3–12.

83. Hughes, T. R., Mao, M., Jones, A. R., et al. (2001) Expression profiling using microarrays fabricated by an ink-jet oligonucleotide synthesizer. *Nat. Biotechnol.* **19,** 342–347.

84. Xia, Y. N. and Whitesides, G. M. (1998) Soft lithography. *Angew. Chem. Int. Ed. Engl.* **37,** 550–575.

85. Ermantraut, J., Wölfl, S., and Saluz, H. -P. (1997) Herstellung einer Matrix-gebundenen miniaturisierten kombinatorischen Poly- and Oligomerbibliothek, *Ger. Patent DE 19543232 A1.*

86. Xiao, P. F., He, N. Y., Liu, Z. C., He, Q. G., Sun, X., and Lu, Z. H. (2002) In situ synthesis of oligonucleotide arrays by using soft lithography. *Nanotechnology* **13,** 756–762.

87. Gillespie, D. and Spiegelman, S. (1965) Quantitative assay for DNA-RNA hybrids with DNA immobilized on a membrane. *J. Mol. Biol.* **12,** 829–842.

88. Belosludtsev, Y., Iverson, B., Lemeshko, S., et al. (2001) DNA microarrays based on noncovalent oligonucleotide attachment and hybridization in two dimensions. *Anal. Biochem.* **292,** 250–256.

89. Dequaire, M. and Heller, A. (2002) Amperometric detection of nucleic acids in 25 µL droplets on screen printed electrodes. *Anal. Chem.* **74,** 4370–4377.

90. Avidin-Biotin Technology. (1990) *Meth. Enzymol.* **184,** (Wilchek, M. and Bayer, E. A., eds.) Academic Press, Inc. San Diego, CA.

91. Southern, E. M. (1975) Detection of specific sequences among DNA fragments separated by gel electrophoresis. *J. Mol. Biol.* **98,** 503–517.

92. Alwine, J. C., Kemp, D. J., and Stark, G. R. (1977) Method for detection of specific RNAs in agarose gels by transfer to diazobenzyloxymethyl-paper and hybridization with DNA probes. *Proc. Natl. Acad. Sci. USA* **74,** 5350–5354.

93. Kafatos, F. C., Jones, C. W., and Efstratiadis, A. (1979) Determination of nucleic acid sequence homologies and relative concentrations by a dot hybridization procedure. *Nucleic Acids Res.* **7,** 1541–1552.

94. Seliger, H., Hinz, M., and Happ, E. (2003) Arrays of immobilized oligonucleotides—contributions to nucleic acids technology. *Curr. Pharm. Biotechnol.* **4,** 379–395.

95. Kusnezow, W. and Hoheisel, J. D. (2003) Solid support for microarray immuno-assays. *J. Mol. Recognit.* **16,** 165–176.
96. Benters, R., Niemeyer, C. M., Drutschmann, D., Blohm, D., and Wöhrle, D. (2002) DNA microarrays with PAMAM dendritic linker systems. *Nucleic Acids Res.* **30,** E10.
97. Ameringer, T., Hinz, M., Mourran, C., Seliger, H., Groll, J., and Moeller, M. (2005) Ultrathin functional star PEG coatings for DNA microarrays. *Biomacromolecules* **6,** 1819–1823.
98. Riepl, M., Enander, K., Liedberg, B., Schäferling, M., Kruschina, M., and Ortigao, F. (2002) Functionalized surfaces of mixed alkanethiols on gold as a platform for oligonucleotide microarrays. *Langmuir* **18,** 7016–7023.
99. Li, Z., Jin, R., Mirkin, C. A., and Letsinger, R. L. (2002) Multiple thiol-anchor capped DNA–gold nanoparticle conjugates. *Nucleic Acids Res.* **30,** 1558–1562.
100. Prokein, T., Hinz, M., and Seliger, H. (2004) Immobilisation of oligonucleotides to gold surfaces via chain extension with lipoic acid residues. Poster 15/I, *Oligonucleotide and peptide technology conferences (TIDES)*, Las Vegas, USA.
101. Gabig, M. and Węgrzyn, G. (2001) An introduction to DNA chips: principles, technology, applications and analysis. *Acta Biochim. Pol.* **48,** 615–622.
102. Nakauchi, G., Ohtani, Y., Inaki, Y., and Miyata, M. (2002) DNA microarray fabrication by photo-sensitive polyvinyl alcohol. *J. Photopolymer Sci. Technol.* **15,** 109–110.
103. Glazer, M., Fidanza, J., McGall, G., and Frank, C. (2001) Colloidal silica films for high-capacity DNA probe arrays. *Chem. Mater.* **13,** 4773–4782.
104. Shamansky, L. M., Davis, C. B., Stuart, J. K., and Kuhr, W. G. (2001) Immobilization and detection of DNA on microfluidic chips. *Talanta* **55,** 909–918.
105. Beier, M., Baum, M., Rebscher, H., Mauritz, R., Wixmerten, A., and Stähler, P. F. (2001) Exploring nature's plasticity with a flexible probing tool, and finding new ways for its electronic distribution. *Biochem. Soc. Trans.* **30,** 78–82.
106. Yuen, P. K., Li, G., Bao, Y., and Müller, U. R. (2003) Microfluidic devices for fluidic circulation and mixing improve hybridization signal intensity on DNA arrays. *Lab. Chip* **3,** 46–50.
107. Peterson, D. S. (2005) Solid supports for microanalytical systems. *Lab. Chip* **5,** 132–139.
108. Elghanian, R., Xu, Y., McGowen, J., et al. (2001) The use and evaluation of 2 + 2 photoaddition in immobilization of oligonucleotides on a three dimensional hydrogel matrix. *Nucleosides Nucleotides Nucleic Acids* **20,** 1371–1375.
109. Birch-Hirschfeld, E., Egerer, R., Striebel, H. M., and Stelzner, A. (2002) New ways to immobilize oligonucleotides on DNA-Chips. *Collect. Czech. Chem. Commun., Symp. Ser.* **5,** 299–303.
110. Schena, M., Shalon, D., Davis, R. W., and Brown, P. O. (1995) Quantitative monitoring of gene expression patterns with a complementary DNA microarray. *Science* **270,** 467–470.
111. Eisen, M. B. and Brown, P. O. (1999) DNA arrays for analysis of gene expression. *Methods Enzymol.* **303,** 179–205.

112. Dolan, P. L., Wu, Y., Ista, L. K., Metzenberg, R. L., Nelson, M. A., and Lopez, G. P. (2001) Robust and efficient synthetic method for forming DNA microarrays. *Nucleic Acids Res.* **29,** E107.

113. Prokein, T. (2004) Dithiolanderivate zur Immobilisierung von Nukleinsäuren an Goldoberflächen, *Thesis, Univ. Ulm*, Germany.

114. Beier, M. and Hoheisel, J. D. (2002) Analysis of DNA-microarrays produced by inverse in situ oligonucleotide synthesis. *J. Biotechnol.* **94,** 15–22.

115. Albert, T. J., Norton, J., Ott, M., et al. (2003) Light-directed 5′—>3′ synthesis of complex oligonucleotide microarrays. *Nucleic Acids Res.* **31,** E35.

116. Wang, H. -Y., Malek, R. L., Kwitek, A. E., et al. (2003) Assessing unmodified 70mer oligonucleotide probe performance on glass-slide microarrays. *Genome Biol.* **4,** R5/1–R5/6.

117. Kane, M. D., Jatkoe, T. A., Stumpf, C. R., Lu, J., Thomas, J. D., and Madore, S. J. (2000) Assessment of the sensitivity and specificity of oligonucleotide (50 mer) microarrays. *Nucleic Acids Res.* **28,** 4552–4557.

118. Lemeshko, S. V., Powdrill, T., Belosludtsev, Y. Y., and Hogan, M. (2001) Oligonucleotides form a duplex with non-helical properties on a positively charged surface. *Nucleic Acids Res.* **29,** 3051–3058.

119. Földes-Papp, Z., Baumann, G., Birch-Hirschfeld, E., et al. (1998) Quantitative analysis of oligonucleotide preparations by fractal measures. *Biopolymers* **45,** 361–379.

120. Jobs, M., Fredriksson, S., Brookes, A. J., and Landegren, U. (2002) Effect of oligonucleotide truncation on single-nucleotide distinction by solid-phase hybridization. *Anal. Chem.* **74,** 199–202.

121. Stillman, B. A. and Tonkinson, J. L. (2001) Expression microarray hybridization kinetics depends on length of the immobilized DNA but are independent of immobilization substrate. *Anal. Biochem.* **295,** 149–157.

122. Belosludtsev, Y., Belosludtsev, I., Iverson, B., et al. (2001) Nearly instantaneous, cation-independent, high selectivity nucleic acid hybridization to DNA microarrays. *Biochem. Biophys. Res. Comm.* **282,** 1263–1267.

123. Peterson, A. W., Heaton, R. J., and Georgiadis, R. M. (2001) The effect of surface probe on DNA hybridization. *Nucleic Acids Res.* **29,** 5763–5768.

124. Vainrub, A. and Pettitt, B. M. (2002) Coulomb blockage of hybridization in two-dimensional DNA arrays. *Phys. Rev.* **66,** 041905/1–041905/4.

125. Dixon, A. E. and Damaskinos, S. (2001) Confocal scanning of genetic microarrays. *Meth. Mol. Biol.* **170,** 237–246.

126. Epstein, J. R., Biran, I., and Walt, D. R. (2002) Fluorescence-based nucleic acid detection and microarrays. *Anal. Chim. Acta* **469,** 3–36.

127. Frutos, A. G., Pal, S., Quesada, M., and Lahiri, J. (2002) Method for detection of single-base mismatches using bimolecular beacons. *J. Am. Chem. Soc.* **124,** 2396–2397.

128. Csáki, A., Maubach, G., Born, D., Reichert, J., and Fritzsche, W. (2002) DNA-based molecular nanotechnology. *Single Mol.* **3,** 275–280.

129. Taton, T. A., Mirkin, C. A., and Letsinger, R. L. (2000) Scanometric DNA array detection with nanoparticle probes. *Science* **289,** 1757–1760.

130. Shchepinov, M. S., Udalova, I. A., Bridgman, A. J., and Southern, E. M. (1997) Oligonucleotide dendrimers: synthesis and use as polylabelled DNA probes. *Nucleic Acids Res.* **25,** 4447–4454.

131. Stears, R. L., Getts, R. C., and Gullans, S. R. (2000) A novel, sensitive detection system for high-density microarrays using dendrimer technology. *Physiol. Gen.* **3,** 93–99.

132. Marti, G. E., Gaigalas, A., and Vogt, R. F. (2000) Recent developments in quantitative fluorescence calibration for analyzing cells and microarrays. *Cytometry* **42,** 263.

133. Cheung, V., Morley, M., Aguilar, F., Massini, A., Kucherlapati, R., and Childs, G. (1999) Making and reading microarrays. *Nat. Genet.* **21,** 15–19.

134. Mac Beath, G. and Schreiber, S. L. (2000) Printing proteins as microarrays for high-throughput function determination. *Science* **289,** 1760–1762.

135. Salin, H., Vujasinovic, T., Mazurie, A., et al. (2002) A novel sensitive microarray approach for differential screening using probes labeled with two different radioelements. *Nucleic Acids Res.* **30,** E17.

136. Battaglia, C., Salani, G., Consolandi, C., Bernardi, L. R., and De Bellis, G. (2000) Analysis of DNA microarrays by non-destructive fluorescent staining using SYBR green II. *BioTechniques* **29,** 78–81.

137. Nelson, B. P., Grimsrud, T. E., Liles, M. R., Goodman, R. M., and Corn, R. M. (2001) Surface plasmon resonance imaging measurements of DNA and RNA hybridization adsorption onto DNA microarrays. *Anal. Chem.* **73,** 1–7.

138. Stomakhin, A. A., Vasiliskov, V. A., Timofeev, E., Schulga, D., Cotter, R. J., and Mirzabekov, A. D. (2000) DNA sequence analysis by hybridization with oligonucleotide microchips: MALDI mass spectrometry identification of 5 mers contiguously stacked to microchip oligonucleotides *Nucleic Acids Res.* **28,** 1193–1198.

139. Edman, C. F., Raymond, D. E., Wu, D. J., et al. (1997) Electric field directed nucleic acid hybridization on microchips. *Nucleic Acids Res.* **25,** 4907–4914.

140. Thewes, R., Hofmann, F., Frey, B., et al. (2002) Sensor arrays for fully-electronic DNA detection on CMOS, in *Proceedings of International Solid-State Circuits Conference (ISSCC),* IEEE International 2002, pp. 350–351.

141. Drummond, T. G., Hill, M. G., and Barton, J. K. (2003) Electrochemical DNA sensors. *Nat. Biotechnol.* **21,** 1192–1193.

142. Lassalle, N., Roget, A., Livache, T., Mailley, P., and Vieil, E. (2001) Electropolymerisable pyrrole-oligonucleotide: synthesis and analysis of ODN by fluorescence and QCM. *Talanta* **55,** 993–1004.

143. Marquette, C. A., Lawrence, I., Polychronakos, C., and Lawrence, M. F. (2002) Impedance based DNA chip for direct T_m measurement. *Talanta* **56,** 763–768.

144. Kerman, K., Kobayashi, M., and Tamiya, E. (2004) Recent trends in electrochemical DNA biosensor technology. *Meas. Sci. Technol.* **15,** R1–R11.

145. Sterrenburg, E., Turk, R., Boer, J. M., van Ommen, G. B., and den Dunnen, J. T. (2002) A common reference for cDNA microarray hybridizations. *Nucleic Acids Res.* **30,** E116.

146. Hu, L., Cogdell, D., Jia, Y., Hamilton, S. R., and Zhang, W. (2002) Monitoring of microarray production with a common oligonucleotide and specificity with selected targets. *BioTechniques* **32,** 528–534.
147. Tahi, F., Achddou, B., Decraene, C., et al. (2002) Automatic quantitation of hybridization signals on cDNA arrays. *BioTechniques* **32,** 1386–1397.
148. Chuaqui, R. F., Bonner, R. F., Best, C. J., et al. (2002) Post-analysis follow-up and validation of microarray experiments. *Nat. Genet.* **32,** 509–514.
149. Wang, Y., Wang, X., Guo, S. W., and Ghosh, S. (2002) Conditions to ensure competitive hybridization in two-color microarray: the theoretical and experimental analysis. *BioTechniques* **32,** 1342–1346.
150. Sinibaldi, R., O'Connell, C., Seidel, C., and Rodriguez, H. (2001) Gene expression analysis on medium-density oligonucleotide arrays. *Meth. Mol. Biol.* **170,** 211–222.
151. Heller, M. J. (2002) DNA microarray technology: devices, systems, and applications. *Annu. Rev. Biomed. Eng.* **4,** 129–153.
152. Khan, J., Bittner, M. L., Chen, Y., Meltzer, P. S., and Trent, J. M. (1999) DNA microarray technology: the anticipated impact on the study of human disease. *Biochim. Biophys. Acta* **1423,** M17–M28.
153. Fathman, C. G., Soares, L., Chan, S. M., and Utz, P. J. (2005) An array of possibilities for the study of autoimmunity. *Nature* **435,** 605–611.
154. Boothroyd, J. C., Blader, I., Cleary, M., and Singh, U. (2003) DNA microarrays in parasitology: strengths and limitations. *Trends Parasitol.* **19,** 470–476.
155. Shioda, T. (2004) Application of DNA microarray to toxicological research. *J. Environ. Pathol. Toxicol. Oncol.* **23,** 13–31.
156. Koizumi, S. (2004) Application of DNA microarrays in occupational health research. *J. Occup. Health* **46,** 20–25.
157. Pattnaik, P. and Jana, A. M. (2005) Microbial forensics: application in bioterrorism. *Environ. Forensics* **6,** 197–204.
158. Draghici, S., Chen, D., and Reifman, J. (2004) Applications and challenges of DNA microarray technology in military medical research. *Mil. Med.* **169,** 654–659.
159. Marx, J. (2000) DNA microarrays reveal cancer in its many forms. *Science* **289,** 1670–1672.
160. Mocellin, S., Provenzano, M., Rossi, C. R., Pilati, P., Nitti, D., and Lise, M. (2005) DNA array-based gene profiling: from surgical specimen to the molecular portrait of cancer. *Ann. Surg.* **241,** 16–26.
161. Mocellin, S., Wang, E., Panelli, M., Pilati, P., and Marincola, F. M. (2004) DNA array-based gene profiling in tumor immunology. *Clin. Cancer Res.* **10,** 4597–4606.
162. Wadlow, R. and Ramaswamy, S. (2005) DNA microarrays in clinical cancer research. *Curr. Mol. Med.* **5,** 111–120.
163. Nambiar, S., Mirmohammadsadegh, A., Bär, A., Bardenheuer, W., Roeder, G., and Hengge, U. R. (2004) Applications of array technology: melanoma research and diagnosis. *Summary Expert Rev. Mol. Diagn.* **4,** 549–557.
164. Ooslander, A. E., Meijer, G. A., and Ylstra, B. (2004) Microarray-based comparative genomic hybridization and its application in human genetics. *Clin. Genet.* **66,** 488–495.

165. Tejedor, D., Castillo, S., Mozas, P., et al. (2005) Reliable low-density DNA array based on allele-specific probes for detection of 118 mutations causing familial hypercholesterolemia. *Clin. Chem.* **51,** 1137–1144.
166. Snijders, A. M., Pinkel, D., and Albertson, D. G. (2003) Current status and future prospects of array-based comparative genomics by hybridisation. *Briefings Funct. Genomics Proteomics* **2,** 37–45.
167. Mann, M. (1999) Quantitative proteomics? *Nat. Biotechnol.* **17,** 954–955.
168. Kodadek, T. (2001) Protein microarrays: prospects and problems. *Chem. Biol.* **8,** 105–115.
169. Stoll, D., Templin, M. F., Schrenk, M., Traub, P. C., Vöhringer, C. F., and Joos, T. O. (2002) Protein microarray technology. *Front. Biosci.* **7,** C13–C32.
170. Haab, B. B., Dunham, M. J., and Brown, P. O. (2001) Protein microarrays for highly parallel detection and quantitation of specific proteins and antibodies in complex solutions. *Genome Biol.* **2,** research 0004.1–0004.12.
171. Gygi, S. P., Rist, B., Gerber, S. A., Turecek, F., Gelb, M. H., and Aebersold, R. (1999) Quantitative analysis of complex protein mixtures using isotope-coded affinity tags. *Nat. Biotechnol.* **17,** 994–999.
172. Frank, R. (1992) Spot-synthesis: an easy technique for the positionally addressable, parallel chemical synthesis on a membrane support. *Tetrahedron* **48,** 9217–9232.
173. MacBeath, G. and Schreiber, S. L. (2000) Printing proteins as microarrays for high-throughput function determination. *Science* **289,** 1760–1763.
174. Moerman, R., Frank, J., Marijnissen, J. C. M., Schalkhammer, T. G. M., and van Dedem, G. W. K. (2001) Miniaturized electrospraying as a technique for the production of microarrays of reproducible micrometer-sized protein spots. *Anal. Chem.* **73,** 2183–2189.
175. Ouyang, Z., Takats, Z., Blake, T. A., et al. (2003) Preparing protein microarrays by soft-landing of mass-selected ions. *Science* **301,** 1351–1354.
176. Paweletz, C. P., Charbonneau, L., Bichsel, V. E., et al. (2001) Reverse phase protein microarrays which capture disease progression show activation of pro-survival pathways at the cancer invasion front. *Oncogene* **20,** 1981–1989.
177. Shin, I., Park, S., and Lee, M. R. (2005) Carbohydrate microarrays: an advanced technology for functional studies of glycans. *Chem. Eur. J.* **11,** 2894–2901.
178. Wang, D. (2003) Carbohydrate microarrays. *Proteomics* **3,** 2167–2175.
179. Zähringer, H. (2005) Glykomik: Vom lästigen Anhängsel zum Molekül-Star. *Labor J.* 18–21.
180. Werz, D. B. and Seeberger, P. H. (2005) Carbohydrates as the next frontier in pharmaceutical research. *Chem. Eur. J.* **11,** 3194–3206.
181. Plante, O. J., Palmacci, E. R., and Seeberger, P. H. (2001) Automated solid-phase synthesis of oligosaccharides. *Science* **291,** 1523–1527.
182. Love, K. H. and Seeberger, P. H. (2002) Carbohydrate arrays as tools for glycomics. *Angew. Chem. Int. Ed. Engl.* **41,** 3583–3586.
183. Chen, X., Liu, Z. -Y., Zhang, J. B., Zheng, W., Kowal, P., and Wang, P. G. (2002) Reassembled biosynthetic pathway for large-scale carbohydrate synthesis: α-gal epitope producing "superbug". *Chembiochem* **3,** 47–53.

184. Ratner, D. M., Adams, E. W., Disney, M. D., and Seeberger, P. H. (2004) Tools for glycomics: mapping interactions of carbohydrates in biological systems. *Chembiochem* **5,** 1375–1383.

185. Khraltsova, L. S., Sablina, M. A., Melikhova, T. D., et al. (2000) An enzyme-linked lectin assay for alpha1, 3-galactosyltransferase. *Anal. Biochem.* **280,** 250–257.

186. Park, S., Lee, M., Pyo, S., and Shin, I. (2004) Carbohydrate chips for studying high-throughput carbohydrate-protein interactions. *J. Am. Chem. Soc.* **126,** 4812–4819.

187. Wang, P., Liu, S., Trummer, B. J., Deng, C., and Wang, A. (2002) Carbohydrate microarrays for the recognition of cross-reactive molecular markers of microbes and host cells. *Nat. Biotechnol.* **20,** 275–281.

188. Willats, W. G., Rasmussen, S. E., Kristensen, T., Mikkelsen, J. D., and Knox, J. P. (2002) Sugar-coated microarrays; a novel slide surface for the high-throughput analysis of glycans. *Proteomics* **2,** 1666–1671.

189. Lundquist, J. J. and Toone, E. (2002) The cluster glycoside effect. *Chem. Rev.* **102,** 555–578.

190. Ratner, D. M., Adams, E. W., Su, J., O'Keefe, B. R., Mrksich, M., and Seeberger, P. H. (2004) Probing protein-carbohydrate interactions with microarrays of synthetic oligosaccharides. *Chembiochem* **5,** 379–383.

191. Schwarz, M., Spector, L., Gargir, A., et al. (2003) A new kind of carbohydrate array, its use for the profiling of antiglycan antibodies, and the discovery of a novel human cellulose-binding antibody. *Glycobiology* **13,** 749–754.

192. Fazio, F., Bryan, M. C., Blixt, U., Paulson, J. C., and Wong, C. H. (2002) Synthesis of sugar arrays in microtiter plate. *J. Am. Chem. Soc.* **124,** 14,397–14,402.

193. Houseman, B. T. and Mrksich, M. (2002) Carbohydrate arrays for the evaluation of protein-binding and enzymatic modification. *Chem. Biol.* **4,** 443–454.

194. Köhn, M., Wacker, R., Peters, C., et al. (2003) Staudinger ligation: a new immobilization strategy for the preparation of small-molecule arrays. *Angew. Chem. Int. Ed. Engl.* **42,** 5829–5834.

195. Park, S. and Shin, I. (2002) Fabrication of carbohydrate chips for studying protein-carbohydrate interactions. *Angew. Chem. Int. Ed. Engl.* **41,** 3180–3182.

196. Fukui, S., Feizi, T., Galustian, C., Lawson, A. M., and Chai, W. (2002) Oligosaccharide microarrays for high-throughput detection and specificity assignments of carbohydrate-protein interaction. *Nat. Biotechnol.* **20,** 1011–1017.

197. Feizi, T., Fazio, F., Chai, W., and Wong, C. H. (2003) Carbohydrate microarrays—a new set of technologies at the frontiers of glycomics. *Curr. Opin. Struct. Biol.* **13,** 637–645.

198. Shin, I., Cho, J. W., and Boo, D. W. (2004) Carbohydrate arrays for functional studies of carbohydrates. *Combin. Chem. and High Throughput Screen.* **7,** 565–574.

199. Adams, E. W., Ratner, D. M., Bokesch, H. R., McMahon, J. B., O'Keefe, B. R., and Seeberger, P. H. (2004) Oligosaccharide and glycoprotein microarrays as tools in HIV glycobiology: glycan-dependent gp120/protein interactions. *Chem. Biol.* **11,** 739–740.

200. Fourmy, D., Recht, M. I., Blanchard, S. C., and Puglisi, J. D. (1996) Structure of the A site of *Escherichia coli* 16S ribosomal RNA complexed with an aminoglycoside antibiotic. *Science* **274,** 1367–1371.
201. Griffey, R. H., Hofstadler, S. A., Sannes-Lowery, K. A., Ecker, D. J., and Crooke, S. T. (1999) Determinants of aminoglycoside-binding specificity for rRNA by using mass spectrometry. *Proc. Natl. Acad. Sci. USA* **96,** 10,129–10,133.
202. Magnet, S., Lambert, T., Courvalin, P., and Blanchard, J. S. (2001) Kinetic and mutagenic characterization of the chromosomally encoded Salmonella enterica AAC(6′)-Iy aminoglycoside-N-acetyltransferase. *Biochemistry* **40,** 3700–3709.
203. Hedge, S. S., Javid-Majd, F., and Blanchard, J. S. (2001) Overexpression and mechanistic analysis of chromosomally encoded aminoglycoside 2′-N-acetyltransferase (AAC(2′)-Ie) from *Mycobacterium tuberculosis. J. Biol. Chem.* **276,** 45,876–45,881.
204. Disney, M. D. and Seeberger, P. H. (2004) The use of carbohydrate microarrays to study carbohydrate-cell interactions and to detect pathogens. *Chem. Biol.* **11,** 1701–1707.
205. Mellet, C. O. and Fernández, J. M. G. (2002) Carbohydrate microarrays. *Chembiochem.* **3,** 819–822.
206. Horan, N., Yan, L., Isobe, H., Whitesides, G. M., and Kahne, D. (1999) Nonstatistical binding of a protein to clustered carbohydrates. *Proc. Natl. Acad. Sci. USA* **96,** 11,782–11,786.
207. Dupnik, K. (2004) Support materials for cell arrays. *Thesis, Univ. Ulm.* Germany.
208. Xu, C. W. (2002) High-density cell microarrays for parallel functional determinations. *Genome Res.* **12,** 482–486.
209. Yamamura, S., Kishi, H., Tokimitsu, Y., et al. (2005) Single-cell microarray for analyzing cellular response. *Anal. Chem., ASAP Article,* A–G.
210. Schwenk, J. M., Stoll, D., Templin, M. F., and Joos, T. O. (2002) Cell microarrays: an emerging technology for the characterization of antibodies, *Biotechniques* **33,** S54–S61.
211. Choi, J. W., Park, K. W., Lee, D. B., Lee, W., and Lee, W. H. (2005) Cell immobilization using self-assembled synthetic oligopeptide and its application to biological toxicity detection using surface plasmon resonance. *Biosens. Bioelectron.* **20,** 2300–2305.
212. Wheeler, D. B., Bailey, S. N., Guertin, D. A., Carpenter, A. E., Higgins, C. O., and Sabatini, D. M. (2004) RNAi living-cell microarrays for loss-of-function screens in Drosophila melanogaster cells. *Nat. Methods* **1,** 127–132.
213. Ziauddin, J. and Sabatini, D. M. (2002) Microarrays of cells expressing defined cDNA's. *Nature* **411,** 107–110.
214. Wu, R. Z., Bailey, S. N., and Sabatini, D. M. (2002) Cell-biological applications of transfected cell microarrays. *Trends Cell. Biol.* **12,** 485–488.
215. Blagoev, B. and Pandey, A. (2001) Microarrays go live—new prospects for proteomics. *Trends Biochem. Sci.* **26,** 639–641.
216. Bailey, S. N., Wu, R. Z., and Sabatini, D. M. (2002) Applications of transfected cell microarrays in drug discovery. *Drug Discov. Today, High-throughput Technol. Suppl.* S113–S118.

217. Delehanty, J. B., Shaffer, K. M., and Lin, B. (2004) Transfected cell microarrays for the expression of membrane-displayed single-chain antibodies. *Anal. Chem.* **76,** 7323–7328.
218. Baghdoyan, S., Roupioz, Y., Pitaval, A., et al. (2004) Quantitative analysis of highly parallel transfection in cell microarrays. *Nucleic Acids Res.* **32,** e77.
219. Packeisen, J., Korsching, H., Herbst, H., Böcker, W., and Bürger, H. (2003) Demystified tissue microarray technology. *J. Clin. Pathol./Mol. Pathol.* **56,** 198–204.
220. Braunschweig, T., Chung, J. Y., and Hewitt, S. M. (2004) Perspectives in tissue microarrays. *Comb. Chem. High Throughput Screen.* **7,** 575–585.
221. Shergill, I. S., Shergill, N. K., Arya, M., and Patel, H. R. H. (2004) Tissue microarrays: a current medical research tool. *Curr. Med. Res. Opin.* **20,** 707–712.
222. Jacquemier, J., Ginestier, C., Charafe-Jauffret, E., et al. (2003) Small but high throughput: how "tissue-microarrays" became a favorite tool for pathologists and scientists. *Ann. Pathol.* **23,** 623–632.
223. Watanabe, A., Cornelison, R., and Hostetter, G. (2005) Tissue microarrays: application in genomic research. *Expert Rev. Molec. Diagn.* **5,** 171–181.
224. Rao, J. S. and Bond, M. (2001) Microarrays: managing the data deluge. *Circ. Res.* **88,** 1226–1227.
225. Zhou, Y. and Abagyan, R. (2003) Algorithmus for high-density oligonucleotide array. *Curr. Opin. Drug Discov. Dev.* **6,** 339–345.
226. Suyama, A., Nishida, N., Kurata, K., and Omagari, K. (2000) Gene expression analysis by DNA computing, in *Computational Molecular Biology,* (Miyano, S., Shamir, R. and Takagi, T., eds.), Universal Academy Press, Tokyo, pp. 12–13.
227. Nishida, N., Wakui, M., Tokunaga, K., and Suyama, A. (2001) Highly specific and quantitative gene expression profiling based on DNA computing. *Genome Inf. Series* **12,** 259–260.
228. Sakakibara, Y. and Suyama, A. (2000) Intelligent DNA chips: logical operation of gene expression profiles on DNA computers, *Genome Inf.* **11,** 33–42.
229. Baldi, P. and Long, A. D. (2001) A Bayesian framework for the analysis of microarray expression data: regularized *t*-test and statistical inferences of gene changes. *Bioinformatics* **17,** 509–519.
230. Mills, J. C. and Gordon, J. I. (2001) A new approach for filtering noise from high-density oligonucleotide microarrays datasets. *Nucleic Acids Res.* **29,** E72.
231. Yang, I. V., Chen, E., Hasseman, J. P., et al. (2002) Within the fold: assessing differential expression measures and reproducibility in microarray assays. *Genome Biol.* **3,** research 0062.10062.12.
232. Relógio, A., Schwager, C., Richter, A., Ansorge, W., and Valcárcel, J. (2002) Optimization of oligonucleotide-based DNA microarrays. *Nucleic Acids Res.* **30,** E51.
233. Chittur, S. V. (2004) DNA microarrays: tools for the 21st century. *Combin. Chem. High Throughput Screen.* **7,** 531–537.
234. Gerhold, D., Rushmore, T., and Caskey, C. T. (1999) DNA chips: promising toys have become powerful tools. *Trends Biochem. Sci.* **24,** 168–173.

235. Paul, H. (2001) Qualitätskriterien bei der Herstellung von DNA-Mikroarrays. *Transcript Laborwelt* **3**, 12–16.
236. Simon, R., Radmacher, M. D., Dobbin, K., and McShane, L. M. (2003) Pitfalls in the use of DNA microarray data for diagnostic and prognostic classification. *J. Natl. Cancer Inst.* **95**, 14–18.
237. Ziebolz, B. (2005) Genom-sequenzierung—schnell und kostengünstig. *Bio. Tech.* **9–10**, 32–35.
238. Mills, J. C., Roth, K. A., Cagan, R. L., and Gordon, J. I. (2001) DNA microarrays and beyond: completing the journey from tissue to cell. *Nat. Cell Biol.* **3**, E175–E178.
239. Panda, S., Sato, T. K., Hampton, G. M., and Hogenesch, J. B. (2003) An array of insights: application of DNA chip technology in the study of cell biology. *Trends Cell Biol.* **13**, 151–156.
240. Ng, J. H. and Ilag, L. L. (2003) Biochips beyond DNA: technologies and applications. *Biotechnol. Annu. Rev.* **9**, 1–149.
241. Moch, H., Kononen, T., Kallionemi, O. P., and Sauter, G. (2001) Tissue microarrays: what will they bring to molecular and anatomic pathology? *Adv. Anat. Pathol.* **8**, 14–20.
242. Forthcoming (2007) The Second World Congress on Gender-Specific Medicine and Aging: The Endocrine Impact, Rome.
243. Neumaier, M. and Funke, H. (2005) Ethik und Qualitätsmanagement genetischer Untersuchungen. *Klinikarzt* **34**, 66–70.
244. Gershon, D. (2005) More than gene expression. *Nature* **437**, 1195–1200.
245. Bowtell, D. D. L. (1999) Options available from start to finish—for obtaining expression data by microarray. *Nat. Genet. Suppl.* **21**, 25–32.
246. Gwynne, P. and Page, G. (1999) Microarray analysis: the next revolution in molecular biology. *Science* **288**, 911, 914, 918, 922, 926, 932, 938.
247. Marktübersicht Microarray-Reader (2002). *Transcript Laborwelt* **III/2002**, Verlag der Biocom A. G., Berlin 32–34.
248. Wheeler, D. L., Chappey, C., Lash, A. E., et al. (2000) Database resources of the National Center for Biotechnology Information. *Nucleic Acids Res.* **28**, 10–14.
249. Vente, A., Korn, B., Zehetner, G., Poustka, A., and Lehrach, H. (1999) Distribution and early development of microarray technology in Europe. *Nat. Genet.* **22**, 22.

2

Current Microarray Surface Chemistries

David W. Grainger, Charles H. Greef, Ping Gong, and Michael J. Lochhead

Summary

In almost all microarray technologies that are currently used, some type of surface chemistry serves as the interface between immobilized biomolecules and the solid support. Factors such as probe loading, spot morphology, and signal-to-noise ratio are all intimately linked to surface chemistry. Surface chemistry also significantly impacts important performance parameters such as three-dimensional structure of the immobilized biomolecules and nonspecific assay backgrounds. Here, an overview of the major types of surface chemistries currently used in printed microarrays is provided, with an emphasis on standard glass slide formats. The first part of this chapter focuses on DNA array surface chemistries, including both commercially fabricated and custom-made arrays. The second part of the chapter focuses on emerging protein, peptide, and carbohydrate array techniques. The intent is to provide the molecular biology researcher and bioanalytical or diagnostic specialist with a guide to the surface chemistry state-of-the-art for established and emerging array technologies.

Key Words: Bioassay; carbohydrate array; diagnostics; DNA array; glass slide; hydrogel; microarray; nonspecific binding; peptide array; polymer coating; protein array; silane; surface chemistry; surface immobilization.

1. Introduction

Microarrays have emerged over the last decade as one of the most important new tools in molecular biology. In addition to their now well-established role in gene expression analysis, microarray technologies have been expanded to include a range of biomolecules in a variety of assay formats. As is the case with any solid-phase assay, the nature of the immobilization surface plays a critical role in microarray function. Most commonly used microarray technologies are based on planar substrates, typically glass, with dozens to tens of thousands of individual features arrayed on the surface. Therefore, microarray surface chemistries are

From: *Methods in Molecular Biology, vol. 381: Microarrays: Second Edition: Volume 1: Synthesis Methods*
Edited by: J. B. Rampal © Humana Press Inc., Totowa, NJ

directed at converting a planar solid surface into an assay and biologically friendly interface for assessing complex biological interactions.

Microarray surface chemistry has been the subject of several comprehensive reviews. For a historical perspective on the field, readers are referred to the Southern chapter in the first edition of this book *(1)*. General reviews of microarray surface chemistry can be found in **refs.** *2* and *3*. Topical reviews have targeted specific areas such as protein arrays *(4–7)*, peptide arrays *(8)*, and carbohydrate arrays *(9)*.

Given progress in the field over the last several years, individual researchers are now less likely to build microarrays entirely from the beginning, instead relying on commercially manufactured microarray slides for self-spotting applications or even moving to manufactured, complete array platforms entirely. This is particularly true for DNA arrays, where from a surface chemistry perspective, most of the activity has consolidated to a relatively narrow, well-defined set of surface technologies. However, this is not true for emerging protein, peptide, and carbohydrate array formats that are both less mature and technically more challenging than DNA arrays. Consequently, diverse surface chemistry strategies are used in the protein, peptide, and carbohydrate, with new approaches continuously appearing in the literature, attempting to improve assay performance for these formats.

2. Surface Chemistry and Microarray Performance

Surface chemistry directly affects several important mechanistic aspects of microarraying, including probe loading, spot morphology, backgrounds, and hybridization yields. **Figure 1** illustrates several of the processing parameters and nonidealities of microarraying that are directly impacted by surface chemistry.

Probe loading refers the amount of probe immobilized in an array spot. Optimal loading can increase assay dynamic range and also improve the capability to detect low copy-number gene products. *Significantly, optimal loading is not necessarily the same as maximum loading.* Experimental and theoretical studies have shown that excess immobilized DNA can produce electrostatic repulsion between surface-immobilized probe and target during hybridization, therefore, lowering hybridization efficiency *(10–12)*. Reducing probe density on surfaces eventually compromises on assay sensitivity. Hence, an optimum probe immobilization process is sought that eliminates both of these problems. Therefore, the goal of an engineered surface chemistry is to provide high signal, large dynamic range, and high hybridization efficiency.

Probe-spot morphology in all printed microarray formats is critically dependent on surface chemistry. Whether contact or noncontact printing is used, spot morphology and the efficiency of any surface-probe coupling reactions both depend on an intimate balance between the probe-drop delivery

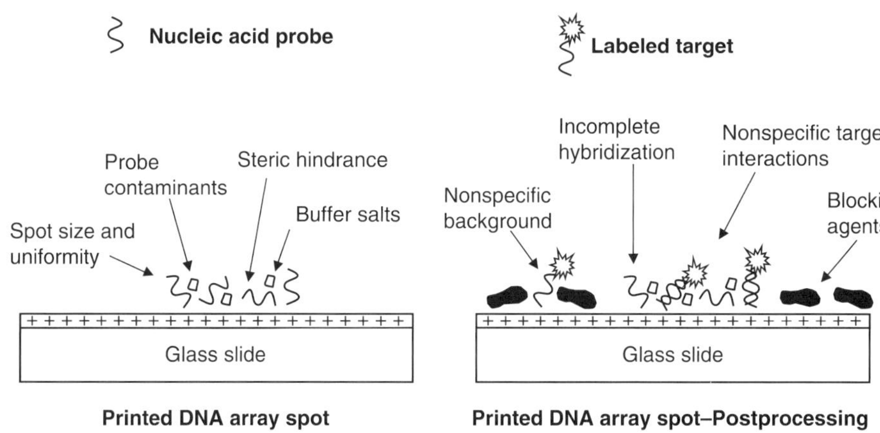

Fig. 1. Processing parameters and nonidealities of microarraying directly related to surface chemistry; issues surrounding the reliable, reproducible printing of microarrayed probes (left) and target hybridization (right). The surface chemistry, immobilized probes, blocking agents, targets, and contaminants together consist of a complex interfacial environment within which accurate and reproducible hybridization must occur.

system, spotting solution properties (surface tension, viscosity, ionic strength), and the surface energy of the slide chemistry. For this reason, every commonly used slide surface chemistry has an optimized spotting protocol that matches print buffer and conditions to the specific surface chemistry characteristics. Diverging from the optimized printing can dramatically lower the quality of printed arrays. Many core labs have invested substantial effort to establish well-developed protocols (usually empirically based) for a given slide type or vendor, which makes these users less likely to try alternative surface chemistries. Many commercial slide vendors sell optimized reagent buffers as part of a microarraying slide kit.

A general comment on probe-spot morphology is that surface chemistries that are relatively hydrophobic (apolar) yield uniform, small, reliably sized spot morphologies, a highly desired property for assay. Surface hydrophobicity reduces printed spot size, allowing higher density arrays, reduced drying artifacts, and more uniform spots. Slide surface chemistry vendors will often advertize "high water-contact angle" (equated to low surface polarity, wettability, or interfacial tension in this context) as a performance feature of their particular array surface chemistry.

Background is typically defined in terms of off-feature (off-spot) fluorescence signal that interferes with on-spot signal contrast, assay signal:noise ratios, and dynamic range and sensitivity. Two sources of off-feature background in a microarray experiment are commonly recognized and directly associated with surface properties: intrinsic fluorescence of the coated slide itself and nonspecific binding of assay components. Until recently, intrinsic surface fluorescence was an important contributor to array assay backgrounds. Now nearly all slide vendors have consolidated onto high quality, low-fluorescence glass-substrate selections, and have developed low-fluorescence surface chemistry coating formulations.

However, nonspecific binding (NSB) remains an important contributor to background in many array experiments. Nonspecific surface binding of biological components is most commonly dealt with using surface blocking strategies as part of the microarray slide processing protocol. Blocking is typically a postprint, prehybridization step in which the regions between arrayed spots are masked with a surface-active blocking agent that adsorbs irreversibly to the off-spot array features to prevent undesired adsorption. Common blocking agents include bovine serum albumin, sodium dodecyl sulfate, and salmon-sperm DNA. Proprietary blocking formulations have also been developed specifically for microarrays (e.g., Pronto! system, Promega Corp. [Madison, WI] and StabilGuard™, Surmodics, Inc. [Eden Prairie, MN]). A general disadvantage of surface-specific adsorptive strategies is that they can be difficult to perform consistently. Despite relatively well-developed products and protocols for masking, background signal variations remain a source of both intrinsic variability and frustration to the microarraying community.

Surface chemistries engineered to have low intrinsic nonspecific background without masking or blocking offer a potential solution to this blocking problem. For example, Schott Nexterion (Jena, Germany) Slide H™ is a polyethylene glycol (PEG)-based polymer coating designed to have extremely low nonspecific binding backgrounds without any adsorptive blocking requirement.

Finally, surface chemistry plays a critical role in defining the interfacial physico–chemical environment of the immobilized biomolecule (e.g., probe). As the ultimate goal of the microarray experiment is to represent a biologically relevant interaction with reliability, sensitivity, and fidelity, the microarray should ideally mimic a biological environment as closely as possible in order to preserve the native physical chemistry responsible for the capture interaction. This is particularly true for protein arrays, where preservation of native structure postprinting is a fundamental requirement for optimal assay function. The special requirements of protein arrays are discussed (*see* **Subheading 4.**). Interfacial physical chemistry is detailed in more detail in the accompanying Chapter 3.

3. DNA Microarray Surface Chemistry

Current DNA microarray technology can be divided into two major categories.

1. Commercially produced arrays from industrial vendors.
2. Custom-made arrays produced by genomic core facilities and individual research labs.

For commercially produced DNA arrays, nucleic acid probes are either synthesized directly on the solid support or by traditional methods and then printed onto the substrates. Custom-made arrays are produced within individual labs or institutional core labs in which the arraying facility selects the print technology, array substrates, and detection technologies. Nucleic acid content is typically purchased in library form from oligo-DNA vendors. A large number of core labs have evolved into relatively sophisticated operations with well-established array manufacturing and quality-control procedures. Many offer nucleic acid arrays for sale within the research community, functioning as quasi-commercial entities.

3.1. Commercial DNA Arrays

A large fraction of microarray-based gene expression research reported in the literature is based on commercially produced arrays from vendors such as Affymetrix (Santa Clara, CA), Agilent (Palo Alto, CA), and GE Healthcare (Chal Font St. Giles, UK). All commercial array vendors have array products based on propriety surface chemistries. Commercial DNA arrays can be divided into two general categories based on a specific production method.

1. On-chip oligonucleotide synthesis.
2. Printed, presynthesized oligo-DNAs.

Affymetrix, Agilent, and Nimblegen (Madison, WI) all utilize on-chip oligonucleotide synthesis to build the array. The Affymetrix GeneChip® microarrays consists of 25-mer probes synthesized directly on the solid support using a sequential, light-directed photolithography process *(13)*. The surface chemistry exploited for this strategy is a proprietary hydroxy-terminated organosilane layer on fused silica *(13,14)*. Nimblegen uses a light-directed, non-photolithographic approach based on an array of micromirrors *(15)*. The substrate is a silanized glass *(15)*. Agilent performs an *in situ* synthesis of 60-mer probes on a silanized glass slide surface using a high-fidelity inkjet process. All three major on-chip synthesis vendors use proprietary surface chemistries optimized for their particular synthesis approach. From the perspective of the end-user, the surface chemistry used by the on-chip synthesis vendors is a two-dimensional (2D) organic-based linker layer that covalently attaches the *in situ* synthesized oligonucleotide probes to the solid substrate. Significantly, this surface is also then present during the biological assay, and therefore, must serve a second important function to facilitate sufficient target capture while avoiding NSB.

The second major category of commercial microarray vendors prints presynthesized oligonucleotides onto solid substrates coated with specific organic capture chemistry. Examples of this approach include GE Healthcare (formerly Amersham) and Mergen (San Leandro, CA), both of which use the traditional glass slide substrate format with proprietary linker coatings. The GE Healthcare

CodeLink™ platform is a three-dimensional (3D), crosslinked polyacrylamide matrix on the glass slide, onto which amine-terminated oligonucleotide 30-mers are printed in a noncontact mode *(16,17)*. Mergen's arrays are also based on amine-modified 30-mers printed onto a 3D polymer-coated glass slide *(18)*.

3.2. Custom-Made DNA Microarrays

DNA microarray researchers who do not select a commercial array platform typically build custom arrays within their own lab or utilize the expertise of a local core facility to manufacture arrays. Two types of nucleic acid content are commonly used in the fabrication of these arrays: cDNA and oligonucleotide probes. The latter category can be further broken down into short (<40 bases) and long (>40 bases) oligonucleotides. The intent of this section is to highlight the major classes of surface chemistries currently used on glass supports by the custom-made DNA microarray community. **Table 1** provides a 2005 survey of vendors that supply coated-glass slides for custom array printing applications.

3.2.1. Aminosilane Surfaces

The first function of glass slide surface chemistry is to provide robust immobilization of nucleic acid probes. Given that nucleic acids are polyanions of high-polyphosphate charge density, electrostatic interactions with a positively charged surface chemistry have been the most commonly used mechanism for DNA surface immobilization. Amine-terminated organosilane-coated slides are the dominant cationic surface in practice for both cDNA and oligoarraying. Silanization of glass is a well-established and extensively reviewed technique *(19)*, wherein the native, negatively charged, silanol-terminated surface of inorganic silicate glass is derivatized with organosilanes. Despite its popularity, the exact mechanisms and precise control of this chemistry remains unclear.

In the aminosilane treatment, clean glass surfaces are converted to a primary amine-containing organo overlayer surface through either solution- or vapor-phase deposition. Because of the terminal amine functional group, the aminosilane surface carries a net positive charge, when exposed to aq solutions approx <pH 9.0. During microarray printing, anionic oligonucleotides or cDNA in the aq print buffer are electrostatically immobilized to this basic, cationic surface. A postprint ultraviolet or thermal crosslinking step is often used to covalently attach the printed molecules to the surface. Organosilane adlayer films on glass are typically on the order of about 1–2 nm thick, and can thus be considered ultrathin organic surface-modifying films. **Figure 2** depicts the relative sizes of a silane film and commonly printed biomolecules.

Despite the long history of silanization of glass surfaces, generating silane films with the uniformity and consistency required for microarray applications is generally not practical for most individual labs. Therefore, DNA microarray researchers most often utilize commercially produced aminosilane slides that

Table 1
Example Commercial Microarray Slide Vendors

Vendor (listed alphabetically)	Example products
Asper Biotech (www.asperbio.com)	Genorama™ (aminosilane)
BioCat (www.biocat.de)	EasySpot™ (epoxide)
Bioslide (www.bioslide.com)	Precision CT (aminosilane, epoxide, aldehyde, PLL)
Corning (www.corning.com)	GAPS II (aminosilane); UltraGAPS (aminosilane); Epoxide
Erie Scientific (www.eriemicroarray.com)	SuperChip™ (aminosilane, epoxide, aldehyde, PLL)
Full Moon Bio (www.fullmoonbio.com)	cDNA Slides; PowerMatrix (3D, oligo); Protein slides
GE Healthcare (formerly Amersham) (www1.amershambiosciences.com)	CodeLink™ (amine-reactive, 3D polymer thin film)
Genetix (www.genetix.com)	ArrayXL cDNA; ArrayXL 3D; Epoxy; amine; Aldehyde slides
GenTel Biosurfaces (www.gentel.com)	PATH™ slides (ultrathin nitrocellulose)
Matrix Technolgies (www.matrixtechcorp.com)	ez-rays (3D polymer, activatable)
Matsunami Glass (www.matsunami-glass.co.jp)	Various
Nunc (www.nalgenunc.com)	ArrayCote™ MaxiSorp™ NucleoLink™
Pall, Inc. (www.pall.com)	Vivid™ Gene Array (nylon membrane)
Perkin Elmer (www.perkinelmer.com)	MICROMAX™ (various); Hydrogel™
Sigma-Aldrich (www.sigmaaldrich.com)	SigmaScreen® slides
Schott Nexterion (www.schott.com/nexterion)	Slide A; A+ (aminosilane); Slide E (epoxide); Slide AL (aldehyde); Slide H (amine-reactive, 3D polymer thin film)
Takara (www.takara-bio.co.jp)	Takara-Hubble™ slide
TeleChem, Inc. (www.arrayit.com)	SuperAmine 2; SuperEpoxide 2; SuperAldehyde 2; SuperProtein; SuperAvidin; Streptavidin
Whatman Schleicher & Schuell (www.schleicher-schuell.com)	FASTslides (nitrocellulose membrane)
Xenopore (www.xenopore.com)	Various

have undergone product development and are subjected to manufacturing quality assurance and quality control. Commercially produced aminosilane slides are available from a large number of vendors, including Corning, TeleChem, Schott Nexterion, and Erie Scientific (*see* **Table 1**).

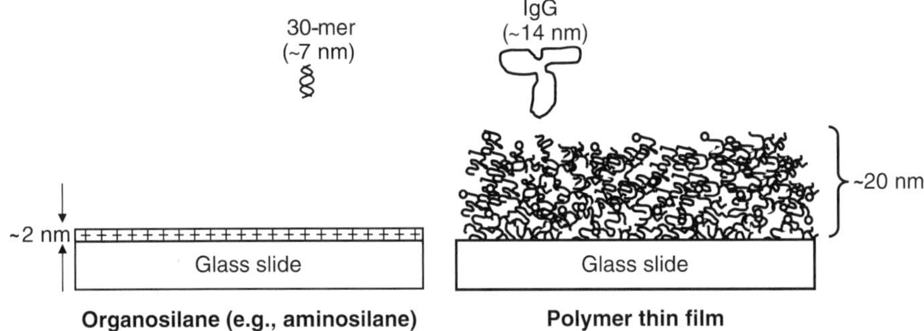

Fig. 2. Issues of scale in surface coatings, bioimmobilization into array formats and capture of target molecules. Ideal 2D-organosilane coatings (e.g., aminosilane) on glass supports (left) are impermeable and nonporous, exhibiting thicknesses smaller than the molecules being printed. This is a distinctly different chemical and physical environment than thicker polymer coatings (right) with permeable architectures and thicknesses greater than the molecules on array. (30-mer dimension assumes duplex with 0.23 nm rise per basepair.)

Aminosilane array slides are relatively inexpensive and a growing legacy dataset has been built on this substrate. Well-developed protocols and reagent kits can deliver high quality arrays. A downside of the aminosilane format is that it lacks proper capabilities to reduce NSB during assay: careful blocking strategies are required to reduce off-feature backgrounds during hybridization. Blocking strategies vary from lab to lab and several microarray slide vendors now offer packaged array processing kits that include blocking reagents. An additional concern for aminosilane slides is that electrostatic and subsequent photo- or thermal-immobilization reactions between DNA and the surface are condition- and probe sequence-specific. And this could lead to steric hindrance of the immobilized probe during hybridization, particularly for short-oligo probes highly reacted to the surface and for low copy-number targets.

3.2.2. Poly-L-Lysine Surfaces

Like the aminosilane surface, poly-L-lysine (PLL)-modified slides also provide cationic immobilization of nucleic acid probes. PLL is a basic, synthetic, positively charged, poly(aminoacid) polyelectrolyte that adsorbs from aq solution to the negatively charged surface of glass. PLL surfaces have historically been used to facilitate cell-adhesion for histology and cell culture applications. Consequently, PLL solutions formulated specifically for glass-slide treatments are available commercially (e.g., through Sigma-Aldrich, St. Louis, MO). These commercial formulations are typically 0.01% aq solutions of 75,000–150,000 mol wt PLL (e.g., Sigma-Aldrich P-4707). Protocols for PLL slide preparation for DNA microarraying are readily available, and all are closely related to the protocol originally described by Brown *(20)*.

The primary advantage of PLL slides is that they are inexpensive and simple to produce within individual labs. The main disadvantage of PLL-coated slides is that intraslide uniformity and batch-to-batch performance can be inferior to other slide formats. The preparation is particularly sensitive to the quality of the glass-cleaning process and because the PLL surface ages with time, most protocols recommend a 2-wk aging period between preparation and use. Because the PLL adheres to the underlying glass through electrostatic interactions, these surfaces are also subject to delamination under high salt conditions.

3.2.3. Epoxide and Aldehyde Surfaces

Silanized slides that provide active-binding chemistries are becoming increasingly popular within the custom-made arraying community. This is particularly true for labs migrating from cDNA or polymerase chain reaction-probe product platforms to those based on oligonucleotide probes. The most popular active silane chemistry is epoxide, which provides covalent immobilization of both amine-modified and unmodified oligos. The epoxide surfaces are typically based on a glycidoxy-functionalized organosilane or related epoxide molecule derivative. The main advantages of epoxide chemistry are its relative stability and generally lower background than aminosilane *(21)*. Several commercial vendors of epoxide slides are listed in **Table 1**. Aldehyde-terminated silanes are another class of active chemistries used for DNA immobilization. Aldehyde-functionalized surfaces selectively react with amine-terminated oligos through a well-known Schiff base reaction, which is often driven to completion through the addition of a reducing agent to the reaction solution.

3.2.4. Polymer Thin Films

Glass slides treated with polymer coating formulations are designed to provide a degree of three-dimensionality and theoretically higher probe-loading capacities. Polymer-coated slides are also designed to provide additional performance features such as tethered-probe immobilization with a more benign, "solution-like" environment, and low nonspecific binding backgrounds. The term "polymer film" has been used here to generally describe polymer thin films residing on glass slides, ranging in thickness from about 20 to 2000 nm. These supported films are treated separately from thicker polymer monolithic membrane materials such as nitrocellulose and nylon, which are discussed in more detail in **Subheading 3.2.5.**

Commercial polymer film formulations generally consist of hydrophilic polymer formulations activated with reactive chemistries that provide chemical capabilities for covalent immobilization of appropriately modified oligonucleotides.

Polyacrylamides and PEG-derivatives are common polymer components in this category. Because of their hydrophilic nature, coated slides in this category are often referred to as "hydrogels" or "hydrogel-like" materials with porous, 3D character and some water imbibation. Hydrogel™ is also the trade name of the Perkin Elmer (Wellesley, MA) protein array slide product.

Here, we first consider thin polymer films, which are typically transparent-coated layers less than a few hundred nanometers thick immobilized on a glass slide. **Figure 2** provides an illustration of a 20-nm-thick polymer film in size-scale relation to a silane film and some molecular dimensions. This thickness disparity distinguishing the silane from the polymer films is readily apparent. A well-known formulation in this category underlies the CodeLink™ microarray platform (GE Healthcare), exploiting a proprietary crosslinked and chemically activated polyacrylamide formulation to enable covalent immobilization of amine-modified oligonucleotides in a 3D solution-like environment *(16)*. Another example is Schott Nexterion Slide H™, a crosslinked, multicomponent PEG-coated glass slide activated with succinimidyl esters to provide covalent immobilization of amine groups. Both CodeLink™ and Slide H™ exhibit extremely specific reactivity for the amine groups on amine-modified probes. Other commercial polymer thin film formulations include the PowerMatrix™ slides from FullMoonBio (SunnyVale, CA) and ez-rays™ Universal™ slides from Matrix Technologies (Hudson, NH).

Thicker (>1000 nm) polymer films have also been demonstrated, the most celebrated of which is the gel-pad technology described by Mirzabekov et al. *(22,23)*. Commercial applications of thick hydrogel technologies have been marketed primarily for protein applications, as discussed in more detail below.

3.2.5. Membranes

DNA microarray analysis has been described as a natural progression from dot-blot analysis in which the DNA probe is adsorbed in a thick membrane for analysis *(1)*. Proceeding from traditional dot-blot analysis to microarray formats, new products have been introduced that provide dot-blot-like membrane layers immobilized onto glass slides, giving the functionality of membranes but in the standard glass slide microarray format. These are particularly useful for DNA array analysis that uses radioactive labeling and chemiluminescence, where high probe loading is needed for good signal-to-noise generation. For example, Pall Vivid™ gene array slides feature a nylon membrane immobilized on a glass slide. Whatman Schleicher & Schuell (Keene, NH) offers nitrocellulose membrane slides (FASTslides), but these are marketed primarily for protein arrays and are discussed in **Subheading 4.2.**

In conclusion, both commercially produced and custom-made arrays have seen significant development over the last several years. In the context of surface chemistry, much of the glass slide DNA array work has consolidated onto a relatively well-defined set of surface chemistries, dominated by organosilane-coatings (aminosilane and epoxide) and a subset of 3D, active polymers. The driving forces of data quality (signal-to-noise, reproducibility, and so on) and biological accuracy will lead to continued improvements in DNA microarray surface chemistries.

4. Protein, Peptide, and Carbohydrate Arrays

A growing number of researchers and commercial organizations are expanding microarray technology into non-DNA applications such as protein, peptide, and carbohydrate arrays. This section highlights the surface chemistry performance considerations for these systems, again focusing primarily on glass slide formats.

4.1. Fundamental Surface Considerations for Non-DNA Arrays

It is fair to say that protein array technology has been more problematic and slow to develop in contrast to the anticipation of many researchers. Much of this difficulty can be attributed to the fact that proteins are more difficult to work with in a solid-phase format than their DNA counterparts *(24)*. Issues such as native conformation, denaturation and stability, molecular orientation, and cross-reactivity have proven to be major challenges. This is particularly true for protein capture arrays, which many envisioned as microenzyme-linked immunosorbent assay (ELISA) with thousands of capture antibodies immobilized on a surface, but in reality has been limited to relatively small panels of capture molecules.

Surface chemistry is and will be a fundamental aspect of protein, peptide, and carbohydrate array development. There exists no analog to the aminosilane-DNA system in the protein format and because of the diversity of biomolecular structures and the unique challenges of protein-based assays, the concept of a universal protein array surface is unlikely to become a reality. Instead, sets of engineered surfaces will be applied for specific array applications. In particular, these surface chemistries will ideally provide the following four performance features:

1. Stability of the immobilized biomolecule (i.e., retained native conformation, tertiary structure, and so on).
2. Retained specificity of the immobilized biomolecule for its cognate binding partner (e.g., antibody–antigen, enzyme–substrate, and so on).
3. Mitigation of surface-induced, on-feature (on-spot) crossreactivity.
4. Minimization of off-feature nonspecific binding backgrounds.

Carefully designed surface chemistry intends to satisfy these performance requirements by simultaneously endowing the assay surface with both specific and nonspecific performance features.

4.1.1. Surface Preservation of Biomolecule Native Structure and Bioactivity

Protein interfacial behavior is a well-studied phenomenon with substantial literature detailing the formidable challenges remaining unsolved *(25–30)*. On top of the established challenges associated with protein interfacial behavior, the physical act of printing microarray presents a unique set of new challenges. Protein microarrays are manufactured using contact (e.g., quill pin) and noncontact (e.g., piezo or inkjet-style) printing methods. Typically, nanoliter to picoliter print solution drop sizes are dispensed with µg/mL protein concentrations in aq print media (buffer), producing functionally "dried" spots on the surface. The drying process is highly nonequilibrium, with spots drying within 2–3 s on the >60% relative humidity print deck. This evaporative loss of water concentrates the printing salts to extremely high ionic strength and high protein concentrations within the spot, producing conditions highly favorable to both protein aggregation and structural loss (denaturation). Further drying on the surface and storage under desiccating conditions on-slide promotes time-dependent loss of structure and bioactivity, regardless of low temperature storage or desiccant additives. Hence, many printed arrays exhibit suboptimal capture bioactivities and limited shelf-life in storage, restricting their ultimate practical utility. Fundamental protein–protein and protein–surface interactions, long known and described in the biomedical device and biosensing literature, represent constraints on improved protein array performance.

Arraying surfaces have continually been developed in attempts to minimize these structural losses to the "dry-down" immobilization process postprint. To date, hydrogel and organosilane immobilization chemistries have had only incremental reported success to date to significantly alter this problem. Soluble excipients added to the print buffer, including well-known desiccant saccharide protectants trehalose and sucrose frequently used to stabilize pharmaceutical protein-lyophilized formulations, exhibit limited capability to preserve immobilized biomolecule structure or bioactivity significantly postprint.

Although significant native structure challenges remain, the number of protein array-based publications continues to increase and commercial products are beginning to emerge. In an academic setting, the long-term storage stability problem is not a factor and assays can be performed in well-controlled conditions soon after protein printing. In the commercial space, capture arrays based on antibodies immobilized in nitrocellulose membranes (e.g., FASTslides[™]; Whatman Schleicher & Scheull, Keene, NH) have demonstrated shelf life stability. Several other immobilized protein formats have appeared on the market (*see* **Table 2**), although the fraction of immobilized proteins retaining native structure in these products is not well established.

Table 2
Survey of Commercial Protein and Peptide Microarray Products

Vendor	Product description	Web site
BD-Clontech	Antibody (Ab) Microarray 500	www.clontech.com
EMD Bioscience	ProteoPlex™	www.emdbiosciences.com
	16-Well Human Cytokine Array	
	16-Well Murine Cytokine Array	
Whatman Schleicher & Schuell (S&S)	Serum Biomarker Chip	www.schleicher-schuell.com
Invitrogen	PROTOArray™	www.invitrogen.com
	Human and Yeast Protein Microarrays for kinase substrate identification	
Sigma-Aldrich (& Procognia)	Panorama™ Human p53 Protein Function Microarray	www.sigmaaldrich.com
Pepscan Systems	PepChip® Kinase Microarray of 1200 kinase substrate peptides	www.pepscan.com

4.1.2. On-Feature Nonspecific Binding (Cross-Reactivity)

On-feature nonspecific binding or cross-reactivity has been an onerous problem for protein microarrays, and in particular capture arrays. Part of this problem is a fundamental property of proteins, and antibodies in particular: there are measurable cross-reactivities in solution phase that will also manifest on a surface. Surface chemistry cannot alter these types of interactions. A fraction of the on-feature cross-reactivity; however, can be attributed to surface immobilization. On a given microarray spot, it is inevitable that some fraction of immobilized protein will be in a denatured, non-native state. It is reasonable to assume that denatured proteins can be a significant contributor to nonspecific interactions. Even in well-developed, microplate-based ELISA formats, literature suggests that only a fraction of immobilized capture antibodies retain activity *(25)*. But the ELISA format has the advantage that different samples or analytes are physically isolated from each other by the plastic wells. Microarray systems have no such advantage, as all spots interact with the sample. So if each spot contains a fraction of denatured protein, the likelihood of on-feature nonspecific interactions is dramatically increased. An outcome of this has been that capture arrays, instead of having hundreds or thousands of features, have largely been limited to only a few dozen features.

Surface chemistry improvements have the potential to partially mitigate this on-feature cross-reactivity problem by creating a local environment that helps to preserve native structure. Improved surface chemistries are expected to continue to appear in the literature and from commercial sources.

4.1.3. Off-Feature Nonspecific Binding Backgrounds

NSB background is an age-old problem in solid-phase assays, and one that often sets the sensitivity limit of the assay. In a microarray experiment, off-feature NSB is typically manifested as a high background fluorescence. In formats with large specific signal such as that observed on nitrocellulose membranes, high off-feature backgrounds often render assay signal-to-noise ratios similar or even inferior to their low specific signal counterparts. These nonspecific binding effects, collectively consisting of acid/base, hydrophobic, hydration, dispersion, electrostatic, and hydrogen bonding forces at the surface, can be irreversible with increasing surface residency time, resistant to stringency rinsing. Therefore it is always a problem in ultimate assay sensitivity and fidelity.

Off-feature NSB is most commonly addressed by surface blocking (described in **Subheading 2.**), or time-consuming prepurification of the sample to remove background proteins, often impractical and sometimes impossible. Ideally, microarray surfaces should be resistant to adsorbing this nonspecific protein background from complex samples, permitting direct sample assay without blocking, analyte abundance amplification (e.g., by polymerase chain reaction), or purification. In the absence of suitable chemistry coating solutions that perform with requisite properties, assay formats, and integrated devices must rely on either (1) compromised, substandard detection limits and dynamic range, (2) extensive sample prepurification before assay, and (3) analyte amplification schemes to boost signal over noise.

Surface chemistry efforts have been directed continuously at this problem and although significant work remains, some progress has been made. The unique challenge of the microarray format is to provide high specific (on-spot) immobilization, while minimizing or eliminating all off-feature interactions. Authors of this chapter (Accelr8 Technology) have developed an activated PEG-based microarraying surface chemistry, engineered specifically to simultaneously provide on-spot covalent immobilization and extremely low off-feature NSB, without required adsorptive blocking. This surface is commercially available as Schott Nexterion Slide H™.

In summary, challenges in preservation of stability of printed capture-protein molecular structure on arrays, especially in desiccated storage states, and in minimizing nonanalyte protein adsorption both to on- and off-array features, remains a primary obstacle to improve the assay performance. Coatings have provided notable, incremental progress to improved-signal and reduced-background, but not a complete, general solution to long-standing two-body adsorption problems that reduces signal and increases background in microarray formats.

4.2. Protein and Peptide Arrays

The concepts and practices of microarraying as a high-throughput multiplexing experimental method were initially proposed as miniaturized multi-ELISA-type protein assays, as an extension of a well-developed field of antibody-based ligand-binding assay development *(31)*. The implementation of protein arrays was preceded by nucleic acid arrays because of the generally simpler aspects of working with nucleic acids. But advances in the understanding of proteomics, and also the increased availability of capture and detection reagents, has spurred the growth of many new methods that allow unprecedented analyses of protein content and interactions within biological samples.

As discussed in the previous section, protein array performance has a greater dependence on surface chemistry considerations than nucleic acids arrays, because proper function of protein arrays depend on maintenance of tertiary structure, especially when the screening test involves activity, or the binding event requires tertiary and quaternary interactions. Further, detection schemes for protein arrays often require signal generation through labeled proteins, which can result in high background and nonspecific signal generation with suboptimal surface chemistries. Finally, shelf-life requirements of commercial products dictate that surface chemistries of protein arrays must support maintenance of structure and activity for extended periods of time. Protein microarrays can be categorized into three groups based on their format and function:

1. Capture arrays.
2. Reverse-phase arrays.
3. Interaction arrays.

Protein microarray substrate types can be categorized according to dimensionality at the surface (2D vs 3D), and whether the binding to the surface occurs through covalent chemistry or nonspecific adsorption. Considerations include gross loading and degree of retention of printed material, maintenance of tertiary structure and/or activity, propensity toward nonspecific signal generation, and shelf life.

The term *capture array* is used to describe systems in which capture elements, typically antibodies, are printed onto a microarray surface. A complex sample is then incubated with the array, and target analytes (e.g., antigens) are "captured" on their cognate spots. The antibody capture array is an attractive and intuitive extension of the well-known ELISA format to a highly multiplexed small-scale assay. Initial efforts at capture arrays transferred known methods from nucleic acids arrays and, hence, used 2D surfaces and nonspecific adsorptive-binding mechanism *(32)*. Subsequent efforts have tended more toward surface chemistries that are thought to promote retention of tertiary structure of capture molecules in 3D polymer matrices (e.g., PerkinElmer Hydrogel™, GE Healthcare CodeLink™, Schott Nexterion Slide H™) or membrane slides (e.g., S&S FASTslide™). Film surfaces can load higher mass amounts on a unit area basis, and though the binding mechanism is noncovalent and some signal loss can be expected, the greater loading and activity retention of the 3D environment has afforded good signal-response in many applications *(33)*.

Reverse-phase arrays are based on spotting complex samples onto a surface that is then probed with a single-detection molecule *(34,35)*. The format is conceptually an inverted form of a capture array, hence the name reverse-phase. From a surface chemistry perspective, reverse-phase arrays require the printing of complex biological samples with subsequent probing for often sparsely distributed targets. Hence, mass loading in this method is a primary concern in order to maximize density of targets of interest. Therefore, membrane-types surfaces (e.g., nitrocellulose) have commonly been used for these assays. It is expected that ultralow nonspecific binding surfaces and high-sensitivity detection methods can expand the functional utility of this method *(34,35)*.

Interaction arrays, also called functional arrays, represent the broadest category of protein arrays because of the diversity of molecules available to array and assay. As such, surface requirements are very much system specific and should be considered an integral part of the experimental design. An outstanding example of interaction arrays is the seminal paper by MacBeath and Schreiber *(36)*, which studied three different protein-interaction systems in the array format: protein–protein, kinase substrate, and protein–peptide interactions *(36)*. Another example is the yeast proteome array developed by Snyder et al., where >5800 recombinant yeast proteins were immobilized on a surface and then probed for specific protein–protein and protein–lipid interactions *(37)*.

From a surface chemistry perspective, protein-interaction arrays can be conceived with orientation-specific immobilization in mind. For example, the recombinant yeast proteins in the Snyder work *(37)* were tagged with HIS-6 peptides in order to direct the attachment and orientation of the proteins to Nickel-modified glass slides allowing optimal presentation of the bound proteins *(37)*. Procognia, Ltd. (Maidenhead, UK) uses a streptavidin-derivatized surface to build their recombinant p53 mutant arrays *(38)*. MacBeath's protein–protein studies were based on aldehyde-silane slides *(36)*. Alternatively, designs with less stringent orientation requirements can utilize nonspecific immobilization mechanisms, for which silanized glass is a convenient solution *(39)*.

Peptide arrays are a subset of the interaction array category and have been the subject of recent reviews *(8)*. Because of their small size, peptides are less amenable to nonspecific adsorptive-binding strategies and consequently, specific chemical-binding strategies are generally required.

4.2.1. Commercial Protein and Peptide Arrays

There has been a concerted effort by industrial microarray manufacturers to transfer protein array technology from the research world to the commercial market. The commercial offerings available at the present time fall under the capture array and interaction array categories, and to date they offer a limited-range of applications and target sets. The long-term success of any of these offerings has yet to be established, but it is anticipated that a growing market for both types of platforms will provide sufficient demand for continued development. Although many commercial offerings of protein arrays have been derived from research generated in academic settings, development efforts, and proprietary concerns often preclude disclosure of the exact nature of surface chemistries used. However, some companies offer products based on known surface technologies and thus, the surface chemistry component can be evaluated on known parameters. **Table 2** provides a partial listing of currently available commercial protein and peptide array products.

4.3. Carbohydrate Arrays

Carbohydrate arrays are a rapidly emerging area of microarray research, providing rapid and diverse screening of carbohydrate–protein interactions in multiplex form *(9,40–42)*. These arrays operate analogously to other array types, frequently using fluorescence-detection of labeled proteins or matched antibodies. Affinity for various saccharide types and sequences using proteins or lectins is achieved using array screening, and receiving increasing attention for several bioanalytical goals. Array binding of targets (typically carbohydrate-recognizing lectins or other sugar-binding protein targets) assumes that, like any other array, the target interacts specifically with spotted array features and minimally with off-spot features. As higher order structures in polysaccharides have not been observed beyond intra- and intermolecular hydrogen bonding and network formation in aq solution (e.g., gelation), specific protein- or lectin-binding to array oligosaccharides is thought currently to be determined simply through saccharide primary sequence specificity. Hence, structural perturbations in oligosaccharide ligands, on printing or drying that alter target binding are not currently a concern. However, surface- and time-dependent ligand accessibility issues involving the carbohydrate-decorated arraying base protein and its surface interactions (e.g., surface-induced denaturation) on-array are likely the same as for any other arrayed protein, and therefore, a current concern for variability and storage issues. Additionally, as complexity of microarrayed saccharide chemistry presentations increases, surfaces perturbations of their structures and effects on target capture are likely to become more evident.

From the surface chemistry perspective, carbohydrate immobilization strategies that provide specific binding are most desirable. Surfaces that provide amine-reactive chemistries coupled to biocompatible polymer films are well-suited for synthetically generated libraries of peptides or carbohydrates, and are attracting more interest as new synthetic methods and reagent libraries become available *(43,44)*.

In summary, the technologies and variety of applications for protein, peptide, and carbohydrate arrays continue to grow and evolve. As new methods are conceived and executed, enterprising researchers are adapting available surface chemistries for substrate functionalization and implementing new chemistries tailored for specific uses. It is clear that this will be a most fascinating field of research to observe and participate in the near future.

5. Emerging Microarray Applications

In coming years, microarray technology is expected to expand significantly beyond the 25 × 75-mm glass slide format. Array integration within microfluidic cartridges, microtiter plates, and diagnostic platforms are already beginning to emerge. In addition to alternative detection formats, arrays will be used with increasingly complex samples, including direct assay of biological fluids, environmental samples, food products, and so on. This combination of new formats and complex samples will undoubtedly place enormous demands on the biomolecule–synthetic surface interface. Therefore, surface chemistry can be viewed as an important and enabling technology for the continued development and expansion of the microarray field.

References

1. Southern, E. M. (2001) DNA microarrays: History and overview, in *DNA Arrays,* (Rampal, J. B., ed.), Humana, Totowa, NJ, pp. 1–15.
2. Aboytes, K., Humphreys, J., Reis, S., and Ward, B. (2003) Slide coating and DNA immobilization chemistries *in A Beginner's Guide to Microarrays* (Blalock, E., ed.), Kluwer, Norwell, MA, pp. 1–42.
3. Schena, M. (ed.) (2003) *Microarray Analysis,* John Wiley & Sons, Hoboken, NJ.
4. Sobek, J. and Schlapbach, R. (2004) Substrate architecture and functionality. *Microarray Technology (September)* 32–44.
5. Templin, M. F., Stoll, D., Schwenk, J. M., Potz, O., Kramer, S., and Joos, T. O. (2003) Protein microarrays: promising tools for proteomic research. *Proteomics* **3,** 2155–2166.

6. Angenendt, P., Glokler, J., Sobek, J., Lehrach, H., and Cahill, D. J. (2003) Next generation of protein microarray support materials: evaluation for protein and antibody microarray applications. *J. Chromatogr. A* **1009,** 97–104.

7. Kusnezow, W. and Hoheisel, J. D. (2003) Solid supports for microarray immunoassays. *J. Mol. Recognit.* **16,** 165–176.

8. Panicker, R. C., Huang, X., and Yao, S. Q. (2004) Recent advances in peptide-based microarray technologies. *Comb. Chem. High Throughput Screen.* **7,** 547–556.

9. Wang, D. (2003) Carbohydrate microarrays. *Proteomics* **3,** 2167–2175.

10. Peterson, A. W., Heaton, R. J., and Georgiadis, R. M. (2001) The effect of surface probe density on DNA hybridization. *Nucleic Acids Res.* **29,** 5163–5168.

11. Vainrub, A. and Pettitt, B. M. (2002) Coulomb blockage of hybridization in two-dimensional DNA arrays. *Phys. Rev. E Stat. Nonlin Soft Matter Phys.* **66,** 041905.

12. Vainrub, A. and Pettitt, B. M. (2003) Surface electrostatic effects in oligonucleotide microarrays: control and optimization of binding thermodynamics. *Biopolymers* **68,** 265–270.

13. McGall, G. H. and Fidanza, J. A. (2001) Photolithographic synthesis of high density oligonucleotide arrays, in *DNA Arrays,* (Rampal, J. B., ed.), Humana, Totowa, NJ, pp. 71–101.

14. McGall, G. H. and Christians, F. C. (2002) High-density genechip oligonucleotide probe arrays. *Adv. Biochem. Eng. Biotechnol.* **77,** 21–42.

15. Nuwaysir, E. F., Huang, W., Albert, T. J., et al. (2002) Gene expression analysis using oligonucleotide arrays produced by maskless photolithography. *Genome Res.* **12,** 1749–1755.

16. Ramakrishnan, R., Dorris, D., Lublinsky, A., et al. (2002) An assessment of Motorola CodeLink microarray performance for gene expression profiling applications. *Nucleic Acids Res.* **30,** E30.

17. Shippy, R., Sendera, T. J., Lockner, R., et al. (2004) Performance evaluation of commercial short-oligonucleotide microarrays and the impact of noise in making cross-platform correlations. *BMC Genomics* **5,** 61.

18. Mergen, Ltd. (San Leandro, CA), www.mergen.ltd.com

19. Mittal, K. L. (ed.) (1992) *Silanes and Other Coupling Agents*, VSP International Science Publishers, *VSP International Science Publishers*, Utrecht, The Netherlands.

20. Patrick Brown Lab, Department of Biochemistry, Stanford, http://cmgm.stanford.edu/pbrown/protocols/1_slides.html.

21. Taylor, S., Smith, S., Windle, B., and Guiseppi-Elie, A. (2003) Impact of surface chemistry and blocking strategies on DNA microarrays. *Nucleic Acids Res.* **31,** E87.

22. Vasiliskov, A. V., Timofeev, E. N., Surzhikov, S. A., et al. (1999) Fabrication of microarray of gel-immobilized compounds on a chip by copolymerization. *Biotechniques* **27,** 592–594, 96–98, 600 passim.

23. Zlatanova, J. and Mirzabekov, A. (2001) Gel-immobilized microassays of nucleic acids and proteins, in *DNA Arrays,* (Rampal, J. B., ed.), Humana, Totowa, NJ, pp. 17–38.

24. Metzger, S., Lochhead, M. J., and Grainger, D. W. (2002) Surface technologies to improve performance in protein microarray based molecular diagnostics. *IVD Technol.* **8,** 39–45.

25. Andrade, J. D., Hlady, V., Feng, L., and Tingey, K. (1996) Proteins at interfaces: principles, problems, and potential. *Bioprocess Technol.* **23,** 19–55.

26. Hlady, V. and Buijs, J. (1996) Protein adsorption on solid surfaces. *Curr. Opin. Biotechnol.* **7,** 72–77.

27. Kleijn, M. and Norde, W. (1995) The adsorption of proteins from aqueous solution on solid surfaces. *Heterogeneous Chem. Rev.* **2,** 157–172.

28. Malmsten, M. (1999) Protein adsorption at the solid-liquid interface. *Protein Architecture: Interfacial Molecular Assemblies and Immobilization Biotechnology* (LVOV, Y. and Möhwald, eds. Marcel Dekker, Inc., New York, pp. 1–23.

29. Norde, W. (2000) Proteins at solid surfaces, in *Physical Chemistry of Biological Interfaces* (Baszkin, A., Norde, W., eds.), Marcel Dekker, New York, NY, pp. 115–135.

30. Wahlgren, M. and Arnebrant, T. (1991) Protein adsorption to solid surfaces. *Trends Biotechnol.* **9,** 201–208.

31. Ekins, R., Chu, F., and Biggart, E. (1990) Fluorescence spectroscopy and its application to a new generation of high sensitivity, multi-microspot, multianalyte, immunoassay. *Clin. Chim. Acta* **194,** 91–114.

32. Haab, B. B., Dunham, M. J., and Brown, P. O. (2001) Protein microarrays for highly parallel detection and quantitation of specific proteins and antibodies in complex solutions. *Genome Biol.* **2,** Research0004.

33. Miller, J. C., Zhou, H., Kwekel, J., et al. (2003) Antibody microarray profiling of human prostate cancer sera: antibody screening and identification of potential biomarkers. *Proteomics* **3,** 56–63.

34. Espina, V., Mehta, A. I., Winters, M. E., et al. (2003) Protein microarrays: molecular profiling technologies for clinical specimens. *Proteomics* **3,** 2091–2100.

35. Liotta, L. A., Espina, V., Mehta, A. I., et al. (2003) Protein microarrays: meeting analytical challenges for clinical applications. *Cancer Cell* **3,** 317–325.

36. MacBeath, G. and Schreiber, S. L. (2000) Printing proteins as microarrays for high-throughput function determination. *Science* **289,** 1760–1763.

37. Zhu, H., Bilgin, M., Bangham, R., et al. (2001) Global analysis of protein activities using proteome chips. *Science* **293,** 2101–2105.

38. Boutell, J. M., Hart, D. J., Godber, B. L., Kozlowski, R. Z., and Blackburn, J. M. (2004) Functional protein microarrays for parallel characterisation of p53 mutants. *Proteomics* **4,** 1950–1958.

39. Robinson, W. H., DiGennaro, C., Hueber, W., et al. (2002) Autoantigen microarrays for multiplex characterization of autoantibody responses. *Nat. Med.* **8,** 295–301.

40. Drickamer, K. and Taylor Maureen, E. (2002) Glycan arrays for functional glycomics. *Genome Biol.* **3,** Reviews 1034.

41. Ratner, D. M., Adams, E. W., Su, J., O'Keefe, B. R., Mrksich, M., and Seeberger, P. H. (2004) Probing protein–carbohydrate interactions with microarrays of synthetic oligosaccharides. *Chem. BioChem.* **5,** 379–382.

42. Feizi, T., Fazio, F., Chai, W., and Wong, C. H. (2003) Carbohydrate microarrays - a new set of technologies at the frontiers of glycomics. *Curr. Opin. Struct. Biol.* **13,** 637–645.

43. Blixt, O., Head, S., Mondala, T., et al. (2004) Printed covalent glycan array for ligand profiling of diverse glycan binding proteins. *Proc. Natl. Acad. Sci. USA* **101,** 17,033–17,038.

44. Disney, M. D., Magnet, S., Blanchard, J. S., and Seeberger, P. H. (2004) Aminoglycoside microarrays to study antibiotic resistance. *Angew. Chem. Int. Ed. Engl.* **43,** 1591–1594.

3

Nonfouling Surfaces

A Review of Principles and Applications for Microarray Capture Assay Designs

Ping Gong and David W. Grainger

Summary

Microarray technology, like many other surface-capture diagnostic methods, relies on fidelity of affinity interactions between a surface-bound probe (e.g., nucleic acid or antibody) and its target in the sample milieu to produce an assay signal specific to analyte. These interfacial interactions produce the assay result with the associated assay requirements for sensitivity, specificity, reproducibility, and ease-of-use. For surface-capture assays, surface properties play a critical role in this performance. Microarray surfaces are routinely immersed into aqueous target solutions of varying complexity, from simple saline or buffer solutions to serum, tissue, food, or microbiological lysates involving thousands of different solutes. The surface chemistry must not only be capable of immobilizing probes at high density in microscale patterned spots, retaining probe affinity for target within these spots, reducing target capture outside of these spots, but also be efficient at eliminating nontarget capture anywhere else on the surface. Historically, the development of surface chemistry with these specific "nonfouling" properties has been an intense interest for bioassays, with many types of architectures, molecular compositions, and performance capabilities across many different surface-capture assays. The unique environment of the bioassay, including the long-standing problems associated with high concentrations of "nontarget" proteins and other surface-active biomolecules in the assay milieu, has proven to be quite challenging to surface chemistry performance. Microarray technology designs with microspotted patterns must address these problems in these challenging dimensions in order to improve signal:noise ratios for captured target signals on surfaces. This chapter reviews principles of protein–surface interfacial physical chemistry, protein adsorption as a source of assay noise, and various approaches to control this interface in the context of surface-capture assay fabrication and improving assay performance from complex milieu. Practical methods to modify surfaces for biological assay are presented. Polymer substrate coating methods, including "grafting from" and "grafting to" strategies, polymer brushes, and alternative surface modification methods are reviewed. Methods to assess biological "fouling" in the bioassay format are also discussed.

From: *Methods in Molecular Biology, vol. 381: Microarrays: Second Edition: Volume 1: Synthesis Methods*
Edited by: J. B. Rampal © Humana Press Inc., Totowa, NJ

Key Words: Bioassay; diagnostics; microarrays; nonspecific binding; PEG; polyethylene glycol; protein adsorption; surface chemistry; surface immobilization.

1. Introduction

Biomedical diagnostic assay device performance often relies on deliberate and precise modification of device surface properties attempting to control interfacial behavior with multiple biological components. Typically, two surface properties are sought: (1) selective control of desired surface binding or capture of molecular and cellular species from solution and (2) elimination of nonspecific adsorption or "fouling" of surfaces by any undesired solute. Most often, in complex, "real" biological samples, prevention of nonspecific platelet, protein, lipid, cell, and bacterial surface attachment (nonspecific binding or NSB) is desired to improve device performance both in vivo and in vitro *(1–5)*. Surfaces that *eliminate* all nonspecific protein adsorption and cell binding are a common desirable end point for many diagnostic devices, surgical implants, biological reactors, maritime hardware, and pharmaceutical packaging. Surfaces that *selectively bind* certain desired solutes, whereas resisting other nonspecific biological component adsorption are a compelling need for technologies spanning diagnostics, sensors, and tissue-engineered devices. Both performance goals exploit coating chemistries that resist NSB. Additionally, selective surface capture technologies also incorporate biological recognized or synthetically designed surface-bound ligands with high affinity for desired targets. This is required for many biosensors, diagnostic assays, and for some implant applications, where specific interactions are required to improve performance through capture, integration, or surface interfacial response. All surface-design approaches require a base surface coating to resist NSB events together with a coupled ability to further tailor specific surface-binding interactions with desired solutes. Importantly, in some cases, surface binding should, maintain captured molecules native structure and essential bioactivity (e.g., growth factor immobilization in tissue engineering and enzyme-antibody or nucleic acid immobilization in sensing and diagnostics).

Unfortunately, many of these NSB and specific surface-performance requirements have not been met to date, providing plenty of opportunity to innovate improved surface-performance across many technology bases using surface modification: no surface reliably exhibits either perfect nonfouling or specific capture capabilities to date. Surfaces and coatings comprising only 100% bound ligands at immobilized densities sufficient to cover the entire surface without defects are not yet synthetically realizable. Any uncoated material surface left exposed, defects, and most ligand chemistries produce measurable NSB of undesired biological components. Improved performance in sensing and sophisticated diagnostics, therefore, requires a technological basis in improved surface designs that reliably resist NSB of biological molecules to preclude adverse performance in each context. Diagnostic signal-to-noise frequently depends critically on surface reactivity differences to specific and nonspecific surface–milieu interactions. Given that most bioanalytical identification technologies involve exposure of biological or environmental fluids containing trace analyte with capture surfaces (enzyme-linked immunosorbent assay, microarrays, beads, chromatography, gels, and membrane sorbents), and that these biotechnological assay materials are pervasive with markets exceeding billions of USD annually world wide, examination of the principles driving rational surface design to improve microarray and diagnostic capture surfaces motivates this review.

2. Protein Adsorption to Diagnostic and Bioassay Surfaces

Most surface-capture bioassays are ideally intended to discriminate target analyte from complex biological milieu (serum, tissue lysate, food, and environmental samples). Serum and cell lysates contain thousands of nontarget proteins, peptides, and interferants; other samples can be equally complex. Hence, a first-priority for assay effectiveness is to implement a surface that effectively avoids nonspecific protein adsorption in order to capture analyte. From first, principles for designing improved surfaces effective in these technologies, all protein–surface interactions might be considered as two-body problems. Attraction, adsorption, and desorption of solutes at surfaces involve complex physical and chemical, time- and space-dependent properties of both the solute and the interfacial chemical binding partners. Adsorption of water, ions, and small molecule solutes remains a challenging scenario to model and understand *(6,7)*. Protein adsorption, subject to intensive study *(8–10)*, remains a more complex and largely uncontrolled phenomena addressed by largely empirical approaches in most technologies. As protein fouling remains a primary obstacle to performance across many diagnostic applications, improved surfaces that control protein adsorption remain a major research focus. Imposed upon equilibrium protein-surface thermodynamic considerations are more complex nonequilibrium thermodynamics and kinetic factors for competitive surface affinity, occupancy, displacement, and protein conformational changes. Hence, rational design features for capture surfaces—both chemical and physical interfacial properties—that reliably control protein interfacial behavior are not obvious or intuitive, and in fact have proven literally impossible to realize or perfect in most biological applications to date.

2.1. The Case for Diagnostic Surface Selectivity

In serum, 90% of the proteome mass content comprises albumin, transferrin, haptoglobin, and immunoglobulins (totaling about 50 mg/mL), few of which are ever interesting analytes. Typical target protein analytes for clinical diagnostics, proteomics, microarraying, and surface capture purposes are low abundance, usually pg/mL or less, embedded within the 80 mg/mL proteome background. Critically, diagnostic or clinical assay signal generation for these analytes must be readily and reliably perceived through high abundance serum or cell-protein noise, particularly for small, highly hydrophobic or highly basic protein analytes, or other trace targets with difficult assay parameters.

Reduced assay noise (NSB) remains a critical challenge to capture surface contributions to improving both signal and noise performance in bioassays. In molecular isolations, purifications, manipulations, and assays, the technological benchmarks for proteins contacting surfaces are similar, specifically described in **Table 1**.

None of the thousands of polymer films and organic modifications proposed to date fulfill these important requirements for optimal surface capture from complex milieu. To understand these limitations critical to assay performance, some appreciation of the complexity of the generic protein adsorbate, as both signal and noise on surfaces, is required. Adsorption of globular serum proteins in particular to bioassay surfaces is most relevant to biomedical and biotechnological applications. On exposure to aqueous milieu posttranslational in the cell, proteins as copolymers of some 22 amino acids of varying hydrophobicity, polarity, acidity, positive- or negative-charged side groups, frequently and spontaneously fold into compact, often globular, domain-containing native (organized) structures. These globular folded native states represent local structural free energy minima and metastability, often only a few kcal/mol deep, permitting reversible unfolding to nonnative structures given sufficient energetic drivers (e.g., thermal, mechanical, and chemical). Their metastable, amphiphilic structural nature makes most proteins intrinsically highly surface active, promoting their spontaneous, nonspecific surface adsorption and unfolding. In aqueous solution, compact globular protein native conformations tend to bury nonpolar amino acid groups into their interior core region through hydrogen bonding, weak acid–base and hydrophobic interactions. Packing of these nonpolar groups away from external bulk water results in a dramatic decrease in protein conformational entropy, which is energetically unfavorable, but compensated by collective hydrophobic associations, increased hydration forces on the resulting, more polar protein hydration surface, and other enthalpic contributions to structural folding and hydration. Protein globular conformational entropy loss is outweighed by the presence of favorable hydrophobic interactions. Dehydration of apolar amino acids by placement in the globular core away from the aqueous environment increases the aqueous entropic part of the system, lowering the overall Gibbs free energy for a folded state. Besides conformational entropy and hydrophobic interactions, collective electrostatic

Table 1
**Performance Benchmarks for Assay Capture Surfaces
in Complex Biological Milieu**

- Reliable and consistent surface performance without surface defects or interfacial variability.
- Substantial reductions in nonspecific protein surface interactions (NSB) to <1% of a monolayer (<3 ng/cm^2) in solutions of cell or tissue lysates, serum, homogenized food, or blood.
- Elimination of "blocking" steps using sacrificially adsorbed globular, surface-active proteins such as albumin, lactoglobulin, or casein, or adsorption of amphiphilic surface-active synthetic polymers (e.g., Pluronics™) to proactively mask surfaces against NSB from other sources.
- Effective, direct surface immobilization of capture ligands to substrates to facilitate analyte capture, stringency washing, multiple-step processing under high shear and extreme pH or solvent conditions.
- Stabilization of immobilized affinity ligands in surface-capture environments to effectively preserve assay bioactivity under long residence times and lyophilized storage conditions.
- Surface immobilization treatments and protocols compatible with substrates comprising metals, plastics, metal oxides, and silicates with disposable economy.
- High-density affinity captures ligand immobilization on surfaces with minimal lateral interaction, competitive interference in assay or steric hindrance issues.
- Lot-lot and areal uniformity of affinity ligand immobilization and assay performance (signal:noise).
- Direct assay from complex milieu (serum, cell lysate, food, and environmental samples) without sample prepurification or analyte amplification or enrichment steps that cost time and money.
- Quantitative, direct correlation of assay analyte signal with sample analyte abundance.
- Reduction in capture feature sizes, surface patterning fidelity, lateral feature spacing for miniaturized formats and integrated approaches to achieve sufficiently discriminating signal:noise performance.

(Coloumbic) and van der Waals interactions, polar hydrogen and acid–base residue bonding, and effects of protein folding on bond lengths and angles are also important factors that affect protein folding to least-energy conformations, although to a much lesser extent. Therefore, with protein folding-based hydrophobic effects outweighing protein conformational entropy by small amounts, proteins tend to fold into metastable globular conformations often stabilized by multiple hydrogen bonds (each worth a few kcal/mol). Relatively low-energy barriers between various protein domain conformational states make the overall native conformation highly susceptible to local structural changes with any environmental disturbance, including introduction of an interface (e.g., solid surface or air), and a protein's intrinsic surface free energy driver for adsorption owing to intrinsic amphiphilicity. Protein-surface adsorption is often spontaneous—an energetically favored event with both enthalpic and entropic driving forces. Hence, protein nonspecific adsorption on assay capture surfaces is generally thermodynamically favored and difficult to prevent by using "simplistic" surface modification approaches.

Within the unstirred fluid boundary layer adjacent to any liquid–solid interface, protein interfacial adsorption is often transport- or diffusion-limited. Recognizing this kinetic constraint to rates of adsorption, proteins diffusing close to a hydrophobic interface adsorb spontaneously regardless of the protein's own hydrophobicity. The same hydrophobic dehydration forces promoting native protein folding in solution work in reverse at the solid–liquid interface to denature the protein over time at the interface (surface residence time), releasing surface-bound water molecules, increasing aqueous entropy of the system *(4,11–14)*. Surface adsorption energetics are generally sufficient to disrupt metastable protein globular structure, increasing both its conformational and surface hydration entropy. This collective entropy gain might be sufficiently favorable to produce spontaneous adsorption of the protein under otherwise adverse conditions, i.e., even at a hydrophilic, electrostatically repelling surface, or unfavorable adsorption enthalpy. Hydration energy at hydrophobic surfaces is unfavorable. Replacement of surface-water molecule contacts with protein-surface contacts can often be energetically favorable if accompanied by sufficient protein conformational loss on adsorption. These requirements are usually met for most globular proteins beyond a few kDa in size. The end result in all protein-surface systems, regardless of either protein or surface chemistry, is the adsorption of finite amounts of protein through these nonspecific, collective, and universal thermodynamic effects. Generally, hydrophobic surfaces adsorb much higher amounts of globular proteins than highly hydrated, polar hydrophilic surfaces *(8)*. A second important consequence is that the adsorption event is accompanied by sufficient free energy change to produce surface-, time-, and protein-dependent unfolding transitions on the surface. Significantly, this leads to several technologically important end points ubiquitously observed in "real systems" that involve surfaces and biological milieu:

- Spontaneous adsorption of proteins to all surfaces, but with increasing affinity (reduced off-rate) for increasingly hydrophobic surfaces.
- Time-dependent denaturation of proteins on surfaces regardless of either protein or surface chemistry.
- Highly irreversible protein-adsorbed states, even if physically adsorbed (no covalent bonds).
- Loss of protein structure and bioactivity accompanying surface residence time.

From the standpoint of designing diagnostic surfaces, this means that two general but important first-principle consequences hinder current surface-capture assay performance:

- Nonanalyte proteins present as spectators in the assay milieu will adsorb to the bioassay surface, producing assay background noise and blocking desired analyte–surface interactions.
- Capture agents deliberately immobilized on a surface (nucleic acid probes, affinity reagents, or antibodies) lose desired native structure and function on surfaces simply as a result of the heterogeneous interphase energetics and surface residency effects on proximal proteins.

This translates to challenges to limit NSB and capture agent inactivation using new surface features that will improve surface capture performance. With more than 200 known proteins in serum and thousands more in cell and tissue lysates, the intention to directly assay trace analytes at femtomolar or less abundances from these impure, relevant samples is severely limited by generic interfacial nonspecific adsorption dynamics that generate excess noise. Current approaches either prepurify samples to remove nonspecific fouling species and enrich the sample in analyte, or exploit surface designs that select analyte over background fouling species. Surface modification strategies must address the source of the adsorption phenomena with appropriate chemical and physical methods to limit nonspecific protein adsorption problems and enhance surface capture bioassay capabilities.

2.2. Surface Modification Strategies to Limit Protein Fouling

Improved surface performance in biological environments often reduces NSB of biologicals to surfaces by exploiting "interfacial energy matching" that typically increases surface hydration stability using polar chemistry applied to diagnostic surfaces by organic coatings *(15)*. A review of current commercial approaches for microarraying surface chemistries and coatings appears elsewhere in this book (*see* Chapter 2). Surface chemistry polar-enough to hydrate, heavily reduces surface free energy gradients to adsorption of generally highly hydrated protein "surfaces." Surface hydrophilicity is a critical parameter used to produce interfacial energy matching in aqueous milieu. "Hydrophilic" in the context of a surface functionally refers to surfaces that readily adsorb, hydrate, swell, or imbibe water, using polar or charged groups that interact strongly with water. Stable surface hydration minimizes interaction energies with other hydrated biological chemistries. Because most biological molecule surfaces are also highly hydrated as result of surface structures with polar groups. At the molecular level, native protein structures exhibit highly hydrated surfaces contributing to their folded stability. Empirically, material surface resistance to protein adsorption by most criteria appears to correlate directly (although not completely) with the surface's degree of hydrophilicity and stable, structured water. Bulk water exhibits a strong self-association through multiple clathrate hydrogen bonds. Any disturbance of this self-associated network increases its entropy, decreases enthalpy, and increases water-free energy *(6,16,17)*. Importantly, some polar, uncharged hydrophilic surfaces such as polyethylene glycol (PEG) and polysaccharides *(18,19)* exhibit Lewis acid/base strength comparable with that of water. This facilitates energetically favorable donation/accepting of electron density between water and these hydrophilic surface moieties (hydration). Anionic and cationic groups on surfaces also cluster water around each charged group based on ion-dipole forces. These cases of surface hydration are spontaneously stable. On the other hand, hydrophobic surfaces provide little chemical or energetic grounds for hydrogen bond formation or other dipole interactions with contacting water molecules, and thereby, forcing water self-network breakdown and restructuring at hydrophobic surfaces with unfavorable thermodynamic consequences.

By interacting with water through multiple interactions, hydrophilic surfaces reduce their surface energy over hydrophobic surfaces that interact with water only through pair-wise dispersion forces *(20)*. More specifically, a hydrophilic, hydrated interface creates a low interfacial tension between the material and the aqueous solution where it resides. This leads to a low Gibbs free energy change for solute adsorption to the surface from biological fluids. In this case, highly hydrated solutes perceive little energetic difference between the solution where they reside fully solvated and the surface to which they might potentially adsorb. Hence, reduced biological adsorption is often observed on highly hydrated surfaces over more polar or hydrophobic surfaces, where energies of hydration are less stable. Hydrophobic surfaces exhibit unfavorable hydration enthalpy and entropy, where contacting water is more liable to be displaced by other more favorable surface contacts through adsorption. Therefore, energetics favor interfacial adsorption, particularly with proteins that dynamically restructure to expose core hydrophobic residues to contact hydrophobic surfaces, and thereby displacing surface-structured water through protein adsorption and denaturation. This mechanism has been reviewed extensively in **refs. 8–10,13,17,21–32**.

3. Polyethylene Glycol Surfaces to Limit Protein Adsorption

In the context of exploiting stable surface hydration as a mechanism to prevent adsorption, surface modification by polar hydrophilic chemistry, or grafted or coated polymers, is widely used to achieve low adsorbing properties for biomedical and biotechnological applications *(1,5,33–38)*. Although the scope of specific, coated surface chemistries is enormous, surfaces bearing PEG or polyethylene oxide (PEO) species by a wide variety of chemical and physical methods are of particular technological and scientific interest. The PEG polymer is a commercial synthetic linear, neutral polyether of general structure, $RO-(CH_2-CH_2-O)_n-CH_2-CH_2-OH$, where R is often either CH_3 or H, but can also be many other chemically reactive functional groups *(39,40)*. The distinction between PEO and PEG is not chemical because their structures are identical, but lies in the value of n, the number of polymer repeat units. PEG is generally noncrystalline, lower molecular weight wax-like polymer, where $n \cong 4 - 40$. PEO is the higher molecular weight crystalline polymer solid. PEG and PEO are both commercially derived from the anionic polymerization of gaseous ethylene oxide, available in a wide range of molecular weights. As a Lewis basic unit, each ethylene oxide repeat unit presents a polar but uncharged hydrogen bond accepting group. Crystalline PEG is a helix with torsion angles in a trans-gauche(+)-trans conformation about the $-O-C-C-O$ bonds, shown by computer simulation to exhibit very rapid rates of conversion between trans and gauche states *(41)*. The unusually flexible ether $C-O$ bond confers unique chain rotational and conformational mobility *(2,42)*. Under aqueous environments, these repeat units readily form multiple hydrogen bonds with water molecules, creating highly hydrated, flexible polymer chains, essentially infinitely soluble in water *(1)*. PEG also exhibits wide solubility in most solvents usually adopted in organic reactions as well, making chemical modification and manipulation readily achievable on its terminal hydroxyl groups or during polymerization termination. This in turn opens up diverse possibilities for specific end group modifications to facilitate direct, specific biological attachment or surface grafting *(43)*. Commercial availability of multifunctional PEG derivatives also allows construction of multiple different surface architectures, coatings, and applications of crosslinked PEG hydrogels or tethered conjugates *(44)*. Both PEG physical adsorption on and chemical immobilizations to surfaces, in most cases through chain end point attachment, have been used to fabricate "biotolerable" surface coatings from PEG derivatives. The covalent chemical grafting approach is more robust and perceived more versatile and chemically reliable (i.e., a covalently grafted surface was shown to resist deleterious oxidative biodegradation compared with a physically grafted counterpart *[45]*). Nonetheless, noncovalent PEG graft coatings are a popular surface modifying alternative, as physisorption of modified PEGs is convenient *(46–48)*. Regardless of the surface method, effort has focused for decades on improving PEG coating performance by tailoring graft molecular weight, chain graft density on the surface, and to understand mechanisms underlying its demonstrated, consistent nonfouling interfacial properties.

The origin of PEG's uniquely effective protein and cell-resistant capabilities in coatings has long been the subject of study and discussion. Molecular attributes contributed by PEG's formal electrostatic neutrality, unique electronic/hydration structure and stable reduction in surface interfacial energy, with accompanying entropically driven polymer steric exclusion and osmotic effects, are common foci to explain its unique surface properties *(1,39)*. Based on experimental performance for PEG brush-like surfaces in blood, a "hydrated dynamic surface" model for PEG's interfacial properties was introduced in 1987 *(2,49)*. The hypothesized rapid movement of hydrated long PEG end point grafted chains on surfaces was initially proposed to resemble active cilia in some microorganisms. Microstreams of water generated by the random movements of PEG "molecular cilia" exploring their local energy space on surfaces, combined with the flexibility of single-point immobilized PEG chains, were proposed to repel proteins approaching these surfaces, schematically shown in **Fig. 1**. A quantitative model for protein adsorption on end-grafted brush-like PEG surfaces *(50,51)* compares repulsive steric exclusion forces against attractive van der Waals forces and hydrophobic effects as a function of PEG chain length (ranging from 80 to 120 residues long), grafting density, and adsorbing protein sizes. This study concluded that a weak, long-range hydrophobic attraction between PEG and protein competes with PEG's repulsive steric exclusion entropically derived force, determining the surface resistance to protein adsorption. The magnitude of repulsion is shown to be greatest for the longest grafted PEG chain length. A higher surface density of PEG is proposed to best resist smaller protein interactions. However, PEG surfaces with high grafting density, consisting of densely packed, nearly crystalline chains, might not be effective for protein resistance *(41,50–52)*.

Jeon's protein-resistant surface model *(50,51)* has been widely appreciated experimentally. For years, the general understanding of PEG's nonfouling surface properties is credited to its large exclusion volume of heavily hydrated, randomly oriented polymer chains as well as rapid segmental motion of the chain as proposed in the "molecular cilia" model *(2)*. In both cases, the protein repellant power of these PEG surfaces derives from the unfavorable entropy loss on compressing and restricting the grafted hydrated polymer layer by adsorbing protein. Szleifer *(53)* pointed out one concern with Jeon's theory *(50,51)*, its theoretical foundation in the Alexander-de Gennes theory of polymer brushes *(50,51,54)* in which the approach is only valid for very long chains in the so-called "dense brush regime" (i.e., at relatively high surface coverages). Conditions for most experimental observations of "grafted to" PEG surfaces are not within this regime. Szleifer presented a general theoretical framework for studying protein adsorption onto grafted polymer surfaces based on single-chain mean-field theory, in which the grafted polymer-protein solvent layer is assumed to be inhomogeneous in the direction perpendicular to the grafting surface. Grafted polymer chain length, density, and the polymer's attraction to the underlying substrate surface are monitored for their impact on the surface's

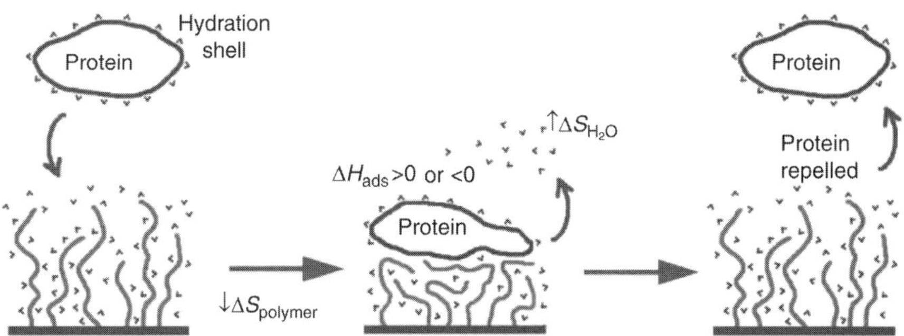

Fig. 1. The proposed "exclusion volume" mechanism *(50,51)* for hydrated polymer (e.g., PEG) surface-immobilized brushes. Protein adsorption onto surface-grafted dense PEG brush releases hydrating water molecules on both protein surface and PEG brush, creates favorable entropy gain for protein adsorption. Compression and steric hindrance of extended PEG brush by protein adsorption restrains free mobility of the polymer chains, creating unfavorable entropy loss, which compensates the entropy gain from the released water molecules, making this protein adsorption entropically unfavorable. Few solvent-cast (nonbrush) polymer films can produce brush-like thermodynamics as shown, and hence, despite similar hydrophilic surface properties, do not perform as well in biological milieu. Protein-polymer adsorption enthalpy (ΔH_{ads}) can be either favorable or unfavorable, depending on specific protein–surface pairing as described in **ref. 8**.

protein adsorption behavior *(53)*. This model indicates that proteins absorb to grafted surfaces by changing the grafted polymer structure to optimize the energetic compromise between polymer conformational entropy, protein-polymer repulsions, and surface–protein attractions. On absorbing to polymer-grafted surfaces, a simplified protein conformational change from spherical to pancake was also included in their model *(53)*. Kinetics of adsorption were shown to play a significant role in judging a surface's protein-resistant properties based on these model parameters *(53,55–57)*. To validate these interpretations, many experiments have attempted to create PEG surfaces, both for modeling purposes and practical biomedical applications, using various grafting chemistries with PEG molecular derivatives under different conditions. A primary objective has been to attempt defect-free high PEG grafting densities and consistent chain physical states, commonly called "PEG brushes," with varying chain lengths effective to limit biofouling *(58–62)*.

4. Polymer Brush Coatings as Alternatives to Solvent-Applied Polymer Coatings in Aqueous Systems

Interfacial energy matching, hydration, and hydrophilic chemistry alone have not proven sufficient to produce completely nonfouling surfaces, especially in anything but simple buffer systems. Surface coating architecture has shown to provide additional enhancements in performance in biological milieu. Polymer tethered chains on surfaces are attractive alternatives to alkylsilanes and spin-coated polymer films as interfaces for reducing adsorption events. In an end point grafted, immobilized polymer surface layer, the actual polymer chain configuration at the surface depends on its tethered density, length, and interaction forces between the polymer and itself, the solvent, and the surface. If the chain segmental chemistry has no affinity for the surface but some affinity for solvent (e.g., water), the chain adopts a coiled configuration stretching out into the solution away from the surface grafting points. Neighboring chains repel one another entropically and this repulsion influences the coil conformation when the lateral surface grafting density exceeds R_g^{-2}, where R_g is the polymer's radius of gyration, used as a measure of solvated polymer coil dimension. Many polymer molecules attached to a surface in this way produce a "polymer brush," a unique grafted coating architecture with free chain ends imparting interesting properties. The relationship between chain grafting density and polymer radius of gyration is a benchmark parameter used to describe different brush regimes. Swollen tethered chains extension lengths *(l)* from surfaces results from lateral density influences are called "dense brushes" *(54,63–66)*. Less densely grafted brushes preserve random coil polymer conformations attached to surfaces (e.g., mushroom regime) as shown in **Fig. 2**. Brushes may be attached chemically or physically, grafted by single-point or multipoint attachment, producing a wide array of different tethering structures on surfaces *(63,67)*. Many biologically relevant surface modifications utilize brush technology in attempts to block unwanted biological adsorption events. However, practical experimental fabrication limitations define the performance limits of real systems over models; polymer brush synthetic methods currently do not approximate ideal brush models in controlling density of homogeneity on surfaces.

4.1. Polymer Brush Fabrication

Surface grafting approaches with hydrophilic brushes can be roughly divided between two fabrication methods: "grafting to" and "grafting from." The "grafting to" method modifies coating substrates by polymer electrostatic, hydrophobic, or covalent multipoint surface bonds, attaching a preformed polymer to surfaces covalently through active functional coupling groups, or physically by adsorption of multiple polymer surface-active domains. Although "grafting to" methods immobilize polymer brushes robustly on surfaces, resistant to removal by common chemical environmental conditions, the resulting immobilized polymer chain surface density is generally poorly organized owing to the random, statistical nature of the polymer–surface interactions produced by diffusion-limited, stochastic polymer immobilization reactions at surface sites. Polymer

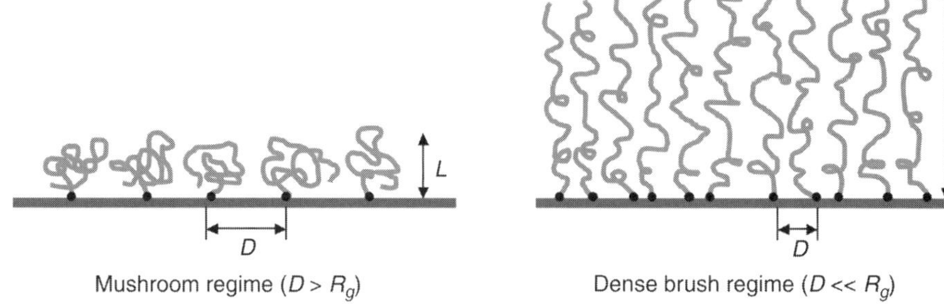

Fig. 2. Polymer brush grafting density, D, compared with radius of gyration, R_g, as an important parameter to describe polymer brush regimes. Thickness of the immobilized polymer brush depends both on D and R_g with a relationship of thickness, $L \cong (R_g^{5/3})(D^{-2/3})$. When the density of grafted points is increased beyond a certain limit, the grafted polymer coils begin to overlap, leading to polymer stretching, and increasing L as described in **ref. 65**. Such surface designs remain largely ideal and theoretical constructs because control of polymer chain immobilization density and chain length independently and reliably has proven elusive for chemistries interesting to diagnostics.

chains tethered to a surface by these random reactions at average statistical density (R_g^{-2}) form mushroom-like grafted coil structures, occupying a local volume in the surface region that sterically hinders further chain attachment *(68)*. These grafted polymer mushrooms attain limiting surface densities proportional to their solvated size (R_g) in solution (a function of their chemistry, solvation, molecular weight, and surface interactions), occupying a surface area that sterically shadows this portion of the surface and adjacent area from incoming polymer chains, shown in **Fig. 3**. Targeted ideally to react with each other for immobilization, reactive sites on both surface and incoming polymer molecules may or may not be exposed or properly reactive during transient diffusion-limited surface-polymer encounters, often limiting reaction yields, lateral homogeneity, and immobilization efficiency. For end-grafting, polymers may be only semitelechelic, averaging one possible reactive chemistry site per polymer chain. Immobilization statistics are poor for surface reactivity when a single polymer reactive group must encounter a single surface-bound reactive group along the appropriate reaction coordinate to produce a covalent bond by diffusion-limited kinetics (**Fig. 3**). Therefore, several solution-based conditions such as stirring or concentration effects are often employed to increase rates of diffusion, reaction encounters, and thereby, increase polymer surface reaction efficiencies. Often, ability to control "grafted to" immobilized chain densities is poor, fraught with lateral heterogeneities and defects, and frequently targeted as a reason for less-than-optimal biological passivation behaviors observed in applications. Kingshott and coworkers *(69)* studied the effects of "cloud point" (collapsed

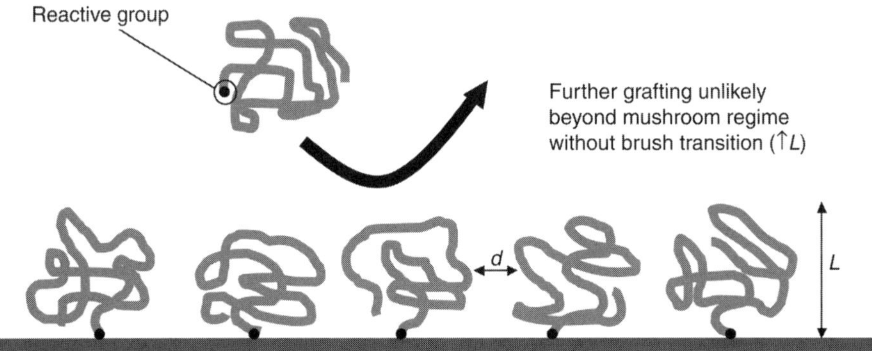

Fig. 3. Polymer chains immobilized to a surface by stochastic "grafting-to" reactions at average statistical density (R_g^{-2}) form mushroom-like grafted coil structures occupying a local volume in the surface region that sterically hinders further chain attachment. Surface reaction is density-limited by steric factors that preclude high-density grafting. Intergraft spacing of d < R_g prohibits higher grafting density without mushroom-brush transition by chain extension away from surface (increase thickness by l). Chain extension is entropically unfavored in most graft systems. The resulting mushroom regime often has surface heterogeneity and spatial defects that promote nonspecific adsorption of adsorbates that fit between the mushroom grafts.

chain, reduced R_g) against typical PEG grafting conditions to increase polymer densities and improve the ability to immobilized PEG dense brush layers to minimize adsorption from multicomponent protein solutions. Use of a marginal PEG solvent compacts PEG chains to poorly solvated collapsed coils, lowering PEG R_g and chain-chain repulsion during solution "grafting to" reactions, facilitating greater PEG chain packing density at interfaces without large typical steric hindrance complications *(69)*. Nevertheless, despite the promising approach, similar protein adsorption problems persist. For example, grafted polymers establish an initial surface occupancy that inhibits further diffusion and accessibility of additional PEG chains for surface immobilization, leading to incomplete and generally insufficient polymer tethering densities. Resulting surface defects are difficult to remedy, creating subsequent NSB problems in biological milieu resulting from lack of adequate PEG brush coverage.

The antiadsorptive effects of PEG chain length and grafting density in the "grafting to" method, as well as interfacial forces involved have been extensively investigated using model proteins, cells, and bacteria *(45,59,60,70–72)*. An interpenetrating network of poly(acrylamide-co-ethylene glycol) (p[AAm-co-EG]) was applied as a hydrogel covalently grafted onto polyethylene terephthalate angioplasty balloons to increase surface hydrophilicity *(73,74)*. N-hydroxyl-succinimide end-functionalized PEG (PEG-NHS) as a masking agent has been employed to form a protein-resistant background surrounding DNA arrays in SPR sensor fabrication *(75)*. PEG-grafted polysiloxane copolymers exhibit spontaneous surface assembly as polymer monolayers and exhibited resistance to protein adsorption in a surface plasmon resonance analysis *(76)*. Functionalized PEG monolayers *(77)* and mixed PEG brushes made up of both methoxy-terminated PEG and terminally functionalized PEG *(76)* have been explored to provide both nonspecific resistance, whereas, allowing direct covalent immobilization of other species to functionalized PEG ends. PEG-grafted poly-L-lysine (PLL-g-PEG), a polycationic copolymer positively charged at neutral pH, adsorbs spontaneously onto negatively charged surfaces (e.g., TiO_2, Ta_2O_5, Nb_2O_5, and SiO_2), forming stable, protein-resistant polymeric monolayers *(78–80)*. This surface has shown protein NSB resistance to <1% of an adsorbed monolayer from serum in vitro. More oxidation-tolerant than popular alkanethiol monolayers on gold and adsorbed films of triblock copolymer PEG-PPS-PEG on gold substantially reduced protein adsorption, even after exposure to whole blood and cell adhesion over long culture durations *(47)*. Silane-modified PEG with hydrophobic spacers attached to metal oxide substrates have been reported to reduce protein adsorption from serum *(81)*. Surface adsorption of fluoroalkyl-terminated PEG onto poly(tetrafluoroethylene) (PTFE) effectively modifies this surface from hydrophobic to hydrophilic, providing a potentially practical way to produce nonfouling surfaces on difficult to modify PTFE *(82)*.

4.2. Other Grafted Brush Chemistries

Many other "grafting to" chemistries reported involve virtually all the hydrophilic polymers (both synthetic and natural) known. Synthetic polymers includes poly(2-hydroxyethyl methacrylate) (PHEMA) *(83,84)*, polyacrylic acid *(85)*, polyvinyl alcohol *(86,87)*, polyvinyl pyrolidone, polyacrylamide-based polymers *(74,88,89)*. Natural polymers including dextrans *(90–93)*, heparin *(86,94,95)*, hyaluronic acid *(94,96)*, alginic acid *(96)*, and carboxylated celluloses. The polysaccharide dextran, activated by oxidation of anhydroglucopyranoside subunits with sodium periodate, to convert glucopyranoside subunits to cyclic hemiacetals, can then be reacted with primary amine groups on substrates *(91,97,98)*.

Alternatively, thiolated dextran (2-mercaptoethyl-carbamoyl-dextran) can be directly chemisorbed onto coinage metal substrates *(90)*. Both approaches produce multipoint attachment, not end point polymer grafting. Heparin, partially degraded with nitrous acid (HNO_2), yields fragments with reactive aldehyde groups at the reducing terminal that are then covalently linked with primary amine groups on substrate surfaces in aqueous solution through reductive amination using cyanoborohydride *(94,95)*. Hyaluronic acid and alginic acid both have multiple carboxyl groups readily coupled with surface amino groups using carbodiimide condensation *(94,96)*. A photoimmobilization approach for hyaluronic acid was also reported, using polysaccharides derivatized with azidophenylamine and subsequently ultraviolet irradiated onto *poly(ethylene)* terephthalate (PET) substrates *(99)*. This method enables surface patterning of hyaluronic acid moieties, opening up new possibilities for these coatings in microarray applications. Extensive theoretical and experimental studies have been conducted for decades with the "grafting to" method because of the wide availability of hydrophilic polymers to attach to surfaces and their empirical improvements in reducing but never eliminating biofouling. However, the many different architectures and strategies used to immobilize large hydrophilic polymers to surfaces have achieved only a fraction of the performance sought for nonfouling applications in biological milieu. Seemingly, grafting-to approaches, regardless of the polymer chemistry, have limitations experimentally that preclude true NSB resistance in relevant physiological systems.

4.3. Nonfouling Polymer Brush Surfaces Using "Grafting From" Strategies

A less-developed class of direct covalent polymer modification methods, collectively known as "grafting from," has gained substantial interest for achieving predictable and more reliable, high polymer surface grafting densities. As the method's name implies, small molecule polymer initiator precursors are first covalently attached to a surface at high, small molecule densities (monolayer, pmol/cm^2). *In situ* polymer initiation is prompted from these surface-bound initiator species using bulk-phase monomer reactions, producing grafted polymer chains that propagate from the surface but remain tethered to the substrate by single point attachment. Importantly, the "grafting from" strategy yields reliably higher, single point grafting densities, both theoretically and experimentally, relative to the "grafting to" technique *(100)*. However, "grafting from" suffers from problems of bulk-phase chain initiation and competitive propagation, monomer diffusion-limited chain growth, surface recombination reactions, and nonhomogeneous lateral brush chain length distributions from less- or nonreactive surface initiators. Although significant work has been focused on hydrophobic brushes *(101,102)*, less attention has been paid to more biologically relevant hydrophilic "grafting from" surfaces. Patten and coworkers *(101)* have employed atom transfer radical polymerization (ATRP) to graft a number of different hydrophilic monomers from surfaces of silica particles. Typically, rapid rates of polymerization are observed even under remarkably mild reaction conditions *(103–107)*. Jones and Huck *(108)* have also reported ATRP of hydrophilic monomers such as 2-hydroxyethyl methacrylate from planar gold surfaces in water–methanol mixtures at room temperature to produce

Table 2
Comparison of the Polymer "Grafting To" With "Grafting From" Techniques

Grafting to	Grafting from
• Polymer obtained prior to surface immobilization	• Polymer grows from the surface
• No initiator needed on the surface. Matched functional reactive groups required on both polymer and surface	• Initiator is first immobilized onto the substrate surface
• Immobilized polymer occupies a mushroom conformation on the surface	• Polymer chains are more brush-like
• Low surface coverage	• High surface coverage
• Low lateral density control	• High lateral density control
• Low surface density achieved	• High surface density achieved
• Thin film thickness (30–50 monolayers)	• Thickness controlled by polymerization conditions (2–200 nm)
• Heterogeneous coverage problems (defects)	• Polymer growth kinetic problems produce chain length heterogeneity (defects)

hydrophilic polymer brushes of up to 125 nm thickness. Chilkoti and coworkers *(109)* employed the ATRP method in grafting PEG-containing comb polymer to gold to control adsorption of proteins at surfaces *(109)*. Surface grafting of hydrophilic acrylamide monomers was also reported by Lopez to tune transitions in surface energy *(110)*. Synthesis of unusually thick PHEMA films on gold surfaces by surface-initiated ATRP was reported by Bruening and coworkers *(111)*. Film thickness up to 700 nm can be achieved with loosely packed brush-like chains that become increasingly extended as the PHEMA hydroxyl groups are derivatized as esters, carbamates, and similar groups. Gradient brush surfaces of grafted polyacrylic acid formed on gold SPR surfaces have been fabricated by Bohn using electrochemical initiation from the electrode surface as a potential artificial cell adhesion surface with extracellular matrix mimics. Both chemical (polymer composition) and physical (thickness) properties of this model anisotropic film may be varied spatially to modify biological interactions *(112)*. **Table 2** compares the two surface grafting methods for forming grafted polymer brushes.

5. Alternative Well-Controlled Interfacial Chemistries to Brushes

Producing homogeneous, controlled density, long chain PEG grafting on different commercially important surfaces is synthetically challenging. As reviewed elsewhere in this text (*see* Chapter 2), many current polymer coatings, applied as spin- or dip-coated films, do not perform sufficiently to satisfy current design or performance (i.e., signal:noise) requirements in microarray assays, particularly from anything but simple buffers. In order to better elucidate physical mechanisms underlying PEG's superior nonfouling properties in more detail, model systems comprising self-assembled monolayers (SAMs) of oligoethylene glycol (OEG) alkanethiols on gold, first introduced by Prime and Whitesides in 1993 *(113)* and extensively developed over the subsequent decade *(114–119)*, have become increasingly popular. SAMs offer a molecularly well-defined model surface suitable for precise physical measurements at interfaces and systematically tailorable using routine organic synthetic methods. Although conformational freedom of the OEG end groups is restricted in these densely packed films compared with brush-like polymeric PEG (i.e., OEGs have little unfavorable conformational entropy loss upon protein adsorption), they exhibit high protein resistant properties in various dilute protein conditions *(113,118,119)*. This observation has prompted the newer hypothesis that observed protein repellent properties of EO surfaces do not necessarily derive from PEG chain conformational energetics, but require further explanation.

Newer detailed molecular conformational studies of OEG-terminated SAMs on gold and silver substrates by Grünze and coworkers *(114–117)* show that methoxy-terminated OEG self-assembled epitaxial monolayers form on both of these metal substrates. However, only SAMs formed on gold demonstrate high nonfouling properties toward proteins. Tilt angles of hydrocarbon SAMs on gold and silver supports are distinctly different (30–35° on Au vs nearly 90° on Ag) *(120–123)*, allowing each to assume different densities and lateral interactions. Areas per molecules for the alkyl SAMs on Au<111> are about 0.184 nm^2, increasing to 0.216 nm^2 on Ag<111> *(121)*. These significant differences in adsorbed molecular densities allow each EO-SAM on these respective supports to hydrate differently and manifest the effects of different hydration structures in their respective protein adsorptive properties. The helical conformer of the EO$_3$-OMe SAM on Au exhibits a favorable epitaxial lattice match spacing on the Au<111> crystal surface that accommodates monolayer hydration and penetration of hydroxide ions *(116)*. This spacing match allows hydrating water molecules to form hydrogen bonding bridges among OEG SAM chains, increasing the OEG SAM water content and stabilizing the overall layer hydration energies and hydrated structure. The levels of incorporated water disorder the OEG component of the SAM and, as a result, increase the overall entropic contribution of the OEG SAM layer. The EO$_3$-OMe SAM with this hydrated conformation on Au substrates presents a stable, hydrated surface with a negative ζ potential from imbibed hydroxyl ions although the SAM molecules are formally neutral *(115,116)*. This observed SAM negative ζ potential was argued to relate to the SAM protein resistance, as most proteins also bear a net negative surface charge (acidic pI) *(124–126)*. Negative surface charging in aqueous solution is by no means unique to these formally neutral EG$_3$-OMe SAMs as it occurs also for other "neutral" SAMs and most polymer surfaces *(127)* that are clearly not protein resistant and are not penetrated by water. Grünze and coworkers *(128)* then assembled alkanethiols of different backbone units, chain lengths, and OEG terminal groups of various hydrophilicities onto both Au and Ag substrates (**Table 3**). SAM surface densities were correlated with their ability to resist protein fouling from solution. Only combinations of several parameters, namely terminal group hydrophilicity, the hydrophilicity of the internal units, and the SAM lateral packing density (Au vs Ag thiolate epitaxy) facilitate full SAM protein resistance. These factors inevitably point to the importance of hydration and water structure at the interface that effectively shield hydrophobic interactions between proteins and surface, and stabilize hydration by increasing overall system entropy through SAM disorder. More recently, molecular simulations of protein resistant behavior of OEG-terminated SAMs were performed by Jiang *(129)*. Simulations involving not only water and SAMs as in the Grünze case but also proteins were included, providing a more comprehensive system representation. Their simulation suggests a correlation between OEG surface resistance to protein adsorption and quantities of tightly bound water molecules surrounding OEG chains. Improved protein resistance was observed for intermediate surface densities, whereas, the too dense and too dilute OEG surfaces exhibited increased protein adsorption. Stable hydration of OEG chains or associated high OEG chain flexibility appears to be linked to their nonfouling properties. This molecular simulation was later confirmed by experimental studies of protein adsorption onto OEG-terminated SAMs, supporting the hypothesis that both chain mobility and hydration are important for surface resistance to protein adsorption *(130)*. Another systematic investigation on surface chemistry parameters involved in reducing protein adsorption, in this specific case, a series of terminal polar groups, was conducted by Whitesides and coworkers *(131)*. SAM surfaces bearing a number of systematically varied polar and apolar functional groups were prepared and examined for protein adsorption behavior *(132)*. Despite practical limitations observed for most polar chemistries in less organized solvent-cast polymer coatings, organized SAM functional groups other than ethylene oxide units have been identified as useful in designing alternative nonfouling surfaces. (**Table 4**). These functional groups were found to share certain molecular characteristics, specifically:

1. Polar functional groups.
2. Hydrogen bond accepting groups.
3. No hydrogen bond donor groups or net charge.

Table 3
Overview of the Synthetic Oligo(ether)Alkanethiols Surveyed
for Protein Interfacial Studies

Oligo(ethylene glycol) termination $HS-(CH_2)_{11}(O-CH_2-CH_2-O)_nR$ $n = 1,2,3,6$	
EG$_1$OMe *(1)*	$HS-(CH_2)_{11}-(O-CH_2-CH_2)_1-OMe$
EG$_2$OH *(2)*	$HS-(CH_2)_{11}-(O-CH_2-CH_2)_2-OH$
EG$_2$OMe *(3)*	$HS-(CH_2)_{11}-(O-CH_2-CH_2)_2-OMe$
EG$_3$OH *(4)*	$HS-(CH_2)_{11}-(O-CH_2-CH_2)_3-OH$
EG$_3$OMe *(5)*	$HS-(CH_2)_{11}-(O-CH_2-CH_2)_3-OMe$
EG$_3$OEt *(6)*	$HS-(CH_2)_{11}-(O-CH_2-CH_2)_3-OEt$
EG$_3$OPr *(7)*	$HS-(CH_2)_{11}-(O-CH_2-CH_2)_3-OPr$
EG$_3$OBu *(8)*	$HS-(CH_2)_{11}-(O-CH_2-CH_2)_3-OBu$
EG$_6$OH *(9)*	$HS-(CH_2)_{11}-(O-CH_2-CH_2)_6-OH$
EG$_6$OMe *(10)*	$HS-(CH_2)_{11}-(O-CH_2-CH_2)_6-OMe$
EG$_6$OEt *(11)*	$HS-(CH_2)_{11}-(O-CH_2-CH_2)_6-OEt$
EG$_6$OPr *(12)*	$HS-(CH_2)_{11}-(O-CH_2-CH_2)_6-OPr$
Oligo(propylene glycol) termination $HS-(CH_2)_{11}(O-CH-CH_2-O)_nCH_3$ with CH_3 side chain $n = 2,3,4$	
PRO$_2$OMe *(13)*	$HS-(CH_2)_{11}-(O-CH(CH_3)-CH_2)_2-OMe$
PRO$_3$OMe *(14)*	$HS-(CH_2)_{11}-(O-CH(CH_3)-CH_2)_3-OMe$
PRO$_4$OMe *(15)*	$HS-(CH_2)_{11}-(O-CH(CH_3)-CH_2)_4-OMe$
Oligo(trimethylene glycol) termination $HS-(CH_2)_{11}(O-CH_2-CH_2-CH_2)_nO-R$ $n = 3$	
TRI$_3$OH *(16)*	$HS-(CH_2)_{11}-(O-CH_2-CH_2-CH_2)_3-OH$
TRI$_3$OMe *(17)*	$HS-(CH_2)_{11}-(O-CH_2-CH_2-CH_2)_3-OMe$

Abbreviations used here (and in the text) refer to the following chemistries: EG$_3$OEt—tri(ethylene glycol) unit with an ethoxy end group; PRO$_3$OMe—tri(propylene glycol) unit with a methoxy end group; TRI$_3$OH—trimethylene glycol unit with hydroxyl termination. All molecules are derivatives of ω-functionalized undecylthiol (for thiolate anchoring on gold and silver surfaces).

Used with permission from the American Chemical Society, *see* **ref. 128**.

Table 4
Overview of SAM Surface Chemistry Derivatives

Entry No.	HNRR'	ML(%) Fib[a,b,c]	ML(%) Lys[a,b,c]	θ_a[d]
1	$H_2N(CH_2)_{11}CH_3$	100	100	163[e]
2	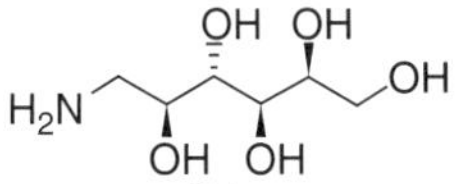 (sorbitol-type structure, H_2N–CH₂...OH OH OH OH–CH₂OH)	68	20	37
3	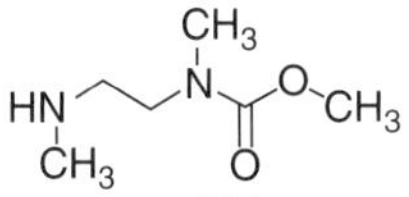 (CH₃ N-carbamate, HN–CH₃)	58	43	75
4	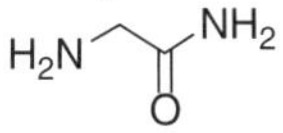 (H_2N–CH₂–C(=O)–NH₂)	58	30	39
5	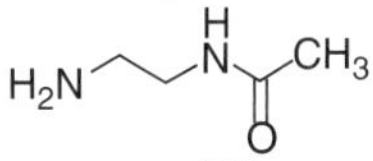 (H_2N–CH₂CH₂–NH–C(=O)–CH₃)	40	5	49
6	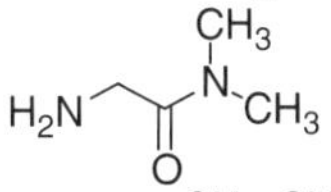 (H_2N–CH₂–C(=O)–N(CH₃)CH₃)	33	15	61
7	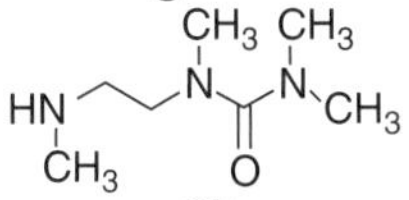 (HN(CH₃)–CH₂CH₂–N(CH₃)–C(=O)–N(CH₃)CH₃)	25	11	62
8	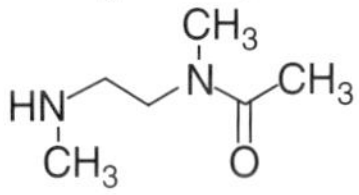 (HN(CH₃)–CH₂CH₂–N(CH₃)–C(=O)–CH₃)	12	4	53
9	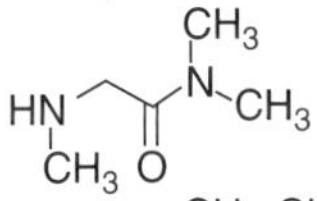 (HN(CH₃)–CH₂–C(=O)–N(CH₃)CH₃)	9	2	65
10	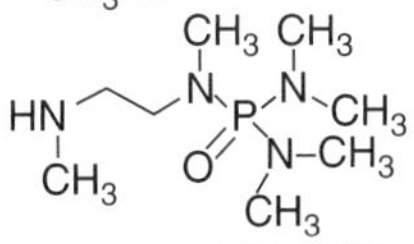 (HN(CH₃)–CH₂CH₂–N(CH₃)–P(=O)(N(CH₃)CH₃)–N(CH₃)CH₃)	4	<1	66
11	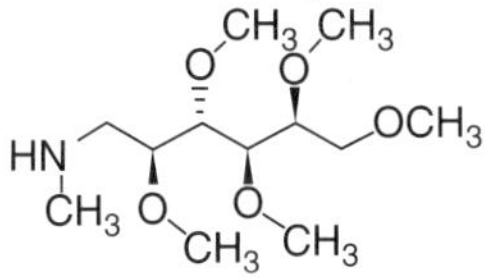 (permethylated sorbitol, HN(CH₃)–CH₂...OCH₃...OCH₃)	3	6	81
12	$H_2N(CH_2CH_2O)_3H$	2	1	54

Used with permission from the American Chemical Society, to study the hypothesized requirements for protein resistance to surface adsorption *see* **ref. *132*.**

Protein-resistant phosphorylcholine coating chemistry *(133,134)* and mannitol-terminated SAM surfaces *(132,135)* are both notable exceptions to this general characterization of protein resistant chemistry, although little data are available to explain their interesting empirical protein interfacial performance. Again, surface hydration and resulting thermodynamic stability must be involved in this behavior.

The practical utility of these general design rules to create chemistries that displace PEG brushes as the functional technological "gold-standard" for coating protein resistance with new alternative chemistries remains to be seen. Further development or identification of such chemistry exhibiting reliable and effective protein nonfouling properties is of great technical importance to provide additional options to suit better individual nonfouling biological applications, including diagnostic assays and clinical sensors. Second, additional examples of biologically inert surfaces would aid in understanding general mechanisms of surface protein resistance, further assisting the design of alternative surface modifications effective for nonfouling applications *(131,135,136)*.

Whereas, surfaces of oligo(ethylene glycol) derivatives have been studied extensively but they are not completely understood. Oligo(EO) SAMs on gold have been shown to be the most effective protein-resistant surface coating to date, but they have certain limitations. Thiolate epitaxial anchoring on gold tends to auto-oxidize, proving problematic in certain applications where metal substrates might not be ideal and where oxidative chemistry might be present. Peroxide formation and chain cleavage are concerns for any PEG-based material in long-term storage, but this usually requires oxidizing conditions, heavy metal ions or thermal catalysis *(40)*. PEG brush coating may also be limited for ligand grafting applications where each surface-immobilized PEG molecule couples only one ligand at <100% yield, producing low surface concentrations of ligands. Fabrication of such PEO ligand-coupled surfaces is also complex, often requiring heterobifunctional coupling group additions to brush (e.g., PEO) molecule termini or the substrate (e.g., NHS, maleimide, acrylate, or silane). Because PEG has a stable, associated hydration difficult to remove except under extreme drying conditions (i.e., azeotropic distillation), hydrolytically labile reactive derivatives (e.g., NHS) can also exhibit stability issues in storage.

Radio frequency glow discharge (RFGD) polymerization has recently proven successful to deposit PEO-like coatings *(137–139)*. Pulsed RFGD deposition of slightly volatile glycol ether ("glyme") compounds was performed with coating conditions selected to minimize monomer fragmentation. Electron spectroscopy for chemical analysis (ESCA) and static time of flight secondary ion mass spectrometry (ToF-SIMS) that are used to characterize the surface chemistry of these glyme coatings exhibit signals similar to PEG brush surfaces. Fibrinogen adsorption to the coated substrate surfaces was very low, <10 ng/cm^2 measured with isotope labeling and ToF-SIMS. The RFGD coating approach allows use of complex substrate geometries and substrate chemistries. The resulting RFGD coating surface chemistry can be controlled using different chemical precursors for plasma deposition and by varying the plasma deposition conditions. It shows considerable practical promise to process automation and reproducibility compared with conventional coating methodologies applicable to PEO and other hydrophilic macromolecules *(5)*.

6. Functional Nonfouling Assessments of Surface Chemistries in Biological Systems

The term "nonfouling" has been empirically defined to represent those surfaces that resist the adsorption of proteins and/or adhesion of cells under biological conditions. No standard technical definition for "protein adsorption resistance," for "NSB," or conditions under which these properties are measured has been developed for functional assays. Whereas, as protein adsorption assays are widely used as a "direct" raw measure of a surface's nonfouling capability. Each different research and application focus arbitrarily selects proteins of different size, charge, and concentrations, sometimes as complex mixtures (e.g., dilute serum) and sometimes alone in buffer. "Protein resistance" is typically reported as a percent reduction in protein adsorption at a specific time as compared with an untreated control, over time-scales of minutes to days. Despite theoretical predictions that PEG brush mechanisms inhibiting fouling are kinetic *(53)*, very few studies actually examine adsorption levels at equilibrium under relevant assay conditions. To be functionally relevant, assay protocols for new applications generally require experimental verification that the surface remains sufficiently inert to adsorption under specific conditions in which it will be used *(140)*. Additionally, equilibrium surface–protein interactions in multicomponent systems are recognized for decades to be a function of competitive conditions distinct from single protein isotherms. The "Vroman effect" *(141,142)* and other kinetic phenomena *(143,144)* occurring in multicomponent competitive protein–surface milieu clearly limits the utility of single protein-adsorption assays in predicting more complex performance. Effects of postassay rinsing, removal of loosely adsorbed proteins and effects of air-drying must also be considered in context of actual use to define protein resistance *(145)*. Single protein adsorption assay from buffer is widely known to produce distinct and often falsely encouraging "NSB" results completely different than either this same protein's surface uptake from binary or more complex solutions, or total protein uptake from relevant biological fluids (cell lysate, serum, or bacterial broths). Inequity of these "nonfouling" claims under widely varying conditions has led to confounding, often misleading conceptions of true performance in "real conditions." In fact, all claims to "complete protein resistance" or "NSB" performance in vitro in simple assay systems have rarely proven valid in more complex biological milieu, either in vitro or in vivo. Lack of surface performance "connectivity" between in vitro and in vivo protein fouling assays is problematic, as little predictive relevance from in vitro simplified assays can be extended to in vivo performance in complex host fluids, in vitro in full serum or cell lysate assays. This has profound implications for *"ab initio"* design of functional surface coatings where interactions of proteins with surfaces might be better understood and thereby controlled for any desired technology.

The current lack of surface adsorption performance correlations and fabrication methods that create surface chemistry reliably and defect-free at sufficient molecular resolution to block protein–surface interactions stymie most efforts to move this field forward rationally and scientifically. It is also critical to note that reported protein resistance performance is also dependent on the surface techniques utilized to measure it. With the development of high-sensitivity surface analytical techniques (e.g., OWLS and ToF-SIMS), lower levels of protein surface deposition are detectable with greater certainty *(79,146)*. Finally, it should be noted that protein adsorption (or nonadsorption) may not correlate to a desired performance endpoint in specific applications. For example, PEO-like plasma polymerized tetraglyme surfaces that reduce fibrinogen adsorption to <10 ng/cm^2 in vitro failed to reduce leukocyte or macrophage adhesion in the presence of plasma, whole blood, or when implanted in mice *(145)*. Several in vivo studies on PEO-containing surfaces even showed increased neutrophil or macrophage adhesion over control surfaces *(147,148)*. Additionally, bacteria tend to adhere and colonize almost any type of surface *(45,149)*. These microorganisms deposit their own binding proteoglycan-adhesives, called adhesins, independent of surface composition or charge. Surface resistance to bacterial adhesion remains an unsolved problem with the same challenges as all other nonfouling surfaces. Therefore, a "low-fouling" designation for these surfaces seems more appropriate since none truly show complete resistance despite reduced protein interfacial activity. Hence, a truly complete "nonfouling" surface is not yet achievable, especially in complex milieu relevant to most biological samples *(1,45)*.

7. Summary: Surface Chemistry Design Principles in Diagnostic Applications

The limitations of surface chemistry in improving the performance of diagnostic capture assays, particularly those operating in complex biological multicomponent systems, are widely acknowledged, and many precedents can be traced back to fundamental mechanisms described for biointerfacial adsorption for 50 yr. Microarray miniaturization does not eliminate any of these mechanisms and, in fact, makes their analysis more difficult. Assay performance is most frequently limited by substandard surface properties that are unable to eliminate nonspecific adsorption of proteins and other components (noise) that compromise capture and detection of analytes (signal). Microarray assay devices compound signal:noise performance issues by reducing platform dimensionality (capture area) without commensurate reduction in confounding noise. Interfacial physical chemical requirements for ideal surfaces for these applications are well-developed from a first-principles perspective: interfacial energy gradients at interfaces that produce adsorption through several types of intermolecular forces can be explained and surfaces that minimize these forces have been described and experimentally attempted. Hydrophilic coatings incorporating high densities of polar, Lewis base chemistry design features that stabilize interfacial hydration are currently a major focus. Immobilized polymer brush coatings fabricated with various surface-grafting methods are also a focus for several technologies, involving mixed immobilized coatings of hydrophilic and reactive functionalized polymer chains. Polyethylene glycol is a prominent chemistry because of its demonstrated success and versatility in reducing NSB in interfacial applications, commercial availability in many different terminal chemistries and highly purified forms. Many architectures and theoretical studies aim to explain its performance in context, microarray, spatially patterned fluidics, and diagnostic technologies will all benefit from improved surface capture chemistries where affinity reagents can be immobilized to operate with high capacity and high selectivity within a nonfouling, low adsorption background surface matrix. Unfortunately, few successful solutions with sufficient performance in "real" biological samples are available for commercial application to immobilize proteins and nucleic acids.

Acknowledgments

Support from National Institutes of Health grants EB1473 and EB726 is gratefully acknowledged. Technical discussions with V. Hlady, D. G. Castner, E. Vogler, M. Lochhead, and C. Greef are appreciated.

References

1. Hoffman, A. S. (1999) Nonfouling surface technologies. *J. Biomat. Sci. Polym. Ed.* **10,** 1011–1014.
2. Andrade, J. D., Nagaoka, S., Cooper, S., Okano, T., and Kim, S. W. (1987) Surfaces and blood compatibility. Current hypotheses. *ASAIO Trans.* **33,** 75–84.
3. Merrill, E. W. (1987) Distinctions and correspondences among surfaces contacting blood. *Ann. NY Acad. Sci.* **516,** 196–203.
4. Malmsten, M. (ed.) (2003) *Biopolymers at Interfaces, 2nd ed., Surfactant Science Series.,* vol. 110, Marcel-Dekkar, NY.
5. Kingshott, P. and Griesser, H. J. (1999) Surfaces that resist bioadhesion. *Curr. Opin. Solid St. M.* **4,** 403–412.

6. Vogler, E. A. (2001) On the origins of water wetting terminology, in *Water in Biomaterials Surface Science,* (Morra, M., ed.), John Wiley & Sons Ltd., West Sussex, UK, pp. 149–182.

7. Garcia, C. A., Hummer, G., and Soumpasis, D. M. (2001) Theoretical and computational methods of biomolecular hydration, in *Water in Biomaterials Surface Science*, (Morra, M., ed.), John Wiley & Sons Ltd., West Sussex, UK, pp. 25–52.

8. Norde, W. (2003) Driving forces for protein adsorption at solid surfaces, in *Biopolymers at Interfaces, 2nd ed., vol. 110, Surfactant Science Series* (Malmsten, M., ed.), Marcel-Dekkar, NY, pp. 21–43.

9. Roth, C. M. and Lenhoff, A. M. (2003) Quantitative modeling of protein adsorption, in *Biopolymers at Interfaces, 2nd ed., vol. 110, Surfactant Science Series* (Malmsten, M., ed.), Marcel-Dekkar, NY, pp. 71–94.

10. Ramsden, J. J. (2003) Protein adsorption kinetics, in *Biopolymers at Interfaces, 2nd ed., vol. 110, Surfactant Science Series* (Malmsten, M., ed.), Marcel-Dekkar, NY, pp. 199–220.

11. Britt, D. W., Jogikalmath, G., and Hlady, V. (2003) Protein interactions with monolayers at the air-water-interface, in *Biopolymers at interfaces, 2nd ed., vol. 110, Surfactant Science Series* (Malmsten, M., ed.), Marcel-Dekkar, NY, pp. 415–434.

12. Horbett, T. A. and Brash, J. L. (1987) "Proteins at interfaces: current issues and future prospects," vol. 343 (Horbett, T. A. and Brash, J. L. ed.), *ACS Sym. Ser.* pp. 1–33.

13. Brash, J. L. and Horbett, T. A. (1995) An overview in *Proteins at Interfaces.* vol. 602 (Horbett, T. A. and Brash, J. L. ed.), *ACS Sym. Ser.* pp. 1–23.

14. Andrade, J. D. (ed.) (1985) *Surface and Interfacial Aspects of Biomedical Polymers*, Plenum Press, New York.

15. Coleman, D. L., King, R. N., and Andrade, J. D. (1974) The foreign body reaction: a chronic inflammatory response. *J. Biomed. Mater. Res.* **8,** 199–211.

16. Andrade, J. D. and Hlady, V. (1986) Protein adsorption and materials biocompatibility: a tutorial review and suggested hypotheses. *Adv. Polym. Sci.* **79,** 1–63.

17. Vogler, E. A. (1999) Water and the acute biological response to surfaces. *J. Biomat. Sci. Polym. Ed.* **10,** 1015–1045.

18. Israelachvili, J. and Wennerstrom, H. (1996) Role of hydration and water structure in biological and colloidal interactions. *Nature (London)* **379,** 219–225.

19. Israelachvili, J. N. (2000) Short-range and long-range forces between hydrophilic surfaces and biopolymers in aqueous solutions. *Hydrocolloids, (based on presentations at [the] Osaka City University International Symposium 98, Joint Meeting with the 4th International Conference on Hydrocolloids), Osaka, October 4–10, 1998* vol. 1 pp. 3–21.

20. Vogler, E. A. (1998) Structure and reactivity of water at biomaterial surfaces. *Adv. Colloid Interface Sci.* **74,** 69–117.

21. Andrade, J. D., Hlady, V., Feng, L., and Tingey, K. (1996) Proteins at interfaces: principles, problems, and potential. *Bioprocess Technol.* **23,** 19–55.

22. Bhaduri, A. and Das, K. P. (1999) Proteins at solid/water interface—a review. *J. Disper. Sci. Technol.* **20,** 1097–1123.

23. Elwing, H., Askenda, A., Ivarsson, B., Nilsson, U., Welin, S., and Lundstroem, I. (1987) Protein adsorption on solid surfaces: physical studies and biological model reactions. *ACS Sym. Ser.* **343,** 468–489.

24. Greig, R. G. and Brooks, D. E. (1981) Protein adsorption. *J. Colloid Interf. Sci.* **83,** 661–662.

25. Hlady, V. and Buijs, J. (1996) Protein adsorption on solid surfaces. *Curr. Opin. Biotechnol.* **7,** 72–77.

26. Kleijn, M. and Norde, W. (1995) The adsorption of proteins from aqueous solution on solid surfaces. *Heterogen. Chem. Rev.* **2,** 157–172.

27. Malmsten, M. (2000) Protein adsorption at the solid-liquid interface in *Protein Architecture*, (Lvov, Y. and Moehwald, H., ed.), Marcel Dekker Inc., New York, NY, pp. 1–23.

28. Nakanishi, K., Sakiyama, T., and Imamura, K. (2001) On the adsorption of proteins on solid surfaces, a common but very complicated phenomenon. *J. Biosci. Bioeng.* **91,** 233–244.

29. Norde, W. (1980) Adsorption of proteins at solid surfaces. *Polym. Sci. Technol. (Plenum)* **12B,** 801–825.

30. Norde, W. (2000) Proteins at solid surfaces, in *Physical Chemistry at Biological Interfaces*, (Baszkin, A. and Norde, W., ed.), Marcel Dekker Inc., New York, NY, pp. 115–135.

31. Ramsden, J. J. (1997) Protein adsorption at the solid/liquid interface. *Conference of the Colloid Chemistry: in Memoriam of Aladar Buzagh, Procceding 7th, Eger, Hungary, September 23–26, 1996*, pp. 148–151.

32. Wahlgren, M. and Arnebrant, T. (1991) Protein adsorption to solid surfaces. *Trends Biotechnol.* **9,** 201–208.

33. LaPorte, R. J., (ed.) (1997) *Hydrophilic Polymer Coatings for Medical Devices: Structure/Properties, Development, Manufacture and Applications*, Technomic, Lancaster, PA.

34. Mrksich, M. and Whitesides, G. M. (1997) Using self-assembled monolayers that present oligo(ethylene glycol) groups to control the interactions of proteins with surfaces. *ACS Sym. Ser.* **680,** 361–373.

35. Andrade, J. D., Hlady, V., and Jeon, S. I. (1996) Poly(ethylene oxide) and protein resistance. Principles, problems, and possibilities. *Adv. Chem. Ser.* **248,** 51–59.

36. Lee, J. H., Lee, H. B., and Andrade, J. D. (1995) Blood compatibility of polyethylene oxide surfaces. *Prog. Polym. Sci.* **20,** 1043–1079.

37. Needham, D., Hristova, K., McIntosh, T. J., Dewhirst, M., Wu, N., and Lasic, D. D. (1992) Polymer-grafted liposomes: physical basis for the stealth property. *J. Lipos. Res.* **2,** 411–430.

38. Castner, D. G. and Ratner, B. D. (2002) Biomedical surface science: foundations to frontiers. *Surf. Sci.* **500,** 28–60.

39. Harris, J. M. (ed.) (1992) *Poly(Ethylene Glycol) Chemistry: Biotechnical and Biomedical Applications*, Plenum Press, New York, NY.

40. Harris, J. M. and Zalipsky, S. (eds.) (1997) *Polyethylene Glycol: Chemistry and Biological Applications*, ACS, Washington, DC.

41. Lim, K. and Herron, J. N. (1992) Molecular simulation of protein-PEG interaction, in *Poly(ethylene Glycol) Chemistry*, (Harris, M., ed.), Plenum Press, New York, NY, pp. 29–56.
42. Valentini, M., Napoli, A., Tirelli, N., and Hubbell, J. A. (2003) Precise determination of the hydrophobic/hydrophilic junction in polymeric vesicles. *Langmuir* **19,** 4852–4855.
43. Harris, J. M. and Chess, R. B. (2003) Effect of pegylation on pharmaceuticals. *Nat. Rev. Drug Discov.* **2,** 214–221.
44. Graham, N. B. (1992) Poly(ethylene glycol) gels and drug delivery, in *Poly(ethylene Glycol) Chemistry*, (Harris, M., ed.), Plenum Press, New York, NY, pp. 263–281.
45. Kingshott, P., Wei, J., Bagge-Ravn, D., Gadegaard, N., and Gram, L. (2003) Covalent attachment of poly(ethylene glycol) to surfaces, critical for reducing bacterial adhesion. *Langmuir* **19,** 6912–6921.
46. Caldwell, K. D. (1997) Surface modifications with adsorbed poly(ethylene oxide)-based block copolymers. Physical characteristics and biological use. *ACS Symp. Ser.* **680,** 400–419.
47. Bearinger, J. P., Terrettaz, S., Michel, R., et al. (2003) Chemisorbed poly(propylene sulphide)-based copolymers resist biomolecular interactions. *Nat. Mater.* **2,** 259–264.
48. Li, J. -T., Carlsson, J., Huang, S. -C., and Caldwell, K. D. (1996) Adsorption of poly(ethylene oxide)-containing block copolymers. A route to protein resistance. *Adv. Chem. Ser.* **248,** 61–78.
49. Merrill, E. W. and Salzman, E. W. (1983) Polyethylene oxide as a biomaterial. *ASAIO J. (1978–1985)* **6,** 60–64.
50. Jeon, S. I., Lee, J. H., Andrade, J. D., and De Gennes, P. G. (1991) Protein–surface interactions in the presence of polyethylene oxide. I. Simplified theory. *J. Colloid Interf. Sci.* **142,** 149–158.
51. Jeon, S. I. and Andrade, J. D. (1991) Protein–surface interactions in the presence of polyethylene oxide. II. Effect of protein size. *J. Colloid Interf. Sci.* **142,** 159–166.
52. Halperin, A. (1999) Polymer brushes that resist adsorption of model proteins: design parameters. *Langmuir* **15,** 2525–2533.
53. Szleifer, I. (1997) Protein adsorption on surfaces with grafted polymers: a theoretical approach. *Biophys. J.* **72,** 595–612.
54. De Gennes, P. G. (1979) *Scaling Concepts in Polymer Physics*, Cornell Univ. Press, Ithaca, NY.
55. McPherson, T., Kidane, A., Szleifer, I., and Park, K. (1998) Prevention of protein adsorption by tethered poly(ethylene oxide) layers: experiments and single-chain mean-field analysis. *Langmuir* **14,** 176–186.
56. Satulovsky, J., Carignano, M. A., and Szleifer, I. (2000) Kinetic and thermodynamic control of protein adsorption. *Proc. Natl. Acad. Sci. USA* **97,** 9037–9041.
57. Fang, F. and Szleifer, I. (2002) Effect of molecular structure on the adsorption of protein on surfaces with grafted polymers. *Langmuir* **18,** 5497–5510.

58. Emoto, K., Harris, J. M., and Van Alstine, J. M. (1996) Grafting poly(ethylene glycol) epoxide to amino-derivatized quartz: effect of temperature and pH on grafting density. *Anal. Chem.* **68,** 3751–3757.
59. Malmsten, M., Emoto, K., and Van Alstine, J. M. (1998) Effect of chain density on inhibition of protein adsorption by poly(ethylene glycol) based coatings. *J. Colloid Interface Sci.* **202,** 507–517.
60. Efremova, N. V., Sheth, S. R., and Leckband, D. E. (2001) Protein-induced changes in poly(ethylene glycol) brushes: molecular weight and temperature dependence. *Langmuir* **17,** 7628–7636.
61. Papra, A., Gadegaard, N., and Larsen, N. B. (2001) Characterization of ultrathin poly(ethylene glycol) monolayers on silicon substrates. *Langmuir* **17,** 1457–1460.
62. Raviv, U., Frey, J., Sak, R., Laurat, P., Tadmor, R., and Klein, J. (2002) Properties and interactions of physigrafted end-functionalized poly(ethylene glycol) layers. *Langmuir* **18,** 7482–7495.
63. Zhao, B. and Brittain, W. J. (2000) Polymer brushes: surface-immobilized macromolecules. *Prog. Polym. Sci.* **25,** 677–710.
64. De Gennes, P. G. (1976) Scaling theory of polymer adsorption. *J. Phys. Paris* **37,** 1445–1452.
65. De Gennes, P. G. (1980) Conformations of polymers attached to an interface. *Macromolecules* **13,** 1069–1075.
66. Alexander, S. (1977) Adsorption of chain molecules with a polar head: a scaling description. *J. Phys. Paris* **38,** 983–987.
67. Halperin, A. (1994) On polymer brushes and biology: an introduction. *NATO ASI Ser. B: Phys.* **323,** 33–56.
68. Stokes, R. J. and Evans, D. F. (eds.) (1996) *Fundamentals of Interfacial Engineering,* VCH, New York, NY.
69. Kingshott, P., Thissen, H., and Griesser, H. J. (2002) Effects of cloud-point grafting, chain length, and density of peg layers on competitive adsorption of ocular proteins. *Biomaterials* **23,** 2043–2056.
70. Sofia, S. J., Premnath, V., and Merrill, E. W. (1998) Poly(ethylene oxide) grafted to silicon surfaces: grafting density and protein adsorption. *Macromolecules* **31,** 5059–5070.
71. Chapman, R. G., Ostuni, E., Liang, M. N., et al. (2001) Polymeric thin films that resist the adsorption of proteins and the adhesion of bacteria. *Langmuir* **17,** 1225–1233.
72. Efremova, N. V., Huang, Y., Peppas, N. A., and Leckband, D. E. (2002) Direct measurement of interactions between tethered polyethylene glycol chains and adsorbed mucin layers. *Langmuir* **18,** 836–845.
73. Bearinger, J. P., Castner, D. G., Golledge, S. L., Rezania, A., Hubchak, S., and Healy, K. E. (1997) P(aam-co-eg) interpenetrating polymer networks grafted to oxide surfaces: surface characterization, protein adsorption, and cell detachment studies. *Langmuir* **13,** 5175–5183.

74. Park, S., Bearinger, J. P., Lautenschlager, E. P., Castner, D. G., and Healy, K. E. (2000) Surface modification of poly(ethylene terephthalate) angioplasty balloons with a hydrophilic poly(acrylamide-co-ethylene glycol) interpenetrating polymer network coating. *J. Biomed. Mater. Res.* **53,** 568–576.

75. Brockman, J. M., Nelson, B. P., and Corn, R. M. (2000) Surface plasmon resonance imaging measurements of ultra thin organic films. *Ann. Rev. Phys. Chem.* **51,** 41–63.

76. Xia, N., Hu, Y., Grainger, D. W., and Castner, D. G. (2002) Functionalized poly(ethylene glycol)-grafted polysiloxane monolayers for control of protein binding. *Langmuir* **18,** 3255–3262.

77. Metzger, S. W., Natesan, M., Yanavich, C., Schneider, J., and Lee, G. U. (1999) Development and characterization of surface chemistries for microfabricated biosensors. *J. Vac. Sci. Technol. A* **17,** 2623–2628.

78. Huang, N. -P., Michel, R., Voros, J., et al. (2001) Poly(l-lysine)-g-poly(ethylene glycol) layers on metal oxide surfaces: surface-analytical characterization and resistance to serum and fibrinogen adsorption. *Langmuir* **17,** 489–498.

79. Pasche, S., De Paul, S. M., Voeroes, J., Spencer, N. D., and Textor, M. (2003) Poly (L-lysine)-graft-poly(ethylene glycol) assembled monolayers on niobium oxide surfaces: a quantitative study of the influence of polymer interfacial architecture on resistance to protein adsorption by TOF-SIMS and in situ OWLS. *Langmuir* **19,** 9216–9225.

80. Tosatti, S., De Paul, S. M., Askendal, A., et al. (2003) Peptide functionalized poly(L-lysine)-g-poly(ethylene glycol) on titanium: resistance to protein adsorption in full heparinized human blood plasma. *Biomaterials* **24,** 4949–4958.

81. Jo, S. and Park, K. (2000) Surface modification using silanated poly(ethylene glycol)s. *Biomaterials* **21,** 605–616.

82. Tae, G., Lammertink, R. G. H., Kornfield, J. A., and Hubbell, J. A. (2003) Facile hydrophilic surface modification of poly(tetrafluoroethylene) using fluoroalkyl-terminated poly(ethylene glycol)s. *Adv. Mater.* **15,** 66–69.

83. Martins, M. C. L., Wang, D., Ji, J., Feng, L., and Barbosa, M. A. (2003) Albumin and fibrinogen adsorption on pu-phema surfaces. *Biomaterials* **24,** 2067–2076.

84. Horak, D., Jayakrishnan, A., and Arshady, R. (2003) Poly(2-hydroxyethyl methacrylate) hydrogels: preparation and properties. *PBM Series* **1,** 65–107.

85. Czeslik, C., Jackler, G., Hazlett, T., et al. (2004) Salt-induced protein resistance of polyelectrolyte brushes studied using fluorescence correlation spectroscopy and neutron reflectometry. *Phys. Chem. Chem. Phys.* **6,** 5557–5563.

86. Rollason, G. and Sefton, M. V. (1989) Inactivation of thrombin in heparin-pva coated tubes. *J. Biomat. Sci. Polym. Ed.* **1,** 31–41.

87. Evangelista, R. A. and Sefton, M. V. (1986) Coating of two polyether-polyurethanes and polyethylene with a heparin-poly(vinyl alcohol) hydrogel. *Biomaterials* **7,** 206–211.

88. Proudnikov, D., Timofeev, E., and Mirzabekov, A. (1998) Immobilization of DNA in polyacrylamide gel for the manufacture of DNA and DNA-oligonucleotide microchips. *Anal. Biochem.* **259,** 34–41.

89. Zlatanova, J. and Mirzabekov, A. (2001) Gel-immobilized microarrays of nucleic acids and proteins, in *DNA Microarrays*, (Rampal, J. B., ed.), Humana Press, Totowa, NJ, pp. 17–38.

90. Frazier, R. A., Matthijs, G., Davies, M. C., Roberts, C. J., Schacht, E., and Tendler, S. J. B. (2000) Characterization of protein-resistant dextran monolayers. *Biomaterials* **21,** 957–966.

91. Massia, S. P. and Stark, J. (2001) Immobilized rgd peptides on surface-grafted dextran promote biospecific cell attachment. *J. Biomed. Mater. Res.* **56,** 390–399.

92. Massia, S. P., Stark, J., and Letbetter, D. S. (2000) Surface-immobilized dextran limits cell adhesion and spreading. *Biomaterials* **21,** 2253–2261.

93. Zhou, Y., Andersson, O., Lindberg, P., and Liedberg, B. (2004) Protein microarrays on carboxymethylated dextran hydrogels: immobilization, characterization and application. *Microchimica Acta* **147,** 21–30.

94. Larsson, R. (1987) Biocompatible surfaces prepared by immobilized heparin or hyaluronate. *Acta Otolaryngol. Suppl.* **442,** 44–49.

95. Larm, O., Larsson, R., and Olsson, P. (1983) A new nonthrombogenic surface prepared by selective covalent binding of heparin via a modified reducing terminal residue. *Biomater. Artif. Cell* **11,** 161–173.

96. Morra, M. and Cassineli, C. (1999) Nonfouling properties of polysaccharide-coated surfaces. *J. Biomat. Sci. Polym. Ed.* **10,** 1107–1124.

97. Dai, L. T., St. John, H. A. W., Bi, J., Zientek, P., Chatelier, R. C., and Griesser, H. J. (2000) Biomedical coatings by the covalent immobilization of polysaccharides onto gas-plasma-activated polymer surfaces. *Surf. Interfacial Anal.* **29,** 46–55.

98. Oesterberg, E., Bergstroem, K., Holmberg, K., et al. (1993) Comparison of polysaccharide and poly(ethylene glycol) coatings for reduction of protein adsorption on polystyrene surfaces. *Colloid Surf. A* **77,** 159–169.

99. Chen, G., Ito, Y., Imanishi, Y., Magnani, A., Lamponi, S., and Barbucci, R. (1997) Photoimmobilization of sulfated hyaluronic acid for antithrombogenicity. *Bioconj. Chem.* **8,** 730–734.

100. Lemieux, M., Minko, S., Usov, D., Stamm, M., and Tsukruk, V. V. (2003) Direct measurement of thermoelastic properties of glassy and rubbery polymer brush nanolayers grown by\grafting-from\approach. *Langmuir* **19,** 6126–6134.

101. Patten, T. E. and Matyjaszewski, K. (1998) Atom-transfer radical polymerization and the synthesis of polymeric materials. *Adv. Mater.* **10,** 901–915.

102. Matyjaszewski, K. and Xia, J. (2001) Atom transfer radical polymerization. *Chem. Rev.* **101,** 2921–2990.

103. Robinson, K. L., de Paz-Banez, M. V., Wang, X. S., and Armes, S. P. (2001) Synthesis of well-defined, semibranched, hydrophilic-hydrophobic block copolymers using atom transfer radical polymerization. *Macromolecules* **34,** 5799–5805.

104. Wang, X. S. and Armes, S. P. (2000) Facile atom transfer radical polymerization of methoxy-capped oligo(ethylene glycol) methacrylate in aqueous media at ambient temperature. *Macromolecules* **33,** 6640–6647.

105. Wang, X. S., Jackson, R. A., and Armes, S. P. (2000) Facile synthesis of acidic copolymers via atom transfer radical polymerization in aqueous media at ambient temperature. *Macromolecules* **33,** 255–257.

106. Wang, X. S., Lascelles, S. F., Jackson, R. A., and Armes, S. P. (1999) Facile synthesis of well-defined water-soluble polymers via atom transfer radical polymerization (atrp) in aqueous media at ambient temperature. *Chem. Commun.* **18,** 1817–1818.

107. Ashford, E. J., Naldi, V., O'Dell, R., Billingham, N. C., and Armes, S. P. (1999) First example of the atom transfer radical polymerization of an acidic monomer: direct synthesis of methacrylic acid copolymers in aqueous media. *Chem. Commun.* **14,** 1285–1286.

108. Jones, D. M. and Huck, W. T. S. (2001) Controlled surface-initiated polymerizations in aqueous media. *Adv. Mater.* **13,** 1256–1259.

109. Chilkoti, A., Ma, H., Hyun, J., and Nath, N. (2003) Passive and active nonfouling polymer grafts. *Polymer Preprints (Am. Chem. Soc. Div. Polym. Chem.)* **44,** 455–456.

110. Mendez, S., Ista, L. K., and Lopez, G. P. (2003) Use of stimuli responsive polymers grafted on mixed self-assembled monolayers to tune transitions in surface energy. *Langmuir* **19,** 8115–8116.

111. Huang, W., Kim, J. -B., Bruening, M. L., and Baker, G. L. (2002) Functionalization of surfaces by water-accelerated atom-transfer radical polymerization of hydroxyethyl methacrylate and subsequent derivatization. *Macromolecules* **35,** 1175–1179.

112. Wang, X. and Bohn, P. W. (2004) Anisotropic in-plane gradients of poly(acrylic acid) formed by electropolymerization with spatiotemporal control of the electrochemical potential. *J. Am. Chem. Soc.* **126,** 6825–6832.

113. Prime, K. L. and Whitesides, G. M. (1993) Adsorption of proteins onto surfaces containing end-attached oligo(ethylene oxide): a model system using self-assembled monolayers. *J. Am. Chem. Soc.* **115,** 10,714–10,721.

114. Feldman, K., Haehner, G., Spencer, N. D., Harder, P., and Grunze, M. (1999) Probing resistance to protein adsorption of oligo(ethylene glycol)-terminated self-assembled monolayers by scanning force microscopy. *J. Am. Chem. Soc.* **121,** 10,134–10,141.

115. Kreuzer, H. J., Wang, R. L. C., and Grunze, M. (2003) Hydroxide ion adsorption on self-assembled monolayers. *J. Am. Chem. Soc.* **125,** 8384–8389.

116. Pertsin, A. J., Hayashi, T., and Grunze, M. (2002) Grand canonical monte carlo simulations of the hydration interaction between oligo(ethylene glycol)-terminated alkanethiol self-assembled monolayers. *J. Phys. Chem. B* **106,** 12,274–12,281.

117. Kreuzer, H. J., Wang, R. L. C., and Grunze, M. (1999) Effect of stretching on the molecular conformation of oligo(ethylene oxide): a theoretical study. *New J. Phys.* **1,** 21.1–21.16.

118. Harder, P., Grunze, M., Dahint, R., Whitesides, G. M., and Laibinis, P. E. (1998) Molecular conformation in oligo(ethylene glycol)-terminated self-assembled monolayers on gold and silver surfaces determines their ability to resist protein adsorption. *J. Phys. Chem. B* **102,** 426–436.

119. Herrwerth, S., Rosendahl, T., Feng, C., et al. (2003) Covalent coupling of antibodies to self-assembled monolayers of carboxy-functionalized poly(ethylene glycol): protein resistance and specific binding of biomolecules. *Langmuir* **19,** 1880–1887.

120. Ulman, A. (1989) Ultrathin organic films: from langmuir-blodgett to self-assembly. *J. Mater. Edu.* **11,** 205–280.

121. Ulman, A. (1996) Formation and structure of self-assembled monolayers. *Chem. Rev.* **96,** 1533–1554.

122. Grunze, M. (1993) Preparation and characterization of self-assembled organic films on solid substrates. *Phys. Scripta T* **T49B,** 711–717.

123. Nemetz, A., Fischer, T., Ulman, A., and Knoll, W. (1993) Surface-plasmon-enhanced-raman spectroscopy with 21-hydroxyheneicosanethiol (HS(CH2)21-OH) on different metals. *J. Chem. Phys.* **98,** 5912–5919.

124. Horsley, D., Herron, J., Hlady, V., and Andrade, J. D. (1991) Fluorescence quenching of adsorbed hen and human lysozymes. *Langmuir* **7,** 218–222.

125. Kawasaki, K., Kambara, M., Matsumura, H., and Norde, W. (2003) Protein adsorption at polymer-grafted surfaces: comparison between a mixture of saliva proteins and some well-defined model proteins. *Biofouling* **19,** 355–363.

126. Norde, W. (1998) Driving forces for protein adsorption at solid surfaces, *"Biopolymers At Interfaces,"* 2nd ed., *Surfactant Sci. Ser.* **75,** 27–54.

127. Van Wagenen, R. A., Coleman, D. L., King, R. N., et al. (1981) Streaming potential investigations: polymer thin films. *J. Colloid Interf. Sci.* **84,** 155–162.

128. Herrwerth, S., Eck, W., Reinhardt, S., and Grunze, M. (2003) Factors that determine the protein resistance of oligoether self-assembled monolayers—internal hydrophilicity, terminal hydrophilicity, and lateral packing density. *J. Am. Chem. Soc.* **125,** 9359–9366.

129. Zheng, J., Li, L., Chen, S., and Jiang, S. (2004) Molecular simulation study of water interactions with oligo (ethylene glycol)-terminated alkanethiol self-assembled monolayers. *Langmuir* **20,** 8931–8938.

130. Li, L., Chen, S., Zheng, J., Ratner, B. D., and Jiang, S. (2005) Protein adsorption on oligo(ethylene glycol)-terminated alkanethiolate self-assembled monolayers: the molecular basis for nonfouling behavior. *J. Phys. Chem. B* **109,** 2934–2941.

131. Ostuni, E., Chapman, R. G., Liang, M. N., et al. (2001) Self-assembled monolayers that resist the adsorption of proteins and the adhesion of bacterial and mammalian cells. *Langmuir* **17,** 6336–6343.

132. Chapman, R. G., Ostuni, E., Takayama, S., Holmlin, R. E., Yan, L., and Whitesides, G. M. (2000) Surveying for surfaces that resist the adsorption of proteins. *J. Am. Chem. Soc.* **122,** 8303–8304.

133. Tegoulia, V. A., Rao, W., Kalambur, A. T., Rabolt, J. F., and Cooper, S. L. (2001) Surface properties, fibrinogen adsorption, and cellular interactions of a novel phosphorylcholine-containing self-assembled monolayer on gold. *Langmuir* **17,** 4396–4404.

134. Kitano, H., Kawasaki, A., Kawasaki, H., and Morokoshi, S. (2005) Resistance of zwitterionic telomers accumulated on metal surfaces against nonspecific adsorption of proteins. *J. Colloid Interf. Sci.* **282,** 340–348.

135. Luk, Y. -Y., Kato, M., and Mrksich, M. (2000) Self-assembled monolayers of alkanethiolates presenting mannitol groups are inert to protein adsorption and cell attachment. *Langmuir* **16,** 9604–9608.

136. Deng, L., Mrksich, M., and Whitesides, G. M. (1996) Self-assembled monolayers of alkanethiolates presenting tris(propylene sulfoxide) groups resist the adsorption of protein. *J. Am. Chem. Soc.* **118,** 5136–5137.
137. Lopez, G. P., Ratner, B. D., Tidwell, C. D., Haycox, C. L., Rapoza, R. J., and Horbett, T. A. (1992) Glow discharge plasma deposition of tetraethylene glycol dimethyl ether for fouling-resistant biomaterial surfaces. *J. Biomed. Mater. Res.* **26,** 415–439.
138. Shen, M., Pan, Y. V., Wagner, M. S., et al. (2001) Inhibition of monocyte adhesion and fibrinogen adsorption on glow discharge plasma deposited tetraethylene glycol dimethyl ether. *J. Biomat. Sci. Polym. Ed.* **12,** 961–978.
139. Wu, Y. J., Timmons, R. B., Jen, J. S., and Molock, F. E. (2000) Nonfouling surfaces produced by gas phase pulsed plasma polymerization of an ultra low molecular weight ethylene oxide containing monomer. *Colloid Surf. B: Biointerfaces* **18,** 235–248.
140. Griffith, L. G. (2000) Polymeric biomaterials. *Acta Mater.* **48,** 263–277.
141. Ball, V., Schaaf, P., and Voegel, J. -C. (2003) Mechanism of interfacial exchange phenomena for proteins adsorbed at solid-liquid interfaces. "Biopolymers at interfaces," 2nd ed., *Surfactant Sci. Ser.* **110,** 295–320.
142. Slack, S. M. and Horbett, T. A. (1995) The Vroman effect. A critical review. *ACS Sym. Ser.* **602,** 112–128.
143. Krishnan, A., Sturgeon, J., Siedlecki, C. A., and Vogler, E. A. (2004) Scaled interfacial activity of proteins at the liquid-vapor interface. *J. Biomed. Mater. Res., Part A* **68A,** 544–557.
144. Krishnan, A., Siedlecki, C. A., and Vogler, E. A. (2004) Mixology of protein solutions and the vroman effect. *Langmuir* **20,** 5071–5078.
145. Shen, M., Martinson, L., Wagner Matthew, S., Castner David, G., Ratner Buddy, D., and Horbett Thomas, A. (2002) Peo-like plasma polymerized tetraglyme surface interactions with leukocytes and proteins: in vitro and in vivo studies. *J. Biomat. Sci. Polym. E.* **13,** 367–390.
146. Wagner, M. S., McArthur, S. L., Shen, M., Horbett, T. A., and Castner, D. G. (2002) Limits of detection for time of flight secondary ion mass spectrometry (ToF-SIMS) and X-ray photoelectron spectroscopy (XPS): detection of low amounts of adsorbed protein. *J. Biomat. Sci. Polym. Ed.* **13,** 407–428.
147. Ronneberger, B., Kao, W. J., Anderson, J. M., and Kissel, T. (1996) In vivo biocompatibility study of aba triblock copolymers consisting of poly(l-lactic-co-glycolic acid) blocks attached to central poly(oxyethylene) b blocks. *J. Biomed. Mater. Res.* **30,** 31–40.
148. Suggs, L. J., Krishnan, R. S., Garcia, C. A., Peter, S. J., Anderson, J. M., and Mikos, A. G. (1998) In vitro and in vivo degradation of poly(propylene fumarate-co-ethylene glycol) hydrogels. *J. Biomed. Mater. Res.* **42,** 312–320.
149. Johnston, E. E., Bryers, J. D., and Ratner, B. D. (2005) Plasma deposition and surface characterization of oligoglyme, dioxane, and crown ether nonfouling films. *Langmuir* **21,** 870–881.

4

Optimization of Oligonucleotide DNA Microarrays

Martin Dufva and Claus B. V. Christensen

Summary

Optimization of oligonucleotide DNA microarrays is a multiparametric problem. The goal of the optimization process is to get conditions that capture target DNA with high sensitivity and selectivity. Parameters determining the performance of the microarray are spot morphology, probe and target density, background, and selectivity. More than 10 variables can be adjusted to obtain a well-optimized protocol. However, some variables only affect spot morphology and other factors affect for instance hybridization and selectivity, which can limit the optimization work considerably. This chapter suggests an outline on how an optimization procedure is made. Moreover, a simple method for absolutely quantify the number of hybridized target to the spots is given. The latter is important because it gives the possibilities to compare results with those obtained in the literature.

Key Words: DNA microarray; hybridized density; hybridization; optimization; probe density; quantification; spot morphology; spotting.

1. Introduction

Optimization of DNA microarrays is a multiparametric problem with many unknown factors. The parameters can be divided into different functional groups according to **Table 1**. The aim of the optimization procedure is to have spots that are homogenous, yields high hybridization density and with low surrounding background. Furthermore, the immobilized DNA should specifically capture the target DNA. In principle, an optimization process can use fluorescence intensity as readout but it is convenient to measure the performance of the DNA microarray by absolute numbers. Determining the probe and hybridization density is easily done using standard curves of fluorescently labeled molecules and gives the better understanding of the probe–solid support and probe–target interactions. Furthermore, it gives the possibility to compare the performance of the specific experimental set up with those obtained in the literature. DNA has been

From: *Methods in Molecular Biology, vol. 381: Microarrays: Second Edition: Volume 1: Synthesis Methods*
Edited by: J. B. Rampal © Humana Press Inc., Totowa, NJ

Table 1
Factors That Affect Different Parameter of Microarray Performance

	Spot quality	Probe density	Hybridized density	Background	Specificity	Recommendations[a]
Spotter type (pin, inkjet)	Yes	No	No	No	No	–
Pin type	Yes	No	No	No	No	Reservoir
Humidity	Yes	Yes	Yes	No	No	60–70%
Temperature at spotting	Yes	Yes	Yes	No	No	Keep constant
Probe concentration	Yes	Yes	Yes	No	No	10–20 μM (*11,12,15–17*)
Spotting buffer	Yes	Yes	Yes	No	Yes	Depends on substrate (*20*)
Microarray substrate (type of slide)	Yes	Yes	Yes	Yes	No	CodeLink, agarose, dendrimeric linkers (*15,17,19*)
Immobilization chemistry	Yes	Yes	Yes	Yes	No	Amine reactive (*21*)
Blocking technique	No	No	No	Yes	No	Depend on slide type
Spacers	Yes	Yes	Yes	No	No	15–20 poly dT or other spacers (*16,17,22*)
Probe sequence	No	No	Yes	No	Yes	Test a panel (*23*)
Stringency during hybridization/washing	No	No	Yes	No	Yes	Depend of probes/application
Target preparation	No	No	Yes	No	No	Short single stranded DNA (*24*)

[a]Some of the factors are affecting microarray performance is based on unpublished observations. The references given are only examples and many more exist in the literature.

94

successfully attached to glass *(1–6)*, oxidized silicon *(7)*, silicon wafers *(8)*, and gold surfaces *(9,10)* and polymers like poly(methylmethacrylate) *(11,12)* and SU-8 polymers *(13)*. Naturally quite different optimal conditions are required in each report mainly owing to the large variety of substrates and chemistries used.

Probe density is defined as the number of probes that are immobilized on a given surface area. In many cases only 10–20% of the probes are involved in the hybridization reaction, which can be calculated as a ratio between the hybridized target density and immobilized probe density. This ratio is also termed as hybridization efficiency. Often a maximum in hybridization efficiency is observed at relatively low probe densities indicating that too high probe density negates hybridization probably owing to electrostatic repulsion of immobilized probes *(14)*. Despite this immobilization, probes are often chosen at a higher probe density than that giving the highest hybridization efficiency because the highest signal possible is desired. That is, although the hybridization efficiency decreases the signal increases *(11,12)*. The first parameter to tune is the probe density on the surface by varying the spotting concentrations. For many substrates spotting at 10–20 μ*M* *(11,12,15–17)* gives high signals without using too much of the relatively expensive probes.

The background defines the sensitivity of the system meaning that the background should be as low as possible and with little variation. As mentioned earlier, the probe density should be chosen that yields the highest hybridization signal, because the span between the maximum signal and the background defines the dynamic range of what the spot can measure.

The spot quality is important for several reasons. First, the spots should be perfectly rounded as it will simplify for quantification programs to analyze and find the spots. The spot should also be homogenous meaning that all the probes on the spot should be evenly spaced. This is important as high probe densities sometimes lead to a reduction in hybridization efficiency. A general problem with pin spotting is the so-called coffee-ring spot phenomena, in which the spots have extremely high probe density in the perimeter with little or no probes in the middle of the spot. This leads to uneven hybridization properties over the spot and spots with high signal in the periphery and low signal in the middle of the spot. Such spots are difficult to quantify correctly (*see* example in **Fig. 1** where a simple spotting buffer optimization is shown).

2. Materials

2.1. Construction of a Standard Curve

1. Target molecules can be of different kinds and may be changed depending on their specific needs. For instance, finding immobilization condition for gene expression array many use 60 bp long probes *(18)*, whereas, optimization for SNP detection

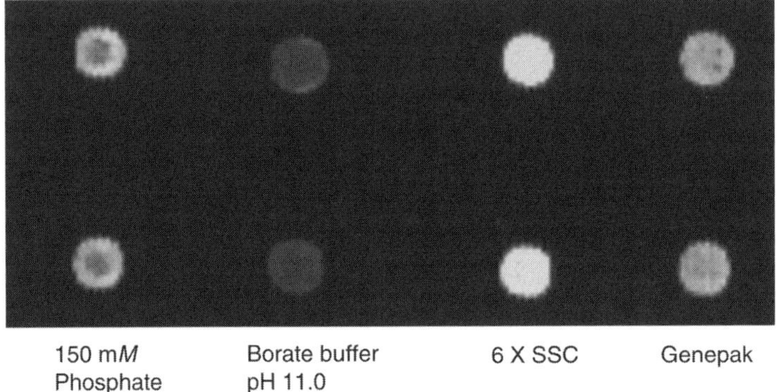

Fig. 1. The goal is to identify spots that are perfectly rounded and homogenous. The signal from the spots should also be strong. A small experiments where different spotting solutions has been used for spotting 20 μM DNA probes onto agarose film coated slides as described in **ref. 19**. Looking at the spots by eye shows that the phosphate buffer yield strong signals but have a tendency toward coffee-spot phenomena, where there are strong signal in the periphery of the spots and weak in the middle. The borate buffer resulted in very little immobilized DNA as well as coffee ring phenomena. The 6X SSC buffer yield homogenous spots and high signals whereas the Genpak buffer yield homogenous spots and lower signals. From this experiments is appears that the 6X SSC buffer gave the best results.

it may better to use probes between 13 and 25 bp. For these experiments we use W54-T for construction of the standard curve *(15)* 5′-Cy3-GATGACTTCGCA-GAGACAAGCTTGTCGATGAGCGTGGCAGCATCTTCAACTGTCGCAATT-3′ (high-performance liquid chromatography purified). The Cy3 label and high-performance liquid chromatography purification of the DNA is made by the vendor. Dilute the DNA in 150 mM phosphate buffer (pH 8.3) supplemented with 20-μM unlabeled DNA (*see* **Note 1**). It is earliest to make a 2X spotting buffer before to making twofold dilution of the target DNA. A procedure for spotting a dilution series of probes is given in **Subheading 2.2.**, and could be followed when spotting the standards curve for quantification of hybridized density although when spotting the standard curve, labeled target is spotted.

2. Prepare the agarose slides as described in **refs. *15*** and ***19***. We use aldehyde-coated slides (CSS-100, CEL Associates, Pearland, TX) for many application as they are reasonably inexpensive.

2.2. Spotting the Slides

1. Probes W45R: 5′-amino-AATTGCGACAGTTGAAGATGCTGCCACGC-TCATCG-ACAAGCT-TGTCTCTGCGAAGTCATC-3′, LucP 5′-amino-GCGTCGAG-TTT-TCCG-GTAAG-ACCTTTCGGTACTTCGTCCACAAACACAACTCCT CCGCGC-3′ is diluted in spotting buffer. Use a 2X spotting buffer and make a

30 µL solution consisting by adding 15 µL 2X spotting buffer and 15 µL of a 40 µ*M* probe solution. The mixing is made directly in the microtiter plate well (384-well plate with cone bottom). Prepared a two dilution series by adding 15 µL 2X spotting buffer to four other wells. Take 15 µL of the 20 µ*M* probe solution to the first well consisting only 15 µL spotting solution. Mix by pipetting and transfer 15 µL to the next well and so on.

2. 0.1X Standard saline citrate (SSC) is prepared from stock solution of 20X SSC (Promega, Madison, WI) diluted in Milli-Q or distilled water. Sodium dodecyl sulfate (SDS; Sigma-Aldrich, Steinheim, Germany) is added to a final concentration of 0.5% (w/v). A 5 L 0.1X SSC + 0.5% SDS solution is made by adding 25 g SDS and 25 mL 20X SSC. Fill up with water and mix thoroughly.

3. Treat the slides in sodium borohydride solution for 5 min at room temperature to reduce free aldehydes. Dissolve 1 g NaBH$_4$ (1 g tablet from Fluka [Sigma-Aldrich]) in 300 mL phosphate-buffered saline (PBS; 150 m*M* Sigma Aldrich, pH 7.4), and then add 100-mL 100% ethanol to reduce bubbling and prepare just before to use. NaBH$_4$ is reactive so work in a fume hood.

4. SMB3, split-pin (TeleChem, Sunnyvale, CA).

5. QArray™ robot (Genetix, New Milton, Hampshire, UK).

2.3. Hybridizations

1. Hybridization buffer are easiest prepared by making a 2X hybridization buffer by mixing 1% w/v SDS and 10X SSC. Note that the salt will precipitate the SDS immediately, insoluble at room temperature. Before to use, put the flask with the 2X hybridization buffer in the microwave oven and heat *gently* until a clear solution is obtained. Dilute the target to 20 n*M* in Milli-Q water. Mix one volume of prewarmed 2X hybridization buffer with one volume of 20 n*M* target solutions.

2. Microarray Wash Station (Arrayit, TeleChem).

3. ScanArray™ Lite, microarray scanner (Packard Biosciences, Billerica, MA).

3. Methods

As apparent from **Table 1**, a complete optimization procedure is not practical to show here. A small optimization procedure that concerns finding optimal probe densities for two different slides will be used as an example for how an optimization procedure can look like. Finally, the hybridization density of the optimal probe density will be quantified using a standard curve.

The hybridized density is however, difficult to compare between different reports as it is heavily dependent on the sequence of the probe as well as the length and sequence of the target. The parameter is correctly measured if the probes in a spot are saturated with target molecules. For several reasons, the concentration of target where saturation is reached is highly variable. In most cases, a short fully complementary target is used to determine the hybridization density, because these are easy to produce and are easy to use experimentally to saturate the spots. It should be noted, however, that using PCR products or long cDNA as target, which are more relevant targets, will usually result in less hybridization density.

3.1. Construction of a Standard Curve for Quantification

1. Make a dilution series of fluorescently labeled target molecules in spotting buffer. The target molecule should be labeled with an appropriate fluorochrome. Modifications are added directly during synthesis of the probes. The dilution series should cover a fairly large spectrum and start at a target dilution (*see* **Note 1**) of 10 µ*M* and end at 10 n*M*. One nanomolar fluorescently labeled target is barely detected because only 1 nL is deposited on the surface (10 amol).
2. Transfer the dilution series to a microtiter plate and put the microtiter plate in the spotter. The humidity during spotting should be 60–70% and the temperature at 20°C. Make 10 replicates of each dilution on each slide type (*see* **Note 3**). Use at least two slides of each slide type you want to test (agarose- and aldehyde-coated glass slides). Do not wash these slides before scanning.
3. Scan the slides so that none of the signals is saturated (*see* scanner manual for how to change the scanning settings). Quantify the spots and plot the spots against spotting concentrations. However, it is more convenient to plot fluorescence toward amount spotted DNA. Because 1 nL is deposited (assumed) a fluorescent value corresponding to a specific amount of DNA can be calculated and plotted **(Fig. 2)**.

A similar approach can be taken to construct a standard curve for measuring probe density (*see* **Note 2**).

3.2. Spotting the Slides for Probe Optimization

1. Place the agarose- and aldehyde-coated slides in the spotter. Make a dilution series of probes (between 20 and 1.2 µ*M*) and transfer the probe solution to a microtiter plate. Put the microtiter plate in the spotter and spot the slides using 70% humidity. Spot replicates to get statistics (*see* **Note 3**).
2. Ultraviolet crosslink the agarose slides for 4 min in a Stratalinker 2400 (Stratagene, La Jolla, CA) equipped with 254-nm bulbs *(15)* or similar. Remove unbound probes by washing the slides in $0.1 \times$ SSC + 0.5% SDS. Place the slides in a microarray wash station and put the washing station in a 400-mL beaker. Add a stirring magnet and 400 mL $0.1 \times$ SSC + 0.5% SDS. Stir vigorously using a magnetic stirrer for 10 min. Rinse the slides under water briefly and dry by centrifuge or by blowing air over the slide.
3. The aldehyde-coated slides are incubated overnight at room temperature and washed in 0.2% SDS for 10 min and in water for 5 min. Block the slides by washing them in freshly made $NaBH_4$/PBS solution for five min. Wash the slides with $1 \times$ SSC + 0.5% SDS followed by a rinse in water. Dry the slides. All the washing steps follow those in **Subheading 3.2.2.**

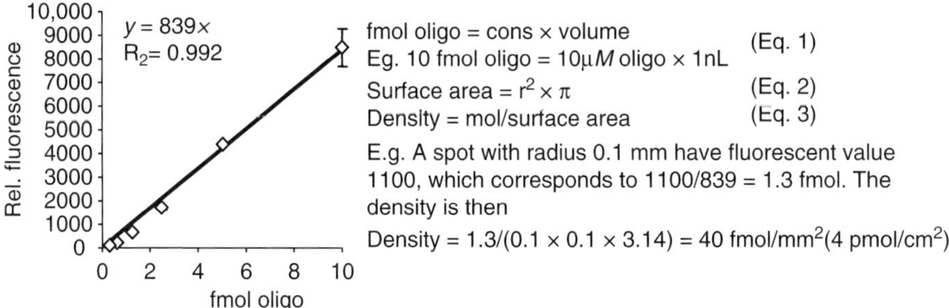

fmol oligo = cons × volume

Eg. 10 fmol oligo = 10µM oligo × 1nL (Eq. 1)

Surface area = r^2 × π (Eq. 2)

Density = mol/surface area (Eq. 3)

E.g. A spot with radius 0.1 mm have fluorescent value 1100, which corresponds to 1100/839 = 1.3 fmol. The density is then

Density = 1.3/(0.1 × 0.1 × 3.14) = 40 fmol/mm^2(4 pmol/cm^2)

Fig. 2. Standard curve for determination of hybridization densities. A dilution series of target is spotted on the slides that are going to be tested. The slides are dried and scanned in a microarray scanner (ScanArray Lite). The spots were quantified using the freeware (for academics) ScanAlyze v2.5 *(25)* and analyzed using Excel (Microsoft, Redmond, WA). The linear regression equation is convenient to use for quantification of hybridized density as indicated in example in the figure.

3.3. Hybridizations

Dilute target and hybridization buffer to a final concentration of 10 nM target (*see* **Note 4**), 5X SSC, and 0.5% SDS. Denature the target molecules by heating the target solution for 4 min at 95°C. Place 25 µL target hybridization solution where the microarray is located on the slide. Place a 20 × 20-mm^2 cover slip on the microarray avoiding air-bubbles. Put the slide in a hybridization chamber with humidity and place the chamber in oven at 37°C, and hybridize for 1 h (*see* **Note 5**).

1. Wash the slides at room temperature in 0.1X SSC + 0.5% SDS for 10 min with vigorous stirring (*see* **Subheading 3.2.2.**). Rinse the slides for a few second in water to remove buffers and spin dry (*see* **Note 6**).
2. Scan the slide using scanner settings so that the strongest spots in the series. Quantify the spots. A representative result of a quantified hybridization reaction is shown in **Fig. 3**. The fluorescent values can be converted to absolute values using the standard curve and calculation in **Fig. 2**.

4. Notes

1. The spotting buffer used has large influence on spot morphology and the amount of liquid that is deposited on the spot. However, also the DNA content in the probe solution can modulate the performance of the spotting procedure and the spot morphology where high-DNA concentrations (≥10 µM) generally gives better spot morphology and better reproducibility in the depositing process that probe solution in which the DNA content is less than 10 µM. To avoid such differences and unlabeled unrelated carrier DNA is spiked into the spotting solution so that the final concentration is at least 10 µM.

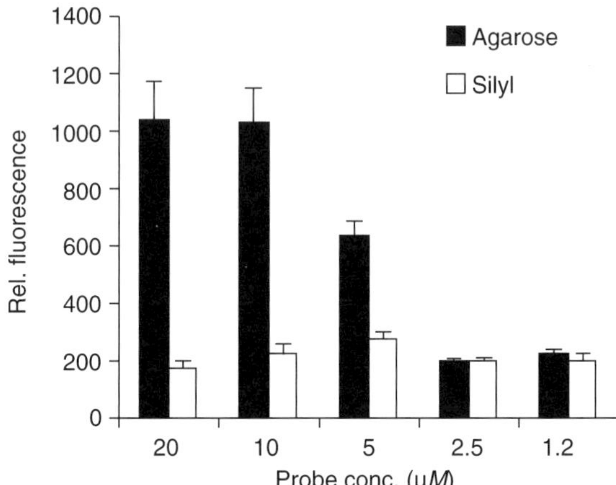

Fig. 3. Probe concentration optimization. Probes were spotted in different concentration by contact printing using a split-pin controlled by a QArray robot. The slides was processed as described earlier and hybridized with 10 n*M* target (Cy3-W54-T) for 1 h at 37°C. The spots was scanned and quantified as described in **Fig. 2**. The results shows that the aldehyde coated slides have a maximum hybridization signal when spotted at 5 µ*M* probe solution whereas three-dimensional agarose slides have maximum hybridization signals from 10 and 20 µ*M* probe solutions. Under the condition used the unspecific probe (LucP) has 1:100 of the signal as compared with the specific probe. The hybridized density can be quantified using the standards curve in **Fig. 2**. The relative fluorescence of about 1000 is converted into an absolute number as described in **Fig. 2**.

2. A similar approach can be made to quantify the density of immobilized probes on the surface *(11,12,15)*. However, in many protocols for immobilizing DNA to the solid support involves steps that can potentially destroy the fluorochrome on the surface. For instance, ultraviolet light induced crosslinking and $NaBH_4$ treatment efficiently destroys the fluochromes, which will underestimate the number of probes on the surface. Therefore, make sure that the immobilization protocol does not destroy the fluorochrome. If so, use probes that are radioactively labeled.

3. The slides should be spotted asymmetrically so that the spots can be identified by the pattern. Use if necessary a labeled but nonspecific probe (cannot participate in hybridizations reactions) to print orientation spots in the microarray. Replication of spots is easily done using microarray spotter. However, 10 spots in one slide is not really statistically sound approach. It is far better to have two spots of each probe solution on five different slides and make five different hybridizations. We usually spot a probe once per subarray as in production scale arrays but we make four subarrays per slides and test three slides (12 spots). If the four subarrays are small enough four subarrays were printed at the top of the slide and two at the bottom of the slide (just above the frosted part). By that way we can make two independent hybridizations using 20×20-mm^2 cover slips saving half of the slide costs.

4. The condition used for saturating the spots is dependent on the probe-target configuration. These condition is suitable for the 60-bp probes described here and will suffice for many conditions even when high hybridized densities is expected to be observed (like agarose slides or other three-dimensional slides). However, certain novel chemistries and solid support might capture such large quantities of targets so that 10 nM target is not enough to saturate the spots. It is recommended that microarrays are hybridized with a dilution series of target to find condition that saturates the spots. It is however, desirable to minimize target concentration because it is expensive and too high target concentration will results in undesirable background.

5. It is very important that the array is kept wet during hybridization otherwise undesirable background will be obtained. There are many ways to obtain this and the solution is depending on the hybridization time. For short incubation like 1–2 h the cover slip will reduce evaporation enough if the slide is placed in a humid chamber. Such chambers can easily be constructed using a closed box with wet paper towels in the bottom and small rods atop the towels to avoid the slides being contact with the towels. Other solution (more expensive) is to use Geneframes (Abgene, Epsom, UK) where the microarray is hermetically closed into a plastic structure that is taped onto the slide. Even more expensive solution involves dedicated hybridization machines that also provide mixing in small volumes.

6. Wash the slides using nonstringent washing condition to remove unhybridized DNA from the slide while leaving the hybrids intact. Rinsing in water must be very short (s), otherwise the target will be removed form the probes

References

1. Beier, M. and Hoheisel, J. D. (1999) Versatile derivatisation of solid support media for covalent bonding on DNA-microchips. *Nucleic Acids Res.* **27,** 1970–1977.

2. Britcher, L. G., Kehoe, D. C., Matisons, J. G., Smart, R. S. C., and Swincer, A. G. (1993) Silicones on glass surfaces 2: coupling agent analogs. *Langmuir* **9,** 1609–1613.

3. Guo, Z., Guilfoyle, R. A., Thiel, A. J., Wang, R., and Smith, L. M. (1994) Direct fluorescence analysis of genetic polymorphisms by hybridization with oligonucleotide arrays on glass supports. *Nucleic Acids Res.* **22,** 5456–5465.

4. Janolino, V. and Swaigood, H. E. (1982) Analysis and optimisation of methods using water-soluble carbidiimide for immobilization of biochemicals to porous glass. *Biotechnol. Bioeng.* **24,** 1069–1080.

5. Joos, B., Kuster, H., and Cone, R. (1997) Covalent attachment of hybridizable oligonucleotides to glass supports. *Anal. Biochem.* **247,** 96–101.

6. Rogers, Y. H., Jiang-Baucom, P., Huang, Z. J., et al. (1999) Immobilization of oligonucleotides onto a glass support via disulfide bonds: a method for preparation of DNA microarrays. *Anal. Biochem.* **266,** 23–30.

7. Chrisey, L. A., Lee, G. U., and O'Ferrall, C. E. (1996) Covalent attachment of synthetic DNA to self-assembled monolayer films. *Nucleic Acids Res.* **24,** 3031–3039.

8. Strother, T., Hamers, R. J., and Smith, L. M. (2000) Covalent attachment of oligodeoxyribonucleotides to amine-modified Si (001) surfaces. *Nucleic Acids Res.* **28,** 3535–3541.

9. Steel, A. B., Herne, T. M., and Tarlov, M. J. (1998) Electrochemical quantitation of DNA immobilized on gold. *Anal. Chem.* **70,** 4670–4677.

10. Steel, A. B., Levicky, R. L., Herne, T. M., and Tarlov, M. J. (2000) Immobilization of nucleic acids at solid surfaces: effect of oligonucleotide length on layer assembly. *Biophys. J.* **79,** 975–981.

11. Fixe, F., Dufva, M., Telleman, P., and Christensen, C. B. (2004) Functionalization of poly(methyl methacrylate) (PMMA) as a substrate for DNA microarrays. *Nucleic Acids Res.* **32,** E9.

12. Fixe, F., Dufva, M., Telleman, P., and Christensen, C. B. (2004) One-step immobilization of aminated and thiolated DNA onto poly(methylmethacrylate) (PMMA) substrates. *Lab. Chip* **4,** 191–195.

13. Marie, R., Schmid, S., Johansson, A., et al. (2006) Immobilisation of DNA to polymerised SU-8 photoresist. *Biosens. Bioelectron.* **21,** 1327–1332.

14. Vainrub, A. and Pettitt, B. M. (2002) Coulomb blockage of hybridization in two-dimensional DNA arrays. *Phys. Rev. E Stat. Nonlin. Soft Matter Phys.* **66,** 041905.

15. Dufva, M., Petronis, S., Bjerremann J. L., Krag, C., and Christensen, C. (2004) Characterization of an inexpensive, nontoxic and highly sensitive microarray substrate. *Biotechniques* **37,** 286–296.

16. Benters, R., Niemeyer, C. M., Drutschmann, D., Blohm, D., and Wohrle, D. (2002) DNA microarrays with PAMAM dendritic linker systems. *Nucleic Acids Res.* **30,** E10.

17. Le Berre, V., Trevisiol, E., Dagkessamanskaia, A., et al. (2003) Dendrimeric coating of glass slides for sensitive DNA microarrays analysis. *Nucleic Acids Res.* **31,** E88.

18. Hughes, T. R., Mao, M., Jones, A. R., et al. (2001) Expression profiling using microarrays fabricated by an ink-jet oligonucleotide synthesizer. *Nat. Biotechnol.* **19,** 342–347.

19. Afanassiev, V., Hanemann, V., and Wolfl, S. (2000) Preparation of DNA and protein micro arrays on glass slides coated with an agarose film. *Nucleic Acids Res.* **28,** E66.

20. Hessner, M. J., Wang, X., Hulse, K., et al. (2003) Three color cDNA microarrays: quantitative assessment through the use of fluorescein-labeled probes. *Nucleic Acids Res.* **31,** E14.

21. Zammatteo, N., Jeanmart, L., Hamels, S., et al. (2000) Comparison between different strategies of covalent attachment of DNA to glass surfaces to build DNA microarrays. *Anal. Biochem.* **280,** 143–150.

22. Shchepinov, M. S., Case-Green, S. C., and Southern, E. M. (1997) Steric factors influencing hybridisation of nucleic acids to oligonucleotide arrays. *Nucleic Acids Res.* **25,** 1155–1161.

23. Keramas, G., Bang, D. D., Lund, M., et al. (2003) Development of a sensitive DNA microarray suitable for rapid detection of *Campylobacter* spp. *Mol. Cell Probes* **17,** 187–196.

24. Southern, E., Mir, K., and Shchepinov, M. (1999) Molecular interactions on microarrays. *Nat. Genet.* **21,** 5–9.

25. Eisen, M. B. and Brown, P. O. (1999) DNA arrays for analysis of gene expression. *Methods Enzymol.* **303,** 179–205.

5

Detection of DNA Copy-Number Alterations in Complex Genomes Using Array Comparative Genomic Hybridization

Wei-Wen Cai

Summary

Array-based comparative genomic hybridization (array CGH) is becoming a prominent genomic technology with many important applications in biomedical research. Although several platforms of this technology have been published, successful implementation of this technology still demands technical expertise. Here, we describe the technology that has been developed and improved in the past few years are described. Our array CGH technology is primarily based on robust and readily implemented array production method. We also developed related protocols for using our bacterial artifical chromosomes CGH microarrays. This technology was successfully used to detect DNA copy-number alterations in various mouse and human samples.

Key Words: Bacterial artificial chromosomes; comparative genomic hybridization; DNA copy-number alteration; DNA microarray; DNA silanization.

1. Introduction

Microarray-based comparative genomic hybridization (array CGH) is a maturing high resolution technology to identify DNA copy-number alterations *(1)* in complex genomes. Array CGH borrows the idea of comparative hybridization from the conventional CGH *(2)* in which differentially labeled "test" and "reference" DNA are competitively hybridized to normal metaphase chromosomes and the copy number changes are detected by significant drift of fluorescence ratio along the metaphase chromosomes. In array CGH, DNA fragments spanning known genomic regions are arrayed on a surface and CGH is performed on the array, and thus, allowing the entire genome to be scanned in a single experiment. Copy number changes are detected by analysis of the fluorescence ratio variation of the arrayed spots.

From: *Methods in Molecular Biology, vol. 381: Microarrays: Second Edition: Volume 1: Synthesis Methods*
Edited by: J. B. Rampal © Humana Press Inc., Totowa, NJ

Array CGH has the potential to overcome all the limitations in conventional CGH such as low resolution and technical difficulties. The theoretical resolution of array CGH is only limited by the density of probes representing or sampling the genome. Clone-based arrays such as bacterial artifical chromosomes (BAC) or fosmid clones CGH arrays have a resolution only limited by the clone insert size. Normally, BAC clones have an average insert size of 100–160 kb and fosmid clones are usually around 40 kb. The resolution limit for deletion–detection is about half of the clone insert size. Thus, a genome tiling path arrays *(3)* consisting of BAC clones have the ability to detect 50 kb or larger heterozygous deletion in the genome. Detection limit for sequence insertion or amplification depends on the copy number of the amplified sequences. Theoretically, when the amplified sequences contribute signals equivalent to 30% of the whole clone, the gain of signal can be reliably detected. For example, a 100-kb BAC tiling path array can detect 3 kb sequence amplification if the amplification level is >10 copies. Such high resolution detection of gain across the genome is difficult to achieve with other array platforms without full sequence coverage of the genome.

Although the concept of array CGH is straightforward, successful implementation of the technology need to overcome a battery of technical challenges, which include:

1. The high sequence complexity of mammalian genomes dilutes the signals from any specific region to such low level that large amount of probe must be used.
2. The requirement for reliable detection of heterozygous deletion in diploid genome demands highly quantitative hybridization. Because of this quantitative requirement, methods that can introduce nonlinear amplification during manipulation of genomic DNA could lead to false detection.
3. The presence of large amount of repetitive sequences in the complex genome causes significant problems for clone-based CGH arrays. In high-repetitive genomic region (repeat content >75%) clone arrays might generate significant false-negative detection.

To overcome some of the technical difficulties in array CGH, an array production method was developed based on a novel concept *(4)*. Instead of printing DNA on positively charged or reactive surfaces, a crosslinker-like silane compound with an epoxide group was used that can covalently attach to DNA at slightly alkaline pH to modify the DNA first and then print the modified DNA on to neutral or hydrophobic surfaces (**Fig. 1**). Using this compound, DNA can be modified in a controlled manner so that the extent of modification (modified bases per kilobase) is optimal. It was found that the modified DNA retained normal hybridization specificity and could be purified after modification using standard procedures.

DNA arrays can be conveniently made from modified DNA by simple deposition of the samples onto clean natural glass surfaces. Once in contact with

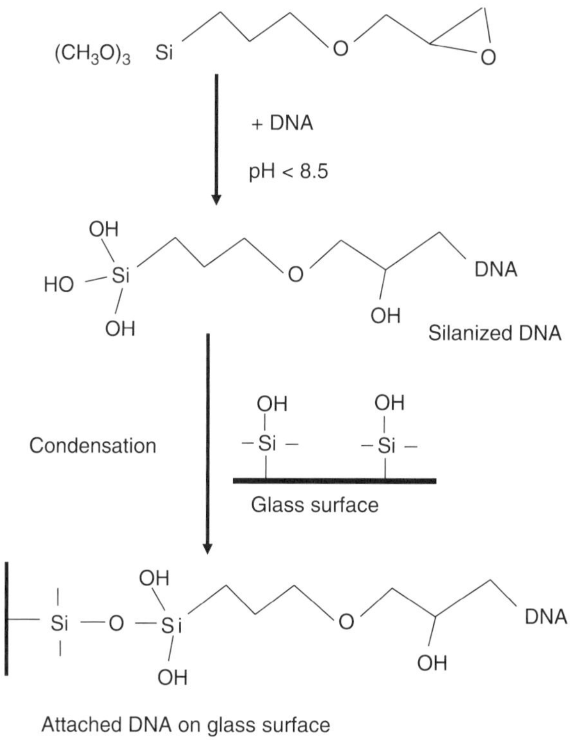

Fig. 1. Silanization of DNA. DNA reacts readily with epoxide compound in alkaline solution. Modification of DNA with epoxysilane activates DNA into a form that can specifically bind to surfaces with excess silanol groups. Covalent attachment occurs through condensation between silanol groups under dehydrating condition. The bound DNA retains hybridization specificity and can sustain stringent hybridization and wash conditions.

the glass covalent attachment occurs at neutral pH after some simple curing treatment. The arrays are permanently stable in dry state. It can sustain overnight stringent washes at 65°C without obvious loss of DNA. Because natural glass surfaces are slightly negatively charged and thus repulsive to DNA. DNA arrays prepared by this method have low-background noise after hybridization washes as expected (*4*).

A distinct advantage of our DNA premodification method over all other methods using activated surfaces is that all activated surfaces are unstable, difficult to prepare reproducibly and sensitive to impurities in the samples. Because of limited DNA binding capacity, activated surfaces require careful calibration of DNA concentration before printing to avoid overspill of access DNA. In contrast, in the premodification method, it is not necessary to calibrate

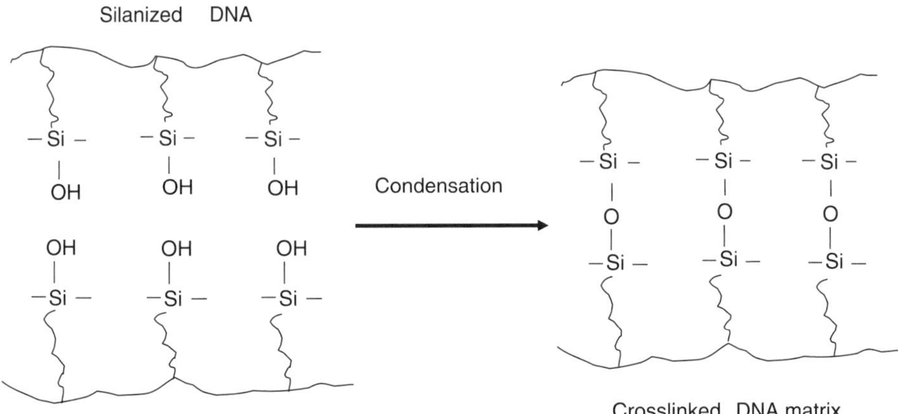

Fig. 2. Self-crosslinking of silanized DNA. Silanized DNA molecules, when dried on surfaces, can crosslink with each other through silanol condensation to produce a multilayer three-dimensional matrix with enhanced hybridization efficiency because of more available hybridization targets. This self-crosslinking property of silanized DNA offers unique advantages for making sensitive DNA microarrays.

DNA concentration because modified DNA molecules can crosslink with each other after drying to form three-dimensional matrix (**Fig. 2**). This has improved hybridization efficiency because of more available probe targets within spots. It was found that modified DNA can be printed at any concentration without saturating the surface thus causing spills. For example, the authors were able to print DNA at a concentration of 2 µg/µL and all the spots were perfectly confined but when the same samples were printed without modification on commercial slides at a concentration exceeding 300 ng/µL, significant DNA spills were noticed. Because of this crosslinking property of silanized DNA, it can be arrayed on other surfaces such as metal and plastics, which are difficult to activate for DNA binding.

This DNA modification chemistry can be applied to make DNA array of different size ranging from 200 kb BAC DNA to plain oligos of 20 of bases. When this premodification chemistry is applied to make oligoarrays, the advantage of this method is obvious. When oligos are attached by single site on either end to a surface they tend to take a brush conformation to minimize repulsive intermolecular electrostatic interaction. This conformation is not favorable for probe hybridization because the probe molecules have to creep in to hybridize to the target sequence and the unhybridized probe end near the surface must overcome certain level of steric hindrance *(5)*. However, when oligos are silanized, multiple attachments on each molecule are possible and they can form three-dimensional fabric structure through crosslinking (**Fig. 2**). Moreover, oligos attached by single sites are not very stable on glass surfaces. Stringent washes can result in significant loss as detected by staining assay using a fluorescent DNA binding dye. In contrast, arrays of silanized oligos are perfectly stable in hybridization and stringent washes, and appear to have much higher hybridization efficiency.

In this chapter, the procedures for preparing BAC DNA arrays using our premodification method and the related protocols for using such arrays are provided for detection of chromosomal abnormalities in complex genomes. The array production method using premodified DNA can be directly applied to make oligo or cDNA arrays.

2. Materials

2.1. Silanization of DNA

1. 0. 1 *M* NaOH.
2. Isopropanol.
3. 3-Glycidoxylpropyltrimethoxysilane (glycidsilane or epoxysilane) (Sigma-Aldrich, St. Louis, mo; cat. no. 440167).
4. Neutralization solution: 0.5 *M* Tris-HCl, 1 *M* NaCl, pH 3.7.

2.2. Surface Cleaning and Coating

1. Plain microscope slides (Fisher Scientific, Pittsburg, PA; cat. no. 12-550A).
2. 2.5 *M* NaOH, 50% ethanol (v/v) surface cleaning solution.
3. 3 *M* HCl surface activating solution.
4. 1,8-Bis(chlorodimethylsilyl)-octane (BDSO) (Gelest, Morrisville, PA; cat. no. SIB1048.0).
5. 65°C Incubator.

2.3. Array Production

1. MicroGrid printer from Genomic Solution.
2. 80% Ethanol.
3. Kapak heat sealable pouches.

2.4. DNA Labeling

2.4.1. Chemical Activation Method

1. 0.5 *M* NaOH.
2. 0.5 *M* NaHCO$_3$.
3. 0.2 *M* 2,2′-(Ethylenedioxy)bis(ethylamine) (EDBE) (Aldrich, cat. no. 929-59-9).
4. 2 *M* NaH$_2$PO$_4$.
5. 0.2 *M* NaHCO$_3$.
6. 0.1 *M* Tris-HCl base.
7. N-Bromosuccinimide (NB) at 1 mg/mL (Aldrich, cat. no. 128-08-05).
8. Cy3, Cy5 NHS monoreactive dye (GE Healthcare Bio-Sciences, Piscataway, NJ).

2.4.2. Random Priming Method

1. Bioprime DNA labeling kit (Invitrogen, Carlsbad, CA).
2. 5 X AA-dUTP-dNTP solution (2 mM dATP, 2 mM dGTP, 2 mM dCTP, 1.4 mM AA-dUTP, 0.6 mM dTTP Aminoally-dUTP), AA-dUTP (Sigma-Aldrich).
3. Anhydrous dimethyl sulfoxide.
4. 0.5 M NaOH.
5. 0.5 M Ethylenediaminetetraacetic acid.
6. 0.2 M NaHCO$_3$, pH 9.6.
7. 80% Ethanol (v/v).
8. Sheared salmon testes DNA at 2 µg/µL.

2.5. Array Hybridization

1. Hybridization chamber from Corning.
2. 22×60 mm^2 cover glass (Fisher Scientific).
3. Humidifying buffer: 2X SSC + 50% Formamide.
4. Hybridization buffer: 60% formamide, 12% polyethylene glycol-17,000, 5% sodium dodecyl sulfate, 5% Triton X-100.
5. Cot DNA from Invitrogen or Roche (Palo Alto, CA).
6. Wash buffer: 0.5X SSC, 0.25% sodium dodecyl sulfate, 0.5% Triton X-100.
7. Deionized water.
8. 20X SSC: dissolve 175.3 g of NaCl and 88.2 g of sodium citrate in 800 mL, adjust the pH to 7.0 and bring up the volume to 1 L.
9. Boekel shaker incubator model 136400 (VWR, West Chester, PA).

2.6. Array Scanning

1. ScanArray 5000 (Perkin Elmer, Wellesley, MA).

2.7. Image Quantation

1. Imagene 6.0 or higher (Biodiscovery, EL Segundo, CA).

3. Methods

3.1. Silanization of DNA in 96-Well Format

1. Dissolve DNA pellet (1–5 µg) in 100 µL of 0.1 M NaOH. Incubate at 65°C for 30 min to ensure complete dissolving (*see* **Note 1**).
2. Dispense 10 µL of prepared 150 mM 3-glycidoxypropyltrimethoxysilane (*see* **Note 2**) to the dissolved sample and tap the block to mix well.
3. Seal the blocks in a Kapak plastic bag with wet paper towel and transfer them back to the 65°C incubator and continue to incubate for 2 h.
4. Add 100 µL of neutralization solution (0.5 M Tris-HCl + 1 M NaCl, pH 3.7). Mix by tapping. To make 1 L of this neutralization solution, mix 500 mL of 1 M Tris base and 500 mL of 1 M HCl, add 58 g of NaCl, shake to dissolve into clear solution.

5. Add 200 µL isopropanol and tap to mix well. Keep the block at room temperature (RT) for 15 min.
6. Spin down the DNA at 4000*g* for 15 min. Discard the supernatant by inverting the block.
7. Rinse the DNA pellet with 200 µL of 80% ethanol. Use a pipet to add ethanol slowly to avoid loosening the pellet.
8. Discard ethanol by inverting the block in the tray. Blot dry on a piece of paper towel.

3.2. Slide Cleaning and Coating

3.2.1. Slide Cleaning

1. Load individual slides on glass racks. Remove any visible dusts using compressed air or simply wipe off the dusts with soft tissue paper such as Kimwipes (Kimberly-Clark Global Sales Inc., Roswell, GA).
2. Wash the slides two times using deionized water by plunging in a glass jar.
3. Drain away the water. Add 2.5 *M* NaOH in 50% ethanol and soak at RT overnight. To make the 2.5 *M* NaOH solution, mix 500 mL of 5 *M* NaOH with 500 mL ethanol.
4. Collect the NaOH solution in a glass bottle. The NaOH can be reused two to three times.
5. Rinse the slides with deionized water two times.
6. Add 2 *M* HCl solution in 50% ethanol and incubate at RT for 2 h or longer. To make 1000 mL of 2 *M* HCl solution in 50% ethanol, dilute 167 mL of concentrated HCl in 333 mL deionized water and then add 500 mL ethanol.
7. Collect the HCl–ethanol solution in a bottle. This solution can be reused five times. Rinse the slides with deionized water. First, flush the jar with deionized water and then plunge the slide rack in the jar, and discard water. Repeat the cycle at least five times.
8. Rinse slides in ethanol three times. To save ethanol, put ethanol in three separate jars, all the slides racks can go through the three jars without ethanol changes.
9. Dry the slides completely at 65°C for 2 h or longer.
10. Store the slides in a clean slide box until use.

3.2.2. Coating Slides With BDSO

1. Load slides in small containers.
2. Make enough fresh 0.15–0.20% BDSO in 100% ethanol in a dedicated plastic bottle (*see* **Note 3**).
3. Wrap several containers together with plastic wrap and incubate them at 65°C for exactly 3 h.
4. Take out containers from the 65°C incubator and cool down to RT.
5. Discard the BDSO solution and immediately fill the container with 100% ethanol. Put the container in a separate group.
6. Keep going through all the containers. Avoid drying any part of the slide in these two steps (**steps 5** and **6**).

7. Shake the container vigorously several times to rinse off the residual BDSO solution. Discard ethanol and refill immediately as **step 5**.
8. Repeat **steps 5–7** four times.
9. Take the containers filled ethanol to the bench.
10. Pour ethanol in three 200-mL beakers and put them in a fixed order.
11. Using a tweezers, take slides out from the container and immediately dip into beaker no. 1. Rinse the slide several times in the beaker. Also let go the tweezers and rinse it to clean.
12. Rinse the slide in beaker nos. 2 and 3 the same way as in **step 11**.
13. Stand the slide on paper towel with one edge of the slide against a clean surface to blot away the ethanol. Do not touch the surface even with soft material.
14. After 3–5 min, inspect the slide surface. Blow off any visible liquid droplets on the surface or close to the end of the slides using compressed air.
15. Dry the slides for 20 min at RT.
16. Collect the coated slides in a clean slide box and label. Keep the slides in dry state until use by sealing the slide box in a plastic bag. Clean coated slides are completely clear and transparent. No specks should be visible to naked eyes.

3.3. Array Printing and Processing

1. Dissolved the silanized DNA in 30–50 µL of 0.2 M NaOH + 0.3 M NaCl at RT overnight. (Seal the plate or block in a bag with wet paper towel.)
2. Print the samples on slightly hydrophobic glass slides (*see* **Note 4**) using a MicroGrid printer (Genomic Solutions, Ann Arbor, MI). Use 0.2 M NaOH for pin washing in static baths.
3. Dry the printed slides in a 65°C incubator overnight.
4. Transfer the slides to small container and fill the container with 80% ethanol.
5. Keep the containers in a 65°C incubator overnight.
6. Discard the 80% ethanol and rinse the slides one time with 100% ethanol.
7. Dry the slides at 65°C for 1 h.
8. Put the arrays in a clean slide box and seal it in a plastic bag. Store the array at RT in a desiccator until use.
9. Print as many slides as practically possible (*see* **Note 5**).

3.4. Genomic Probe Labeling

3.4.1. Chemical Activation Method

1. Take 6 µg genomic DNA and dilute to 200 µL with 0.6 M NaCl in a 1.5-mL microtube. Add 150 µL isopropanol and mix by tapping the tube (*see* **Note 6**).
2. At **step 1**, if the DNA is high-molecular weight you will observe the DNA come out as a piece of floccule. Otherwise the solution is still clear. Spin down the DNA at maximum speed for 5 min.
3. Remove the liquid by aspiration. Add 500 µL of 80% ethanol to wash the pellet. Remove the ethanol completely. Dry the pellet at 65°C for 10 min.
4. Add 50 µL 0.5 M NaOH solution and incubate at 50°C for 10 min. Pipet to dissolve the DNA.

5. Transfer the tube to a 100°C heating blocked and incubate for 10 min if the DNA is high molecular weight, for 5 min if the DNA is fragmented, which will not show floccule in **step 1**.
6. Cool down on ice for 10 min. Add 150 µL 0.5 M NaCO$_3$ solution and keep on ice for 10 min.
7. Add 3 µL of 0.5X NB solution (always keep this solution on ice), mix well, and incubate on ice for strictly 10 min.
8. Add 20 µL 0.2 M EDBE solution and mix by vortexing. Spin down briefly.
9. Transfer the tube to a 50°C incubator and incubate for 60 min.
10. Add 10 µL of 2 M NaH$_2$PO$_4$ solution. Mix well. Add 300 µL isopropanol and mix.
11. Wait for 3 min then spin down at maximum speed for 5 min.
12. The liquid will now turn into two layers. The bottom layer contains the DNA and is about 15–20 µL.
13. Carefully remove the top layer by aspiration. Do not attempt to remove the top layer clean as you may easily take away the DNA in the bottom.
14. Add 500 µL of 80% ethanol and turn or invert the tube slowly. The DNA solution will now become murky.
15. Wash two more times with 500 µL of 80% ethanol.
16. Carefully remove the ethanol. Stop aspiration when there is only little ethanol left. Use a pipet to remove the ethanol as much it can be removed.
17. Dry the DNA at 100°C for 10–15 min. When the DNA is dry it will rise as a thin layer.
18. Add 500 µL of 80% ethanol to rehydrate the DNA. Wash the DNA by tapping or vortexing for 1 or 2 min.
19. Spin at maximum speed for 2 min.
20. Remove the ethanol by aspiration. The DNA at this step will stick to the bottom of the tube.
21. Completely dry the DNA at 100°C. The DNA will rise when it is completely dried. It usually takes 15 min or more to get the DNA dried completely.
22. Add 40 µL 0.2 M NaHCO$_3$ (pH 9.6) solution and pipet to assist dissolving. Make sure the pellet dissolve completely.
23. Split the sample into two aliquots, 20 µL for Cy5 labeling and 20 µL for Cy3 labeling.
24. Add 3 µL of NHS monoreactive Cy5 to label one aliquot (20 µL) and 3 µL of NHS monoreactive Cy3 to label the other. Mix by stirring with a pipet tip.
25. Keep the reaction at RT for overnight or at 37°C for 2–3 h.
26. Add 160 µL of Sheared salmon testes (SST) DNA (1 µg/µL), 10 µL of 5 M NaCl, and mix.
27. Add 100 µL isopropanol and mix.
28. Spin down the DNA at 13,000g for 10 min.
29. Wash the pellet with 500 µL 80% ethanol. Aspirate to remove the ethanol.
30. Dry the DNA at RT for 30 min. Keep the label DNA as dry pellet at −20°C until use.

3.4.2. Random Priming Labeling Using Bioprime Kit

1. Mix 200 ng high-molecular weight DNA and water and 20 µL 2.5X random primer solution to final 50 µL, heat for 5 min at 100°C (*see* **Note 7**).
2. Cool on ice for 5 min.

3. Add 10 µL AA-dUTP-dNTP mix and 2 µL Klenow (40 U/µL), mix well, and spin briefly.
4. Incubate at 37°C for 3 h.
5. Measure DNA concentration with Fluorometer. The reading should reach 50 ng/µL or higher. This DNA reading will be used as empirical concentration for taking probe amount for hybridization.
6. Add 10 µL of 0.5 mM ethylenediaminetetraacetic acid.
7. Add equal volume of 0.5 M NaOH and mix well.
8. Heat for 15 min at 100°C (*see* **Note 8**).
9. Add equal volume of isopropanol and mix well, keep at room temperature for 20 min.
10. Spin at maximum speed 10 min to precipitate DNA.
11. Wash the DNA pellet with 300 µL of 80% ethanol.
12. Remove ethanol and dry DNA pellet at RT.
13. Completely dissolve DNA pellet with 40 µL 0.2 M NaHCO$_3$ (pH 9.6).
14. Separate DNA solution in two equal aliquots.
15. Add 5 µL of Cy3 or Cy5 to DNA solution (*see* **Note 9**).
16. Incubate at 37°C for 3 h or keep on the bench for overnight.
17. Add 20–40 µg of SST DNA and mix well.
18. Add 5 M of NaCl to final 0.5 M and mix well.
19. Add isopropanol and wait at RT for 10 min.
20. Spin down DNA at maximum speed 10 min.
21. Remove isopropanol and wash the pellet with 500 µL 80% ethanol.
22. Remove ethanol and dry DNA pellet at 65°C for 10 min.
23. Keep the probe pellet at –20°C until use.

3.5. Hybridize Labeled DNA to the Array

1. Dissolve the probe pellet with TE buffer.
2. Combine 2 µg sample DNA labeled with Cy3 and equal amount control DNA labeled with Cy5, conversely sample DNA labeled with Cy5 and control DNA labeled with Cy3, mix well.
3. Add 100 µg of Cot1 DNA (amount by Fluorometer reading).
4. Add 5 M NaCl to final 500 mM and mix well.
5. Add equal volume isopropanol and mix well, keep mixture on bench for 10 min.
6. Spin at 13,000g for 10 min to pellet DNA.
7. Remove the supernatant and wash with 500 µL 80% ethanol (the pellet should have purplish hue).
8. Remove ethanol and dry pellet on 65°C for 10 min.
9. Dissolve DNA pellet and mouse Cot1 DNA pellet with 10 µL of water.
10. Add 50 µL hybridization buffer, mix by pipeting.
11. Heat at 100°C for 3 min and then incubate at 37°C for 2 h.
12. Load probe onto the slide and cover with 22 × 60-mm^2 cover slip to spread it out. Make sure there is some solution running around the edges of the cover slip.
13. Add 30 µL 2X SSC/50% formamide to either side of the Corning hybridization chamber (the wells at the ends Corning hybridization chamber need to be expanded to hold more volume), then put the slide in the chamber.

14. Close chamber and put all the chambers in a block and wrap with aluminum foil. Seal the chamber block in a plastic bag together with wet paper.
15. Put the bag in 37°C shaker incubator. Hybridize for 18–20 h at the lowing shaking speed (*see* **Note 10**).

3.6. Posthybridization Wash

1. Put the slide in a Petri dish and soak it with 30-mL 2X SSC buffer for 2 min. Gently slide off the cover slip using a clean forceps.
2. Transfer the slide to a clean Petri dish and wash with slide wash solution at 67°C in a shaking incubator for 30 min.
3. Briefly rinse slide with deionized water and immediately dry the slide with nitrogen. The slide is ready for scanning. Scan the slide as soon as possible.

3.7. Array Scanning

Arrays after washing should be scanned as soon as possible. ScanArray 5000 was used to scan our CGH arrays but other two-color microarray scanner work equally well. Attention should be paid to proper selection of scanner setting to acquire high-quality image data as they directly affect the final CGH results. In order to optimize the signal intensity in both channels, the scanner should be adjusted to the proper setting for each array. The signals from both Cy3 and Cy5 channel should be roughly balanced. Normally, a fast scan should be performed on a small area for each array to find out the proper setting. It is not necessary to scan the whole slide each time because the signal intensity should be uniform across the whole array. The laser can be set at 100% if the signals are not very strong. The scanner photomultiplier tube (PMT) gain should be adjusted to achieve the proper signal intensity in both channels such that saturated pixels or white pixels on array spots are absent or minimized.

3.8. Image Quantitation

Imagene program was used for image data quantitation. This program seems to generate consistent results for variable quality of image data. Grid aligning is semi-automatic with this program. For high-throughput batch mode image processing the Bluefuse program from BlueGnome was preferred, which is specifically designed for analysis of CGH images. There are also other attractive features with this program such as normalization and chromosome profile display.

3.9. Fluorescence Ratio Normalization

Normalization of the fluorescence ratios between the sample and control DNA on array spots is critical for sensitive and reliable detection of DNA copy-number differences. In a self vs self hybridization the log fluorescence ratios of Cy5 over Cy3 follow a normal distribution. Ratios drift from the mean is highly correlated with the spot fluorescence intensity for reasons currently unclear. Another dominant factor affecting the fluorescence ratios is the spot positions. The normalization process is thus involved in finding the baseline ratios for different groups of spots with different intensity and positions. After normalization against intensity and position, spots indicating chromosomal abnormalities should be in outliers including spots drifting away from a set cutoff value. An in-house excel macro program was used for array CGH ratios normalization and chromosome ratio profile display.

For whole genome arrays (21 k arrays) the spots are first divided into 100 groups based on spot total fluorescence intensity (Cy3 + Cy5). The fluorescence ratios are then ranked within each subgroup. Twenty percent of the spots the low end and high end are taken out and the mean value of the 60% of the midrange ratios is calculated and used as the normalization factor for the subgroup. After normalization against intensity, the ratios are further normalized using local spot position information. We take 40–50 spots in each subarray printed by the same pin and rank them by intensity normalized ratios. Again, 20% of each spots at the low and high end are excluded and the rest are taken to calculate the normalization factor.

After normalization, chromosomal abnormalities are scored by outlier detections using a set cutoff value. The cut off normally used for detecting chromosomal loss or gain is 2.5 standard deviation (SD) units. Our self vs self experiments indicate that if dye reversal experiment is performed the average false-positive rate for our whole genome array is around 0.3%. To detect smaller scale of deletion or gain this 2.5 SD cut off can be relaxed. But other clues must be used to exclude the false-positive detections.

4. Notes

1. Tris interferes with silanization reaction. If the DNA is not in a Tris-based buffer NaOH can be added directly to a final concentration of 0.1 *M*. The DNA premodification method will work with DNA prepared by PCR amplification of purified clone DNA or direct extraction of cloned DNA from bacterial culture. However, less fragmented cloned DNA (>2 kb) offers better hybridization efficiency.
2. Glycidsilane solution should be freshly prepared by diluting the stock (>98%) in isopropanol.
3. BDSO is very sensitive to humidity. Take the required amount from the bottle as quickly as possible and immediately cap the bottle and wrap the bottleneck with parafilm. BDSO should be stored in a desiccator. The ethanol used for treating the slides must be 200 proof. The slide containers for slide coating must be completely dry. Coating will fail if the containers are not dry. The containers should be dried and kept in a 65°C incubator.
4. The surface hydrophobicity is an important factor controlling the spot size and morphology. Surface hydrophobicity can be adjusted by treating the slide at different concentration of BDSO. Treating acid clean slide in 0.02–0.15% of BDSO in

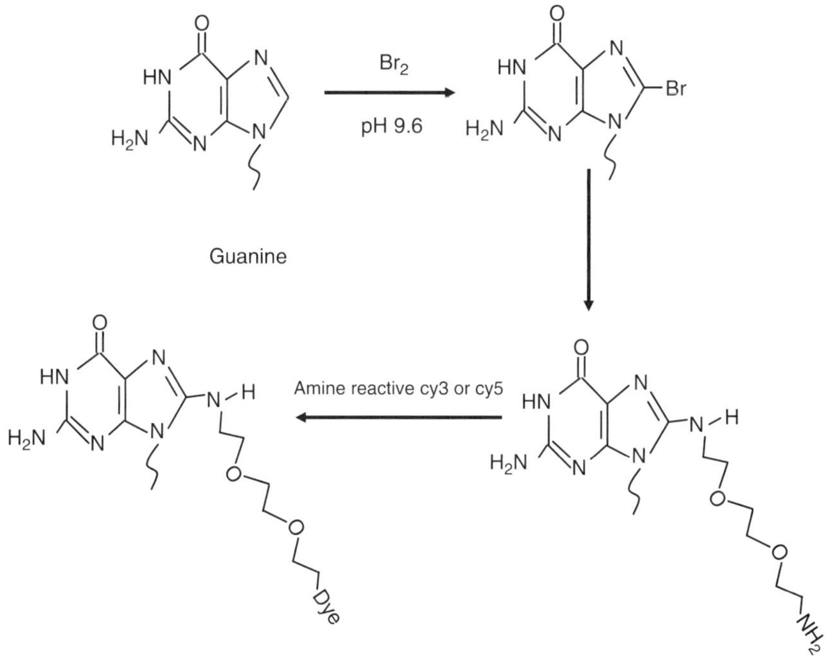

Fig. 3. Chemical activation labeling of DNA. Bromine primarily activates guanine on DNA at the C7 site. The bromine activated DNA is further functionalized to contain active amino group that can be readily labeled by amino-reactive fluorescent dyes.

ethanol for 2 h can give different degrees of reproducibly hydrophobic surfaces. To get the desired spot size other factors such as humidity, size and type of pins for printing, and the pin contact time in printing must match with the surface hydrophobicity. Insufficient hydrophobicity can cause spot merging. Too much hydrophobicity will generate tiny spots and poor sample deposition. Matching all these factors requires experience and titration test printing.

5. Silanized DNA is not good for storage. It slowly aggregates as the pH drops when the sodium hydroxide solution is gradually neutralized by CO_2 in the air. When printing lasts longer than 2-wk fresh NaOH should be added every 2 wk to keep pH >12.0.

6. Chemical activation provides uniform labeling of DNA without the concern for biased amplification as in the enzymatic labeling methods. The disadvantage of this method is relatively large quantity of DNA must be available. The protocol described here is modified from an original protocol (*6*). The reaction scheme is shown is **Fig. 3**. Critical improvements include fragmentation of DNA using NaOH and the use of EDBE to enhance the solubility of amine modified product. N-Bromosuccinimide can be made into stock solution and stored at –20°C. To get consistent results, the starting DNA should be of high molecular weight DNA (>5 kb).

7. The Bioprime kit from Invitrogen is specifically designed for nonradioactive labeling of DNA. The kit can also amplify the input DNA through strand displacement reaction. In the method described here genomic DNA is first amplified and labeled with aminoally-dUTP. The amine activated product is then labeled with Cy3 or Cy5 monoreactive NHS ester. This two-step indirect labeling guarantees efficient amplification and uniform dye incorporation. The method is also more cost effective than direct incorporation of dye labeled nucleotide. In 3 h, 1000-fold amplification can be readily achieved. Input DNA amount as low as 12 ng does not cause significant amplification bias detectable by array CGH. Impurity in the genomic DNA is the most frequent cause of failure in this method. If the amplification is slow, the input DNA should be purified by one more round of phenol/chloroform extract. Extending the amplification time to achieve high DNA yield is not recommended as this will generate significant bias in amplification. The protocol also works with formalin-fixed/parafilm-embedded DNA. But normal control DNA must be extracted from the same fixed tissue. Otherwise, consistent artificial gains or losses can be observed as a result of nonuniform fixation of genomic region in the source tissue.

8. Treating DNA with 0.5 M NaOH at 100°C is a very convenient way of fragmenting the probe of small size, which is critical for quantitative hybridization. The optimal probe size for array CGH is in the range of 100 to 400 bp.

9. Amine monoreactive Cy3 or Cy5 are supplied as small package in dry form for protein labeling. Each vial is sufficient for labeling 1 mg protein. For DNA labeling, each vial should be reconstituted with 30 μL high-quality anhydrous dimethyl sulfoxide. Unused dye solution can be kept at –20°C.

10. Active agitation of hybridization mix can significantly enhance the hybridization efficiency. Some commercial hybridization stations such as the MAUI microarray hybridization system offer both sensitive and uniform hybridization. Hybridization time should be longer than 20 h. Extended hybridization time does increase the signals but the ratios are most often dampened as a result of Cy5 degradation and complex probe reassociation kinetics in the hybridization solution.

References

1. Solinas-Toldo, S., Lampel, S., Silgenbauer, S., et al. (1997) Matrix-based comparative genomic hybridization: biochips to screen for genomic imbalances. *Genes Chromosomes Cancer* **20,** 399–407.

2. Kallioniemi, A., Kallioniemi, O. P., Sudar, D., et al. (1992) Comparative genomic hybridization for molecular cytogenetic analysis of solid tumors. *Science* **258,** 818–821.

3. Ishkanian, A. S., Malloff, C. A., Watson, S. K., et al. (2004) A tilling resolution DNA microarray with complete coverage of the human genome. *Nat. Genet.* **36,** 299–303.

4. Cai, W. W., Mao, J. H., Chow, C. W., Damani, S., Balmain, A., and Bradley, A. (2002) Genome-wide detection of chromosomal imbalances in tumors using BAC microarrays. *Nat. Biotechnol.* **20,** 393–396.
5. Southern, E., Mir, K., and Shchepinov, M. (1999) Molecular interaction on microarrays. *Nat. Genet. Suppl.* **21,** 5–9.
6. Keller, G. H., Cumming, C. U., Huang, D. P., Manak, M. M., and Ting, R. (1988) A chemical method for introducing haptens onto DNA probes. *Anal. Biochem.* **170,** 441–450.

6

Evaluating the Quality of Data From DNA Microarray Measurements

Lili. Wang, A. K. Gaigalas, M. B. Satterfield, M. Salit, Y. Zong, and J. Noble

Summary

Gene expression technology offers great potentials to generate new insights into human disease pathogenesis; however, the data quality remains a major obstacle for realizing its potentials. In the present study 60-mers oligonucleotide target immobilized on coated glass slides were utilized as a model system to investigate parameters, such as target concentration, retention, signal linearity, and fluorescence properties of fluorophores, which likely affect the quality of microarray results. An array calibration slide was used to calibrate an Axon GenePix 4000A scanner and ensure the dynamic range of the instrument. The work is a first step toward our goal of quantitative gene expression measurements.

Key Words: Cy3; Cy5; oligonucleotide microarray; retention; scanner calibration; signal linearity; spectral properties; target concentration.

1. Introduction

The use of DNA microarrays is growing rapidly in searching biomarkers for cancer diagnostic *(1,2)*. DNA microarray experiments consist of multiple steps, from extraction and PCR amplification of mRNA to robotic printing of purified cDNA oligonucleotides on coated glass slides, hybridization with reverse-transcribed and fluorescent-labeled test cDNA, fluorescence detection with array scanners, and finally microarray data analysis. The complexity of the microarray processes enhances the potential for variability; however, with proper control of each step, DNA microarrays will eventually become an important tool for clinical applications.

Owing to various advantages of oligonucleotide microarrays over cDNA arrays *(3,4)*, the use of oligo-microarrays has gained momentum recently. There

From: *Methods in Molecular Biology, vol. 381: Microarrays: Second Edition: Volume 1: Synthesis Methods*
Edited by: J. B. Rampal © Humana Press Inc., Totowa, NJ

is great need in optimizing parameters and improving efficiencies of various steps in oligoarrays to make the technology applicable for clinical diagnostics and drug discovery. The usefulness of all bioinformatics tools is ultimately dependent upon the quality of the data generated from microarray measurements. In this work, the 60-mer oligonucleotide target immobilized on coated glass slides were used as a model system to investigate parameters, such as target concentration, retention, signal linearity, and fluorescence properties of fluorophores, likely affecting qualities of data from microarray experiments. The study is a part of the National Institute of Standards and Technology (NIST) effort aimed toward quantitative gene expression measurements.

2. Materials

1. Two oligonucleotide targets (60-mer), defined as oligonucleotides immobilized on glass slides, and two probes (60-mer) were synthesized and high-performance liquid chromatography purified by integrated DNA technologies (Coralville, IA). Both targets have a 5′ amine linker modification and target 1 is further labeled with a 3′ Cy3 fluorophore for determination of target retention. Probes 1 and 2 are labeled with a 5′ Cy3 dye molecule and a 5′ Cy5 fluorophore, respectively. Their sequences are a part of rat tubulin genome and given in **Table 1**. The oligonucleotide targets were dissolved in 150 mM sodium phosphate buffer, pH 8.5, to make 20 µM stock solutions and the probes were dissolved in RNase-free water to make 2 µM stock solutions.
2. PowerMatrix oligo slides and oligohybridization buffer from Full Moon BioSystems (Sunnyvale, CA).
3. Gen III spotter from Amersham Biosciences, Piscataway, NJ.
4. 2X Standard saline citrate (SSC)/0.1% (w/w) bovine serum albumin/0.2% (w/w) sodium dodecyl sulfate (SDS) solution (pH 7.0) prewarmed at 55°C.
5. Cover slips (22 × 30 mm^2) prepared from 22 × 60 mm^2 HybSlip (Molecular Probes, Eugene, OR) and cleaned with 70% ethanol and blow-dried with nitrogen.
6. 0.2X SSC/0.2% (w/w) SDS and 0.2X SSC solutions (pH 7.0) prewarmed at 55°C.
7. Raman microscope operating software (Wire V1.3 and Grams/32) from Renishaw, Gloucestershire, UK, model RM1000.
8. Argon ion laser 514 nm (Spectra-Physics, Mountain View, CA).
9. HeNe laser 633 nm (Renishaw, Gloucestershire, UK).
10. Axon GenePix 4000A scanner (Union City, CA).

3. Methods

3.1. Array Fabrication, Hybridization, and Posthybridization Washing

3.1.1. Array Fabrication

1. Print the consecutive seven twofold dilutions of the target stock solutions (10, 5, 2.5, 1.25, 0.625, 0.312, 0.156, and 0.078 µM, respectively) onto PowerMatrix Oligo Slides using Gen III.
2. Place the spotted slides in a dark chamber with 75% humidity for 18 h.

Table 1
Oligonucleotide Sequences

Name	Sequence	Study
Target 1	5′-/AminoC6/CTG CTG ATG GAG AGG CTC TCT GTC GAC TAC GGA AAG AAG TCC AAG CTG GAG TTC TCT TTT /Cy3/-3′	Retention
Target 2	5′-/AminoC6/CTG CTG ATG GAG AGG CTC TCT GTC GAC TAC GGA AAG AAG TCC AAG CTG GAG TTC TCT TTT-3′	Target, concentration, spectral properties
Probe 1	5′-/Cy3/TTA AAA GAG AAC TCC AGC TTC GAC TTC TTT CCG TAG TCG ACA GAG AGC CTC TCC ATC AGC-3′	
Probe 2	5′-/Cy5/TTA AAA GAG AAC TCC AGC TTC GAC TTC TTT CCG TAG TCG ACA GAG AGC CTC TCC ATC AGC-3′	

Target and probe are defined as oligonucleotides immobilized on glass slides and hybridized to immobilized nucleotides, respectively.

3. Scan each slide using an Axon GenePix 4000A scanner with 10 μm/pixel resolution at a PMT setting of 500 V for both Cy3 and Cy5 channels.
4. Treat the printed slides with prewarmed (55°C) 2X SSC/bovine serum albumin/SDS for 20 min in the dark to wash off unbound excess targets and block nonspecific bonding sites on the slides.
5. Rinse the slides extensively with deionized water and dry with compressed N_2 at 20 psi.
6. Scan the slides with the same scanner at the same photomultiplier tube (PMT) setting as described earlier.

3.1.2. Array Hybridization

1. For a single probe hybridization, a 35 μL of probe solution was prepared by adding a 32 μL of oligohybridization buffer to a 3 μL of the probe stock solution. The probe solution was denatured at 95°C for 5 min and placed on ice for 5 min before use.
2. Place a 25 μL of the probe solution on each slide and carefully lay down a cover slip.
3. Incubate slides in a humidified chamber with 100% humidity at 42°C for 16 h.

3.1.3. Posthybridization Treatment

1. Rinse off the cover slips with deionized water and place slides in a slide rack.
2. Immerse slides in a prewarmed 0.2X SSC/SDS solution and place on an orbital shaker for 20 min at ambient temperature.

3. Immerse slides in a prewarmed 0.2X SSC solution and gently dip slides up and down for 1 min.
4. Repeat the **steps 2** and **3** twice. Use fresh 0.2X SSC solutions every time.
5. Rinse slides extensively with Milli-Q water at ambient temperature.
6. Dry slides with a gently stream of nitrogen (20 psi) immediately.

3.2. Calibration of the Microarray Scanner

An Axon GenePix 4000A scanner was used to detect fluorescence signals and data analysis was performed using the scanner data analysis software, GenePix Pro. To ensure the performance of the scanner, a FMB scanner calibration slide with a series of known concentrations of either Cy3 or Cy5 embedded in a polymer matrix *(5)* was used to calibrate the scanner in terms of the dynamic range, detection limit, and scanning uniformity. The signal-to-noise ratios (SNR) generated for Cy3 and Cy5 fluorophores are shown in **Fig. 1A** using the calibration slide. SNR is commonly defined as average signal intensity with average background intensity subtracted divided by standard deviation of the background signals *(5–7)*. The average background signal was obtained using multiple printed blank spots in the absence of fluorophores. For Cy5 channel, a linear relationship is observed between 9 fluorophores/μm^2 and 9100 fluorophores/μm^2 as shown in **Fig. 1A** by a linear fitting curve (solid black line). Similarly, a linear relationship is observed between 5 fluorophores/μm^2 and 4590 fluorophores/μm^2 for Cy3 channel. The linear dynamic ranges for both channels are close to three orders of magnitude.

The limit of detection is determined by locating the dye concentration at which SNR is at least 3. The limit of detection for Cy3 and Cy5 channels are 5 and 9 fluorophores/μm^2, respectively. The data in **Fig. 1A** can be replotted with the SNR as the independent X-coordinate to yield a calibration curve relating a measured SNR value to the number of Cy3 fluorophore, for example, in terms of fluorophore/μm^2 for the scanner shown in **Fig. 1B**. Whereas SNR ≤4000, a linear equation, $y = 2.544x$ was used to fit the data with R^2 of 0.994. When SNR ≥4000, a power equation, $y = 0.0666x^{1.452}$ was applied to fit the data with R^2 of 0.981. In the same fashion, a calibration curve for Cy5 channel of the same scanner was constructed using combined linear and power equations. The two calibration curves allow converting SNRs calculated from intensities of array spots to values in terms of fluorophore/μm^2. This calibration slide is useful for establishing the linearity range of the scanner that is crucial for quality assurance/quality control (QA/QC) of the microarray measurements *(8)* (*see* **Note 1**).

3.3. Retention of the Target Oligonucleotide on Slides

The retention of the target oligonucleotide on PowerMatrix oligo slides was determined by using Cy3-labeled target 1 (**Table 1**) and calculating the ratio of

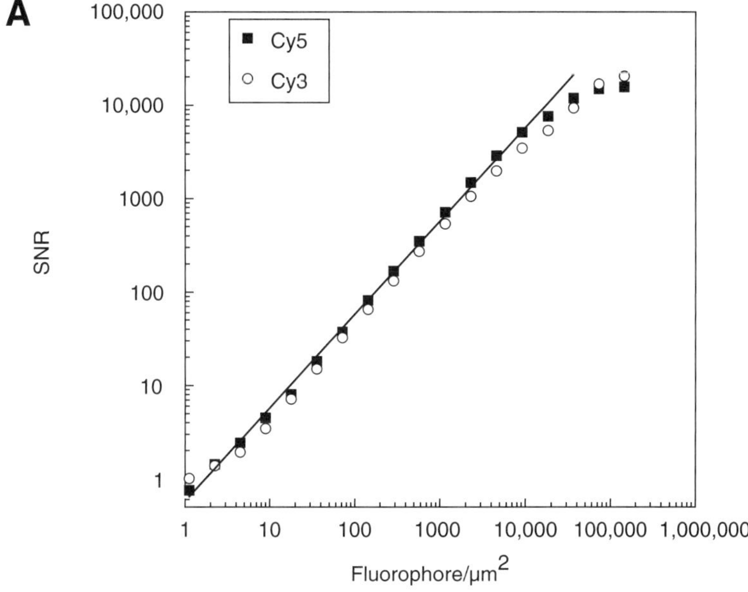

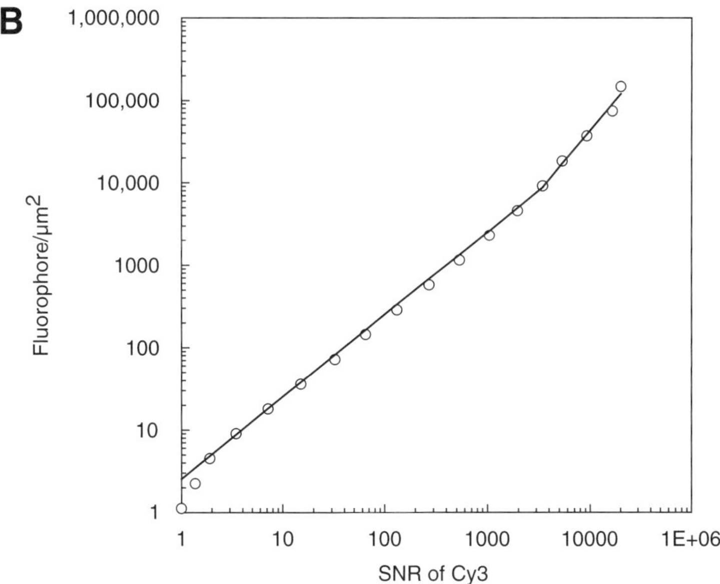

Fig. 1. Signal-to-noise ratios of Cy3 and Cy5 on a FMB scanner calibration slide (**A**) and the resulted calibration curve of Cy3 fluorophore (**B**, as an example) generated on an Axon GenePix 4000A with 10 µm/pixel resolution at the PMT of 500 V for both Cy3 and Cy5 channels. The solid lines are the fitting curves.

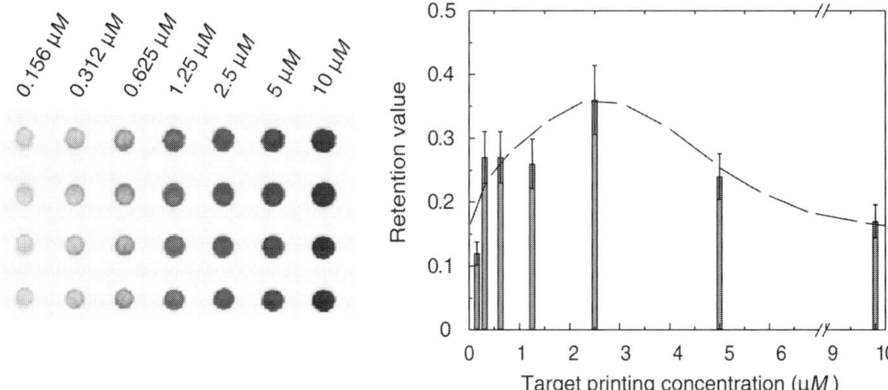

Fig. 2. **(Left)** Images obtained for target 1 after mock hybridization at various printing concentrations. **(Right)** The retention values determined through mock hybridization for different printing concentrations of target 1. The error bar shows the standard deviation of each retention values averaged over multiple experiment repeats.

the signal after mock hybridization and the signal detected after the printing. Mock hybridization refers to hybridization in the absence of the complementary strand. The slides printed with target 1 went through the entire array processes including printing, hybridization, posthybridization washing and scanning, and data analysis. The fluorescence signal from each spot was corrected by subtracting the background signal obtained using an oligonucleotide target (60-mer) whose sequence is very different from the rat tubulin target sequence chose here. Typically, the spot diameter is 180 µm and the center-to-center space between two spots is 340 µm (the image on the left of **Fig. 2**). The retention value as a function of the printing concentration of target 1 is shown on the right of **Fig. 2**. The error bars represent the standard deviations (~15–20%) of the retention values determined through multiple slides in a single experiment and slides from different experiments. The retention values increase before the target concentration reaching 2.5 µM and decrease as the targets become more concentrated. Because the amount of the target attached to the slide surface are limited by the number of binding sites available on the surface, the target retained on the surface will increase with increasing the concentration until the binding sites are saturated. Although the present retention values are reasonable for 60-mer oligonucleotides on this particular kind of coated glass slides, retention values highly depend on the size of the targets, printing buffers, and hybridization conditions. Optimization of the experimental conditions to attain optimal retention values of chosen targets is a perquisite for obtaining high quality array data (*see* **Note 2**).

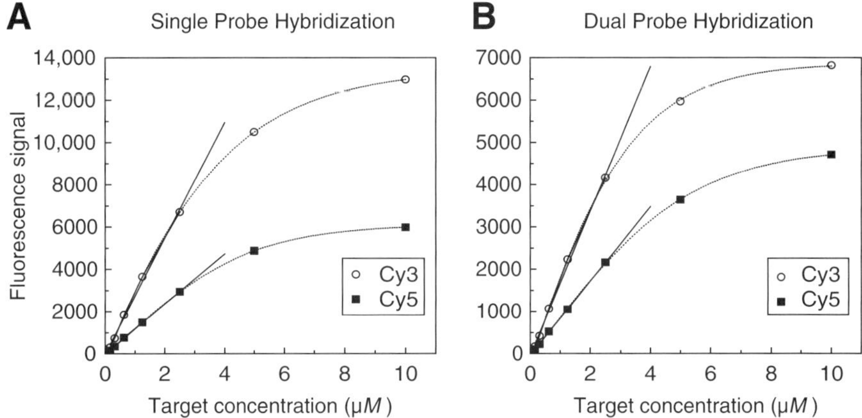

Fig. 3. Fluorescence signal (fluorophores/μm²) as a function of target concentration (Target 2) for single probe hybridization (**A**) and dual probe hybridization (**B**).

3.4. Effect of Target Concentrations on Hybridization Signals

When probe concentration was fixed at 170 nM for single-probe and dual-probe hybridizations (equal molar mixture of two probes, both at 85 nM), we investigated the effect of target concentration on hybridization signal. **Figure 3** shows hybridization signal, expressed in fluorophore/μm² according to the calibration curve, changes with the target concentration. The standard deviations of the measured signals are 15–20% in the present study. In general, hybridization signal has a linear relationship with target concentration when the target concentration is equal or less than 2.5 μM, as shown by the linear fitting lines. Whereas, the target concentration doubled, the fluorescence signal increases by about two times. However, the hybridization signal deviated from the linearity range when the target concentration is more than 2.5 μM and eventually reaches a plateau in all cases. Available binding sites on the slide surface limit the amount of target immobilized. Once target fully occupy all binding sites, any extra probes available in the hybridization solution will not increase the hybridization signals. The results are consistent with those of the retention plot in **Fig. 2**. Within the linear dynamic range (**Fig. 3**) the ratios of fluorescence signals from Cy3 fluorophores in single-probe hybridization and dual-probe hybridization are similar to the ratios from Cy5 fluorophores in the two hybridization conditions. However, these ratios are somewhat less than 2 taking into account that the concentration of Cy3- or Cy5-labeled probe in single-probe hybridization is twice of that in dual-probe hybridization. About 15–20% error in the study may account for the discrepancy. The current investigation reveals that quantitative hybridization results can be obtained with target concentrations falling within the linearity range (*see* **Note 3**).

3.5. Fluorescence Spectral Characterization of Cy3 and Cy5 Fluorophores in Oligoarrays and on Calibration Slides

The instrument for measuring fluorescence spectra of Cy3 and Cy5 fluorophores in microarrays is a dispersive Raman instrument with a spectral resolution of 4/cm. The Raman microscope, optics, stage, and operating software (Wire V1.3 and Grams/32) are from Renishaw. This system is made up of both confocal optics and a CCD detection device allowing acquisition of spectra at discrete areas of a microarray spot. The instrument contains two lasers, an Argon ion laser 514 nm, and a HeNe laser 633 nm. Given that spectral acquisition using this Raman microscope is extremely time consuming, a calibrated fluorimeter *(9)* was employed for measuring emission spectra of Cy3 and Cy5 fluorophores on the FMB calibration slide. The rationale here is that the composed spectrum shape from individual array spots (multiple experimental repeats) should be equivalent to the spectrum averaged over multiple array spots. In the fluorimeter, a water-cooled argon ion laser, Lexel model 95, provides 514 nm light for Cy3 fluorescence measurement, and a HeNe laser (Uniphase model 215-1) supplies a 633-nm beam for Cy5 measurement. The laser beams were used to illuminate the multiple spots with high concentrations of Cy3 or Cy5 fluorophore.

Figure 4 shows the emission spectra of Cy3 (left) and Cy5 (right) in oligonucleotide microarrays (solid) obtained using the Raman microscope and on the FMB calibration slide (dot) taken by the fluorimeter concerning the spectra taken in solutions (dash). The spectrum of Cy5 fluorophore in microarrays matches that on the calibration slide, both are peaked at 670 nm and red-shifted by 7 nm with respect to the spectrum in solutions. On the other hand, the emission spectra of Cy3 fluorophore in oligoarrays and on the calibration slide are red-shifted by 8 and 13 nm, respectively, regarding the spectrum of Cy3 fluorophore in solution. The filter set of Axon's 4000A scanner (575 ± 17.5 nm for Cy3 channel, 670 ± 20 nm for Cy5 detection) is well suited for fluorescence detection. The weaker signals from Cy5 fluorophores than those from Cy3 fluorophores seen in **Fig. 3** are not owing to spectral mismatch between oligoarrays and calibration slide. There may be two reasons for the observation. One is that the fluorescence quantum yield of Cy5 fluorophore drops considerably via hybridization and washing processes, partly owing to atmospheric ozone *(8)*. Another reason is that contaminating fluorescence, which is more severe in Cy3 channel than in Cy5 channel *(10,11)*, is attributed to high fluorescence signals from Cy3 fluorophores. Taking into account that the fluorescence signal corresponded linearly with serial dilutions of the target (**Fig. 3**), particularly at low target concentrations, it is unlikely that strong signals from Cy3 fluorophores are as a result of contaminating fluorescence and/or autofluorescence in the present study (*see* **Notes 4** and **5**).

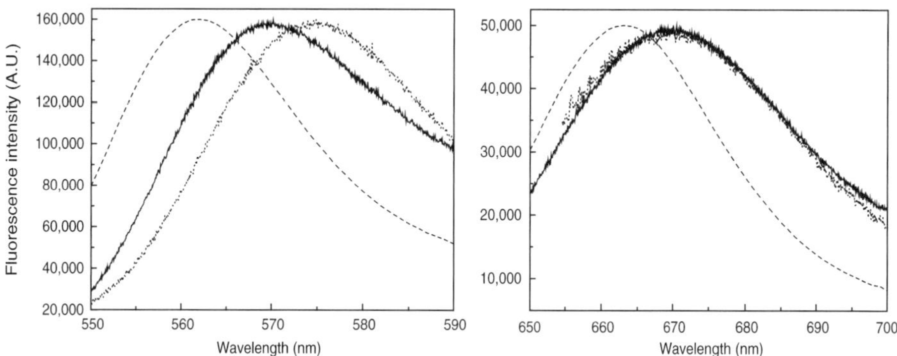

Fig. 4. Emission spectra of Cy3 (left) and Cy5 (right) in oligonucleotide microarrays obtained from the Raman microscope (probes 1 or 2 hybridized to target 2, solid) and on the FMB calibration slide using the fluorimeter (dot) with respect to the spectra of Cy3 and Cy5 in 1X PBS buffer (dash).

4. Notes

1. Users of a Full Moon Biosystems calibration slide should be aware that they are calibrating their scanner to a particular slide and that the measurement is not traceable to known standards or reference material. The FMB slide contains 12 identical rows; determination of the two sigma range associated with row-to-row variability yields a basic understanding of the dispersion of values around the measurement.

2. In this study, the retention value was defined as the ratio of the fluorescence signal after mock hybridization and the signal after printing. In other cases *(12,13)*, retention values were determined by the ratio of the fluorescence signal after washing and blocking and the signal after printing. The retention value determined by this method is higher than ours through mock hybridization by about 10%. Retention values are generally dependent of target size, printing buffer, and hybridization condition. Hessner and coworkers *(13)* reported retention values of cDNA on variety types of commercially available substrates. By and large, cDNA can hold on to slides better than short oligonucleotides. For quality control purpose it is also possible to stain slides with DNA specific dyes *(14,15)*, such as SYBR green II and SYTO 61, before hybridization. However, this step would not produce quantitative retention data, and requires destaining that may cause changes in slide performance. Obtaining retention values ensures reliable and high-quality gene expression data.

3. For quantitative array experiments it is important to optimize the target concentrations on the substrates so that the hybridization signals fall within a linear dynamic range. Wang and coworkers reported that the hybridization signals increased by two- to sixfold for majority of the printing probes when the oligonucleotide printing concentration decreased from 50 to 6 µ*M (4)*. At high printing concentrations low

signals are seen as a result of fluorescence quenching *(16)*. The concentration optimization likely depends on the type of targets, slide surface coating, and printing conditions. Yue and coworkers *(14)* have done excellent work in determining optimal cDNA target concentrations for differentiating expected mRNA expression levels.

4. Fluorescence properties of fluorophores including Cy3 and Cy5 change drastically when they are placed near surfaces. The maximal emission of Cy5 fluorophores immobilized directly was found on UltraGAP slides (Corning, NY) is at 689 nm with half width of 55 nm with respect to emission maximum at 664 nm with half width of 34 nm in solution *(17)*. Given that the emission spectrum of Cy5 fluorophores in oligoarray slides is red shifted only by 7 nm as to that in solution and overlays with the spectrum of Cy5 on the calibration slide, it is apparent that the fluorophore is distanced away from the slide surface and in relatively homogeneous microenvironment. In typical microarray measurements, cDNA labeled with numerous fluorophore molecules are hybridized to targets on slides. Owing to environmental inhomogeneity of the fluorophores, the fluorescence properties, such as emission maximum and spectral shape, may be very different from those seen with singly labeled oligonucleotides in the present study. The spectroscopic investigation is underway with the use of labeled cDNA.

5. The result here is that hybridization signals from Cy5–labeled probe were lower than those from Cy3-labeled probe (*see also* **ref. 8**). It may be wise to implement single-color hybridizations (for instance, Cy3) with housekeeping genes as spike-in controls to eliminate fluorophore biases and hence improve data quality from DNA microarray measurements.

Acknowledgments

Certain commercial equipment, instruments, and materials are identified in this article to specify adequately the experimental procedure. In no case does such identification imply recommendation or endorsement by the NIST, nor does it imply that the materials or equipment are necessarily the best available for the purpose.

References

1. Mischel, P. S., Cloughesy, T. F., and Nelson, S. F. (2004) DNA-microarray analysis of brain cancer: molecular classification for therapy. *Nat. Rev. Neurosci.* **5,** 782–792.

2. Chang, J. C., Hilsenbeck, S. G., and Fuqua, S. A. (2005) Genomic approaches in the management and treatment of breast cancer. *Br. J. Cancer* **92,** 618–624.

3. Kane, M. D., Jatkoe, T. A., Stumpf, C. R., Lu, J., Thomas, J. D., and Madore, S. J. (2000) Assessment of the sensitivity and specificity of oligonucleotide (50 mer) microarrays. *Nucleic Acids Res.* **28,** 4552–4557.

4. Wang, H., Malek, R. L., Kwitek, A. E., et al. (2003) Assessing unmodified 70-mer oligonucleotide probe performance on galss-slide microarrays. *Genome Biol.* **4,** R5.

5. Zong, Y., Wang, Y., Zhang, S., and Shi, Y. (2003) How to evaluate a microarray scanner, in *Microarrays Methods and Applications—Nuts and Bolts*, (Hardiman, G., ed.), DNA Press LLC, pp. 99–114.

6. Worley, J., Bechtol, K., Penn, S., et al. (2000) A systems approach to fabricating and analyzing DNA microarrays, in *Microarray Biochip Technology*, (Schena, M., ed.), Eaton Publishing, pp. 65–85.

7. Basarsky, T., Verdnik, D., Zhai, J. Y., and Wellis, D. (2000) Overview of a microarray scanner: design essentials for an integrated acquisition and analysis platform, in *Microarray Biochip Technology*, (Schena, M., ed.), Eaton Publishing. pp. 265–284.

8. Shi, L., Tong, W., Goodsaid, F., et al. (2004) QA/QC: challenges and pitfalls facing the microarray community and regulatory agencies. *Expert Rev. Mol. Diagn.* **4,** 761–777.

9. Gaigalas, A. K., Wang, L., Schwartz, A., Marti, G. E., and Vogt, R. F. (2005) Quantitating fluorescence intensity from fluorophore: assignment of MESF values. *J. Res. Natl. Inst. Stand. Technol.* **110,** 101–114.

10. Juanita Martinez, M., Aragon, A. M., Rodriguez, A. L., et al. (2003) Identification and removal of contaminating fluorescence from commercial and in-house printed DNA microarrays. *Nucleic Acids Res.* **31,** E18.

11. Raghavachari, N., Bao, Y. P., Li, G., Xie, X., and Müller, U. R. (2003) Reduction of autofluorescence on DNA microarrays and slide surfaces by treatment with sodium borohydride. *Anal. Biochem.* **312,** 101–105.

12. Hessner, M. J., Wang, X., Hulse, K., et al. (2003) Three color cDNA microarrays: quantitative assessment through the use of fluorescein-labeled probes. *Nucleic Acids Res.* **31,** E14.

13. Hessner, M. J., Meyer, L., Tackes, J., Muheisen, S., and Wang, X. (2004) Immobilized probe and glass surface chemistry as variables in microarray fabrication. *BMC Genomics* **5,** 53.

14. Yue, H., Scott Eastman, P., Wang, B. B., et al. (2001) An evaluation of the performance of cDNA microarrays for detecting changes in global mRNA expression. *Nucleic Acids Res.* **29,** E41.

15. Battaglia, C., Salani, G., Consolandi, C., Rossi Bernardi, L., and De Bellis, G. (2000) Analysis of DNA microarrays by non-destructive fluorescent staining using SYBR green II. *Biotechniques* **29,** 78–81.

16. Ramdas, L., Coombes, K. R., Baggerly, K., et al. (2001) Sources of nonlinearity in cDNA microarray expression measurements. *Genome Biol.* **2,** R47.

17. Wang, L., Gaigalas, A. K., and Reipa, V. (2005) Optical properties of Alexa™ 488 and Cy™ 5 immobilized on a glass surface. *Biotechniques* **38,** 127–132.

7

Construction of Oligonucleotide Microarrays (Biochip) Using Heterobifunctional Reagents

Jyoti Choithani, B. Vaijayanthi, Pradeep Kumar, and Kailash Chand Gupta

Summary

A number of hetero- and homobifunctional reagents have been reported to immobilize biomolecules on a variety of supports. However, efforts are on to search for a method, which is relatively simple, involving minimum of steps, cost effective, easy to reproduce, and that produces stable oligonucleotide arrays. Two new reagents, viz., [N-(2-trifluoroethanesulfonatoethyl)-N-(methyl)-triethoxysilylpropyl-3-amine], and [N-(3-trifluoroethanesulfonyloxypropyl)anthraquinone-2-carboxamide] have been designed considering the above points. These reagents contain different functional groups at their two ends. In [N-(2-trifluoroethanesulfonatoethyl)-N-(methyl)-triethoxysilylpropyl-3-amine], one end (triethoxysilyl) is capable of binding to the virgin glass surface and the other one consists of trifluoroethanesulfonate (tresyl) function specific toward aminoalkyl and mercaptoalkyl functionalities, which are easy to introduce at the 3'- or 5'-end of oligonucleotides. Likewise, in [N-(3-trifluoroethanesulfonyloxypropyl)anthraquinone-2-carboxamide], one end consists of photoactivatable moiety (anthraquinone) capable of reacting to a C–H containing surface and the tresyl function at the other end reacts specifically with aminoalkyl and mercaptoalkyl functionalities in modified oligonucleotides. These reagents have successfully been utilized to construct a number of oligonucleotide arrays and subsequently used for the detection of mismatches.

Key Words: Glass surface; heterobifunctional reagents; immobilization; microarray; NTMTA; NTPAC; oligonucleotides.

1. Introduction

Detection, analysis and characterization of oligonucleotide/DNA sequences in the traditional way, i.e., one gene-one experiment, are quite laborious and time consuming. Recently developed microarray technology has proved to be a

From: *Methods in Molecular Biology, vol. 381: Microarrays: Second Edition: Volume 1: Synthesis Methods*
Edited by: J. B. Rampal © Humana Press Inc., Totowa, NJ

boon in this direction. It is a powerful tool that allows simultaneous detection of different target molecules present in a sample.

Microarrays are flat surfaces on which different molecules of oligonucleotides/DNAs have been affixed at discrete locations in an ordered manner *(1,2)*. The simultaneous measurement of expression patterns of thousands of genes *(3–5)* is made possible by microarrays thus making it an effective research tool. The majority of applications are in the area of large-scale screening of mutations, studies of gene polymorphism, differential-gene expression analysis, disease diagnosis, identification and characterization of pathogens, and so on. For the preparation of oligonucleotide arrays, several surface materials have been tried, viz., nylon *(6)*, glass *(7–9)*, polyacrylamide *(10)*, polypropylene *(11)*, polyesters *(12)*, poly(methyl methacrylate) (PMMA) *(13)*, silicon *(14,15)*, optical fibers *(16)*, gold *(17,18)*, polyethylene glycol grafted on silica surfaces *(19)* and so on. The choice of surfaces for microarray preparation depends on their homogeneity, stability upon storage and reactivity toward biomolecules. Based on these criteria, glass and polypropylene could be opted as they can easily be modified to generate functional groups on them. However, glass is most commonly used as it is easily available, inexpensive and can be used in laser scanners in which it offers low background.

Basically, two methods are followed for preparing an oligonucleotide array: (1) *in situ (20–22)* or on chip synthesis of oligonucleotides and (2) arraying of prefabricated oligonucleotides. The former one is based on photolithographic methods, wherein the oligonucleotides are synthesized at the predetermined sites on the solid or polymeric surface, using photolabile protecting groups. Using this technique, a density of 10^6 sequences/cm^2 has been achieved on a biochip *(21)*.

In the second method, oligonucleotides are first prepared and then immobilized on suitable solid surfaces (inorganic and organic polymeric materials). A plethora of reports on the attachment of readily prepared oligonucleotides to a surface have appeared. Noncovalent *(23)* and covalent attachments to surfaces have been analyzed and the latter was found to be superior. In the covalent attachment of immobilization, generally, electrophilic-glass surfaces *(24)* are reacted with nucleophilic oligonucleotides. Many modifications to the surface as well as to the oligonucleotides have been carried out for improving the quality of the microarrays.

To minimize the number of steps in the preparation of arrays, homo- and heterobifunctional reagents (linkers) have been used to couple oligonucleotides directly to virgin/functionalized glass surface *(25–27)*. These linkers have either two similar or different functional groups, one capable of attaching to the surface and the other that can react with a suitably modified oligonucleotide. Heterobifunctional crosslinkers, viz., succinimidyl 4-(maleimidophenyl)butyrate, m-maleimidobenzoyl-N-hydroxysuccinimide ester, succinimidyl-4-(N-male imidomethyl)cyclohexane-1-carboxylate *(SMCC)*, N-(γ-maleimidobutyryloxy) succinimide ester, maleimidopropionic acid N-hydroxysuccinimide ester and N-succinimidyl(4-iodoacetyl)aminobenzoate bearing both thiol and amino reactive moieties have been developed for covalent attachment of thiol-modified oligomers to aminosilane monolayer films *(28)*. In case of succinimidyl 4-(maleimidophenyl)butyrate, a surface density of about 20 pmol of bound oligonucleotide/cm^2 was obtained. This method appeared to be superior to other methods of attachment both in terms of density of immobilized oligonucleotides and also the reduced concentration of oligomers required to achieve this density.

The use of heterobifunctional crosslinkers has proved to be attractive but the hunt is on for a surface which could confer reproducibility to the array preparation. It must also be stable to conditions of hybridization and show no nonspecific binding. Recently, Sulfo-SMCC has been used to attach oligonucleotides on Si(111) and Si(001) surfaces *(14,29)*. However, the surface had to be modified first by reaction with an 11-undecylenic acid methyl or trifluoroethylester under photolight (254 nm), then hydrolyzed to yield a carboxylic acid modified surface that was then reacted with polylysine and Sulfo-SMCC to get a maleimide-activated surface. Fixing of oligonucleotides to virgin glass surface has been made possible by two reagents, viz., 3-mercaptopropyltriethoxysilane and 3-glycidyloxypropyltrimethoxysilane *(25)*. 3-Mercaptopropyltriethoxysilane results in immobilization of oligonucleotides through disulfide linkage, which has proved to be a labile linkage and 3-glycidyloxypropyltrimethoxysilane requires longer reaction time (~8 h).

Recently, a novel heterobifunctional crosslinker, [N-(2-trifluoroethane-sulfonatoethyl)-N-(methyl)-triethoxysilylpropyl-3-amine] (NTMTA) has been developed, which contains a tresyl function on one side and a triethoxysilyl function on the other side *(30,31)*. The tresyl function can couple to a 5′-aminoalkyl or mercaptoalkyl-functionalized oligonucleotide/biomolecules, whereas, the triethoxysilyl function can bind to the virgin glass surface. The method is suitable for the construction of oligonucleotide microarrays on a glass surface without employing any additional coupling reagent. The density of immobilized DNA was found to be comparable or better than the existing glass-specific methods and far superior to these methods in terms of reaction time for construction of microarrays and signal-to-noise ratio (>98). Moreover, the covalently immobilized oligonucleotides were stable to varying heating conditions (encountered during PCR). The thermal stability studies indicated that the microarrays obtained by this method are quite stable and can be exploited successfully in biological research.

Photoreactive groups have also been used for immobilization of biomolecules including oligonucleotides. Recently, a new photolabile heterobifunctional reagent was developed, N-(3-trifluoroethanesulfonyloxypropyl)anthraquinone-2-carboxamide (NTPAC). The idea was to have a reactive functional group (tresyl) at one end, which can be specific toward aminoalkyl- or mercaptoalkyl moiety commonly found in biomolecules and a photoreactive group (anthraquinone) at the other end, which can react with a variety of C–H containing polymer matrices, such as polypropylene, polyethylene, polystyrene, and modified glass surfaces, and so on *(32,33)*. Reagents, based on photoactivable azides and protected carbenes *(34–36)*, have also been initially considered for this purpose but dropped because the generated reactive species (i.e., nitrenes and carbenes, respectively) produce a number of side products leading to lower reaction yields on photo-irradiation.

The immobilization techniques are elucidated using the above two important reagents, namely, NTMTA and NTPAC.

2. Materials

1. Acetonitrile, dichloromethane (DCM), ethylene dichloride (EDC), and toluene are dried by refluxing over calcium hydride, distilled and stored over 4 Å molecular sieves.
2. Methanol, petroleum ether, and diethyl ether are dried over sodium metal, distilled, and stored in dry amber-glass bottles.
3. N,N-dimethylformamide (DMF) is purified by azeotropic distillation with benzene and stored over 4 Å molecular sieves.
4. 1H-tetrazole is purified by sublimation.
5. Pharmacia. LKB Gene Assembler Plus (Uppsala, Sweden).
6. FAM amidite (Applied Biosystems, Foster City, CA).
7. Fluorescein at C-5 position in thymidine-phosphoramidite (Glen Research Inc., USA).
8. Other reagents and chemicals are purchased from reputed manufacturers (e.g., Sigma-Aldrich (St. Louis, MO), Fluka (Chemie AG, Buchs, Switzerland), Merck ltd. (Mumbai, India) and so on). The reagents required in this work are: thioacetic acid, thiobenzoic acid, 6-bromohexanol, 18-crown-6, N,N,N′,N′-tetraisopropyl-2-cyanoethyl-phosphoramidite, N,N′-dicyclohexylcarbodiimide (DCC), controlled pore glass (CPG), dithiothreitol (DTT), 2,2,2-trifluoroethanesulfonyl chloride (tresyl chloride), anthraquinone-2-carboxylic acid, 3-aminopropanol, 5-aminopentanol, N,N-diisopropylethylamine, N-methylaminoethanol, 3-aminopropyltriethoxysilane, octyltrimethoxysilane, 3-chlorotrimethoxysilane, 1-*O*-(4,4′-dimethoxytrityl)-5-aminopentanol *(37)*, 1-*O*-(4,4′,-dimethoxytrityl)-6-mercaptohexanol *(38)* chloro-(2-cyanoethyl)-N,N′,-diispropylaminophosphine, N-hydroxysuccinimide (NHS).
9. Buffers mentioned in this work are as follows:
    a. Buffer 1: 0.1 M triethylammonium acetate, pH 7.1.
    b. Blocking buffer 2: 0.1 M Tris-HCl, pH 8.5.
    c. Reaction buffer 3: 0.1 M sodium phosphate containing 0.5 M NaCl, pH 8.0.
    d. Washing buffer 4: 0.1 M sodium phosphate containing 0.5 M NaCl, pH 7.1.
    e. Hybridization buffer 5: 0.1 M sodium phosphate containing 1 M NaCl, pH 7.1.
    f. Buffer 6: 0.1 M sodium citrate containing 0.5 M NaCl, pH 7.1.

10. Glass microslides are precleaned by immersing in 25% ammonium hydroxide solution overnight and then rinsed with double-distilled (dd) water (3 × 50 mL). Finally, these are submerged in anhydrous ethanol (~95%) for 30 min, dried under vacuum in a desiccator and stored in a dust-free container.

3. Methods

For the immobilization of oligonucleotides using NTMTA and NTPAC reagents, 5′-modified oligonucleotides are required, whereas for quantification, 3′-modified oligonucleotides are required. The following **Subheadings 3.1.** and **3.2.**, give the procedure for the preparation of 5′- and 3′-modified oligonucleotides. The oligonucleotide sequences synthesized in this work along with their deprotection conditions are presented in the **Table 1**.

3.1. Preparation of 5′-Modified Oligonucleotides

A reactive functional group (aminoalkyl or mercaptoalkyl) can be introduced in the protected form with the help of a suitable linker as last coupling step, which during deprotection step generates oligomers with free nucleophilic reactive groups *(39–41)*.

3.1.1. Synthesis of 5′-Mercaptoalkylated Oligonucleotides

Scheme 1 depicts the steps involved in the preparation of reagent, S-acyl-6-mercaptohexyl-2-cyanoethyl-N,N-diisopropylphosphoramidite **3a**, **3b**. This reagent is further used in the last coupling step to obtain 5′-mercaptoalkylated oligonucleotides using Gene Assembler Plus synthesizer.

3.1.1.1. PREPARATION OF POTASSIUM THIOACETATE **1a** OR THIOBENZOATE **1b**

1. To a cold solution of thioacetic or thiobenzoic acid (10 mmol) in methanol, add a methanolic solution of potassium hydroxide (10 mmol) dropwise over a period of 5 min with constant stirring.
2. After complete addition, stir the solution at room temperature for 15 min, filter it to remove any insoluble material, and concentrate on a rotary evaporator to reduce the volume to about 10 mL.
3. Triturate the above solution with diethyl ether (200 mL) to precipitate the potassium salt of thioacetic or thiobenzoic acid.
4. Filter on a sintered-disc glass funnel, and dry under vacuum in a desiccator.

3.1.1.2. PREPARATION OF S-ACYL-6-MERCAPTOHEXANOL **2a**, **2b**

1. Mix potassium thioacetate **1a** (2.4 mmol), 6-bromohexanol (2.0 mmol), and 18-crown-6 (0.2 mmol) in toluene (20 mL) in a round-bottom flask and reflux for 15 min.
2. Filter off the insoluble material and dilute the filtrate with toluene (30 mL).
3. Wash the previously described filtrate successively with saturated aqueous sodium hydrogen carbonate (2 × 20 mL), sodium chloride (1 × 20 mL), and water (1 × 20 mL).

Table 1
Oligonucleotides Synthesized With Their Deprotection Conditions

No.	Oligomer sequence	Deprotection conditions
1.	H_2N-$(CH_2)_5$-OPO_3-d(CAG AGG TTC TTT GAG TCC TT)	Aq NH_4OH, 16 h, 60°C
2.	HS-$(CH_2)_6$-OPO_3-d(CAG AGG TTC TTT GAG TCC TT)	Aq NH_4OH + 50 mM DTT, 16 h, 60°C
3.	HS-$(CH_2)_6$-OPO_3-d(CTC CTG AGG AGA AGG TCT GC)	Aq NH_4OH + 50 mM DTT, 16 h, 60°C
4.	H_2N-$(CH_2)_5$-OPO_3-d(CTC CTG AGG AGA AGG TCT GC)	Aq NH_4OH, 16 h, 60°C
5.	H_2N-$(CH_2)_5$-OPO_3-d(CTC CTG AGG CGA AGG TCT GC)	Aq NH_4OH, 16 h, 60°C
6.	FAM-d(AAG GAC TCA AAG AAC CTC TG)	Aq NH_4OH, 16 h, 60°C
7.	FAM-d(GCA GAC CTT CTC CTC AGG AG)	Aq NH_4OH, 16 h, 60°C
8.	d(T^FTT TTT TTT TTT TTT TTT TT)-OPO_3-$(CH_2)_6$SH	Aq NH_4OH + 50 mM DTT, 16 h, 60°C
9.	H_2N-$(CH_2)_5$-OPO_3-d(CTC CTG CGG AGA ACG TCT GC)	Aq NH_4OH, 16 h, 60°C
10.	H_2N-$(CH_2)_5$-OPO_3-d(CTC CTG CGG CGA ACG TCT GC)	Aq NH_4OH, 16 h, 60°C
11.	H_2N-$(CH_2)_5$-OPO_3-d(TTT TTT TTT TTT TTT TTT TT)	Aq NH_4OH, 16 h, 60°C
12.	H_2N-$(CH_2)_5$-OPO_3-d(AAT CGT TAC TTT TTA TTA TCC)	Aq NH_4OH, 16 h, 60°C
13.	HS-$(CH_2)_6$-OPO_3-d(AAT CGT TAC TTT TTA TTA TCC)	Aq NH_4OH + 50 mM DTT, 16 h, 60°C
14.	HS-$(CH_2)_6$-OPO_3-d(TTG GGT CCG CCA CTC CTT CCC)	Aq NH_4OH + 50 mM DTT, 16 h, 60°C
15.	HS-$(CH_2)_6$-OPO_3-d(TTG GGT CCG CTA CTC CTT CCC)	Aq NH_4OH + 50 mM DTT, 16 h, 60°C
16.	Fluorescein-d(GGA TAA TAA AAA GTA ACG ATT)	Aq NH_4OH, 16 h, 60°C
17.	Fluorescein-d(GGG AAG GAG TGG CGG ACC CAA)	Aq NH_4OH, 16 h, 60°C
18.	Fluorescein-d(TTT TTT TTT TTT TTT TTT TT)OPO_3-$(CH_2)_6$SH	Aq NH_4OH + 50 mM DTT, 16 h, 60°C

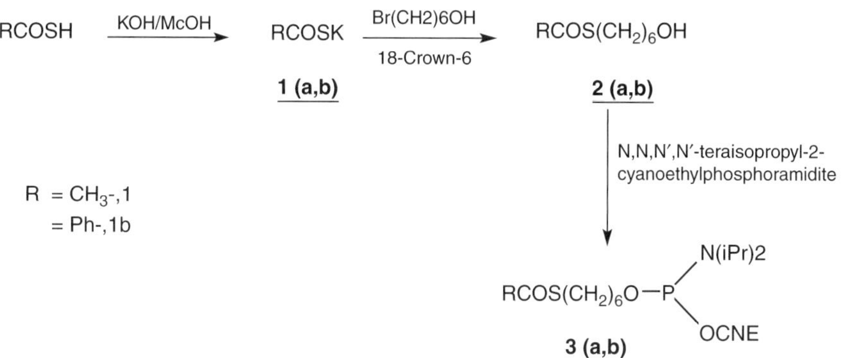

Scheme 1. Preparation of reagents for 5′-mercaptoalkylation of oligonucleotides.

4. Separate the organic layer and dry it over anhydrous sodium sulfate, concentrate and purify by silica gel column chromatography using petroleum ether: ethyl acetate (8:2 v/v) as an eluent.
5. Collect the fractions containing the desired material and concentrate to obtain a yellowish oily material (~92% yield). R_f: 0.45 (hexane: ethyl acetate, 6:4).
6. Characterize the compound S-acetyl-6-mercaptohexanol **2a** by nuclear magnetic resonance (NMR) spectrum: ^{1}H NMR (CDCl$_3$) δ: 1.6 (m, 8H), 2.4 (s, 3H), 2.95 (t, 2H), and 3.8 (m, 2H).

S-Benzoyl-6-mercaptohexanol **2b** can be synthesized in an analogous manner. R_f: 0.52 (hexane: ethylacetate, 6:4). ^{1}H NMR (CDCl$_3$) δ: 1.54 (m, 8H), 3.04 (t, 2H), 3.7 (m, 2H), and 7.6 (m, 5H).

3.1.1.3. PREPARATION OF S-ACYL-6-MERCAPTOHEXYL-2-CYANOETHYL-N,N-DIISOPROPYLPHOSPHORAMIDITE **3a**, **3b**

1. Dissolve the compound **2a** (1 mmol) in anhydrous acetonitrile (10 mL) and add to it N,N,N′,N′-tetraisopropyl-2-cyanoethylphosphoramidite (1.5 mmol).
2. To the previously described mixture, add tetrazole (0.75 mmol) dissolved in acetonitrile (1.5 mL) dropwise at room temperature.
3. Stir the resulting mixture for 2 h and monitor the progress of the reaction on thin layer chromatography (TLC) and after the reaction is complete, quench it by adding dry methanol (0.2 mL).
4. Concentrate the reaction mixture under reduced pressure and take the resulting syrupy mass in ethyl acetate (20 mL).
5. Wash the previously described mixture with sodium chloride solution (2×10 mL). Separate the organic phase and dry over anhydrous sodium sulfate.
6. After filtration, remove the solvent by evaporation and purify the resulting material using silica gel column.

7. Elute the desired compound with hexane: ethylacetate: triethylamine (TEA) (8:1:1, v/v/v), pool together the fractions containing the pure material and concentrate on a rotary evaporator to obtain the title compound **3a** in about 85% yield.

Similarly, prepare the phosphoramidite of compound **2b** and purify to obtain compound **3b** in about 85% yield.

3.1.1.4. Synthesis and Deprotection of 5′-Mercaptoalkylated Oligonucleotides

1. Assemble an oligonucleotide chain at 0.2 µmol scale on a Gene Assembler Plus following manufacturer's recommendations using standard nucleoside containing polymer support.
2. Perform the last coupling step with one of the phosphoramidite reagents **3a** or **3b** in an analogous manner to the normal nucleoside phosphoramidites.
3. Deprotect the mercaptoalkylated oligonucleotides from the support and cleave the protecting groups by treatment with 30% aq ammonia containing 50 m*M* DTT at 60°C for 16 h (*see* **Note 1**).
4. Concentrate the ammoniacal solution in a speed vac and subject to desalting on a C-18 silica gel column.
5. Elute the 5′-mercaptoalkylated oligonucleotide with a solution of 30% aqueous acetonitrile solution and concentrate in a speed vac (*see* **Note 2**).

3.1.1.5. Purification of 5′-Mercaptoalkylated Oligonucleotides by Affinity Chromatography

1. Pack a small chromatography column with 5-nitro-2-thiopyridyl-activated 3-mercaptopropyl-fractosil support and equilibrate with buffer 2 (10 mL).
2. Load the column with the crude mixture of 5′-mercaptoalkylated oligonucleotide dissolved in buffer 2 containing 20% acetonitrile.
3. Allow the solution to react with the column matrix for 1 h and then wash with the buffer 2 containing 20% acetonitrile (10 mL).
4. After all the truncated sequences are removed, elute the desired sequence with buffer 2 containing 50 m*M* DTT and 20% acetonitrile.
5. Concentrate and desalt the oligomer using Sephadex G-50 (Pharmacia Fine Chemicals AB, Uppsala, Sweden) column using buffer 1.

3.1.2. Synthesis of 5′-Aminoalkylated Oligonucleotides

5′-Aminoalkylated oligonucleotide can be prepared by performing the last coupling in a Gene Assembler Plus, using the reagent N-(4,4′-dimethoxytrityl)-5-aminopentyl-2-cyanoethyl-N,N-diisopropylphosphoramidite **5**. **Scheme 2** depicts the synthesis of this reagent.

3.1.2.1. Preparation of N-(4,4′-Dimethoxytrityl)-5-Aminopentan-1-ol **4**

1. To a cold solution of 5-aminopentanol (10 mmol) in dry DCM (50 mL) and TEA (11 mmol), add a solution of 4,4′-dimethoxytrityl chloride (10 mmol), dissolved in DCM (25 mL), dropwise over a period of 30 min with constant stirring.
2. After complete addition, stir the solution at room temperature for 2 h.

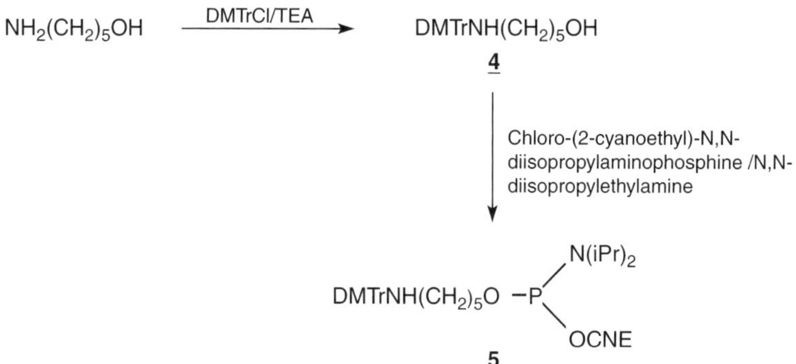

Scheme 2. Preparation of reagent for 5′-aminoalkylation of oligonucleotides.

3. Monitor the reaction on TLC and if it is complete, wash the organic phase sequentially with cold 2.5% aq citric acid (2 × 25 mL), 5% aq sodium bicarbonate (2 × 25 mL), and saturated sodium chloride solutions (2 × 25 mL).
4. Collect the organic phase, dry over anhydrous sodium sulfate, filter, and concentrate on a rotary evaporator to get the desired compound as a syrupy material in about 88% yield.
5. Characterize the desired compound by NMR. ^{1}H NMR (CDCl$_3$) δ: 1.6 (m, 6H), 2.9 (t, 2H), 3.3 (t, 2H), 3.8 (s, 6H), and 6.8–7.5 (m, 13H).

3.1.2.2. Preparation of N-(4,4′-Dimethoxytrityl)-5-Aminopentyl-2-Cyanoethyl-N,N-Diisopropylphosphoramidite 5

1. Dissolve the compound **4** (1 mmol) in anhydrous EDC (10 mL) and add N,N-diisopropylethylamine (4 mmol) to it.
2. Cool the previously described mixture in an ice-bath and add chloro-(2-cyanoethyl)-N,N-diisopropylaminophosphine (1.5 mmol) dropwise.
3. Stir the resulting mixture for 1 h at room temperature, monitor the advancement of the reaction on TLC and after the reaction is complete, quench it by adding dry methanol (0.2 mL).
4. Dilute the reaction mixture with EDC (50 mL) and wash sequentially with 5% aq sodium carbonate (2 × 25 mL) and saturated sodium chloride (2 × 25 mL) solutions.
5. Separate the organic phase and dry over anhydrous sodium sulfate.
6. After filtration, remove the solvent by evaporation on a rotary evaporator and purify the resulting material using silica gel column.
7. Elute the desired compound with hexane: ethylacetate: TEA (8:1:1, v/v/v), pool together the fractions containing the pure material and concentrate to obtain the title compound in about 85% yield.

3.1.2.3. Synthesis and Deprotection of 5′-Aminoalkylated Oligonucleotides

1. Synthesize an oligonucleotide chain at 0.2 μmol scale on a Gene Assembler Plus following manufacturer's recommendations using standard nucleoside containing polymer support.

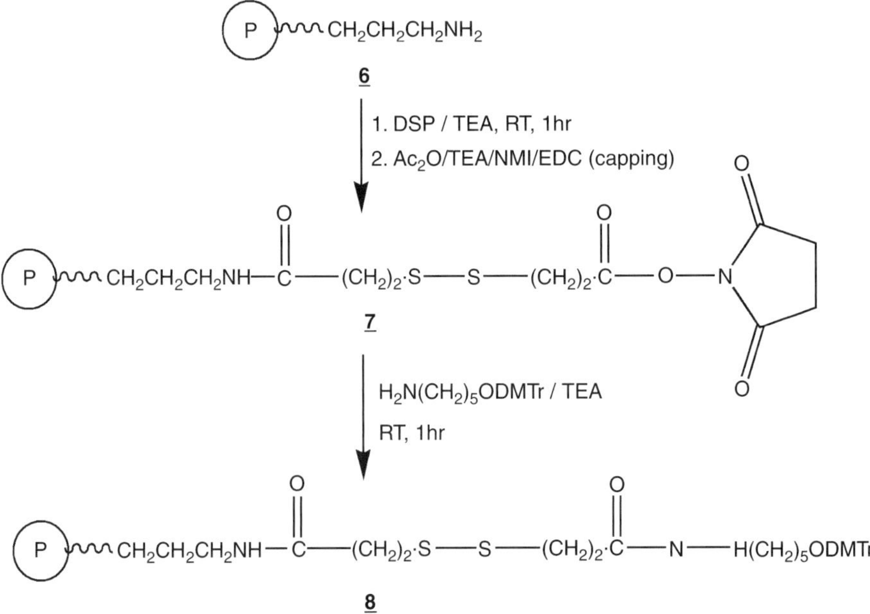

Scheme 3. Preparation of polymer support for 3′-mercaptoalkylated oligonucleotides.

2. Perform the last coupling step with phosphoramidite reagent **5** in an analogous manner to the normal nucleoside phosphoramidites except with an extended coupling time (5 min).
3. Deprotect 5′-aminoalkyl-oligonucleotide from the support and cleave the protecting groups by treatment with 30% aq ammonia at 60°C for 16 h.
4. Concentrate the ammoniacal solution in a speed vac and subject to desalting on a C-18 silica gel column.
5. Elute the 5′-aminoalkylated oligonucleotide with a solution of 30% aqueous acetonitrile solution and concentrate in a speed vac.

3.2. Preparation of 3′-Modified Oligonucleotides

3.2.1. Synthesis of 3′-Mercaptoalkylated Oligonucleotides

Unlike 5′-end modifications, 3′-modifications are not that straightforward. Because 3′- hydroxyl group of the leader nucleoside is not available for manipulation during solid phase synthesis as it is attached to the support through its hydroxyl group, modified supports have been designed to incorporate modifications at the 3′ end of oligonucleotides *(42–46)*.

Out of a number of polymer supports that have been designed for this purpose, universal polymer support *(38)* is of considerable importance. The method used is described in **Scheme 3**. The synthesis can be carried out using a Gene Assembler Plus in the usual way. A coupling efficiency of >98% has been found using these supports.

3.2.1.1. PREPARATION OF DERIVATIZED SUPPORT

1. To a suspension of 3-aminopropylated controlled pore glass **6** (3-AP-CPG) (1 g, 100 µmol amino groups) in dry EDC (5 mL) containing TEA (100 µL), add 10-fold molar excess of 3, 3′-diothio-bis(propionic acid N-hydroxysuccinimide ester) (DSP) (1 mmol).
2. Shake the reaction mixture at room temperature for 1 h.
3. Filter the polymer support on a sintered disc glass funnel and remove the excess reagent by washing sequentially with EDC (3 × 20 mL) and diethyl ether (2 × 10 mL).
4. Cap the residual amino groups on support with a solution of Ac$_2$O:TEA: N-methylimidazole:EDC (1:1:0.4:6, v/v) for 30 min at room temperature followed by washing of the support on the glass funnel with EDC (2 × 20 mL) and diethyl ether (2 × 10 mL).
5. Take a weighed amount of polymer support **7** (500 mg) in dry EDC (5 mL) and add 1-(*O*-4,4′-dimethoxytrityl)-5-aminopentanol (0.5 mmol), dissolved in anhydrous EDC (2 mL) containing TEA (50 µL).
6. Keep the reaction suspension at room temperature for 1 h with occasional shaking.
7. Transfer the support on to a sintered disc glass funnel and wash it with EDC and diethyl ether (2 × 10 mL of each).
8. Dry the fully functionalized support **8** under vacuum and determine its loading.

3.2.1.2. SYNTHESIS OF 3′-MERCAPTOALKYLATED OLIGONUCLEOTIDE

Synthesize the oligonucleotides modified at their 3′-end using the above support **8** in a Gene Assembler Plus. After the synthesis of desired oligonucleotide chain, subject the support to deprotection conditions as given in **Subheading 3.2.1.4.**

3.2.1.3. SYNTHESIS OF 5′-LABELED 3′-MERCAPTOALKYLATED OLIGONUCLEOTIDE

After the synthesis of desired oligonucleotide chain in a Gene Assembler Plus, perform the last coupling with FAM amidite as discussed in **Subheading 3.1.1.4.**

3.2.1.4. DEPROTECTION

1. Treat the support bound oligonucleotide with aq ammonia (30%) containing 50 m*M* DTT in a sealed vial at 60°C for 16 h.
2. Concentrate the ammoniacal solution in a speed vac followed by desalting on C-18 silica gel.
3. Elute the oligomer sequence with 30% aq acetonitrile and again concentrate in a speed vac.

3.3. Preparation of Reagent, NTMTA

3.3.1. Preparation of N-(2-Hydroxyethyl)-N-(Methyl)-Triethoxysilylpropyl-3-Amine **9**

1. Prepare a solution of N-methyl-aminoethanol (0.3 mol) and N,N-diisopropylethylamine (0.1 mol) in hot toluene (100 mL).
2. To this solution, add 3-chloropropyltriethoxysilane (0.1 mol) in one step.

3. Reflux the mixture for 10 h under inert (argon) atmosphere. Cool the mixture to 4°C and keep it for 16 h.
4. Filter the ammonium chloride that is formed in the reaction under an inert atmosphere through a sintered disc glass funnel and wash the residue with dry toluene (2×25 mL).
5. Collect the filtrate and the washings together in a round-bottomed flask and concentrate under vacuum.
6. Distill the syrupy crude material so obtained under vacuum. The title compound is obtained in about 65% yield.
7. Characterize the compound by NMR. ^{1}H-NMR, CDCl$_3$ δ: 0.58 (t, 2H), 1.14 (m, 9H), 1.30 (m, 4H), 2.27 (s, N–CH$_3$), 2.3–2.6 (m, 7H, –CH$_2$), 3.58 (m, 6H, -OCH$_2$), and 3.66 (t, 2H).

3.3.2. Preparation of NTMTA **10**

1. Prepare a mixture of compound **9** (5 mmol) and TEA (6 mmol) in 25 mL of DCM.
2. Maintained at 0°C in an ice bath, add tresyl chloride (6 mmol) dropwise over a period of 5 min (*see* **Notes 3** and **4**).
3. After the addition is complete, take out the mixture from ice-bath and stir at room temperature.
4. Monitor the reaction by TLC and stir until complete conversion takes place (~2 h).
5. After the completion of the reaction, cool the reaction mixture to 0°C in an ice bath and unwanted amine hydrochloride is precipitated.
6. Remove the precipitate by filtration under argon. Concentrate the filtrate under vacuum to obtain a syrupy liquid of the compound **_10_** (*see* **Note 5**).
7. Characterize the product by NMR and MALDI-TOF. ^{1}H NMR, CDCl$_3$ δ: 0.32 (m, 2H), 1.23 (m, 2H), 3.0–3.5 (m, 9H), 4.5 (br, t, 2H). MALDI-TOF: 384 (M + H)$^+$, matrix: 2,5-dihydroxybenzoic acid.

3.4. Preparation of Reagent NTPAC

3.4.1. Preparation of N-(3-Hydroxypropyl)Anthraquinone-2-Carboxamide **11**

1. To a solution of anthraquinone-2-carboxylic acid (2 mmol) and N-hydroxysuccinimide (2.5 mmol) in dry DMF (15 mL), add N,N′-dicyclohexylcarbodiimide (2.5 mmol).
2. Stir the reaction mixture at room temperature for 3 h.
3. After the reaction is complete, add 3-aminopropanol (3 mmol) and TEA (3 mmol) dropwise at room temperature.
4. Stir the reaction mixture at room temperature for 6 h.
5. Remove the precipitated N,N′-dicyclohexylurea by filtration and concentrate the filtrate to syrupy mass.
6. Dissolve the syrupy material in ethyl acetate (100 mL) and wash sequentially with 5% aq citric acid (2×25 mL), 5% aq sodium bicarbonate (2×25 mL), and saturated sodium chloride (1×25 mL) solutions.
7. Collect the organic phase, dry it over anhydrous sodium sulfate, filter, and concentrate under vacuum to obtain the title compound as solid material in 90% yield.
8. Characterize the compound by NMR. ^{1}H NMR, CDCl$_3$ δ: 1.65 (m, 2H), 3.2–3.45 (m, 4H), 7.5–8.5 (m, 8H).

3.4.2. Preparation of NTPAC **12**

1. Take N-(3-hydroxypropyl)anthraquinone-2-carboxamide **11** (1 mmol) in anhydrous DCM (10 mL) containing N,N-diisopropylethylamine (1.2 mmol).
2. Cool the reaction mixture in an ice bath and add tresyl chloride (1.2 mmol) dropwise over a period of 20 min (*see* **Notes 3** and **4**).
3. Stir the reaction mixture at room temperature for another 2 h for completion.
4. Cool the reaction mixture and filter the solution in an inert atmosphere to remove suspended ammonium chloride.
5. Concentrate the filtrate under vacuum and dry in a desiccator to obtain the title compound in quantitative yield as a syrupy material (*see* **Note 5**).
6. Characterize the reagent by MALDI-TOF (456, $[M + H]^+$; matrix: 2,5-dihydroxybenzoic acid) and UV absorptions at 325 and 255.8 nm.

3.5. Immobilization of Oligonucleotides

Immobilization can be effected through two routes. In the first one, 5′-mercapto- or aminoalkylated oligonucleotides react with reagent (NTMTA or NTPAC) to form oligonucleotide-reagent conjugate, which in a subsequent reaction with glass surface (virgin or modified) (*see* **Note 6**) results in surface bound oligonucleotides. In the second route, the reagent (NTMTA or NTPAC) is allowed to react first with glass surface (virgin or modified) to generate tresyl functions on it, which, in a subsequent step, react with 5′-mercapto- or aminoalkylated oligonucleotides to generate surface bound oligonucleotides (*see* **Note 7**).

3.5.1. Immobilization of 5′-Mercapto- and Aminoalkylated Oligonucleotides Using NTMTA

3.5.1.1. Immobilization of 5′-Aminoalkylated Oligonucleotide Through Route A

1. Dissolve 0.15 A_{260} units, 0.85 nmol of oligonucleotide sequence, $H_2N\text{-}(CH_2)_5\text{-}OPO_3\text{-}$d(CAG AGG TTC TTT GAG TCC TT), in 100 µL of reaction buffer 3.
2. Mix it with ethanolic solution of NTMTA (85 µL, 8.5 nmol).
3. Agitate the mixture on an Eppendorf mixer for 1 h at room temperature.
4. Concentrate the solution under vacuum to obtain a semidried residue of oligonucleotide-trimethoxysilyl conjugate.
5. Suspend this residue in 25 µL of dd water.
6. Centrifuge the mixture to get rid of excess of NTMTA as undissolved material.
7. Concentrate the supernatant under vacuum to obtain a residue that is reconstituted in dd water containing 10% DMSO (v/v) to make an oligonucleotide conjugate final concentration to 10 µ*M* (*see* **Note 8**).
8. Spot the oligonucleotide conjugate manually in multiplicates on a glass microslide.
9. Incubate the spotted microslide at 45°C in a humid chamber for 40 min.
10. After 40 min, wash the slide with washing buffer 4 (5 × 10 mL).
11. The slide is ready for hybridization (*see* **Note 7**).

3.5.1.2. Immobilization of 5′-Mercaptoalkylated Oligonucleotide Through Route A

Immobilize the 5′-mercaptoalkylated oligonucleotide in a similar manner as previously described (*see* **Subheading 3.5.1.1.**) except that 30 min are required (**step 3**, **Subheading 3.5.1.1.**) to complete the reaction between mercaptoalkylated oligonucleotide and NTMTA.

3.5.1.3. Immobilization of 5′-Aminoalkylated Oligonucleotide Through Route B

1. Treat the virgin glass microslide with ethanolic solution of NTMTA (85 μg, 0.2 μmol, and 2 mL) by evenly spreading over it.
2. Keep the microslide in a humid chamber maintained at 45°C for 40 min.
3. Wash the slide with ethanol (3 × 15 mL) and dry it under vacuum (*see* **Note 9**).
4. Spot the solution of 5′-aminoalkylated oligonucleotides (10 μ*M*, 0.5 μL) in reaction buffer 3 in triplicates.
5. Keep the spotted microslide in a humid chamber for 1 h at room temperature.
6. Remove the slide and keep it in blocking buffer 2 for 1 h followed by washing with washing buffer 4 (5 × 10 mL).
7. The slide is ready for hybridization (*see* **Note 7**).

3.5.1.4. Immobilization of 5′-Mercaptoalkylated Oligonucleotide Through Route B

Immobilize the 5′-mercaptoalkylated oligonucleotide in a similar manner as previously described (*see* **Subheading 3.5.1.3.**) except that 30 min are required (**Subheading 3.5.1.3., step 5**) for completion of reaction between mercaptoalkylated oligonucleotide and tresyl functions on NTMTA-treated glass microslide.

3.5.2. Immobilization of 5′-Mercapto- and Aminoalkylated Oligonucleotides Using NTPAC

3.5.2.1. Immobilization of 5′-Aminoalkylated Oligonucleotide on C-8 Glass Microslide Through Route A

1. Take 1.5 A_{260} units of the oligonucleotide sequence, $H_2N–(CH_2)_5–OPO_3–d(AAT$ CGT TAC TTT TTA TTA TCC), dissolved in reaction buffer 3 (100 μL).
2. To the above solution, add NTPAC dissolved in DMF (9.1 mg, 0.2 *M*, 100 μL).
3. Agitate the reaction mixture at room temperature for 1 h.
4. After 1 h, concentrate the mixture under vacuum to remove the solvent.
5. Suspend the semisolid residue so obtained in 25 μL of dd water, centrifuge and discard the residue, which is nothing but the excess reagent (it is insoluble in water).
6. Dilute the decanted solution with 1 *M* LiCl to bring the final concentration of LiCl to 0.2 *M*.
7. Spot the anthraquinone-oligonucleotide conjugate so obtained on C-8 glass microslide (obtained after treatment of virgin glass microslide with a 5% solution

of octyltrimethoxysilane in toluene for 2 h at 50°C) in triplicates and allow the spots to dry in the air.

8. Keep the spotted slide in a UV chamber (operating at 365 nm, 150 W) for 30 min.
9. Then wash the slide with washing buffer 4 (3 × 20 mL) and followed by dd water (3 × 20 mL).
10. The slide is ready for hybridization (*see* **Note 7**).

3.5.2.2. IMMOBILIZATION OF 5′-MERCAPTOALKYLATED OLIGONUCLEOTIDE ON C-8 GLASS MICROSLIDE THROUGH ROUTE A

The procedure for the mercaptoalkyl oligonucleotide is similar to the previously described (*see* **Subheading 3.5.2.1.**) except that the reaction time for oligonucleotide HS–$(CH_2)_6$–OPO_3-d(AAT CGT TAC TTT TTA TTA TCC) and NTPAC is 30 min at room temperature (*see* **Subheading 3.5.2.1., step 3**).

3.5.2.3. IMMOBILIZATION OF 5′-AMINOALKYLATED OLIGONUCLEOTIDE ON C-8 GLASS MICROSLIDE THROUGH ROUTE B

1. Spread evenly a solution of the reagent NTPAC (91 mg; 2.0 mL, 0.1 *M*) in DCM on C-8-coated glass microslide.
2. Keep the slide in a fume-cupboard and allow the solvent to evaporate.
3. Place the dried microslide in a UV chamber (operating at 365 nm, 150 W) for 30 min.
4. Wash the slide several times with dry DCM (5 × 20 mL) and dry under vacuum.
5. Then spot a solution of 5′-aminoalkylated oligonucleotide sequence, H_2N-$(CH_2)_5$-OPO_3-d(AAT CGT TAC TTT TTA TTA TCC), (10 µ*M*, 0.5 µL) in reaction buffer 3, in triplicates.
6. Keep the spotted microslide in a humid chamber for 1 h.
7. Then treat the plate with blocking buffer 2 (10 mL) for 1 h.
8. Finally, wash the glass slide with washing buffer 4 (3 × 20 mL) followed by dd water (3 × 20 mL).
9. The slide is ready for hybridization (*see* **Note 7**).

3.5.2.4. IMMOBILIZATION OF 5′-MERCAPTOALKYLATED OLIGONUCLEOTIDE ON C-8 GLASS MICROSLIDE THROUGH ROUTE B

The 5′-mercaptoalkylated oligonucleotide, HS–$(CH_2)_6$–OPO_3-d(AAT CGT TAC TTT TTA TTA TCC), is immobilized on NTPAC-activated glass microslide in a similar fashion as described earlier (*see* **Subheading 3.5.2.3.**) except that the reaction time is 30 min instead of 1 h (**Subheading 3.5.2.3., step 6**).

3.6. General Method for Hybridization Assay

3.6.1. Hybridization of Immobilized Oligonucleotides With Complementary Oligonucleotides

1. Dissolve the complementary oligomer (cf. glass microslide prepared in **Subheadings 3.5.1.** and **3.5.2.**), for example, fluorescein-d(GGA TAA TAA AAA GTA ACG ATT) in hybridization buffer 5 (500 µL).

2. Place a spotted microslide in a Petri dish and paste a hybri-slip (Sigma) on the spotted area of glass microslide.
3. Fill the chamber with complementary oligomer solution (500 µL).
4. Close the holes on the hybri-slip, and place the Petri dish along with microslide in heated chamber at 65°C for 10 min.
5. Allow it to cool slowly to room temperature and then to 10°C with gentle shaking and keep it at that temperature for additional 3–4 h.
6. Remove the complementary oligomer solution with the help of pipetman, remove the hybri-slip and wash the plates with the same buffer 5 and after drying, visualize the fluorescent spots under laser scanner at 570 nm.

3.7. Specificity of Immobilization Chemistry and Detection of Mismatches

1. Dissolve oligonucleotide sequences, H_2N–$(CH_2)_5$–OPO_3-d(CTC CTG AGG AGA AGG TCT GC), H_2N–$(CH_2)_5$–OPO_3-d(CTC CTG AGG CGA AGG TCT GC), H_2N–$(CH_2)_5$–OPO_3-d(CTC CTG CGG AGA ACG TCT GC), H_2N–$(CH_2)_5$–OPO_3-d(CTC CTG CGG CGA ACG TCT GC) and H_2N–$(CH_2)_5$–OPO_3-d(TTT TTT TTT TTT TTT TTT TT), (10 µM each) in reaction buffer 3.
2. Spot these oligomers on an NTMTA-activated glass microslide (Route B), and keep it in a humid chamber at room temperature for 1 h.
3. Place the microslide in blocking buffer 2.
4. Hybridize the spotted oligomers with a complementary FAM-d(GCA GAC CTT CTC CTC AGG AG) as discussed in **Subheading 3.6.**
5. Wash the microslide with buffer 5 and subject to visualization under laser scanner.

3.8. Determination of Optimal Concentration of Oligonucleotide Spots on Surfaces for Fluorescence Detection

3.8.1. Using NTMTA

1. Prepare at least five different concentrations (15, 10, 7.5, 5, and 2.5 µM) of 3′-mercaptoalkyl-oligonucleotide sequence d(T^{F}TT TTT TTT TTT TTT TTT TT)-OPO_3–$(CH_2)_6$SH (T^F is fluorescein attached at C-5 position in thymidine-phosphoramidite) by diluting the stock solution using the reaction buffer 3 containing 10% DMSO (v/v).
2. Spot the glass microslide activated with NTMTA (Route B) with the serially diluted fluorescent oligonucleotide solution (0.5 µL), in duplicate, using a micropipet.
3. Keep the microslide at room temperature for 30 min in a humid chamber.
4. Wash the microslide with washing buffer 4 (5 × 20 mL) and dry under vacuum.
5. Visualize the spots under laser scanner and calculate the optimum concentration.

3.8.2. Using NTPAC

1. Make three different concentrations of fluorescein-d(TTT TTT TTT TTT TTT TTT TT-S-AQ) (fluorescein-labeled anthraquinone-oligonucleotide conjugate) (5, 10, and 15 µM) in 0.2 M LiCl.

2. Spot the oligomer (0.5 µL) manually in triplicates on C-8 glass microslide.
3. Irradiate the slide by placing it in a UV chamber (operating at 365 nm, 150 W) for 30 min.
4. After irradiation, wash the slide with buffer 4.
5. Dry the slide and visualize the spots under laser scanner at 570 nm.
6. Optimum concentration can be calculated by visualizing the bright spot with lowest concentration under laser scanner.

3.9. Quantification of Immobilized Oligonucleotides

1. Prepare various concentrations of d(T^{F}TT TTT TTT TTT TTT TTT TT)-OPO$_3$–(CH$_2$)$_6$SH (0.025, 0.05, 0.10, 0.25, 0.50, 0.75, and 1 µM) by dissolving in buffer 4.
2. Spot each of these solutions on a virgin glass microslide in duplicate.
3. Dry the spots on the microslide and visualize under a laser scanner; measure the fluorescence signal intensity corresponding to each spot.
4. Plot a graph of fluorescence intensity against concentration (*see* **Note 10**).

3.10. Determination of Optimal Time Required for Immobilization of Oligomers on Virgin Glass Surface

In order to determine the optimum time required to immobilize aminoalkyl- and mercaptoalkylated oligonucleotides on the glass surfaces, it is advisable to carry out a model experiment to study the immobilization of 1-O-(4,4′-dimethoxytrityl)-5-aminopentanol or 1-O-(4,4′-dimethoxytrityl)-6-mercapto-hexanol on virgin CPG support (500 Å) using NTMTA by two different routes.

3.10.1. Attachment of O-(4,4′-Dimethoxytrityl)-5-Aminopentan-1-ol on to CPG Through Route A

1. Dissolve *O*-(4,4′-dimethoxytrityl)-5-aminopentan-1-ol (40.5 mg, 100 µmol) in ethanol (1 mL).
2. Similarly, dissolve NTMTA (42.5 mg, 100 µmol) in 1 mL ethanol.
3. Mix the previously listed two solutions and allow the reaction mixture to swirl for 1 h at room temperature in a shaker.
4. After 1 h, add 200 µL of this solution of conjugate to each of the eight Eppendorf tubes containing approx 10 mg of virgin CPG.
5. Place all these Eppendorf tubes in thermo mixer set at 45°C.
6. Remove the tubes one by one at different time intervals (10, 20, 30, 40, 50, 60, 120, and 180 min).
7. Wash the support in each tube with EDC (5 × 1 mL) containing TEA (0.1%), dry it under vacuum in a desiccator and subject it to loading determination.

3.10.2. Attachment of O-(4,4′-Dimethoxytrityl)-6-Mercaptohexan-1-ol on to CPG Through Route A

Mix an ethanolic solution of *O*-(4,4′-dimethoxytrityl)-6-mercapto- hexan-1-ol (43.6 mg, 0.1 mmol, 1.0 mL) with ethanolic solution of NTMTA, allow it to stir for 30 min instead of 1 h in case of *O*-(4,4′-dimethoxytrityl)-5-aminopentan-1-ol and follow rest of steps as discussed in **Subheading 3.10.1.**

3.10.3. Attachment of O-(4,4′-Dimethoxytrityl)-5-Aminopentan-1-ol on to CPG Through Route B

1. Dissolve NTMTA (42.5 mg, 100 μmol) in 1 mL ethanol.
2. Weigh out approx 10 mg of CPG (500 Å) in eight different Eppendorf tubes and add the previously listed solution (100 μL) to each of them.
3. Keep the tubes in a thermo mixer set at 45°C.
4. Take out the tubes at different time intervals, i.e., 10, 20, 30, 40, 50, 60, 120, and 180 min.
5. Wash the support in each tube with ethanol (5 × 1 mL) followed by diethyl ether (2 × 1 mL).
6. After drying the support, add an ethanolic solution of *O*-(4,4′-dimethoxytrityl)-5-aminopentan-1-ol (4.05 mg, 10 μmol, 100 μL) to each of these tubes.
7. Keep the suspension on a mixer at room temperature for 1 h and then wash the contents of each tube with ethanol (5 × 1 mL) and diethyl ether (2 × 1 mL).
8. Dry the support in each tube under vacuum and subject it to loading determination.

3.10.4. Attachment of O-(4,4′-Dimethoxytrityl)-6-Mercaptohexan-1-ol on to CPG Through Route B

For the attachment of O-(4,4′-dimethoxytrityl)-6-mercaptohexan-1-ol on to CPG, first follow **steps 1–5** as given in **Subheading 3.10.3.** Then add an ethanolic solution of *O*-(4,4′-dimethoxytrityl)-6-mercaptohexan-1-ol (4.36 mg, 10 μmol, 100 μL) to each of the eight tubes in **step 6** (*see* **Subheading 3.10.3.**) and allow it to react for 30 min. Follow rest of steps as discussed in **Subheading 3.10.3.**

3.10.5. Determination of Extent of Reaction on the Support

As the dimethoxytrityl cation absorbs at 498 nm, the deoxyribonucleoside loading can be determined by spectrophotometric determination of the released dimethoxytrityl cation on acidic treatment of a sample of support *(47)*.

1. Weigh an accurate amount of support (1–2 mg) and treat it with detritylating solution (a mixture of 51.4 mL of 70% perchloric acid and 46 mL of methanol) for 5 min.
2. Measure the released dimethoxytrityl cation at 498 nm.
3. Calculate the loading on the support in μmol/g using the formula:

$$\text{Loading (μmol/g of support)} = \frac{14.3 \text{X } A_{498} \text{ X vol (mL) of detritylating solution}}{\text{wt. of support (mg)}}$$

4. Plot a graph of loading on the support (μmol/g) vs time of reaction (min).

3.11. Discussion

A number of hetero- and homobifunctional reagents have been reported to immobilize biomolecules on a variety of polymer supports. Most commonly used reagents are either based on thermochemical or photochemical reactive groups. Generally, the thermochemical reactive groups containing reagents require the presence of nucleophilic functions ($-SH$, $-NH_2$) in the ligand molecules, or on the polymer surface. A number of methods are currently available where modified oligonucleotides are immobilized covalently on polymer surfaces for the construction of microarrays. However, looking at the advantages of using heterobifunctional reagents for the construction of microarrays, the following points have been taken into consideration during designing of newer reagents (NTMTA and NTPAC): (1) the preparation of the reagents must be straightforward and involve commonly available reagents and chemicals; (2) these must be sufficiently stable during storage; and (3) one end of the reagents must contain an active chemical group specific toward aminoalkyl and mercaptoalkyl functionalities, easy to introduce at the 5′-end of oligonucleotides during synthesis in the machine itself, whereas, the other end of the reagents should possess either a thermochemical reactive group (triethoxysilyl in case of NTMTA, reactive toward glass surface) or a photochemical reactive group (anthraquinone in case of NTPAC, reactive toward any C–H containing surfaces).

During immobilization using these reagents and for finding the optimal concentration required for the visualization of the oligonucleotide, the oligonucleotide required modification at 3′- and 5′-end. The preparation of reagents and supports for modified oligonucleotides has been depicted in **Schemes 1–3**. These reagents have been used to incorporate modifications in the oligonucleotides in a regioselective manner. The preparation of new heterobifunctional reagents are shown in **Schemes 4** and **5** starting from commercially available chemicals. These are characterized by NMR and mass spectrometry. Subsequently, the utility of these reagents have been demonstrated by immobilizing a number of oligonucleotides on different surfaces following two routes (**Schemes 6** and **7**). The immobilized oligonucleotides can be seen under laser scanner after hybridizing the spotted oligomers with their complementary-labeled oligomers.

In order to arrive at the optimum time required to immobilize the 5′-modified (mercaptoalkyl or aminoalkyl) oligonucleotides on a glass surface using NTMTA, an indirect experiment has been designed using DMTrO–$(CH_2)_5$-NH_2 and DMTrO$(CH_2)_6$-SH on CPG following both the routes. The results are shown in **Fig. 1A,B**. From these data, it can be concluded that the immobilization of mercaptoalkyl and aminoalkylated ligands on CPG using NTMTA essentially

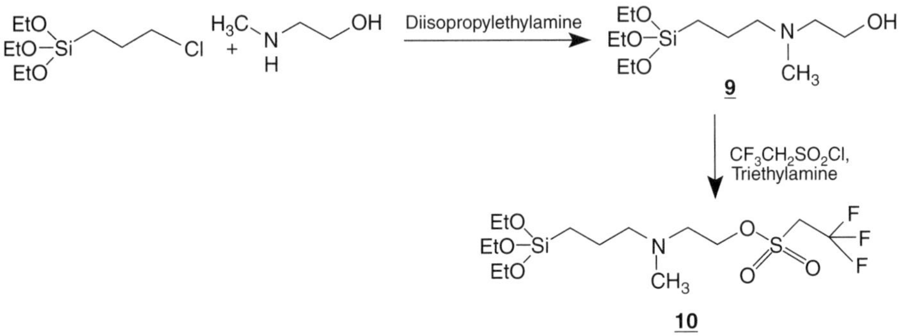

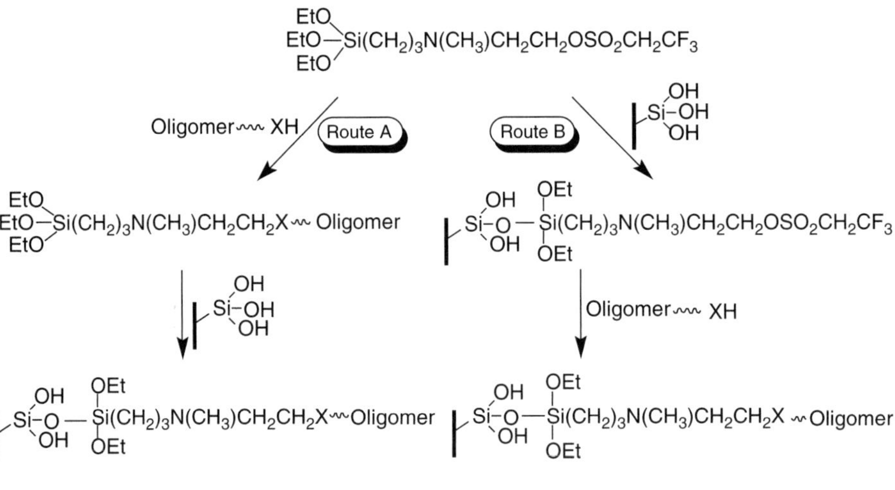

Scheme 4. Synthesis of NTMA reagent.

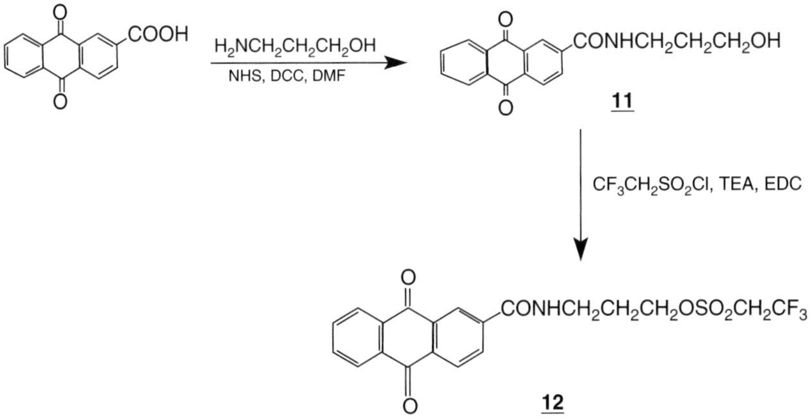

Scheme 5. Synthesis of NTPAC reagent.

X = NH,S

Scheme 6. Immobilization of oligonucleotides using NTMA.

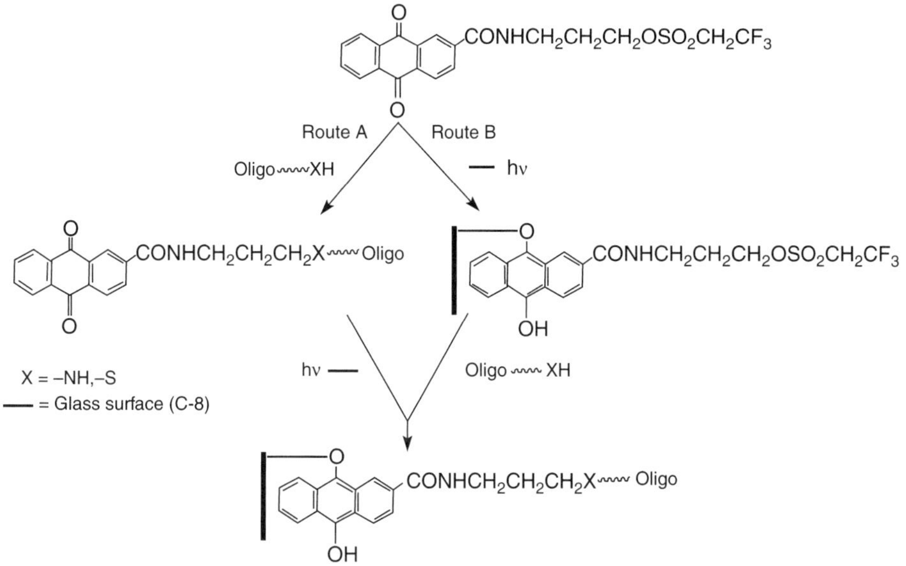

Scheme 7. Immobilization of oligonucleotide using NTPAC.

completes in 40 min at 45°C. Therefore, the same time period has been used for immobilization of oligonucleotides on glass surface through route A or B.

Similarly, to find out the optimal concentration required for visualization of an oligonucleotide on the glass slide, a virgin glass microslide was coated with NTMTA (route B) and spotted an oligomer, d(T^{F}TT TTT TTT TTT TTT TTT TT)-OPO$_3$–(CH$_2$)$_6$–SH), in different concentrations (2.5, 5, 7.5, 10, and 15 μM). After usual washings, the slide was scanned and the spots visualized under a laser scanner set at 570 nm (**Fig. 2, Lane 1**). The immobilization efficiency against each concentration can be determined with the help of the standard curve (**Fig. 3**) and is depicted in **Fig. 2 (Lane 2)**. Though the spots corresponding to 2.5–7.5 μM concentration can be visualized easily, however, to have a good quality bright spots, a 10 μM concentration was selected for construction of microarrays. The same experiment is repeated with NTPAC reagent and the results are shown in **Fig. 4**.

Using the earlier-mentioned parameters, i.e., optimal concentration and time required for immobilization of oligonucleotides through NTMTA, a number of 5′-amino- and mercaptoalkylated oligonucleotides have been spotted following both the routes and, visualized after hybridizing with complementary-labeled oligomers under laser scanner. The spots are shown in **Fig. 5A,B**. The fluorescent signals as well as morphology of the spots on microslides prepared through route A and route B are almost identical. Similarly, 5′-amino- and mercaptoalkylated oligonucleotides have been spotted using NTPAC reagent and

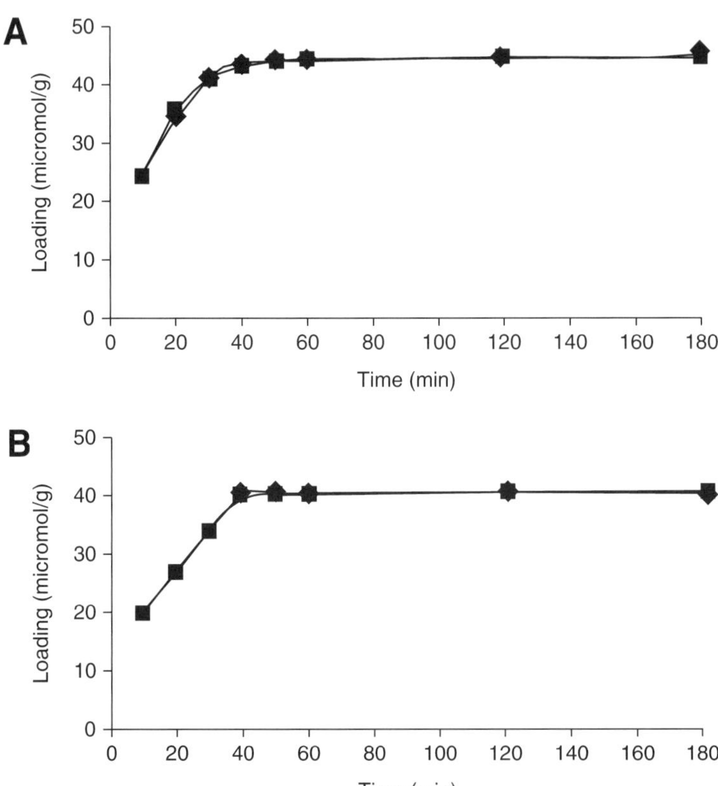

Fig. 1. Time kinetics to determine the optimal time required to immobilize oligo-nucleotides on unmodified glass surface through route A **(A)** and route B **(B)**. Squares, 5′-mercaptoalkylated ligand; diamonds, 5′-aminoalkylated ligand.

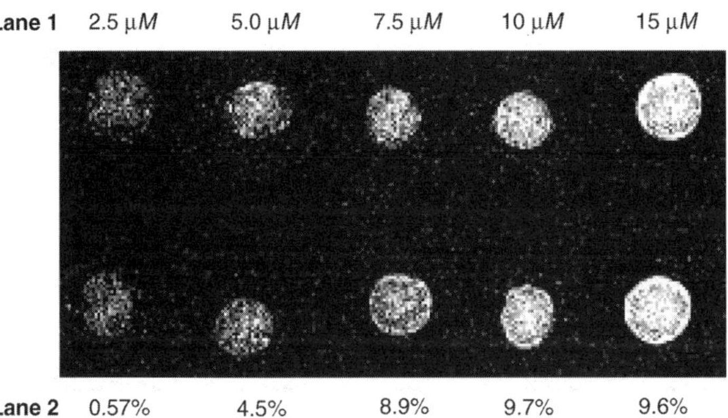

Fig. 2. Threshold concentration of oligonucleotide sequence, d(T^{F}TT TTT TTT TTT TTT TTT TT)-OPO$_3$-(CH$_2$)$_6$SH (entry 8, **Table 1**), required for visualizing fluorescence under a laser scanner. Concentration of spots (2.5, 5, 7.5, 10, and 15 µM) **(Lane 1)**; immobilization efficiency **(Lane 2)**. Reagent: NTMTA.

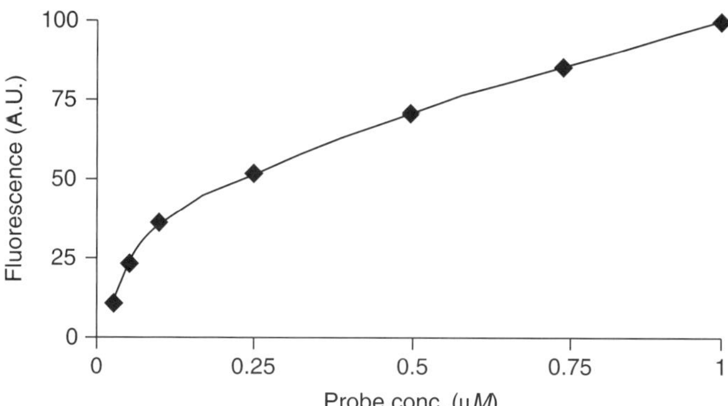

Fig. 3. A correlation sketch between concentration of immobilized probe and fluorescence intensity. Fluorescent oligonucleotide, d(T^{F}TT TTT TTT TTT TTT TTT TT)-OPO$_3$-(CH$_2$)$_6$SH (entry 8, **Table 1**), was spotted in 0.025–1.0 μ*M* concentrations. The spotted microslide was scanned under a laser scanner. Reagent: NTMTA.

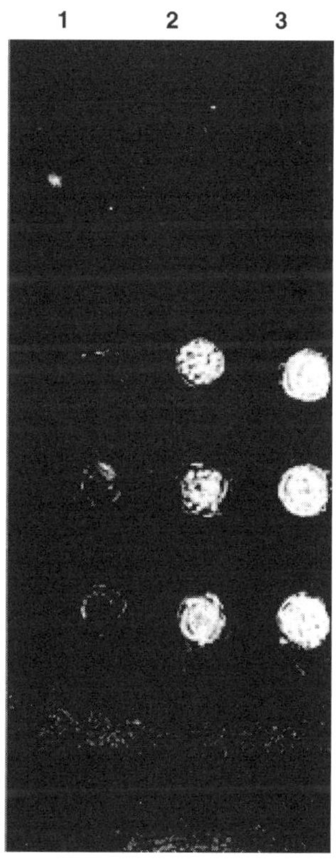

Fig. 4. Threshold concentration of oligonucleotide sequence, fluores-d(TTT TTT TTT TTT TTT TTT TT)OPO$_3$-(CH$_2$)$_6$S-AQ (entry 18, **Table 1**), required for observance of fluorescence under laser scanner. **Lane 1**, 5 μ*M*; **Lane 2**, 10 μ*M*; **Lane 3**, 15 μ*M*. Reagent: NTPAC.

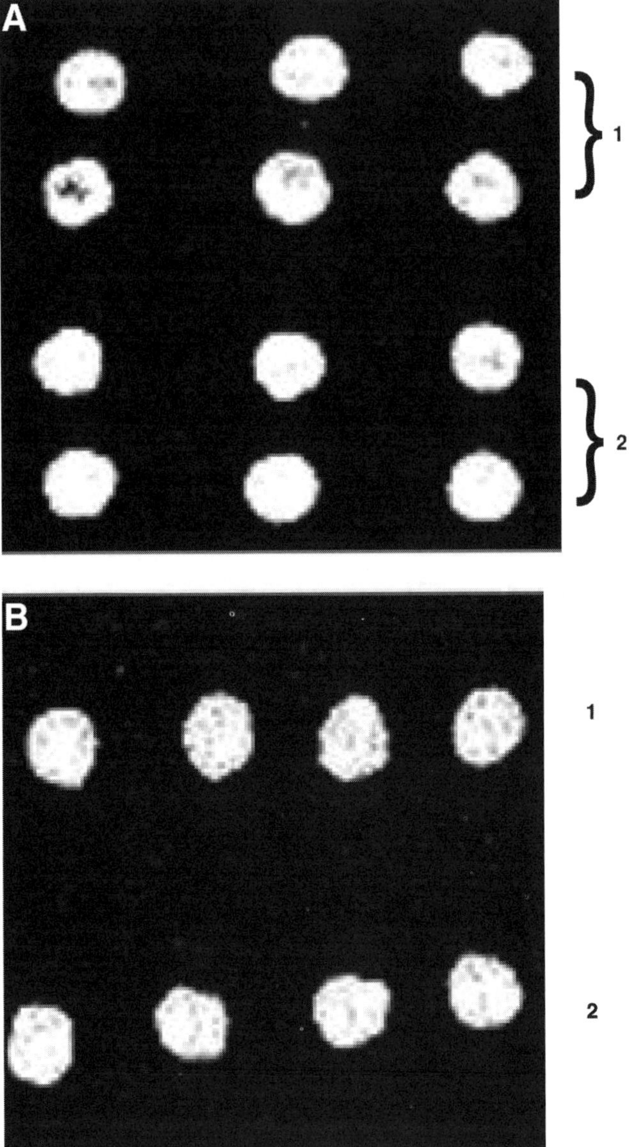

Fig. 5. **(A)** Immobilization of oligonucleotides using NTMTA through route A. **(Lane 1)** H$_2$N-(CH$_2$)$_5$-OPO$_3$-d(CAG AGG TTC TTT GAG TCC TT) (entry 1, **Table 1**) and (**Lane 2**) HS-(CH$_2$)$_6$-OPO$_3$-d(CAG AGG TTC TTT GAG TCC TT) (entry 2, **Table 1**) visualized after hybridization with FAM-d(AAG GAC TCA AAG AAC CTC TG) (entry 6, **Table 1**). **(B)** Immobilization of oligomers through route B. (**Lane 1**) H$_2$N-(CH$_2$)$_5$-OPO$_3$-d(CTC CTG AGG AGA AGG TCT GC) (entry 4, **Table 1**) and (**Lane 2**) HS-(CH$_2$)$_6$-OPO$_3$-d(CTC CTG AGG AGA AGG TCT GC) (entry 3, **Table 1**) visualized after hybridization with FAM-d(GCA GAC CTT CTC CTC AGG AG) (entry 7, **Table 1**).

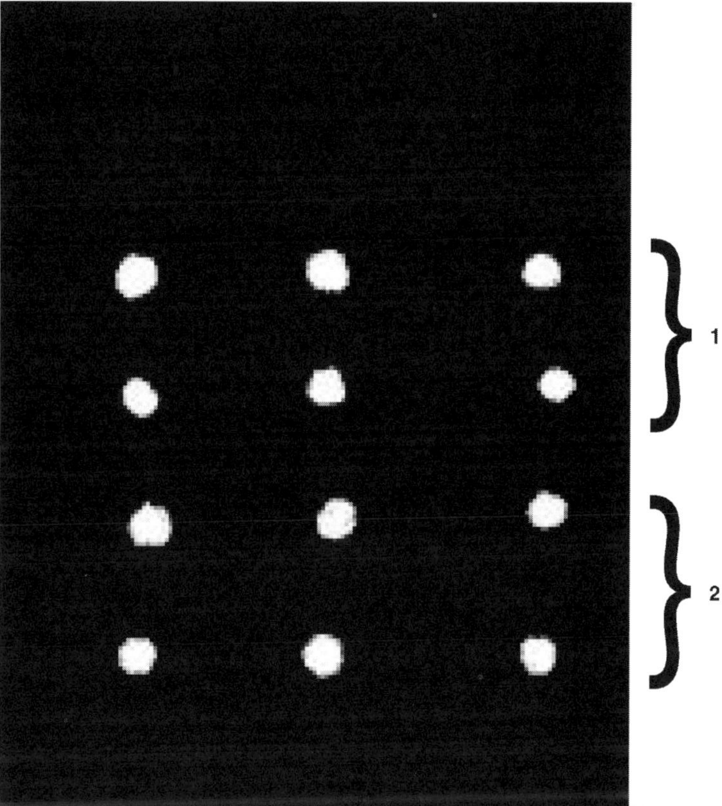

Fig. 6. Immobilization of oligonucleotides H_2N-$(CH_2)_5$-OPO_3-d(AAT CGT TAC TTT TTA TTA TCC) (entry 12, **Table 1**) (**Lane 1**) and HS-$(CH_2)_6$-OPO_3-d(AAT CGT TAC TTT TTA TTA TCC) (entry 13, **Table 1**) (**Lane 2**) in triplicates, using NTPAC and their hybridization with fluorescein–d(GGA TAA TAA AAA GTA ACG ATT) (entry 16, **Table 1**) and visualization under laser scanner at 570 nm.

visualized under laser scanner. **Figure 6** shows the immobilization pattern of oligonucleotides after hybridization with complementary-labeled oligomer.

The specificity of immobilization chemistry and detection of mismatches in case of oligonucleotides immobilized with the help of NTMTA have been depicted in **Fig. 7**. The perfectly matched duplex gives the maximum intensity (**Lane 1**) while the spots having one (**Lane 2**), two (**Lane 3**), and three mismatches (**Lane 4**) show fluorescence intensities in decreasing order and noncomplementary (**Lane 5**) did not emit any signal at all. It is quite evident from these results that the immobilized oligonucleotides are specific toward complementary oligonucleotides and nonspecific hybridization does not occur (**Lane 5**). The signal-to-noise ratio, in all the cases, is >98, as calculated from the signal obtained from complementary hybridized (**Lane 1**) and noncomplementary

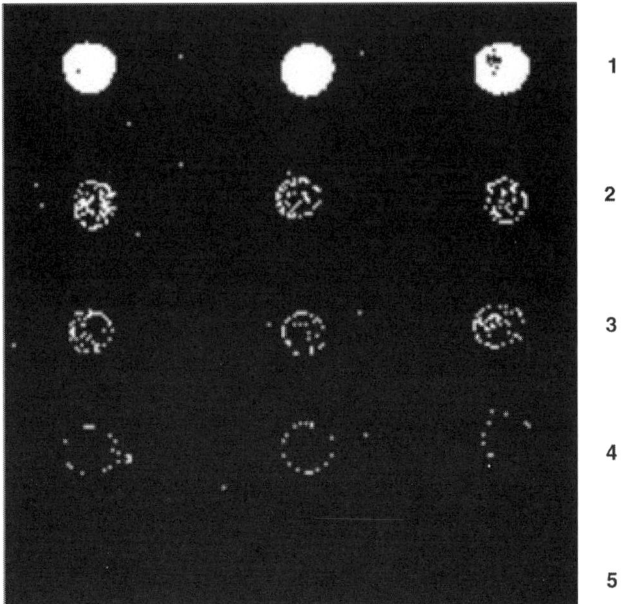

Fig. 7. Detection of nucleotide mismatches and specificity of immobilization through hybridization with FAM labeled complementary oligonucleotides. (**Lane 1**) H_2N-$(CH_2)_5$-OPO_3-d(CTC CTG AGG AGA AGG TCT GC) (entry 4, **Table 1**); (**Lane 2**) H_2N-$(CH_2)_5$-OPO_3-d(CTC CTG AGG CGA AGG TCT GC) (entry 5, **Table 1**); (**Lane 3**) H_2N-$(CH_2)_5$-OPO_3-d(CTC CTG CGG AGA ACG TCT GC) (entry 9, **Table 1**); (**Lane 4**) H_2N-$(CH_2)_5$-OPO_3-d(CTC CTG CGG CGA ACG TCT GC) (entry 10, **Table 1**); (**Lane 5**) H_2N-$(CH_2)_5$-OPO_3-d(TTT TTT TTT TTT TTT TTT TT) (entry 11, **Table 1**). Hybridization was carried out using FAM-d(GCA GAC CTT CTC CTC AGG AG) (entry 7, **Table 1**) followed by laser scanning. Reagent: NTMTA.

hybridized oligomers (**Lane 5**) at 10 µM concentration. The low noise background in this system might be because of the fact that the surface is not aminated (positively charged) and therefore, not susceptible to nonspecific interactions with oligonucleotides (polyanions). A similar pattern has been obtained in case of oligonucleotides (matched and mismatched) immobilized with the help of NTPAC as shown in **Fig. 8**.

The thermal stability of the constructed arrays can be evaluated by exposing them five times to three sets of temperature conditions (encountered during polymerase chain reaction). These consist of 94°C for 30 s (denaturing step), 54°C for 30 s (annealing step), and then 75°C for 30 s (extension step) in buffer 6. **Figure 9** shows the comparison of intensities of fluorescence signals of heat-treated and untreated microarrays after hybridization with complementary-labeled oligomers. The signal intensity of heat-treated microarray decreases by only 1.6% as compared to pretreated microarray.

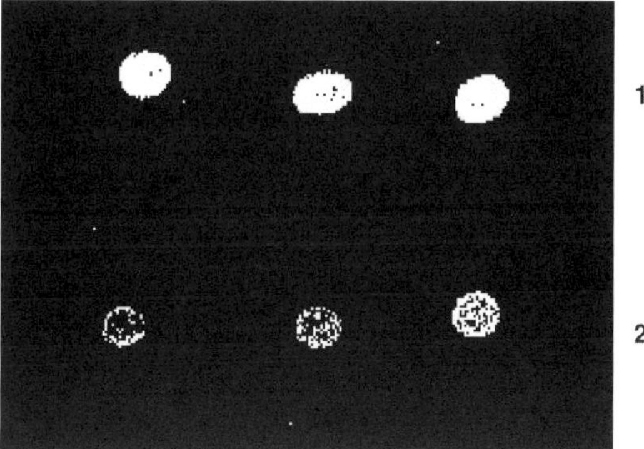

Fig. 8. Detection of single nucleotide mismatch through hybridization with fluorescein-labeled oligonucleotide. (**Lane 1**) HS-$(CH_2)_6$-OPO_3-d(TTG GGT CCG CCA CTC CTT CCC) (entry 14, **Table 1**), and (**Lane 2**) HS-$(CH_2)_6$-OPO_3-d(TTG GGT CCG CTA CTC CTT CCC) (entry 15, **Table 1**). Hybridization was carried out using fluorescein-d(GGG AAG GAG TGG CGG ACC CAA) (entry 17, **Table 1**). Reagent: NTPAC.

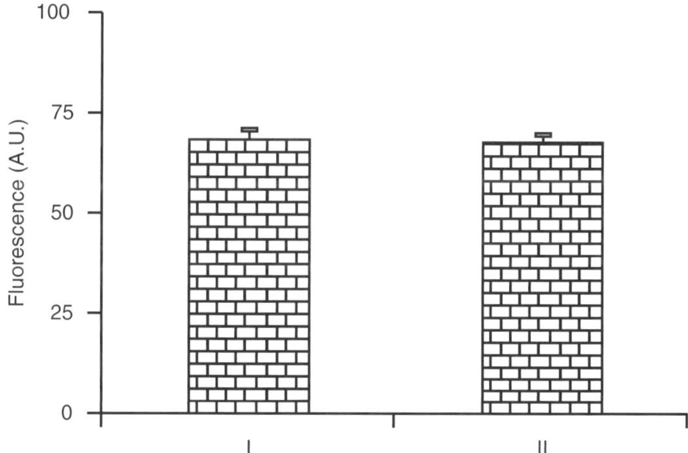

Fig. 9. Thermal stability of immobilized oligonucleotides (microarray). Comparison on the basis of fluorescence signals obtained after hybridization of heat-treated and untreated arrays of oligonucleotides. I, untreated microslide; II, heat-treated microslide.

4. Notes

1. Use freshly opened bottle of aq ammonia (30%) for clean deprotection of oligonucleotides.
2. It is always recommended to treat mercaptoalkylated oligonucleotides with 50 mM DTT solution before their use for immobilization reactions as they are susceptible to arial oxidation on storage.

3. Handle tresyl chloride container with great care as it might develop internal pressure. It is moisture sensitive and reacts violently with water.
4. All the apparatus must be baked in oven and flushed with argon. Because tresylation reaction is extremely moisture sensitive, carry out the reaction in a septum-sealed reaction-vessel fitted with a pressure equalizing syringe.
5. NTMTA and NTPAC should be stored in tightly closed containers under anhydrous conditions at 4°C.
6. Glass microslides must be thoroughly cleaned before their use to remove adsorbed or greasy materials.
7. Spotted microslides can be stored at 4°C and regenerated by placing them in a humid chamber for 1 h before hybridization assay.
8. This oligomer-triethoxysilyl conjugate is stable over several months at 4°C.
9. The tresylated microslides are stable and can be stored for several months at 4°C.
10. The same curve can be used to quantify all the microarrays prepared using the proposed method.

References

1. Pirrung, M. C. (2002) How to make DNA chip. *Angew. Chem. Int. Ed.* **41,** 1276–1289.
2. Seliger, H., Hinz, M., and Happ, E. (2003) Arrays of immobilized oligonucleotides—Contributions to nucleic acids technology. *Curr. Pharm. Biotechnol.* **4,** 379–395.
3. Ramsay, G. (1998) DNA chips: state of the art. *Nat. Biotech.* **16,** 40–44.
4. Nielsen, P. S., Ohlsson, H., Alsbo, C., Andersen, M. S., and Kauppinen, S. (2005) Expression profiling by oligonucleotide microarrays spotted on coated polymer slides. *J. Biotech.* **116,** 125–134.
5. Gerhold, D., Rushmore, T., and Caskey, C. T. (1999) DNA chips: promising toys have become powerful tools. *Trends Biochem. Sci.* **24,** 168–173.
6. Van Ness, J., Kalbfleisch, S., Petrie, C. R., Reed, M. W., Tabone, J. C., and Vermeulen, N. M. (1991) A versatile solid support system for oligodeoxynucleotide probe-based hybridization assays. *Nucleic Acids Res.* **19,** 3345–3350.
7. Beier, M. and Hoheisel, J. D. (1999) Versatile derivatisation of solid support media for covalent bonding on DNA-microchips. *Nucleic Acids Res.* **27,** 1970–1977.
8. Halliwell, C. M. and Cass, A. E. G. (2001) A factorial analysis of silanization conditions for the immobilization of oligonucleotides on glass surfaces. *Anal. Chem.* **73,** 2476–2483.
9. Britcher, I. G., Kehoe, D. C., Matisons, J. G., Smart, R. S. C., and Swincer, A. G. (1993) Silicones on glass surfaces. 2. Coupling agent analogs. *Langmuir* **9,** 1609–1613.
10. Proudnikov, D., Timofeev, E., and Mirzabekov, A. (1998) Immobilization of DNA in Polyacrylamide gel for the Manufacture of DNA and DNA–oligonucleotide microchips. *Anal. Biochem.* **259,** 34–41.

11. Matson, R. S., Rampal, J., Pentoney, S. L., Anderson, P. D., and Coassin, P. (1995) Biopolymer synthesis on polypropylene supports: oligonucleotide arrays. *Anal. Biochem.* **224,** 110–116.

12. Dequaire, M. and Heller, A. (2002) Screen printing of nucleic acid detecting carbon electrodes. *Anal. Chem.* **74,** 4370–4377.

13. Fixe, F., Dufva, M., Telleman, P., and Christensen, C. B. V. (2004) Funtionalization of poly(methyl methacrylate) (PMMA) as a substrate for DNA microarrays. *Nucleic Acids Res.* **32,** E9.

14. Strother, T., Cai, W., Zhao, X., Hamers, R. J., and Smith, L. M. (2000) Synthesis and characterization of DNA-modified Silicon(111) surfaces. *J. Am. Chem. Soc.* **122,** 1205–1209.

15. Chrisey, I. A., O'Ferrall, C. E., Spargo, B. J., Dulcey, C. S., and Calvert, J. M., (1996) Fabrication of patterned DNA surfaces. *Nucleic Acids Res.* **24,** 3040–3047.

16. Healey, B. G., Matson, R. S., and Walt, D. R. (1997) Fiberoptic DNA sensor array capable of detecting point mutations. *Anal. Biochem.* **251,** 270–279.

17. Steel, A. B., Levicky, R. L., Herne, T. M., and Tarlov, M. J. (2000) Immobilization of nucleic acids at solid surfaces: effect of oligonucleotide length on layer assembly. *Biophys. J.* **79,** 975–981.

18. Csáki, A., Möller, R., Straube, W., Köhler, J. M., and Fritzsche, W. (2001) DNA monolayer on gold substrates characterized by nanoparticle labeling and scanning force microscopy. *Nucleic Acids Res.* **29,** E81.

19. Cha, T. -W., Boiadjiev, V., Lozano, J., Yang, H., and Zhu, X. -Y. (2002) Immobilization of oligonucleotide on poly(ethylene glycol) brush-coated Si surfaces. *Anal. Biochem.* **311,** 27–32.

20. Southern, E. M., Maskos, U., and Elder, J. K. (1992) Analyzing and comparing nucleic acid sequences by hybridization to arrays of oligonucleotides: evaluation using experimental models. *Genomics* **13,** 1008–1017.

21. Fodor, S. P. A., Read, J. L., Pirrung, M. C., Stryer, L., Lu, A. T., and Solas, D. (1991) Light directed spatially addressable parallel chemical synthesis. *Science* **251,** 767–773.

22. Pease, A. C., Solas, D., Sullivan, E. J., Cronin, M. T., Holmes, C. P., and Fodor, S. P. (1994) Light-generated oligonucleotide arrays for rapid DNA sequence analysis. *Proc. Natl. Acad. Sci. USA* **91,** 5022–5026.

23. Yong-Sung, C., Do-Kyun, K., and Young-Soo, K. (2002) Development of a new DNA chip microarray by hydrophobic interaction. *Colloids and Surfaces A* **201,** 261–264.

24. Zammatteo, N., Jeanmart, L., Hamels, S., et al. (2000) Comparison between different strategies of covalent attachment of DNA to glass surfaces to build DNA microarrays. *Anal. Biochem.* **280,** 143–150.

25. Kumar, A., Larsson, O., Pardi, D., and Liang, Z. (2000) Silanized nucleic acids: a general platform for DNA immobilization. *Nucleic Acids Res.* **28,** E71.

26. Moller, R., Csaki, A., Kohler, A. M., and Fritzsche, W. (2000) DNA probes on chip surfaces studied by scanning force microscopy using specific binding of colloidal gold. *Nucleic Acids Res.* **28,** E91.

27. Lamture, J. B., Beattie, K. L., Burke, B. E., et al. (1994) Direct detection of nucleic acid hybridization on the surface of a charge coupled device. *Nucleic Acids Res.* **22,** 2121–2125.

28. Chrisey, L. A., Lee, G. U., and O'Ferrall, C. E. (1996) Covalent attachment of synthetic DNA to self-assembled monolayer films. *Nucleic Acids Res.* **24,** 3031–3039.

29. Strother, T., Hamers, R. J., and Smith, L. M. (2000) Covalent attachment of oligodeoxyribonucleotides to amine-modified Si(001) surfaces. *Nucleic Acids Res.* **28,** 3535–3541.

30. Kumar, P., Choithani, J., and Gupta, K. C. (2004) Construction of oligonucleotide arrays on a glass surface using a heterobifunctional reagent, N-(2-trifluoroethane-sulfonatoethyl)-N-(methyl)-triethoxysilylpropyl-3-amine (NTMTA). *Nucleic Acids Res.* **32,** E80.

31. Kumar, P., Agrawal, S. K., Misra, A., and Gupta, K. C. (2004) A new heterobi-functional reagent for immobilization of biomolecules on glass surface. *BioMed. Chem. Lett.* **14,** 1097–1099.

32. Koch, T., Jacobsen, N., Fensholdt, J., Boas, U., Fenger, M., and Jakobsen, M. H. (2000) Photochemical immobilization of anthraquinone conjugated oligo-nucleotides and PCR amplicons on solid surfaces. *Bioconjug. Chem.* **11,** 474–483.

33. Kumar, P., Gupta, K. C., and Gandhi, R. P. (2003) UV light-aided immobilization of oligonucleotides on glass surface using N-(3-trifluoroethanesulfony-loxypropyl)anthaquinone-2-carboxamide (NTPAC) and detection of single nucleotide mismatch. *J. Indian Chem. Soc.* **80,** 1193–1199.

34. Yaqub, M. and Guire, P. (1974) Covalent immobilization of L-asparaginase with a photochemical reagent. *J. Biomed. Mater. Res.* **8,** 291–297.

35. Wilson, D. F., Miyata, Y., Erecinska, M., and Vanerkooi, J. M. (1975) An aryl azide suitable for photoaffinity labeling of amine groups in proteins. *Arch. Biochem. Biophys.* **17,** 104–107.

36. Sigrist, H., Gao, H., and Wegmuller, B. (1992) Light-dependent, covalent immo-bilization of biomolecules on "inert" surfaces. *Biotechnology* **10,** 1026–1028.

37. Gaur, R. K., Sharma, P., and Gupta, K. C. (1989) A simple method for the intro-duction of thiol group at 5′-termini of oligodeoxynucleotides. *Nucleic Acids Res.* **17,** 4404.

38. Gupta, K. C., Sharma, P., Kumar, P., and Sathyanarayana, S. (1991) A general method for the synthesis of 3′-sulfhydryl and phosphate group containing oligonucleotides. *Nucleic Acids Res.* **19,** 3019–3025.

39. Kumar, P., Bhatia, D., Rastogi, R. C., and Gupta, K. C. (1996) Solid phase syn-thesis and purification of 5′-mercaptoalkylated oligonucleotides. *BioMed. Chem. Lett.* **6,** 683–688.

40. Agrawal, S., Christodoulou, C., and Gait, M. J. (1986) Efficient methods for attaching non-radioactive labels to the 5′ ends of synthetic oligodeoxyribo-nucleotides. *Nucleic Acids Res.* **14,** 6227–6245.

41. Connolly, B. A. (1987) The synthesis of oligonucleotides containing a primary amino group at the 5′-terminus. *Nucleic Acids Res.* **15,** 3131–3139.

42. Beaucage, S. L. and Iyer, R. P. (1993) The functionalization of oligonucleotides via phosphoramidite derivatives. *Tetrahedron* **49,** 1925–1963.

43. Beaucage, S. L. and Iyer, R. P. (1992) Advances in the synthesis of oligonucleotides by the phosphoramidite approach. *Tetrahedron* **48,** 2223–2311.

44. Gryaznov, S. M. and Letsinger, R. L. (1993) Anchor for one step release of 3′-aminooligonucleotides from a solid support. *Tetrahedron Lett.* **34,** 1261–1264.

45. Agrawal, S. (ed.) (1994) *Protocols for Oligonucleotides and Analogs, Synthesis and Properties.* Humana Press, Totowa, NJ, pp. 465–496.

46. Truffert, J. C., Lorthioir, O., Asseline, U., Thuong, N. T., and Brack, A. (1994) On-line solid phase synthesis of oligonucleotide-peptide hybrids using silica supports. *Tetrahedron Lett.* **35,** 2353–2356.

47. Atkinson, T. and Smith, M. (1984) Oligonucleotide synthesis, in *A Practical Approach,* (Gait M. J., ed.), IRL Press, Oxford, UK, pp. 35–81.

8

Choice of Polymer Matrix, Its Functionalization and Estimation of Functional Group Density for Preparation of Biochips

Shweta Mahajan, Bhashyam Vaijayanthi, Gopal Rembhotkar, Kailash Chand Gupta, and Pradeep Kumar

Summary

Oligonucleotide microarray has become an important and powerful tool for various genomic analyses, where, unlike conventional methods, one can identify simultaneously a large number of targets in a given sample. Postsynthesis immobilization, the most widely used method, involves the noncovalent and covalent fixing of suitably modified oligonucleotides on the solid supports. Among the various functional groups aminoalkyl, hydroxyalkyl, mercaptoalkyl, aldehyde, epoxy, and carboxyl the most frequently used functional groups on the polymeric surfaces. Because glass and polypropylene, the most widely used materials, are nonporous in nature, the functional groups density on the surface remains very low. In order to know the exact concentration of a ligand to be immobilized, it is essential to estimate the accessible functional groups on these surfaces. For this purpose, sensitive methods are required to estimate exact density of available functional groups on the surfaces. Apart from physical methods, a number of sensitive chemical methods, by making use of high extinction coefficient of 4,4′-dimethoxytrityl cation ($\varepsilon_{498} = 70,000$ Lmol^{-1}cm^{-1}), have been reported to estimate accessible functional groups on the glass based polymer supports. In this chapter, use of these reagents for spectrophotometric determination of functional group density on glass microslides and polypropylene film has been discussed.

Key Words: DMPA; DMTr-Cl; DTNPME; estimation; functional group density; functionalization; glass microslide; polypropylene film; SDTB; spectrophotometric methods.

1. Introduction

DNA microarray has emerged as a powerful tool that allows simultaneous detection of many different target molecules in a sample based on base-pairing rules, and thus, finding wide applications in molecular biology and genomics *(1–4)*. The DNA chip can be used in studies like mutation detection, single

From: *Methods in Molecular Biology, vol. 381: Microarrays: Second Edition: Volume 1: Synthesis Methods*
Edited by: J. B. Rampal © Humana Press Inc., Totowa, NJ

nucleotide polymorphism analysis, differential gene expression analysis, and so on. Currently, two predominant methods of producing oligonucleotide microarrays, viz., (1) *in situ* synthesis of oligonucleotides (Affymetrix approach) on the polymeric surface using photolithographic technique *(5,6)*, and (2) immobilization of presynthesized oligonucleotides on polymer supports (organic or inorganic) are in use *(7–13)*. The former one is advantageous in generating high density (10^6 sequences/cm^2) microarrays but lacks in flexibility. Whereas, the latter one produces low-to-moderate density microarrays and turns out to be suitable for most applications in biological sciences. Therefore, this approach is focused in this chapter. Besides the chemistry involved in fixing oligonucleotide on polymer surface, the polymer matrix plays an important role in influencing the quality of constructed microarrays. A number of polymer matrices *(14–30)* such as nylon, nitrocellulose, polypropylene, polystyrene (organic polymers), glass, silicon wafer, and so on (inorganic polymers) have been examined for this purpose (**Table 1**). Out of these, glass and polypropylene appear attractive because these materials can easily be derivatized generating functional groups such as aminoalkyl, carboxyl, aldehyde, mercaptoalkyl, and so on, on the surface. However, glass is considered to be the best, as it can be used in the laser scanner in order to visualize the fluorescent spots on its surface.

For the introduction of primary functionalities such as –NH$_2$, –SH, and epoxy on glass, the reagents of choice are functionalized trialkoxysilanes such as 3-aminopropyltriethoxysilane (APTS), 3-mercaptopropyltriethoxysilane (MPTS), and 3-glycidyloxypropyltriethoxysilane (GOPTS). Such reagents introduce amino, sulfhydryl, and epoxy groups as primary functional groups, which can be modified into secondary functional groups (–COOH and –CHO), that can further be activated for oligonucleotide attachment. Inert polymeric materials, on the other hand, are functionalized by means of chemical as well as photochemical methods. Functionalization through chemical methods is achieved either by reactive extrusion, in the presence of peroxides, or by plasma-induced surface chemistry. Photochemical generation of functional groups on the surface of inert polymers is relatively a gentle technique and involves the use of light *(31)*. Recently, anthraquinone-2-methanol has been used to generate hydroxyl groups on a variety of inert surfaces using *(32)*.

The immobilization method involves the covalent attachment of modified oligonucleotide sequences at the discrete locations on the polymeric surface of choice. Because the technique deals with the use of expensive material (oligonucleotides) in tiny amounts, it becomes mandatory to know the exact density of the functional groups on the polymer surface, so that, proper concentration of the probe molecules could be determined before immobilization. Functional group loading on polymer supports can be determined by physical as well as chemical methods. Because the physical methods (infrared, ultraviolet, and

Table 1
Types of Polymer Surfaces Used in Oligonucleotide Microarrays

S. No.	Polymer surface	Functional group	Functional group on oligomer	Chemistry	Reference
1.	Glass	Silanol	GOPTS	Covalent (Si-O-Si)	*14*
		$-NH_2$	–	Noncovalent	*15*
		$-NCS$	$-NH_2$	Covalent	*16*
		$-CONHNH_2$	$-CHO$	Covalent	*17*
		$-CHO$	$-NH_2$	Covalent	*18*
		Maleimide	$-SH$	Covalent	*19*
		$-OSO_2CH_2CF_3$	$-SH, -NH_2$	Covalent	*20–23*
		$-SH$	$-S-SR$	Covalent	*24*
		$-SH$	$-SH$	Covalent	*24*
		Epoxy	$-NH_2$	Covalent	*15*
		$-CHO$	$-ONH_2$	Covalent	*25*
		$-NH_2$	$-COOH$	Covalent	*11*
		$-I$	$-SH$	Covalent	*26*
		$-NH_2$	$-PO_4$	Covalent	*27*
2.	Polypropylene	–	$-AQ$	Covalent	*20,21*
		$-OSO_2CH_2CF_3$	$-SH, -NH_2$	Covalent	*20,21*
3.	Polyacrylamide gel pad	$-CHO$	$-NH_2$	Covalent	*28*
4.	Gold layer on glass	Streptavidin	Biotin	Noncovalent	*29*
		–	Lipoic acid	Covalent	*30*

nuclear magnetic resonance [NMR]) involve the use of expensive equipments and are not suitable for the determination of functional groups on polymer supports *(33–37)*, therefore, such methods were not considered appropriate for discussion. Chemical methods have been extensively used, which can further be classified into three broad categories based on: (1) elemental analysis, (2) acid–base titrations, and (3) reactions of functional groups (spectrophotometric method). The methods namely, elemental analysis and acid–base titrations are very tedious as well as time-consuming and often used for quantitative determination of functional groups. Therefore, the methods that could estimate the polymer supported functional groups accessible for further functionalization were considered suitable for this purpose. In literature, several reagents (**Table 2**) have been reported *(38–52)*, some of them from our own laboratory, have been included in this chapter.

Table 2
**Methods for Estimation of Concentration of Functional Groups
on Polymer Surfaces**

S. No.	Functional group	Reagents	Released chromophore	λ_{max} (nm)	Reference
1.	Amino (–NH$_2$)	Ninhydrin	Indoxyl	570	*38*
		TNBS acid	NH$_4$–TNBS	250	*39*
		SDTB	DMTr cation	498	*40*
		3-sulfo-SDTB	DMTr cation	498	*41*
		DMTrCl	DMTr cation	498	*42–44*
		DMPA	DMTr cation	498	*45*
		SPDP	2-Thiopyridone	304	*46*
2.	Mercapto (–SH)	DMTrCl	DMTr cation	498	*44*
		DMPA	DMTr cation	498	*44*
		DTNPME	DMTr cation	498	*47*
		DTNP	5-nitropyridine-2-thione	386	*48*
		DTNB	5-thionitrobenzoic acid	365	*49*
		SPDP	Pyridine-2-thione	304	*46*
3.	Hydroxyl (–OH)	DMTrCl	DMTr cation	498	*44,45*
		DHBA	DMTr cation	498	*50*
		DMPA	DMTr cation	498	*44*
4.	Carboxyl (–COOH)	NP	p-Nitrophenoxide ion	405	*51*
		DTAH	DMTr cation	498	*52*
5.	Aldehyde (–CHO)	DTAH	DMTr cation	498	*52*
6.	Epoxide	DTAH	DMTr cation	498	*52*

SPDP: N-Succinimidyl-3-(2-pyridyldithio) propionate; DTNB: 5,5′-Dithionitrobenzoic acid;
NP: p-Nitrophenol; TNBS: 2,4,6-Trinitrobenzenesulfonic acid.

Although, the polymeric surfaces are used for preparation of biochips are nonporous, the functional group density is extremely low. Therefore, sensitive methods are required to estimate the density accurately. Two reagents, viz., N-succinimidyl-4-*O*-(4,4′-dimethoxytrityl)butyrate (SDTB)/2,4-dinitrophenyl-4-*O*-(4,4′-dimethoxytrityl)butyrate *(40,53)*, and 4,4′-dimethoxytrityloxy-S-(2-thio-5-nitropyridyl)-2-mercaptoethane (DTNPME) *(40)*, have been reported for the estimation of polymer supported amino- and mercaptoalkyl groups, respectively, from the authors' laboratory. The sensitivity of these two reagents relies on the high molar extinction coefficient of the 4,4′-dimethoxytrityl cation (ε_{498} = 70,000 Lmol^{-1}cm^{-1}), which gets released in the solution on an acid treatment after reaction with polymer supported functional groups (–NH$_2$ and –SH). This makes these reagents very sensitive and loadings up to subpicomole level can be calculated. Because microarray surfaces are monofunctional, universal reagents can be designed for estimation of several functional groups. 4,4′-Dimethoxytrityl chloride (DMTr-Cl), a useful reagent for the selective protection of hydroxyl groups in sugar residues *(54)*, has been employed for the estimation of polymer supported amino and hydroxyl groups *(42–44)*. Under similar conditions, this can also be used to estimate density of sulfhydryl groups on polymer supports *(44)*. However, extremely moisture sensitive as well as acid-labile nature of DMTrCl restricts its use in quantitative estimation of functional groups. Recently, a universal reagent, S-(4,4′-dimethoxytrityl)-3-mercaptopropionic acid (DMPA) was developed, which could be used for the estimation of polymer supported functional groups (amino, thiol, and hydroxyl) under microwave irradiation *(44)*. The major advantage of this method is that it is unaffected by the presence of moisture. Therefore, laboratory grade solvents and reagents can be employed for the reactions. The reagent can also be used to monitor the functionalization of polymer supports for peptide and oligonucleotide synthesis.

2. Materials

2.1. Polymeric Supports

1. Glass microslides.
2. Polypropylene films (PP).

2.2. Solvents

N,N-Dimethylformamide (DMF), acetonitrile, dichloromethane (DCM) 1,2-dichloroethane (EDC), methanol, toluene, acetone, diethyl ether, 1,4-dioxane, pyridine, and sym-collidine.

2.3. Miscellaneous Reagents

Tetrabutylammonium nitrate, 4-dimethylaminopyridine (DMAP), 2,2′-dithiobis(5-nitropyridine) (DTNP), 4-hydroxybutyric acid sodium salt, dicyclohexylcarbodiimide, 3-mercaptopropionic acid, bromotrichloromethane (BTCM), triphenylphosphine (TPP), trimethylchlorosilane (TMS-Cl), APTS, MPTS, and GOPTS. All other reagents and chemicals are procured from local vendors (*see* **Note 1**).

3. Methods

3.1. Cleaning and Functionalization of Microslides

3.1.1. Activation of Glass Microslides (see **Note 2**)

1. Immerse virgin glass microslides in 1 *M* NaOH for 2 h at room temperature.
2. Rinse the slides with Milli-Q water (2X 100 mL) followed by placing them in a 2 *M* solution of hydrochloric acid for 30 min.

3. Again rinse the microslides with Milli-Q water (3X 100 mL) and then submerge in approx 95% ethanol for 1 h.
4. Dry under vacuum and store in a dust-free box at room temperature.

3.1.2. 3-Aminopropylation of Glass Microslides

1. Immerse the cleaned microslides in a 5% (v/v) solution of APTS in toluene for 2 h at 50°C with constant agitation at slow speed.
2. After cooling the mixture to room temperature, wash the slides sequentially with toluene (2X 100 mL), acetonitrile (2X 100 mL), and Milli-Q water (2X 100 mL), respectively.
3. Dry these slides under vacuum in a desiccator, and finally bake for 3 h at 100°C.
4. Estimate functional group density on the glass surface.

3.1.3. 3-Mercaptopropylation of Glass Microslides

3-Mercaptopropylation of glass microslides are carried out in the same fashion as the aminopropylation, except that MPTS is employed instead of APTS in **step 1** (*see* **Subheading 3.1.2.**). After washing and drying, 3-mercaptopropylated glass microslides are stored under an inert atmosphere (*see* **Note 3**). Then it is subjected to functional group estimation.

3.1.4. 3-Glycidyloxypropylation of Microslides

3-Glycidyloxypropylation of glass microslides is also carried out as the aminopropylation, except that GOPTS is used instead of APTS in **step 1** (*see* **Subheading 3.1.2.**). After washing and drying, 3-glycidyloxypropylated glass microslides are stored under an inert atmosphere (*see* **Note 4**). Then it is subjected to functional group estimation.

3.1.5. Hydroxyalkylation of Glass Microslides

1. Add a solution of TPP (1 mmol) in DMF (20 mL) to a solution of O-(4,4′-dimethoxytrityl)-4-hydroxybutyric acid (1 mmol) (*see* **Subheading 3.2.**), DMAP (2 mmol), and BTCM (1 mmol) in DMF (20 mL).
2. Agitate the previously described mixture for 30 s and add to 3-aminopropylated glass microslides in a falcon tube. Allow the reaction to stir for 1 h at room temperature.
3. Wash the plates with DMF (2X 50 mL), methanol (2X 50 mL), and diethyl ether (2X 30 mL) and dry them under vacuum in a desiccator.
4. Block the residual amino groups on the surface of microslides with a solution of acetic anhydride:triethylamine (TEA):N-methylimidazole:dichloromethane (DCM) (1:1:0.4:0.6, by vol, 50 mL) for 1 h at room temperature.
5. Wash the plates with DCM (3X 50 mL) and dry in air. Then treat the dried plates with a solution of 3% trichloroacetic acid (TCA) in DCM (3X 50 mL) to remove DMTr group.
6. Wash the plates with DCM (3X 50 mL) and dry under vacuum in a desiccator.
7. Estimate the functional group density on glass surface.

3.1.6. Carboxyalkylation of Glass Slides

1. Immerse 3-aminopropylated glass slides in a 1% (w/v) solution of succinic anhydride in DCM (50 mL) containing TEA (5 mmol) and DMAP (1 mmol) for 2 h at room temperature.
2. Wash the slides with DCM (3X 50 mL), dry under vacuum, and determine the loading of carboxyl groups on the glass surface.

3.1.7. Generation of Aldehyde Groups on Glass Surface

1. Immerse 3-glycidyloxypropylated microslides in a dilute solution of hydrochloric acid (0.2 N, 50 mL) for 3 h at room temperature.
2. Rinse the microslides with Milli-Q water (2X 100 mL) and submerge in an aqueous solution of sodium periodate (0.1 N, 50 mL) for 1 h at room temperature.
3. Rinse the microslides with Milli-Q water (3 × 100 mL) followed by acetone (1X 100 mL).
4. Dry the microslides in a vacuum desiccator, store in dark place and determine the loading of aldehyde groups on the glass surface.

3.1.8. Hydroxylated Polypropylene Film

1. Prepare a 0.1 M solution of anthraquinone-2-methanol in acetonitrile:DMF (1:1, by vol) and coat the polypropylene film (1 × 1 cm^2) with this solution in a Petri dish.
2. Place the coated film in an ultraviolet reactor, operating at 365 nm, for 30 min exposure.
3. Then wash the film with DMF (2X 5 mL) and methanol (2X 5 mL) and dry in air.
4. Treat the air-dried polypropylene film with a solution of 4,4′-dimethoxytrityl chloride (1 mmol) and DMAP (0.1 mmol) in dry pyridine (5 mL) for 4 h at room temperature.
5. Decant off the solution and wash the film with EDC (2X 10 mL), methanol (2X 10 mL), and diethyl ether (1X 10 mL) and dry in air.
6. Determine the extent of photochemical functionalization of the film by spectrophotometric method.

3.2. Preparation of Reagents

3.2.1. Synthesis of S-4,4′-Dimethoxytrityl-3-Mercaptopropionic Acid $\underline{4}$

1. Dry 3-mercaptopropionic acid $\underline{3}$ (10 mmol) by coevaporating twice with anhydrous pyridine (25 mL) and finally take up in dry pyridine (25 mL). To this solution, add DMTrCl $\underline{1}$ (12 mmol) and DMAP (1.0 mmol).
2. Allow the reaction to stir under argon atmosphere for 4 h at room temperature.
3. Monitor the reaction on thin-layer chromatography (TLC) and if it is complete, hydrolyze the excess DMTrCl by adding methanol (0.5 mL).
4. Concentrate the reaction mixture under vacuum and remove the traces of pyridine by coevaporation with toluene (2X 20 mL).
5. Take up the syrupy residue in ethyl acetate (100 mL) and wash it sequentially with 5% aqueous sodium bicarbonate (2X 50 mL), 5% aqueous citric acid (2X 50 mL) and brine (2X 25 mL) solutions.

6. Collect the organic phase and keep it over anhydrous sodium sulfate for 30 min.
7. Filter and concentrate the filtrate to a syrupy residue under vacuum.
8. Purify the crude product by silica gel column chromatography using a stepwise gradient of methanol in EDC (0–5%) containing TEA (1%).
9. Concentrate the pure fractions of DMPA **4** to obtain in 85% yield.
10. Characterize the reagent by NMR. ^{1}H NMR (CDCl$_3$) δ: 1.4 (m, 2H), 2.2 (t, 2H), 2.7 (t, 2H), 3.6 (s, 6H), 6.5–7.8 (m, 13H).

3.2.2. Synthesis of N-Succinimidyl-4-O-(4,4′-Dimethoxytrityl)Butyrate **8**

1. Suspend 4-hydroxybutyric acid sodium salt **6** (20 mmol) in anhydrous pyridine (100 mL), and dry it twice by coevaporation on a rotary evaporator. Finally, resuspend the sodium salt in dry pyridine (100 mL), add DMTrCl **1** (24 mmol) and a catalytic amount of DMAP (2 mmol) with stirring under argon to exclude moisture.
2. Allow the reaction mixture to stir at room temperature for 12 h and then monitor by the TLC.
3. After completion of the reaction, quench the reaction by adding methanol (0.5 mL) and remove the solvent under vacuum to obtain a syrupy residue.
4. Take up the residual mass in DCM (150 mL) and wash the organic phase sequentially with 5% aq. sodium bicarbonate solution (2X 50 mL), 5% aq citric acid solution (2X 50 mL), and brine (2X 50 mL).
5. Collect the organic phase and dry it over anhydrous sodium sulfate followed by filtration and concentration to an oily mass.
6. Purify the crude compound, 4-*O*-(4,4′-dimethoxytrityl)-butyric acid **7**, by column chromatography to get it in approx 82% yield.
7. Dissolve 4-*O*-(4,4′-dimethoxytrityl)-butyric acid **7** (15 mmol) and N-hydroxysuccinimide (15 mmol) in dry DMF (50 mL). Cool this mixture in an ice-water bath and add dicyclohexylcarbodiimide (18 mmol) with stirring.
8. After 2 h, filter off the precipitated dicyclohexylurea and evaporate the filtrate under vacuum to a syrupy material. Dissolve the residue in dry EDC (100 mL), wash it with water (3X 50 mL), and dry over sodium sulfate.
9. Concentrate the organic layer under vacuum to obtain the title compound SDTB **8** in 75% overall yield.

3.2.3. Synthesis of 4,4′-Dimethoxytrityloxy-S-(2-Thio-5-Nitropyridyl)-2-Mercaptoethane *(13)*

1. To a vigorously stirred solution of DTNP **10** (10 mmol) in dry EDC (150 mL), add 2-mercaptoethanol **11** (10 mmol) dropwise over a period of 30 min.
2. Stir the reaction mixture at room temperature for 12 h.
3. Filter off the precipitate of 5-nitropyridine-2-thione and concentrate the filtrate to obtain a yellow oily mass.
4. Purify the product on a silica gel column using a gradient of methanol (0–4%) in EDC.
5. Pool together the fractions containing the compound, S-(2-thio-5-nitropyridyl)-2-mercaptoethanol **12**, and concentrate on a rotary evaporator to obtain an oily product in 61% yield.

6. Characterize the compound by NMR. ^{1}H NMR (CDCl$_3$) δ: 3.0 (t, 2H), 3.7 (t, 2H), 7.5–8.2 (m, 2H), 9.0–9.2 (m, 1H).
7. Dissolve the compound **12** (6 mmol) in dry EDC (50 mL) and add DMTr-Cl **1** (7.2 mmol), DMAP (0.6 mmol), and TEA (7.2 mmol) with stirring under exclusion of moisture.
8. The reaction completes in 3 h; quench the reaction by adding methanol (0.5 mL).
9. Remove the solvent on a rotary evaporator and purify the crude material by silica gel column chromatography using EDC (containing 1% TEA) as an eluent.
10. Concentrate the fractions containing the desired material, DTNPME **13**, to obtain in 80% yield.
11. Characterize the compound by NMR. ^{1}H NMR (CDCl$_3$) δ: 3.0 (t, 2H), 3.5 (t, 2H), 3.9 (s, 6H), 7–7.7 (m, 15H), 9.2–9.3 (m, 1H).

3.2.4. Synthesis of 1-O-(4,4′-Dimethoxytrityl)-6-Aminohexanol **18**

1. Dissolve 6-aminohexanol **15** (10 mmol) in 1,4-dioxane (50 mL) and add TEA (10 mmol), and ethyl trifluoroacetate (20 mmol) at room temperature.
2. Allow the reaction to occur for 12 h followed by removal of solvent on a rotary evaporator at reduced pressure to yield a syrupy residue, N-trifluoroacetyl-6-aminohexan-1-ol **16**, in almost quantitative yield.
3. Dissolve this syrupy material **16** in anhydrous pyridine (50 mL) and evaporate under vacuum to remove traces of moisture. Redissolve the material in anhydrous pyridine (25 mL).
4. To this solution, add 4,4′-dimethoxytrityl chloride **1** (12 mmol) and DMAP (1 mmol). Allow the reaction to proceed at room temperature for 4 h.
5. After monitoring the reaction on TLC, destroy the excess DMTrCl by adding methanol (0.5 mL) and concentrate on a rotary evaporator to a syrupy residue of N-trifluoroacetyl-1-O-(4,4′-dimethoxytrityl)-6-aminohexanol **17**.
6. Dissolve the residue in ethanol (50 mL) and add aqueous ammonium hydroxide (50 mL).
7. Stir the reaction for a period of 12 h at room temperature and concentrated under vacuum to an oily residue.
8. Take up the residue in ethyl acetate (100 mL) and wash it with 5% aqueous citric acid (2X 50 mL) and brine (2X 25 mL), respectively.
9. Collect the organic phase, dry it over sodium sulfate, filter and concentrate to obtain crude 1-O-(4,4′-Dimethoxytrityl)-6-aminohexanol (DTAH) **18**.
10. Purify the crude product by silica gel column chromatography using EDC:methanol (98:2, by vol) containing TEA (0.1%).
11. Collect the fractions containing the desired material, pool together and concentrate to afford the pure material as an oily mass in approx 85% yield.
12. Characterize the compound, DTAH, by NMR. ^{1}H NMR (CDCl$_3$) δ: 1.71 (m, 8H), 2.83 (t, 2H), 3.27 (t, 2H), 3.85 (s, 6H), 6.75–7.54 (m, 13H).

3.3. Preparation of Reagents for Estimation of Different Functionalities

3.3.1. Preparation of Reagent A

Dissolve 4,4′-dimethoxytrityl chloride (1.692 g, 0.1 *M*), tetrabutylammonium nitrate (1.52 g, 0.1 *M*) and sym.-collidine (0.91 g, 0.25 *M*) in 50 mL of dry DMF.

3.3.2. Preparation of Reagent B

Dissolve DMPA (2.04 g, 0.2 *M*) in acetonitrile (25 mL) containing 4-dimethyl-aminopyridine (1.22 g, 10 mmol) and BTCM (1.0 g, 5 mmol).

3.3.3. Preparation of Reagent C

Dissolve TPP (1.31 g, 5 mmol) in dry DMF (25 mL).

3.3.4. Preparation of Reagent D

Dissolve SDTB (2.5 g, 0.1 *M*) in dry DMF (50 mL).

3.3.5. Preparation of Reagent E

Dissolve dithiothreitol (0.385 g, 0.05 *M*) in ethanol (50 mL).

3.3.6. Preparation of Reagent F

Dissolve DTNPME (2.67 g, 0.1 *M*) in dry DMF (50 mL) containing TEA (0.1%).

3.3.7. Preparation of Reagent G

Dissolve DTAH (2.1 g, 0.1 *M*) in dry EDC (50 mL) containing TEA (0.1%) and catalytic amount of DMAP.

3.3.8. Preparation of Reagent H

Prepare detritylating reagent by mixing 70% perchloric acid (51.4 mL) and methanol (46 mL).

3.3.9. Preparation of Reagent I

Prepare 3% (w/v) TCA (by dissolving 3 g of TCA in 100 mL of EDC).

3.4. Determination of Functional Groups on Glass Microslides (–NH$_2$, –OH, –SH, Epoxy, Aldehyde, and Carboxyl Groups)

3.4.1. DMTrCl Method

3.4.1.1. DETERMINATION OF AMINO GROUPS ON GLASS MICROSLIDE

1. Place three to four small pieces of 3-aminopropylated glass microslide (~5 × 5 mm^2) in a reaction vessel and add reagent A (5 mL). Cap the vessel tightly and agitate thoroughly at room temperature for 35 min.
2. Remove the supernatant, wash the glass pieces successively with DMF, methanol and diethyl ether (3X 5 mL of each) containing 0.1% TEA.
3. Dry these pieces under vacuum in a desiccator.
4. Take glass microslide **2** (one to two pieces) in a glass vial and charge it with reagent H (5 mL).

5. Shake the contents of the vial for 2 min and measure the absorbance of the solution at 498 nm using reagent H as a blank.

3.4.1.2. DETERMINATION OF SULFHYDRYL GROUPS ON GLASS MICROSLIDE

Same protocol is followed to determine the contents of sulfhydryl groups on glass microslide as documented in **Subheading 3.4.1.1.,** except taking pieces of 3-mercaptopropylated glass microslide in **step 1**. Moreover, the reaction time required is 1 h at room temperature instead of 35 min.

3.4.1.3. DETERMINATION OF HYDROXYALKYL GROUPS ON GLASS MICROSLIDE

Similarly, the hydroxyl groups content is determined on microslide pieces as given in **Subheading 3.3.1.1.,** except using hydroxyalkylated glass microslide in **step 1**. Allow the reaction to proceed for 2 h at room temperature.

3.4.1.4. DETERMINATION OF HYDROXYL GROUPS ON POLYPROPYLENE FILM

1. Place small portions of PP film (2×2 mm^2) in a glass vial and add pyridine (5 mL), DMTrCl (1 mmol), and DMAP (0.1 mmol).
2. After 2 h of occasional shaking, decant off the solution and wash the film pieces with methanol (2X 5 mL) and diethyl ether (2X 5 mL) containing TEA (0.1%), respectively.
3. Dry these pieces and treat them with reagent I (5 mL) for 10 min and measure the absorbance of the solution at 505 nm using reagent I as blank.

3.4.2. DMPA Method

3.4.2.1. DETERMINATION OF AMINO GROUPS ON GLASS SLIDES

1. Mix reagents B (480 µL) and C (400 µL) in a glass vial, shake the mixture briefly and add it to pieces (three to four) of aminoalkylated glass microslide in a glass vial.
2. Loosely cap the vial and place it inside a microwave oven operating at 800 W and irradiate for 4 min. (No exposure should be longer than 30 s and after each exposure, cool the glass vial in an ice-water mixture to room temperature.)
3. Remove the supernatant, wash the glass pieces successively with DMF, methanol and diethyl ether (3X 5 mL of each) containing TEA (0.1%).
4. Dry these pieces under vacuum in a desiccator.
5. Take glass microslide **5** (one to two pieces) in a glass vial and charge it with reagent H (5 mL).
6. Shake the contents of the vial for 2 min and measure the absorbance of the solution at 498 nm using reagent H as a blank.

3.4.2.2. DETERMINATION OF SULFHYDRYL GROUPS ON GLASS MICROSLIDES

Similar protocol is used for the estimation of sulfhydryl groups on the polymer supports except employing mercaptopropylated polymer surface in **step 1** of **Subheading 3.4.2.1.**

                                                            Mahajan et al.

3.4.2.3. Determination of Hydroxyl Groups on Glass Slides

Similar protocol is used for estimation of hydroxyl groups on the polymer supports except employing hydroxyalkylated polymer surface in **step 1** of **Subheading 3.4.4.1.**

3.4.3. SDTB Method

3.4.3.1. Determination of Amino Groups on Glass Microslide

1. Place three to four small pieces of 3-aminopropylated glass microslide (~5 × 5 mm^2) in a reaction vessel and add reagent D (5 mL), a catalytic amount of DMAP (5 mg) and TEA (100 µL).
2. Cap the reaction vessel tightly and agitate thoroughly at room temperature for 30 min.
3. Remove the supernatant; wash the glass pieces successively with DMF, methanol and diethyl ether (3X 5 mL of each) containing TEA (0.1%).
4. Dry these pieces under vacuum in a desiccator.
5. Take glass microslide **9** (one to two pieces) in a glass vial and charge it with reagent H (5 mL).
6. Shake the contents of the vial for 2 min and measure the absorbance of the solution at 498 nm using reagent H as a blank.

3.4.4. DTNPME Method

3.4.4.1. Determination of Sulfhydryl Groups on Glass Microslide

1. Treat three to four pieces of 3-mercaptopropylated glass microslide (~5 × 5 mm^2) with a solution of reagent E for 15 min at room temperature followed by washings with Milli-Q water (2X 10 mL) and 95% ethanol (2X 10 mL).
2. Dry the glass pieces and add reagent F (5 mL) under argon atmosphere.
3. Cap the reaction vessel tightly and agitate thoroughly at room temperature for 30 min.
4. Remove the supernatant, wash the glass pieces successively with DMF, methanol and diethyl ether (3X 5 mL of each) containing TEA (0.1%).
5. Dry these pieces under vacuum in a desiccator.
6. Take glass microslide **14** (one to two pieces) in a glass vial and charge it with reagent H (5 mL).
7. Shake the contents of the vial for 2 min and measure the absorbance of the solution at 498 nm using reagent H as a blank.

3.4.5. DTAH Method

3.4.5.1. Determination of Epoxide Functions on Glass Microslides

1. Treat 3–4 pieces of 3-glycidyloxypropylated microslide (~5 × 5 mm^2) with a solution of reagent G (1 mL).
2. Cap the reaction vessel tightly and place on a shaker for 12 h at room temperature.
3. Remove the supernatant, wash the glass pieces successively with EDC and diethyl ether (3X 5 mL of each) containing TEA (0.1%).

4. Dry these pieces and take one to two glass pieces **19** in a vial and charge with reagent H (5 mL).
5. Shake the contents of the vial for 2 min and measure the absorbance of the solution at 498 nm using reagent H as a blank.

3.4.5.2. DETERMINATION OF ALDEHYDE FUNCTIONS ON GLASS MICROSLIDES

1. Treat three to four pieces of aldehyde groups bearing glass microslide (~5 × 5 mm^2) with a solution of reagent G (1 mL).
2. Cap the reaction vial tightly and place it on a shaker for 12 h at room temperature.
3. Decant the supernatant, wash the glass pieces successively with DCE and diethyl ether (3X 5 mL of each) containing TEA (0.1%).
4. Dry these pieces and take one to two glass pieces **20** in a vial and charge with reagent H (5 mL).
5. Shake the contents of the vial for 2 min and measure the absorbance of the solution at 498 nm.

3.4.5.3. DETERMINATION OF CARBOXYL FUNCTIONS ON GLASS MICROSLIDES

1. Place three to four pieces of carboxyl group bearing glass microslide, BTCM (0.1 mmol) and DMAP (0.2 mmol) in a glass vial containing DMF (0.5 mL).
2. Add a solution of TPP (0.1 mmol), dissolved in DMF (0.5 mL), to the previously listed contents and swirl for 15–20 s.
3. Add reagent G (0.5 mL) and shake the contents for 30 min at room temperature.
4. Decant the supernatant and wash the pieces with DMF (2X 5 mL) and diethylether (2X 5 mL), respectively, containing TEA (0.1%).
5. Dry these pieces and treat one to two pieces **21** in a glass vial with reagent H (5 mL).
6. After 2 min, measure the absorbance of the solution at 498 nm.

3.5. Estimation of Density of Functional Groups on Polymer Supports

The amount of functional groups on the polymer support can be calculated using the following formula:

Using reagent H

$$\text{Amount of functional groups} \atop (\mu\text{mol/m}^2 \text{ of solid support}) = \frac{14.3 \times \text{Volume (mL)} \times \text{Absorbance at 498 nm}}{\text{Total area (both sides) of glass pieces (mm}^2)}$$

Using reagent I

$$\text{Amount of functional groups} \atop (\mu\text{mol/m}^2 \text{ of film}) = \frac{12.8 \times \text{Volume (mL)} \times \text{Absorbance at 505 nm}}{\text{Total area (both sides) of polymer film (mm}^2)}$$

3.6. Discussion

Estimation of functional groups present on polymer matrices is essentially an important parameter for many biological applications where these surfaces are

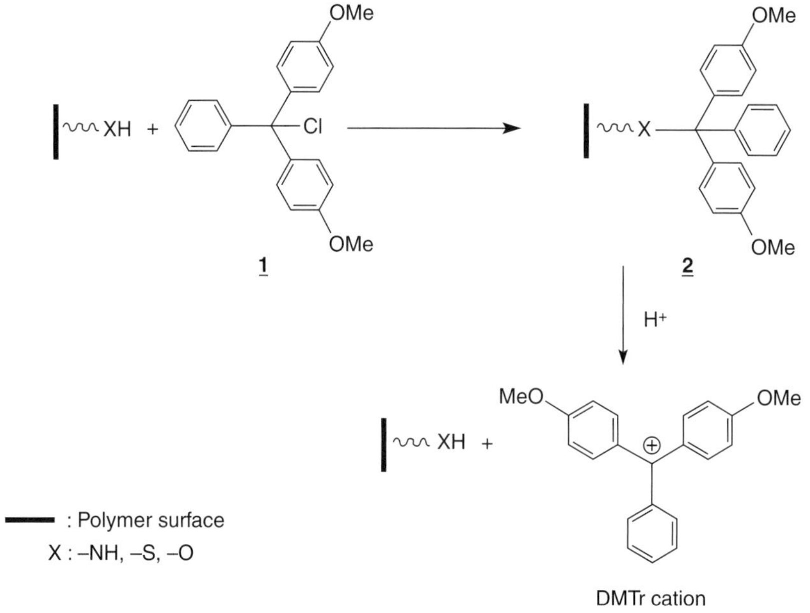

Scheme 1. Spectrophotometric determination of polymer supported functional groups (–NH$_2$, –SH, and –OH).

to be used for further functionalization. Because microarray technology deals with the use of nonporous materials with small surface area, there is a need to design sensitive methods for the estimation of low density of surface functional groups. Not much attention has been paid in this direction, therefore, only a few reagents are available for this purpose. The preparation of these reagents and estimation of functional group density using them are depicted in **Schemes 1–5**.

The first method (*see* **Subheading 3.4.1.**) is based on the reaction of polymer supported amino groups with an excess of DMTrCl in the presence of an efficient catalyst (Q-salt), tetrabutylammonium nitrate **(Scheme 1)**. The same method can be extended to estimate polymer supported mercaptoalkyl and hydroxyl groups. However, the method has some serious limitations, viz., (1) the reagent, DMTrCl, differs in reactivity toward different functionalities in the order –NH$_2$ >–SH >–OH, (2) moisture sensitive and acid labile nature of the reagent, it requires strict anhydrous conditions and careful handling, and (3) being nonspecific in nature, it cannot be used for the estimation of one functional group in the presence of another one.

The limitations of DMTr-Cl reagent have been overcome by developing a new universal reagent, DMPA, which has been used for estimation of polymer supported –NH$_2$, –SH, and –OH groups under microwave irradiation **(Scheme 2)**. The reaction completes in just 4 min and it is hardly affected by presence

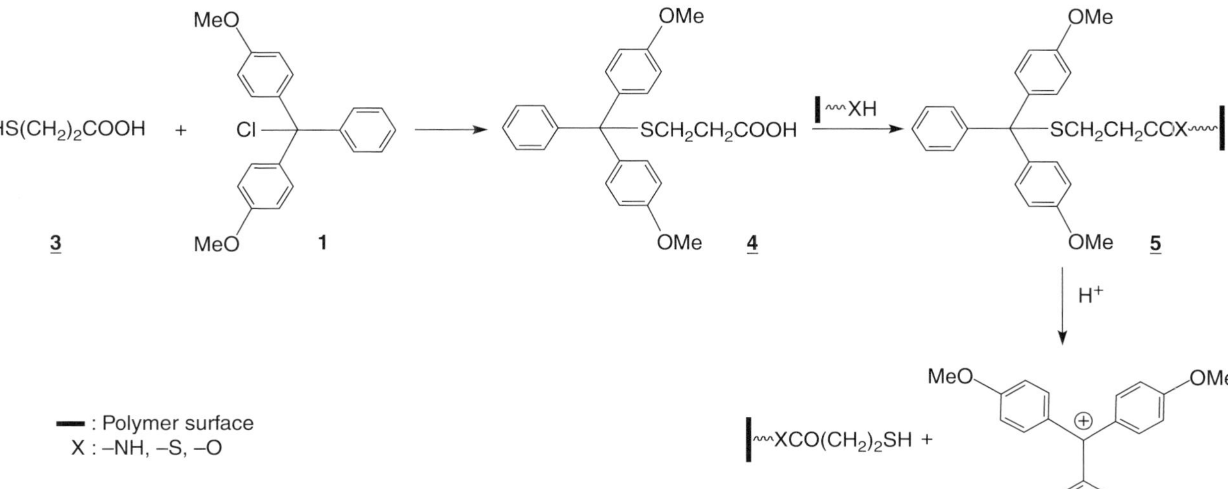

Scheme 2. Synthesis of universal reagent DMPA and estimation of polymer supported functional groups.

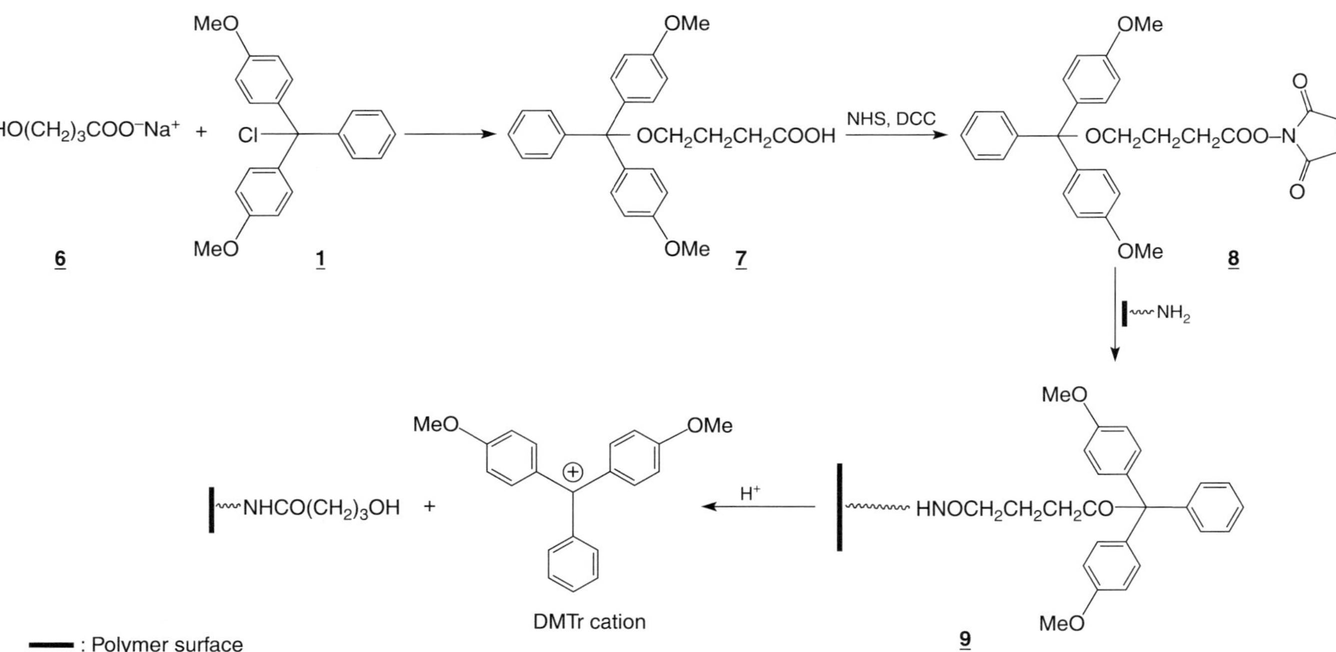

Scheme 3. Synthesis of reagent SDTB and spectrophotometric determination of polymer supported amino groups.

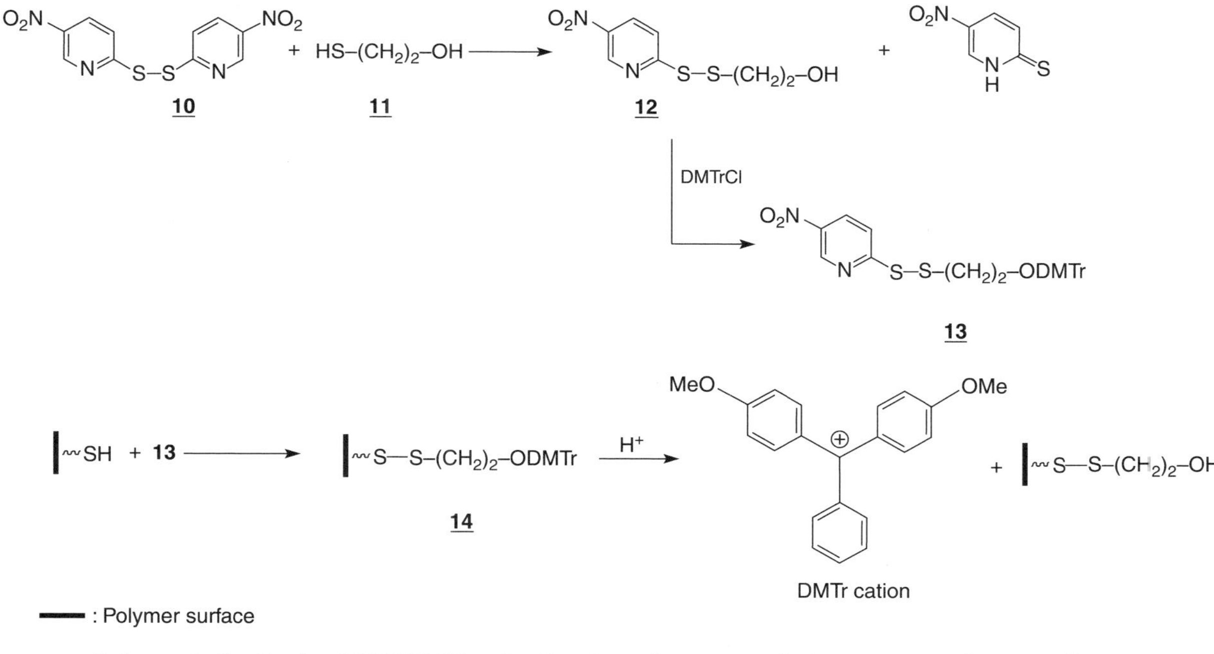

Scheme 4. Synthesis of DTNPME and estimation of mercaptoalkyl groups on polymer surface.

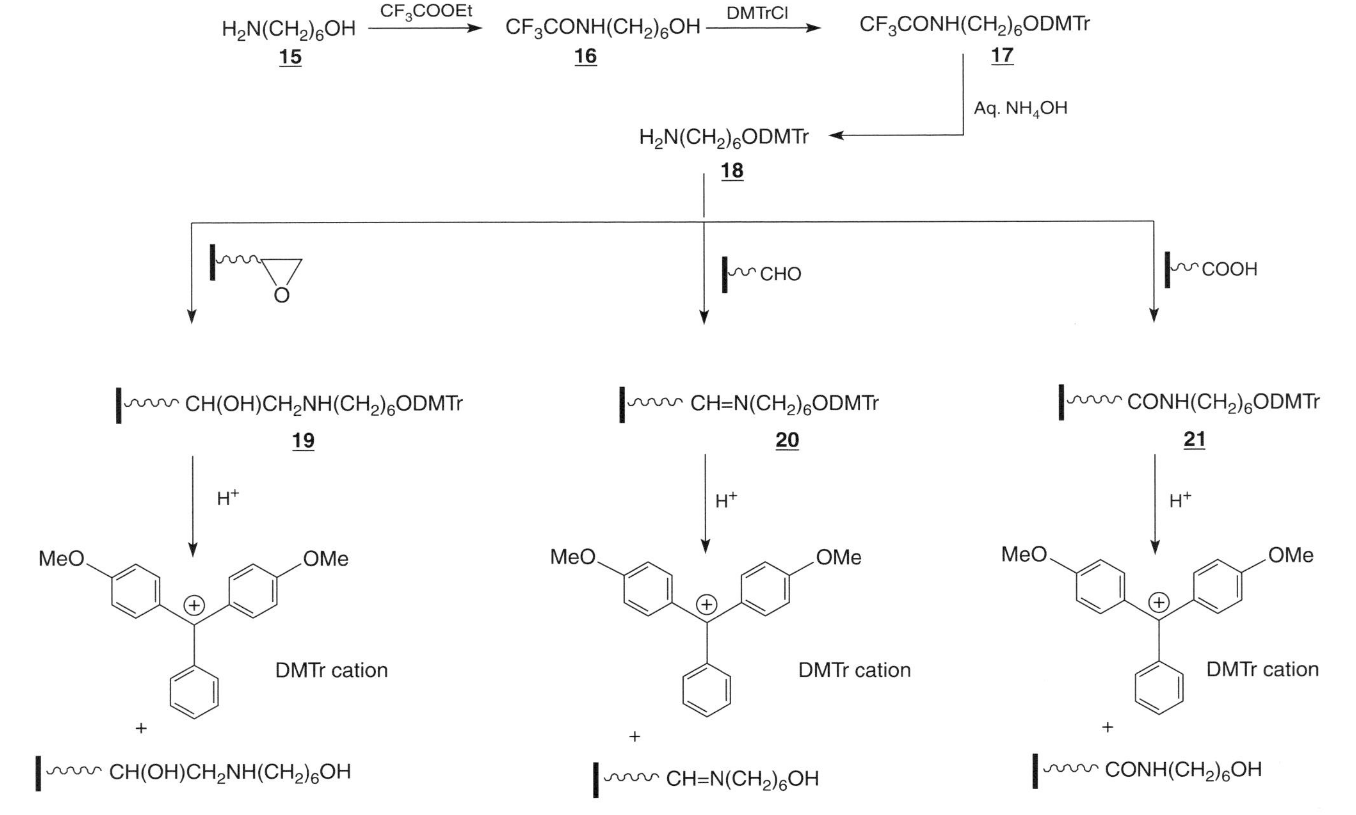

Scheme 5. Synthesis of reagent DTAH and estimation of polymer supported epoxy, aldehyde, and carboxyl groups.

of traces of moisture in the solvents and reagents. Therefore, laboratory grade solvents and reagents can be employed for the reactions. The extent of coupling can be estimated spectrophotometrically by monitoring the released DMTr cation on treatment of polymer surface with an acid.

The previously discussed reagents (DMTr-Cl and DMPA), though very sensitive, are nonspecific for the estimation of polymer supported functionalities. A novel reagent, SDTB, though based on 4,4′-dimethoxytrityl group, has been developed for estimation of polymer supported amino groups (**Scheme 3**). The main advantages of this reagent are that it can selectively be used for the estimation of amino groups in the presence of sulfhydryl and hydroxyl groups. Moreover, the reagent is not affected by the presence of traces of moisture and therefore, allows one to use laboratory grade reagents and is sufficiently stable on storage.

Similarly, a reagent, DTNPME, has been reported (**Scheme 4**) for the estimation of sulfhydryl group selectively on polymer surfaces in the presence of other functional groups (e.g., aminoalkyl or hydroxyalkyl). The reagent possesses good stability in organic solvents over a period of time.

A new reagent has also been designed and developed for estimating polymer supported epoxy, aldehyde, and carboxyl groups. The preparation of the reagent, DTAH, is shown in **Scheme 5**. This is for the first time, a reagent has been developed to determine the surface density of these functionalities. The method involves simple chemical reactions with three different functionalities, through (1) epoxide ring opening by nucleophilic amino groups, (2) condensation of aldehyde with amino groups of the reagent (Schiff's base), and (3) coupling of amino groups with polymer supported carboxyl functionalities in the presence of oxidation-reduction condensation reagent.

The reagents described in this chapter have been selected carefully, which can estimate the functional groups density on the polymer surface available for further functionalization of these supports. Therefore, the number of functional groups determined by the proposed reagents would be an important parameter for calculating the amount of the ligands (biomolecules) required for immobilization.

4. Notes

1. The following reagents may be used as purchased; however, they must be stored in tightly sealed bottles to avoid undue exposure to moisture; APTS, MPTS, GOPTS, trimethylchlorosilane, and DMTr-Cl.
2. This treatment hydrolyzes the strained Si-O-Si bonds and removes adsorbed gases or foreign materials so that a large number of Si-OH groups are available for functionalization process.
3. Avoid over-exposure of mercaptoalkylated polymer supports to air as it generates disulfide linkages. Always store these supports under N_2 or Ar atmosphere. To break these disulfides, treat the support material with ethanolic dithiothreitol

solution (50 m*M*) for 30 min at room temperature followed by washing with ethanol and drying.
4. Avoid exposure of epoxy glass plates to amines.

References

1. Drobyshev, A. N., Mologina, N., Shick, V., Pobedimskaya, D., Yershov, G., and Mirzabekov, A. D. (1997) Sequence analysis by hybridization with oligonucleotide microchip: identification of β-thalassemia mutations. *Gene* **188,** 45–52.
2. Hacia, J. G., Brody, L. C., Chee, M. S., Fodor, S. P. A., and Collins, F. S. (1996) Detection of heterozygous mutations in *BRCA1* using high density oligonucleotide arrays and two–color fluorescence analysis. *Nat. Genet.* **14,** 441–447.
3. Yershov, G., Barsky, V., Belgovskiy, A., et al. (1996) DNA analysis and diagnostics on oligonucleotide microchips. *Proc. Natl. Acad. Sci. USA* **93,** 4913–4918.
4. Cho, R. J., Fromont-Racine, M., Wodicka, L., et al. (1998) Parallel analysis of genetic selections using whole genome oligonucleotide arrays. *Proc. Natl. Acad. Sci. USA* **95,** 3752–3757.
5. Schena, M. (ed.) (2000) *Microarray Biochip Technology*, Eaton Publishing, Natick, MA.
6. Beier, M. and Hoheisel, J. D. (1999) Versatile derivatisation of solid support media for covalent binding on DNA-microchips. *Nucleic Acids Res.* **27,** 1970–1977.
7. Proudnikov, D., Timofeev, E., and Mirzabekov, A. D. (1998) Immobilization of DNA in polyacrylamide gel for the manufacture of DNA and DNA–oligonucleotide microchips. *Anal. Biochem.* **259,** 34–41.
8. Rehman, F. N., Audeh, M., Abrams, E. S., et al. (1999) Immobilization of acrylamide-modified oligonucleotides by co-polymerization. *Nucleic Acids Res.* **27,** 649–655.
9. Raddatz, S., Mueller-Ibeler, J., Kluge, J., et al. (2002) Hydrazide oligonucleotides: new chemical modification for chip array attachment and conjugation. *Nucleic Acids Res.* **30,** 4793–4802.
10. Kumar, A., Larsson, O., Parodi, D., and Liang, Z. (2000) Silanized nucleic acids: a general platform for DNA immobilization. *Nucleic Acids Res.* **28,** E71.
11. Joos, B., Kuster, H., and Cone, R. (1997) Covalent attachment of hybridizable oligonucleotides to glass supports. *Anal. Biochem.* **247,** 96–101.
12. Rasmussen, S. R., Larsen, M. R., and Rasmussen, S. E. (1991) Covalent immobilization of DNA onto polystyrene microwells: the molecules are only bound at the 5′-end. *Anal. Biochem.* **198,** 138–142.
13. Bader, R., Hinz, M., Schu, B., and Seliger, H. (1997) Oligonucleotide microsynthesis of a 200-mer and of one dimensional arrays on a surface of hydroxylated polypropylene tape. *Nucleosides and Nucleotides* **16,** 829–833.
14. Bradley, A., Cai, W. W., and Marathi, U. (2002) PCT Int. Appl. WO 2002 092615 A2 21 Nov. 2002, *Chem. Abstr.* **137,** 366004.
15. Belosludtsev, Y., Iverson, B., Lemeshko, S., et al. (2001) DNA microarrays based on noncovalent oligonucleotide attachment and hybridization in two dimensions. *Anal. Biochem.* **292,** 250–256.

16. Broude, N. E., Woodward, K., Cavallo, R., Cantor, C. R., and Englert, D. (2001) DNA microarrays with stem-loop DNA probes: preparation and applications. *Nucleic Acids Res.* **29,** E92.

17. Podyminogin, M. A., Lukhtanov, E. A., and Reed, M. W. (2001) Attachment of benzaldehyde-modified oligonucleotide probes to semicarbazide-coated glass. *Nucleic Acids Res.* **29,** 5090–5098.

18. Dombi, K. L., Griesang, N., and Richert, C. (2002) Oligonucleotide arrays from aldehyde-bearing glass with coated background. *Synthesis* **6,** 816–824.

19. Yamamoto, N., Okamoto, H., Suzuki, T., and Shimizu, A. (2002) Jpn. Kokai Tokkyo Koho JP 2002 065274 A2 5 Mar. 2002, *Chem. Abstr.,* **136,** 229064.

20. Kumar, P., Agarwal, S. K., and Gupta, K. C. (2004) N-(3-trifluoroethanesulfonyloxy-propyl)anthraquinone-2-carboxamide (NTPAC): a new heterobifunctional reagent for immobilization of biomolecules on a variety of polymer surfaces. *Bioconjugate Chem.* **15,** 7–11.

21. Kumar, P., Gupta, K. C., and Gandhi, R. P. (2003) UV light-aided immobilization of oligonucleotides on glass surface using N-(3-trifluoroethanesulfonyloxypropyl)-anthraquinone-2-carboxamide (NTPAC) and detection of single nucleotide mismatches. *J. Ind. Chem. Soc.* **80,** 1193–1199.

22. Kumar, P., Agarwal, S. K., Misra, A., and Gupta, K. C. (2004) A new heterobifunctional reagent for immobilization of biomolecules on glass surface. *BioMed. Chem. Lett.* **14,** 1097–1099.

23. Kumar, P., Choithani, J., and Gupta, K. C. (2004) Construction of oligonucleotide arrays on a glass surface using a heterobifunctional reagent, N-(2-trifluoroethanesulfonatoethyl)-N-(methyl)-triethoxysilylpropyl-3-amine (NTMTA). *Nucleic Acids Res.* **32,** E80.

24. Riepl, M., Enander, K., Liedberg, B., Schiiferling, M., Krushina, M., and Ortigao, F. (2002) Functionalized surfaces of mixed alkanethiols on gold as a platform for oligonucleotide microarrays. *Langmuir* **18,** 7016–7123.

25. Prokein, T. and Seliger, H. (2002) Dithiolane derivatives for immobilizing biomolecules on noble metals and semiconductors. DE 10251229 (*Ger. Pat.*)

26. Rogers, Y. -H., Jiang-Baucom, P., Huang, Z. -J., Bogbanov, V., Anderson, S., and Boyce-Jacino, M. T. (1999) Immobilization of oligonucleotides onto glass support via disulfide bonds: A method for preparation of DNA microarrays. *Anal. Biochem.* **266,** 23–30.

27. Defrancq, E., Hoang, A., Vinet, F., and Dumy, P. (2003) Oxime bond formation for the covalent attachment of oligonucleotides on glass support. *BioMed. Chem. Lett.* **13,** 2683–2686.

28. Jiang, D., Song, L., Wang, D., and Yuan, C. (2002) Preparation of high-efficiently hybridization DNAchip on glass support. *Chem. Abstr.* **136,** 321400 C; Jinan Dexue Xuebao, Ziran Kexul Yu Yixue ban (2001) **22,** 65–70 (Ch.).

29. Prudnikov, D., Timofeev, E., and Mirzabekov, A. (1998) Immobilization of DNA in polyacrylamide gel for the manufacture of DNA and DNA—ligonucleotide microchips. *Anal. Biochem.* **259,** 34–41.

30. Grabar, K. C., Freeman, R. G., Hommer, M. B., and Natan, M. J. (1995) Preparation and characterization of Au colloid monolayers. *Anal. Chem.* **67,** 735–743.

31. Sigrist, H., Gao, H., and Wegmuller, B. (1992) Light-dependent, covalent immobilization of biomolecules on inert surfaces. *Biotechnology* **10,** 1026–1028.
32. Kumar, P. and Gupta, K. C. (2003) A rapid method for the construction of oligonucleotide microarrays. *Bioconjugate Chem.* **14,** 507–512.
33. Zbinden, R. (ed.) (1964) *Infrared Spectroscopy of High Polymers.* Academic Press, New York.
34. Henniker, J. C. (ed.) (1967) *Infrared Spectroscopy of Industrial Polymers.* Academic Press, New York.
35. Bovey, F. A. (ed.) (1972) *High Resolution NMR of Macromolecules.* Academic Press, New York.
36. Camps, F., Castells, J., Font, J., and Vela, F. (1971) Organic syntheses with functionalized polymers: II. Wittig reaction with polystyryl-*p*-diphenylphosphoranes. *Tetrahedron Lett.* **12,** 1715–1716.
37. Helfferisch, F. (ed.) (1962) *Ion-Exchange.* McGraw Hill Press, New York.
38. Troll, W. and Cannan, R. (1953) A modified photometric ninhydrin method for the analysis of amino and imino acids. *J. Biol. Chem.* **200,** 803–811.
39. Cuatrecasas, P. (1970) Protein purification by affinity chromatography. Derivatizations of agarose and polyacrylamide beads. *J. Biol. Chem.* **245,** 3059–3065.
40. Gaur, R. K. and Gupta, K. C. (1989) A spectrophotometric method for the estimation of amino groups on polymer supports. *Anal. Biochem.* **180,** 253–258.
41. *Applications Handbook and Catalog* (2003–2004), Pierce Biotechnology, Inc., IL.
42. Damha, M. J., Giannaris, P. A., and Zabarylo, S. V. (1990) An improved procedure for derivatization of controlled-pore glass beads for solid-phase oligonucleotide synthesis. *Nucleic Acids Res.* **18,** 3813–3821.
43. Gaur, R. K., Sharma, P., and Gupta, K. C. (1989) 4,4′-Dimethyloxytrityl chloride: a reagent for the spectrophotometric determination of polymer-supported amino groups. *Analyst* **114,** 1147–1150.
44. Rao, N. S., Agrawal, S. K., Chauhan, V. K., et al. (2000) Microwave-assisted spectrophotometric estimation of polymer-supported functional groups using a universal reagent. *Anal. Chimica. Act* **405,** 247–254.
45. Markiewicz, W. T. and Wyrzykiewicz, T. K. (1989) Universal solid supports for the synthesis of oligonucleotides with terminal 3′-phosphates. *Nucleic Acids Res.* **17,** 7149–7158.
46. Ngo, T. T. (1986) A simple spectrophotometric determination of solid supported amino groups. *J. Biochem. Biophys. Methods* **12,** 349–354.
47. Sharma, P., Sathyanarayana, S., Kumar, P., and Gupta, K. C. (1990) A spectrophotometric method for the estimation of polymer-supported sulfhydryl groups. *Anal. Biochem.* **189,** 173–177.
48. Swatditat, A. and Tsen, C. C. (1972) Determining simple sulfhydryl compounds (low molecular weight) and their contents in biological samples by using 2,2′-dithiobis-(5-nitropyridine). *Anal. Biochem.* **45,** 349–356.
49. Ellman, G. L. (1959) Tissue sulfhydryl groups. *Arch. Biochem. Biophys.* **82,** 70–77.
50. Bhatia, D. (1996) Thesis submitted to University of Delhi, Delhi, India.

51. Riordan, J. F., Sokolovsky, M., and Vallee, B. L. (1967) Environmentally sensitive tyrosyl residues. Nitration with tetranitromethane. *Biochemistry* **6,** 358–361.
52. Mahajan, S., Garg, A., Goel, M., Kumar, P., and Gupta, K. C. (2006) Spectrophotometric estimation of functional groups on microslides for preparation of biochips. *Anal. Biochem.* **351,** 273–281.
53. Reddy, M. P., Rampal, J. B., and Beaucage, S. L. (1987) An efficient procedure for the solid phase tritylation of nucleosides and nucleotides. *Tetrahedron Lett.* **28,** 23–26.
54. Gaur, R. K., Paliwal, S., Sharma, P., and Gupta, K. C. (1989) A simple and sensitive spectrophotometric method for the quantitative determination of solid supported amino groups. *J. Biochem. Biophys. Methods* **18,** 323–329.

9

Methods in High-Resolution, Array-Based Comparative Genomic Hybridization

Mark R. McCormick, Rebecca R. Selzer, and Todd A. Richmond

Summary

A method of high resolution, array-based comparative genomic hybridization is described for the mapping of copy-number changes associated with chromosomal amplifications, deletions, and translocations. The method involves the design of whole-genome or targeted, fine-tiling arrays for synthesis on a high-density digital microarray-synthesis platform. The arrays can span entire eukaryotic genomes or be targeted to specific chromosomal regions for high-resolution identification of copy-number changes and the corresponding breakpoint locations. The methods described include the bioinformatics required for array design, and the protocols for DNA fragmentation, dual-color labeling, microarray hybridization, and array scanning. The processes for data extraction, normalization, and segmentation analysis are also described.

Key Words: Amplification; breakpoint; cancer; CGH, comparative genomic hybridization; copy number; deletion; DNA microarray; segmentation; translocation.

1. Introduction

NimbleGen (NimbleGen Systems, Madison, WI) array-based comparative genomic hybridization (CGH) platform is a significant advance in the art of mapping genomic copy-number changes. It can effectively replace metaphase chromosomal hybridization techniques, with greatly enhanced resolution when compared with other array-based CGH methods. The detection of copy-number changes is a rapidly emerging technique in the effort to elucidate the genomic mechanisms of a range of human diseases. Areas of current focus include cancer, autism and congenital defects, and psychiatric disorders. aCGH can effectively map genome-wide copy-number changes associated with a wide range of genomic aberrations including: (1) aneuploidy, (2) HSR amplifications, (3) distributed insertional amplifications, (4) double-minute amplifications, (5) homozygous or

From: *Methods in Molecular Biology, vol. 381: Microarrays: Second Edition: Volume 1: Synthesis Methods*
Edited by: J. B. Rampal © Humana Press Inc., Totowa, NJ

hemizygous deletions, (6) unbalanced (nonreciprocal) translocations, and (7) segmental duplications.

High-resolution mapping of copy-number polymorphisms has a wide range of research applications including: (1) mapping and identifying oncogene and tumor suppressor regions and genes, (2) cohort studies of cancer genomes, (3) assessing tumor progression, (4) assessing genomic status of cultured cell lines, (5) mapping copy-number polymorphisms associated with human diseases, and (6) screening deletion mutants in model organisms.

This chapter will present methodology for aCGH, which has been specifically developed for high-sensitivity detection of DNA copy-number changes in genomic DNA using NimbleGen high-density DNA microarrays. The topics covered include the bioinformatics requirements for array design and data analysis, and also the laboratory techniques for sample labeling and hybridization. One advantage of the protocols described here is that they allow for direct labeling and hybridization of complex genomic DNA samples without any requirement for genomic representation techniques or whole genome amplification methods. This greatly simplifies the process of sample preparation.

1.1. Background

The application of DNA microarrays to the mapping of copy-number changes is an enhancement of earlier methods of comparative genomic hybridization. The aim is to determine the copy-number differences between test- and reference genomic DNA samples. aCGH provides the means for high-resolution mapping of the locations of insertions and deletions of DNA segments in large eukaryotic genomes. One of the goals of these analyses is the identification of genome rearrangements that are signature events in oncogenesis and also implicated in a broad range of other human diseases.

The method relies on competitive hybridization of two dye-labeled samples to a DNA microarray in which one sample is test genomic DNA and the other is a pooled or matched reference sample. By hybridizing both samples to a single array containing probes tiling either the whole genome or focused tiling at higher resolution in target regions, the genomic locations of copy-number differences are readily identified by changes in the ratio of the signals of one dye to the other.

The advent of high-density DNA microarrays has greatly improved the reliability and precision of aCGH by providing the means to tile the entire human genome at a sufficient average probe spacing to allow the mapping of DNA breakpoints to within 25 kbp or less. NimbleGen has developed an array synthesis technology that combines photomediated oligonucleotide synthesis with digital array design to allow for on-demand, limited production run array synthesis without excessive design costs or long-lead times *(1)*. The advantage stems from the use of a digital micromirror array as the method for spatially addressing oligonucleotide synthesis. In the synthesis process, incoming phosphoramidites have a photolabile protecting group attached to their 5′-end to prevent unwanted coupling until it is desired. When a base addition is called for at a specific location in the array, ultraviolet (UV) light is projected at this location to cleave the photolabile group, exposing a hydroxyl group. Although photolithographic approaches have the potential to exceed 385,000 probes in a given array *(2)*, the cost for manufacture of the chromium masks required for spatially addressed light projection must be recovered, contributing to prohibitive costs for limited production runs and significant delays in array availability. In the NimbleGen synthesis process, the use of a digital micromirror array obviates the need for more laborious array manufacturing processes requiring chromium masks, reducing the cost for array design and eliminating any lags in array manufacture. The flexibility in synthesis makes the manufacture of limited production run arrays economically feasible. In addition, the timely availability of *de novo* array designs allows iterative array experimentation in which the results of a given aCGH experiment can be further characterized by fine-tiling arrays in following experiments. Alternative methods for array synthesis, utilizing ink-jet printed phosphoramidites, have been successfully applied to aCGH *(3)*. The main drawback to this approach is that, individual features are >100 µm and the maximum number of features/array is limited to about 45,000.

The NimbleGen technology platform currently produces arrays with a feature capacity of 385,000 probes in a 13×17 mm^2 area. At this probe capacity, the median probe spacing is 6 kbp when tiling the entire, repeat-masked human genome. In addition to standard array designs covering the entire genome, researchers can specify chromosomal regions of interest for focused fine-tiling with the potential for average probe spacing as low as 1 bp. The advantages of a high-density array platform become obvious when trying to tile large genomic regions at high resolution. For example, the entire repeat-masked regions of the human genome were recently tiled on this high density platform at 100 bp spacing in a 38-array set. Achieving this same tiling density on an array format with 45,000 probes would require a 325-array set.

1.2. Important CGH Array Considerations

There are several important array characteristics that greatly enhance performance in CGH. Because this application seeks to measure slight variations in genomic copy-number, the differential level of signal can be difficult to detect unless the probe set, probe characteristics, and sample labeling protocol have been optimized to maximize the signal-to-noise ratio. For example, the detection of single copy amplifications requires the method to differentiate three copies of a DNA fragment in the test sample from two copies in the reference.

In the process of developing aCGH on the NimbleGen platform, several key variables were identified and optimized.

1.2.1. Oligo Length

A comparison of probes of varying length from 25- to 70-mer quickly established that longer oligos are superior to shorter ones in signal-to-noise ratio. However, the probes incorporated in the best-performing arrays were selected not based on length but melting temperature. Standard array designs now incorporate probes that are lengthened or shortened to compensate for shifts in G:C content, to obtain probes whose T_m is as closely matched to 76°C as possible. T_m-matched probes provide better signal uniformity with reduced signal loss and nonspecific binding.

1.2.2. Probe Selection

There are two approaches to the selection of probes for aCGH. One approach is the gene-centric selection of probes derived only from coding regions within a given genome. The advantages of this approach are that the probes selected for expression arrays with known performance characteristics can be incorporated into the CGH array and any identified copy-number changes will necessarily fall within coding regions. Thus identified insertions or deletions provide immediate leads on target disease genes for further study. However, the risk of this approach is that large sections of a given genome will not be represented in the CGH array and the wider distribution of probe spacing can result in uneven probe coverage. Another deficiency of this approach is that it can incorporate only known or putative genes as described in the current annotation. Therefore, it will not sample any regions of potential interest where genes or other functional elements might exist but have not been annotated. The NimbleGen whole-genome array design tiles probes evenly throughout the nonrepetitive regions of the human genome providing coverage of coding and noncoding regions.

1.2.3. Strand Selection

A careful comparison of the results of aCGH with array designs derived from either one or both strands on genomic DNA in a tiled region revealed that probe selection from both strands provides no benefit in terms of the sensitivity of the method for the detection of single copy changes. To maximize the tiling range and minimize the average probe spacing, NimbleGen aCGH array designs are tiled on one strand only without redundant probes derived from the complementary strand.

1.3. Array Design Approaches for CGH

There are two basic formats for CGH using the NimbleGen array platform.

1.3.1. Whole Genome Tiling

This format apportions the available 385,000-probe set in an evenly distributed manner across the nonrepetitive regions of the target genome. As of this writing, designs are available for human, mouse, *Caenorhabditis elegans*, rat, yeast, dog, and *Drosophila* with whole-genome designs anticipated for the full range of available model genomes in the near future.

1.3.2. Fine-Tiling

In this format, probes are distributed across the nonrepetitive regions of researcher-specified subsets of the target genome at high probe density. Tiling probe placement can be as focused as every 1 bp, if desired. Fine-tiling is the preferred method for high-resolution mapping of amplification and deletion breakpoints.

In either approach, the resolution of the method is dependent on the DNA copy-number change and the size of the amplification or deletion. The anticipated resolution is estimated to be approximately four times the average probe interval, thus a whole human genome design with an average spacing of 6 kbp would be expected to map breakpoints to within 25 kbp. One strategy for CGH that takes advantage of the design flexibility of this array synthesis technology is to perform an initial survey to identify regions, where copy-number changes exist in the test genome with the whole genome array, then perform fine-tiling within these regions to fine-map the breakpoints to within the range, where they can be confirmed and screened in other samples by standard methods such as PCR and DNA sequencing.

1.4. Whole Genome Array Design: Bioinformatics

The process for designing probe sets for whole genome CGH array designs is a multistep process that is performed by NimbleGen Bioinformatics to select an optimized, unique set of probes for genome-wide CGH. The following sections describe the process in general terms.

1.4.1. Generate Initial Probe Sets by Tiling

The initial pool of probes for whole-genome designs is generated by selecting a small number of probes at each of a large number of points throughout the genome. Two interval parameters are used:

1. The primary interval specifies the spacing of the clusters of probes.
2. The secondary interval specifies the spacing for the individual probes within each cluster.

These intervals depend on the size of the genome and the amount of repeat sequences and sequence ambiguities present. NimbleGen's CGH designs do not include coverage of repeat regions, as they are uninformative for determining DNA copy-number changes. The genome is surveyed for the presences of repeat sequences and ambiguities and the remaining sequence is used to estimate the interval spacing of the clusters, based on the constraint of 385,000 oligos. The secondary interval is generally set to 100- to 200-fold less than the primary interval for large intervals, though it is rarely set to <10 bp. The estimated primary interval is generally bracketed on both sides by 5–20% to ensure that the target number of clusters is found. At this step, the goal is to end up with 400,000–410,000 clusters so that some filtering can be done at the end of the process, to ensure that only high quality oligos are used in the final design.

The oligos that are selected in each cluster are not of fixed length, but rather are of variable length adjusted to match a target melting temperature of 76°C using the following formula: $T_m = 81.5 + 16.6 \times \{\log_{10}[(Na^+)]\} + 0.41 \times (GC\%) - (600/\text{length})$ *(4)*. There are more-sophisticated T_m-algorithms available, but the primary goal for this step is to balance the GC composition of the oligo, rather than to provide a highly accurate T_m-estimate for the oligo.

The only other constraint placed on probe selection at this point is a restriction placed on the number of synthesis cycles required to manufacture the oligo. Given a defined synthesis order of ACGT, the number of synthesis cycles to manufacture an oligo of N residues can range from N (for $[ACGT]_N$) to N × 4 (for A_N, C_N, G_N, T_N). For oligos of defined length, the normal synthesis cycle threshold is generally N × 3. So for a 36-mer oligo, the normal synthesis cycle threshold is 108 cycles. For designs with variable length oligos, the synthesis cycle threshold is set to 148 cycles. Probe lengths are constrained to be a minimum of 50–75 bp and maximum of 85 bp, and they must meet the synthesis cycle requirements.

1.4.2. 15-Mer Frequency Determination

After generating the oligos by tiling, a measure of low sequence complexity for each oligo is determined. This is done by determining the average 15-mer frequency of each oligo. A window 15 bp in length is moved along the length of the oligo and the frequency of that 15-mer in the genome is determined. For convenience, the 15-mers of both strands of the genome of interest are loaded into hash-table structure for fast look up. The average of the 15-mer frequencies for the oligo is used as a component in the final probe selection (*see* **Subheading 1.4.4.**).

1.4.3. Oligo Uniqueness

After the 15-mer frequency determination, the uniqueness of each full-length oligo in the genome is determined. There are a number of applications available (BLAST, BLAT, FASTA) for sequence comparison. The application that NimbleGen uses is SSAHA *(5)*; an excellent program for quickly determining nearly identical matches, even with large genomes. SSAHA is run using a word size of 12 and a step size of 1, which allows for maximum speed and sensitivity for our purposes. The minimum match parameter is set to 38, the length of the shortest allowable oligo minus the word length. Thus, any match of 38 bp or greater is reported as a match in the genome, regardless of the length of oligo. For the final probe selection, only oligos with a single match are considered for inclusion in the final design. There are more sophisticated measures for uniqueness available, including methods that NimbleGen has developed for selecting oligos for expression arrays. Unfortunately, when dealing with large genomes, the computation time required for more sophisticated measures tends to scale prohibitively.

1.4.4. Probe Selection

After determining the various uniqueness measures, the probe data is loaded into a database for storage and querying. The final probe set is selected by scoring the unique probes within a given probe cluster, and retaining the probe with the highest score. The probe score is based on basepair composition and 15-mer frequency. Starting with an initial score of 1000, a weight penalty of $-50 \times \log_2$ (15-mer frequency) is added. Then, basepair composition penalties are added. Homopolymer C and G stretches of four residues or longer add a penalty of $-100 \times$ homopolymer length. Homopolymer A and T stretches of six residues or longer add a penalty of $-50 \times$ homopolymer length.

After scoring and selection, there is generally an excess of probes because of the oversampling in the initial probe generation (*see* **Subheading 1.4.1.**). Additional filtering is done to bring the number down to the final count of 385,000 probes. First probes with an average 15-mer frequency >50 are eliminated. Then, probes with basepair composition penalties are removed. If there are still too many probes, those probes with the largest deviation from the target T_m are eliminated.

1.4.5. Fine-Tiling Array Design: Bioinformatics

In contrast to whole genome designs, fine-tiling designs are a single-step process. Because fine-tiling designs often focus on relatively small regions, it is often possible to provide 100% coverage of the regions of interest. Thus, little filtering of probes is done at the design stage, although some filtering of probes might be conducted during data analysis. The same application used to generate probes for the whole-genome designs is used to create the probes for fine-tiling designs. Instead of picking probes in clusters, probes are selected at defined interval spacing. Repeat-masked regions and sequence with ambiguities are avoided, and synthesis cycle restrictions are imposed, as for the whole genome designs. Oligo uniqueness is determined as described (*see* **Subheading 1.4.3.**), but depending on the project requirements, filtering by uniqueness might not be done until the data analysis step.

1.5. Preparation of Array Synthesis Masks

The layout of probes within the microarray and generation of the corresponding digital synthesis masks are performed using the proprietary ArrayScribe
software platform (*6*). The software incorporates user-defined probe sets, along
with control and alignment probes, into a set of array manufacturing masks that
are used in the array synthesis process. A demonstration version of the software
is available for download that allows for the preparation of array designs based
on researcher-specified probe sets.

1.6. Accessing aCGH Microarrays

Array manufacture is performed by NimbleGen Systems in their array
manufacturing facility. Arrays are synthesized for use in three primary modes:

1. NimbleGen Service—Researchers can access NimbleGen array technology through
   the NimbleGen microarray services laboratory in Reykjavik, Iceland. Using this
   approach, researchers ship their samples to NimbleGen and the results of their
   aCGH study are returned to them when the analysis is complete.
2. Microarray shipment—Researchers at US Government laboratories and in certain
   countries might opt to have arrays synthesized and shipped to their lab to conduct
   their own aCGH research. Please contact NimbleGen for current information
   regarding array availability.
3. NimbleGen Direct—In this mode, researchers direct the research conducted at the
   NimbleGen microarray services laboratory. Researchers might opt to use the
   trained technical staff of the facility or provide their own personnel.

2. Materials

2.1. Safety Information

1. Wear gloves and take precautions to avoid sample cross-contamination.
2. Purification of samples, apart from ethanol precipitation, is not necessary after the
   labeling reaction.
3. Cy dyes are light sensitive. Be sure to minimize light exposure of the dyes during
   use and store stocks in the dark immediately after use.
4. It was found that the use of commercial deionized water and dithiothreitol (DTT)
   for all posthybridization washes beneficial in preserving Cy5 fluorescence.

2.2. Materials and Equipment

There are several key steps in the labeling and hybridization protocols in which
the protocol steps have been optimized for the indicated equipment or reagents or
for which adequate substitutes have not been evaluated or do not exist. Although
substitutes are possible for the more generic equipment and reagents, the items in
the equipment and reagents list **indicated in bold** should not be substituted. Items
indicated in italics should only be substituted with caution.

- SignalMap, ArrayScribe, and NimbleScan are trademarks of NimbleGen Systems Inc.
- MAUI® is a registered trademark of BioMicro Systems, Inc., Salt Lake City, UT.
- Sonifier is a registered trademark of Branson Ultrasonics Corporation, Danburg, CT.
- GenePix is a registered trademark of Molecular Devices Corporation, Sunnyvale, CA.

2.2.1. Equipment

1. Thermocycler equipped with heated lid (MJ Research, Hercules, CA).
2. Microcentrifuge with 12,000g capability.
3. Heat block(s) at 95°C.
4. NanoDrop ND-1000 Spectrophotometer (NanoDrop Technologies, Wilmington, DE).
5. *Branson Digital Sonifier®* (model no. 450) equipped with microtip (part no. 101-148-062).
6. Electrophoresis apparatus and UV light box.
7. Speed Vac.
8. **Axon GenePix® 4000B Scanner (Molecular Devices).**
9. **MAUI Hybridization System (BioMicro Systems, model no. 02-A002-02).**
10. **MAUI Mixer SL Low Temp. Hybridization Chamber Lids (BioMicro Systems, cat. no. 02-A008-02).**
11. Method of drying arrays after washing (e.g., custom NimbleGen Array-Go-Round unit, slide centrifuge or argon gun).
12. 300-mL Staining dish (VWR Scientific, cat. no. 25608-904).

2.2.2. Reagents

1. **DI water (VWR Scientific, Westchester, PA; cat. no. RC91505).**
2. Cy3, Cy5 5′ Modified Random 9-mers (Tri-Link BioTechnologies, San Diego, CA).
3. 100 mM dNTPs (Invitrogen, Carlsbad, CA; cat. no. 10297-018).
4. 10X TE: 100 mM Tris-HCl, pH 8.0 10 mM EDTA.
5. 1 M Tris-HCl (Sigma-Aldrich Chemical, St. Louis; MO; cat. no. T-2663).
6. 1 $MgCl_2$.
7. 0.1 M DTT, prepared in VWR water and stored in frozen state.
8. *Klenow Fragment (3′→5′ exo-) 50,000 U/mL (New England Biolabs, cat. no. M0212M).*
9. 0.5 M EDTA (Sigma-Aldrich Chemical, cat. no. E-7889).
10. 20X SSC (0.3 M sodium citrate, pH 7.0, 3 M NaCI) (Sigma-Aldrich Chemical, cat. no. S-6639).
11. 10% Sodium dodecyl sulfate (Sigma-Aldrich Chemical, cat. no. L-4522).
12. 10X TAE (0.4 M Tris-agetate, pH 8.3, 0.01 M EDTA) electrophoresis buffer.
13. Agarose.
14. DNA molecular weight markers, 1–12 kbp.
15. **NimbleGen Hybridization Buffer A.**
16. **NimbleGen Hybridization Buffer B.**
17. 0.1 M DTT (Sigma-Aldrich Chemical, cat. no. 150460).
18. 80% Ethanol (stored in freezer).
19. Ethanol (stored in freezer).
20. 5 M NaCl.
21. Isopropanol.

3. Methods

This procedure describes methods for preparing eukaryotic DNA samples for aCGH analysis and hybridization to NimbleGen System's CGH arrays. The method involves shearing unamplified, genomic DNA by sonication and then using a random prime labeling method with Cy dye-labeled random nonamers. Labeled DNA is hybridized to the array using a MAUI hybridization system (**Fig. 1**).

3.1. Sonic Fragmentation of Test- and Reference Genomic DNA Samples

High-molecular weight genomic DNA must be reduced to an average length of between 500 and 2000 bp before the enzymatic labeling (*see* **Subheading 3.3.**) to reduce the size of primer extended products. The lower molecular weight facilitates improved hybridization kinetics and an increased ratio of label to product. Although several methods for cleavage of genomic DNA exist, sonication is the preferred method. The use of restriction enzymes can lead to anomalies in aCGH datasets because of differential cutting between the test- and reference genomes, presumably resulting from differential methylation patterns or other factors, including single nucleotide polymorphisms that can affect digestion and labeling rates differentially between the two samples. As sonication is not a sequence-specific cleavage method, the cleavage products are sheared in the desired, random manner by this method.

1. Dilute 2 µg of each DNA to 80 µL with water in two sterile, nuclease-free 1.5-mL microcentrifuge tubes (*see* **Note 1** regarding DNA requirements).
2. Clean the sonicator tip with isopropanol.
3. Set the Branson sonicator to the following:

Time	10 s
Amplitude	10%
Pulse on	0.5 s
Pulse off	0.5 s

4. Place the sonicator at the bottom of the tube and press Start. Make sure that the sonicator is at the bottom or near the bottom of the tube, or it might splash and the whole sample might not get sonicated.
5. Clean the sonicator tip thoroughly and repeat sonication **steps 1–4** with other sample.

3.2. Electrophoresis Assay of Fragmented Samples

1. Prepare 1% agarose gel with ethidium bromide.
2. Load 250–500 ng each of unsonicated and sonicated samples with loading buffer into separate wells.
3. Run the gel at 100 mV until the loading dye is nearing the bottom of the gel.
4. Transfer gel UV transilluminator and document the gel.

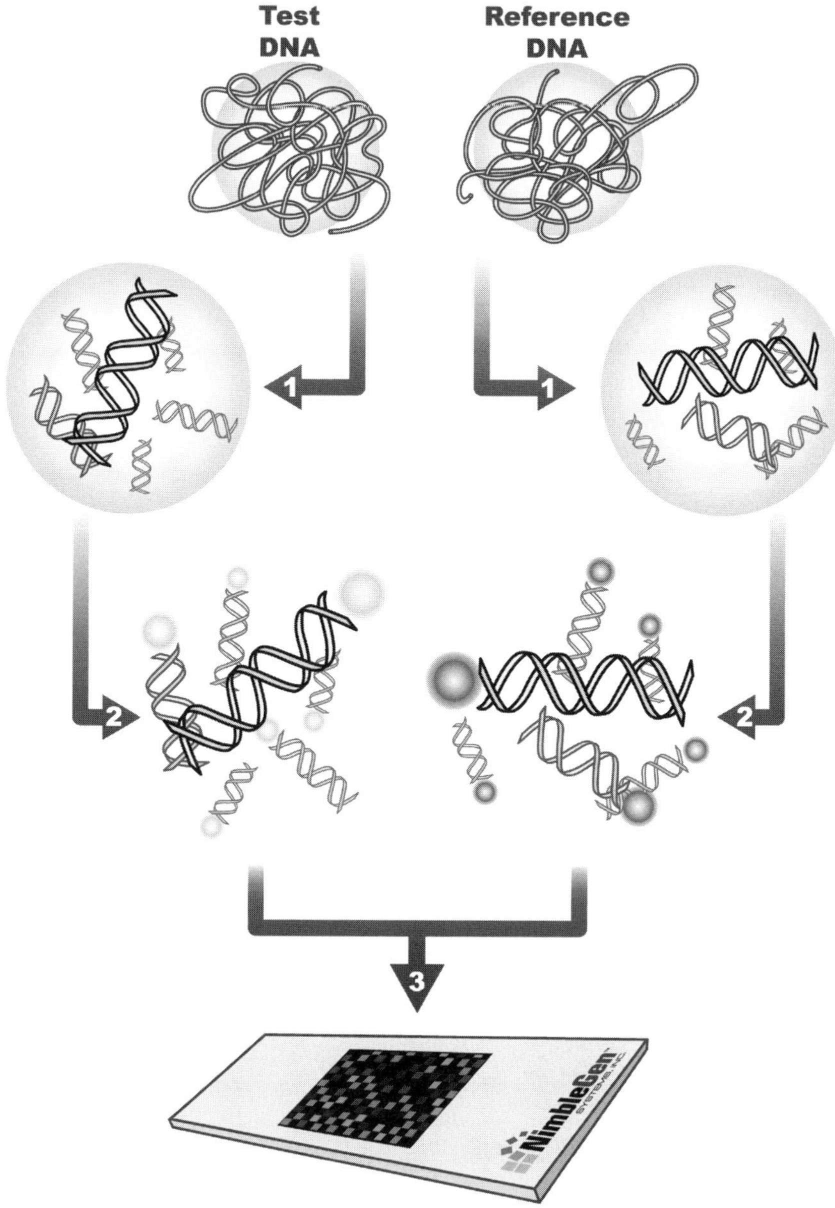

Fig. 1. Graphical depiction of aCGH. Samples of test and reference genomic DNA are sheared to low molecular weight (**step 1**) and separately dye-labeled with cyanine dyes where one sample is labeled green and the other red (**step 2**). The samples are pooled and hybridized to a whole-genome or fine-tiling array (**step 3**). Following washing to remove unbound sample, the array is scanned and the red and green images extracted. The ratios are calculated and mapped back to their genomic position to determine regions, where copy number imbalances exist between the test and reference sample.

5. Genomic samples should appear as a strong band near the top of the gel. Sonicated samples should be a smear with the majority of the fragments between 500 and 2000 bp. If there is more than one discrete band or smearing in the unfragmented samples a contaminant might be present that could affect the labeling procedure.

3.3. Sample Labeling

Important information regarding the handling of labeling reaction components (*see* **Notes 2** and **3**).

1. Prepare the following mixes:

50X dNTP mix:	
Water	250 μL
10X TE	50 μL
100 mM dATP	50 μL
100 mM dGTP	50 μL
100 mM dTTP	50 μL
100 mM dCTP	50 μL
Total volume	500 μL

Random 9-mer buffer	
Water	8.6 mL
1 M Tris-HCl	1.25 mL
1 M MgCl$_2$	125 μL
β-Mercaptoethanol	17.5 μL
Total volume	10 mL

2. Suspend the Cy3 and Cy5 dye-labeled random 9-mers to 1 optical density unit/42 μL in random 9-mer buffer.
3. Assemble the following components in a 0.2 mL thin-walled PCR tube:

	Test	Reference
Fragmented DNA (1 μg)		
Cy3-labeled random primers	40 μL	–
Cy5-labeled random primers	–	40 μL
Water	40 μL	40 μL
Total volume	80 μL	80 μL

4. Heat assembled samples in a thermocycler at 98°C for 5 min. Quick chill in ice-water bath. Add the following (a master mix might be used for this step):

50X dNTP mix	10 μL	10 μL
Water	8 μL	8 μL
Klenow (50 U/mL)	2 μL	2 μL
Total volume	100 μL	100 μL

5. Mix well by pipetting 10 times (do not vortex after addition of Klenow). Spin down briefly to collect contents in the bottom of the tube.
6. Incubate at 37°C for 2 h.
7. Stop reactions by adding 10 µL 0.5 *M* EDTA to each. Mix thoroughly by pipetting or vortexing. The total volume of each reaction is 110 µL.
8. Alcohol precipitate the labeled samples by transferring the reaction to a 1.5-mL tube and adding:

5 *M* NaCl	11.5 µL	11.5 µL
(Mix before addition of isopropanol)		
Isopropanol	110 µL	110 µL
Total volume	231.5 µL	231.5 µL

9. Vortex. Incubate for 10 min at room temperature in the dark.
10. Centrifuge at ≥12,000*g* for 10 min. Carefully remove supernatant with a pipet.
11. Rinse pellet with 500 µL 80% ethanol. Dislodge pellet.
12. Centrifuge at ≥12,000*g* for 2 min. Remove supernatant with pipet.
13. Speed Vac on low heat for until dry, approx 5 min.
14. Resuspend pellets in 25 µL water. Add more water for a very large pellet and add less for a small pellet do not add (15 µL).
15. Measure sample concentration by removing a 1-µL aliquot for NanoDrop A260 measurement.

3.4. Labeled Sample Pooling and Suspension in Hybridization Buffer

1. Combine 15 µg each of the Cy3-labeled test sample and Cy5-labeled reference samples together in a single, nuclease-free, 1.5-mL microcentrifuge tube. Dry the combined samples in Speed Vac on low heat. Protect samples from light.
2. Resuspend the dried pellet in 3.5 µL with VWR water. Vortex to mix and spin down.
3. Prepare a master mix with the following components:

NimbleGen hybridization buffer A	31.5 µL
NimbleGen hybridization buffer B	9 µL
Cy3 CPK6 50-mer Oligo, 50 n*M*	0.5 µL
Cy5 CPK6 50-mer Oligo, 50 n*M*	0.5 µL
Total volume	41.5 µL

4. Add 41.5 µL master mix to each resuspended, pooled sample from **step 2**. Vortex briefly and spin down to gather contents in bottom of the tube.
5. Heat samples at 95°C for 5 min. Protect from light.
6. Vortex, spin down, and place samples at 42°C in the MAUI station for 5–15 min until ready for hybridization setup.

3.5. Hybridization Assembly

1. Preset the MAUI hybridization station temperature to 42°C. Allow adequate time for the system to equilibrate to the target temperature.
2. While the sample is maintained at 42°C on the MAUI station, adhere the MAUI lid to the slide following manufacturers instructions.

3. Once the cover is placed on the slide, rub the gasket brayer over the top of the MAUI lid. Make sure there are no bubbles in the gasket and it is fully adhered to the slide.
4. Pipet all of the sample into one of the two MAUI sample ports, using the manufacturer provided pipetor. Some sample will come out of each port.
5. Make sure there are no bubbles in the lid chamber. If there is, massage any bubbles to either of the ends, away from the center of the array.
6. Blot exterior of cover with a Kimwipe to remove any residual liquid and adhere MAUI stickers to both ports.
7. Place the assembly into one of the four MAUI bays and close the bay clamp.
8. Select mixing mode B and then hold down the mix button to start mixing.
9. Hybridize the sample overnight (16–20 h).

3.6. Stringency Washes

1. Prepare all washes before removing the chip from the MAUI hybridization station. Washes should be in a 300-mL container.

Wash 1 (prepare two containers)	Wash 2	Wash 3
250 mL VWR H_2O	250 mL VWR H_2O	250 mL VWR H_2O
2.5 mL 20X SSC	2.5 mL 20X SSC	625 µL 20X SSC
5 mL 10% Sodium dodecyl sulfate	250 µL 0.1 M DTT	250 µL 0.1 M DTT
250 µL 0.1 M DTT		

2. Remove chip from MAUI hybridization station and immerse in 250 mL Wash 1. While the chip is submerged, carefully peel off the lid. Gently agitate the chip in Wash 1 for 10–15 s.
3. Transfer the slide into a slide rack in a new container of Wash 1 and incubate 2 min with gentle agitation.
4. Transfer to 250 mL of Wash 2 and incubate 1 min with gentle agitation.
5. Transfer to 250 mL Wash 3 and incubate for 15 s with gentle agitation.
6. Remove array and spin dry in array-go-round unit to dry (or use alternate drying method).

3.7. Scanning

1. Place slide in the Axon scanner with the barcode facing down and toward the front.
2. Turn on both lasers and select an initial PMT setting of 600 for both channels.
3. Initiate the scan and zoom the view to the array area.
4. Select the Histogram tab and watch the curve develop.
5. By adjusting the PMT, aim the far right most part of the curve (intensity 65,000) at the y-axis line 10^{-5}.
6. If it appears the curve is above 10^{-5}, stop scan, turn the PMT down and restart. If it appears the curve is below 10^{-5}, turn the PMT up and restart. Finely tuning the PMT by increments of 10 might be required to adjust the curve to this line.
7. After completing PMT settings, scan the entire array area (including corner array fiducials) at 5 µm resolution and save the images as separate files.

3.8. Extraction of Images With NimbleScan™ Software

The extraction in CGH array images is performed with NimbleScan, a proprietary image alignment and extraction software package. The software automatically identifies groups of marker probes, termed fiducials, at the four corners of NimbleGen array images. It also overlays a grid, corresponding to the specific array design file, on the image to establish the appropriate binning of pixel intensities for each probe in the array. The software is designed to function with two-color images, such as those generated by aCGH. There are three primary reporting functions for NimbleScan: Feature Report, Probe Report, and Pair Report. For aCGH data analysis, the required output is the Pair Report.

3.9. Data Analysis

The data analysis procedure for CGH is broken down into four major steps:

1. Combine PAIR data and genome position information.
2. Normalize Cy3 and Cy5 data.
3. Average $\log_2$ ratios in defined window sizes.
4. Perform segmentation analysis on unaveraged and averaged ratio data.

3.9.1. Combining PAIR Data

After image extraction, two PAIR files, one containing the experimental or sample data, and the other containing the reference or control data, are combined into a single file, along with a positions file that contains the genomic coordinates of the individual probes.

A NimbleGen positions file (.pos) contains the following columns:

COLUMN_NAME	DESCRIPTION
PROBE_ID	The name for the individual probe. Generally in the format: chromosome + genome position.
SEQ_ID	The chromosome name for whole genome designs or the chromosome and position range for fine-tiling designs.
CHROMOSOME	The name of the chromosome the probe is located on.
POSITION	Uniqueness. The position of the probe on the chromosome.
COUNT	Information. The number of times this probe appears in the genome. The criteria for this count might vary depending on the type of design.
LENGTH	The length of the probe.

Multiple POS files might exist for each design, so that the probes can be assigned to chromosomal positions for different genome builds. The POS file contains the uniqueness of the probe, so when combining the PAIR files, the researcher can decide whether to include nonunique probes in the subsequent analysis.

The output from the application that combines the PAIR files is a text file with the following columns:

COLUMN_NAME	DESCRIPTION
INDEX	Row number of data
CONTAINER	Name of the NimbleGen ArrayScribe container. Generally just FORWARD for normal CGH designs. If there are replicates within the design, the CONTAINER will differentiate them
SEQ_ID	The chromosome name for whole genome designs or the chromosome and position range for fine-tiling designs
PROBE_ID	The name for the individual probe. Generally in the format: chromosome + genome position
POSITION	The position of the probe in the parent sequence. Will be the same as CHR_POSITION for whole genome designs, and often the same for fine-tiling designs
GC	The GC percentage of the probe
CHROMOSOME	The name of the chromosome the probe is located on
CHR_POSITION	The position of the probe on the chromosome
EXP_CHIPID	Column containing the experimental data, $\log_2$ scale
REF_CHIPID	Column containing the reference data, $\log_2$ scale
RATIO	The $\log_2$ ratio, experimental–reference

3.9.2. Data Normalization

After combining the PAIR reports, the output file is loaded into the R statistical application to normalize the Cy3 and Cy5 data. The current method of normalization used is the qspline method, available as part of the Bioconductor Affy Package. The method has been published *(7)* and is available on-line *(8)*. As part of the normalization process, it is appropriate to make some diagnostic plots to assess the pre- and postnormalization data. Standard two-color microarray diagnostic plots like MvA plots *(9)* will show any dye bias, and whether the normalization was successful in correcting such bias.

After normalization two files are produced: a *normalized.txt file and a *normalized.gff file. The normalized text file contains the same columns as the input file + three additional columns:

EXP_NORM	Column containing the experimental data, $\log_2$ scale, after normalization
REF_NORM	Column containing the reference data, $\log_2$ scale, after normalization
RATIO_CORRECTED	The $\log_2$ ratio, experimental–reference, after normalization

The GFF file contains the normalized $\log_2$ ratio data in a format suitable for loading into SignalMap, NimbleGen's data visualization tool. Information about the GFF format can be found on the Internet *(10)*.

3.9.3. Window Averaging

Window averaging serves two purposes: it reduces the data size, and thereby reducing the computation time and the amount of noise. Window averaging results can be affected by the interval selected, so the data is analyzed unaveraged, and at three intervals: 10X, 20X, and 50X the median probe interval. If the median probe spacing is 60 bp, the data is averaged at 600, 1200, and 3000 bp. The window averaging is done by calculating the biweight mean of the probes in nonoverlapping windows along the chromosome.

The averaged file(s) contain five columns:

CHROMOSOME	The name of the chromosome the window is located on
CHR_POSITION	The midpoint of the window on the chromosome
COUNT	The number of datapoints averaged to produce the RATIO_CORRECTED value
RATIO_CORRECTED	The $\log_2$ ratio, experimental–reference, after averaging
WINDOW_SIZE	The size of the averaging window in base pairs

3.9.4. CGH Segmentation

The final step in CGH analysis is the determination of changes in DNA copy-number based on the $\log_2$ ratio data. A circular binary segmentation algorithm is used to segment the averaged $\log_2$ ratio data *(11)*. An implementation of the algorithm is available as the DNA copy library from the Bioconductor package *(12)*.

The data from the averaged files generated above are loaded into the R analysis software, and the appropriate columns are passed to the DNA copy library. To assess significance, a permutation test is performed by DNA copy. In order to make the analysis run in a reasonable amount of time, the number of permutations is changed from the default number of 10,000 to 1000. One of the options available when doing the segmentation, is the option to undo splits and remove spurious, small segments that are often called in complex genomic data. Typically, the undo.splits = "sd.undo" option is used, with a cutoff of 1.5 standard deviations. The analysis will require from 15 min to several hours, depending on the number of regions, and the window size used to average the data. NimbleGen's software produces four files for each averaged file: a *segtable.txt file, a *.gff file, and two PDF files. The first PDF file, *multi_panel.pdf, contains a chromosome-by-chromosome view of the data and predicted segmentation, three chromosomes per page. The second PDF file, *single_panel.pdf, contains a single plot with all of the data and predicted segmentation. Each chromosome is colored differently, and chromosome boundaries are designated with dashed lines. The GFF file contains the averaged $\log_2$ ratio data plus the predicted segmentation in a format suitable for loading into SignalMap, as described next.

The *segtable.txt file contains the predicted segmentations for the $\log_2$ ratio data in table format. It contains six columns:

SAMPLE_NAME	If a sample key was present, this will be the experimental sample name (Cy3) followed by the reference sample name (Cy5). Otherwise it will be the CHIPID plus the window size
CHROMOSOME	Chromosome containing the segment
START	Genome position of the beginning of the DNA segment
END	Genome position of the end of the DNA segment
DATATPOINTS	Number of data points ($\log_2$ ratio values) in the segment
SEGMENTATION_MEAN	The average $\log_2$ ratio of the data points within the DNA segment

3.9.5. Viewing Data

The GFF file described earlier can be viewed using a NimbleGen data browser, SignalMap. This is an easy-to-use data viewer that allows customized views of individual chromosomes or the entire dataset. Plots can be rescaled and zoomed to analyze specific regions and features an export function to allow the preparation of high-resolution graphics files for publication. With embedded annotation, clicking on specific regions will launch a separate browser with links to the underlying NCBI annotation information for that region of the genome. GFF files can also be added as custom tracks to the UCSC Genome Browser *(13)*.

3.9.6. Example Data

The described aCGH microarray platform and protocols have been applied to genome-wide and targeted high-resolution detection of copy-number changes. An example of genome-wide aCGH, displayed in the SignalMap data browser is shown in **Fig. 2**. The data are sorted in ascending order with the autosomal chromosomes followed by the sex chromosomes. For each chromosome, three tracks are displayed: annotated genes, the segmentation analysis, and the normalized $\log_2$ ratios.

Figure 3 shows a concatenated panel display of a whole-genome aCGH dataset of a human cancer cell line. This perspective provides a clear view of the numerous genome-wide amplifications and deletions identified by aCGH.

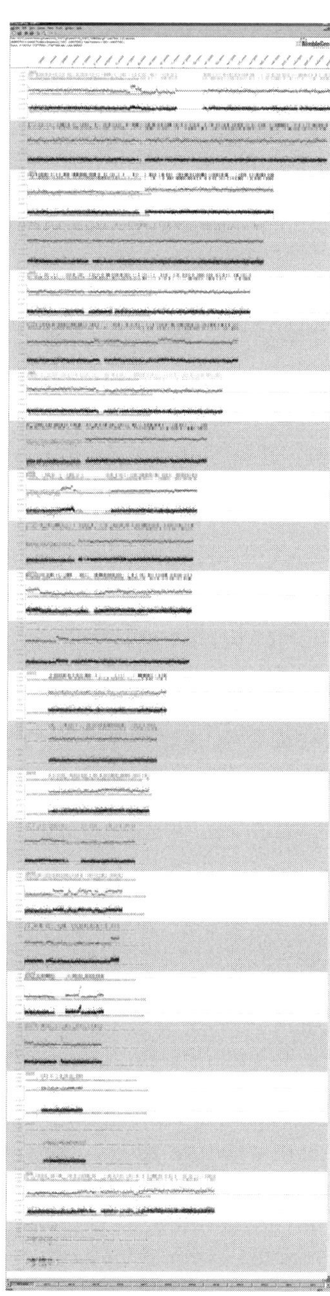

Fig. 2. Whole-genome aCGH dataset displayed in the signalmap data browser.

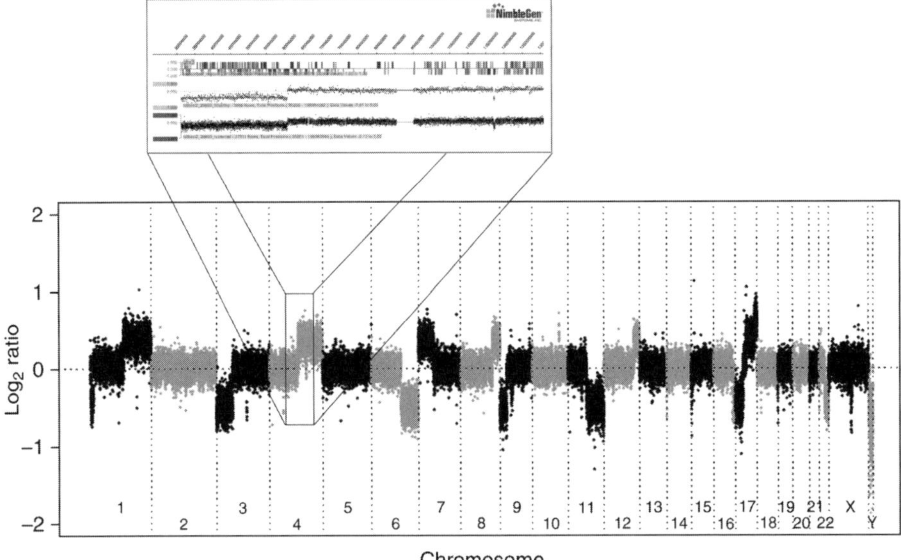

Fig. 3. Whole-genome aCGH dataset displayed in panel view. The results of a whole-genome aCGH analysis of a human cancer cell line are shown in a panel display with each chromosome is displayed in a unique color. The inset is a zoom view of single copy deletion breakpoint found on chromosome 3.

In **Fig. 4**, a high-resolution fine-tiling dataset is shown in the SignalMap data browser. In this experiment, regions from four chromosomes were tiled with 100 bp average probe spacing. The two zoom-views depict the detection of an 80-kbp microdeletion and the breakpoint of single copy deletion.

The example datasets illustrate the advantages of a flexibly deployable array probe set coupled with optimized methods for probe selection, sample labeling, and hybridization for identifying and mapping the copy-number changes at high resolution.

4. Notes

Selection of reference genomic DNA for aCGH. The reference genomic DNA used for comparison is typically a mixture of genomic DNA from several donors, or a matched control. The use of a mixed-genome reference sample reduces the risk of false identification of amplifications or deletions because of copy-number polymorphisms in the reference sample. There are several commercial sources for male, female, or mixed genomic DNA that are suitable for use. It is important to match the sex of the reference to the sex of the test sample in instances where aCGH is focused on the sex chromosomes. For routine aCGH work where the sex of test and reference samples is inconsequential,

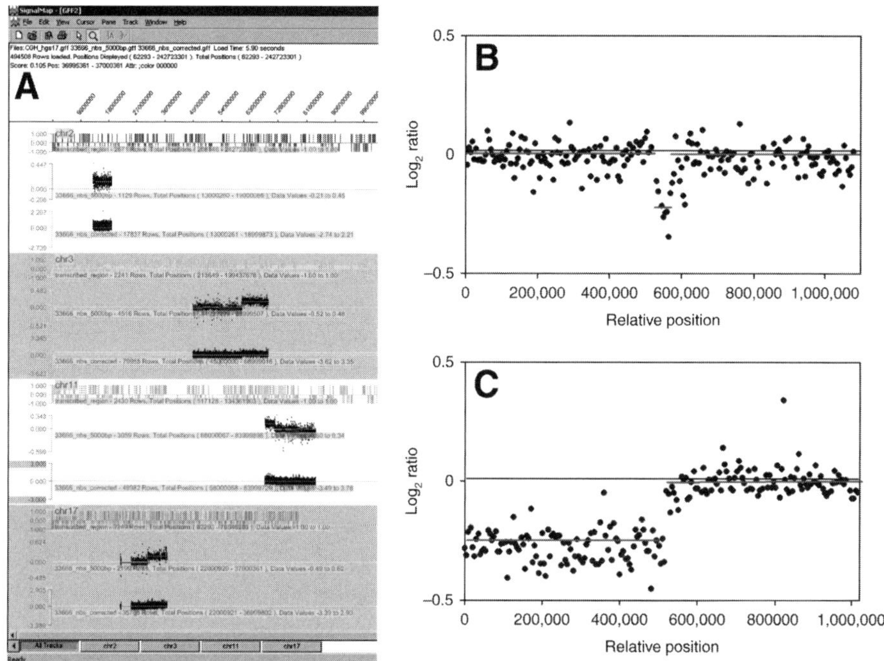

Fig. 4. Fine-tiling aCGH of four selected chromosomal regions The fine-tiling dataset of aCGH analysis performed on a human cancer cell line, displayed in SignalMap, is shown in **A**. The successful detection of an 80-kbp microdeletion in chromosome 3 is shown in the zoom view of the segmentation analysis in **B**. **C** shows the high-resolution mapping of the breakpoint of a single copy deletion.

pooled male genomic DNA was used from a specific vendor to eliminate variables in the process (Human Male Genomic DNA, Promega Corporation, Madison, WI). The labeling protocol requires 0.5 µg or more of test and reference genomic DNA for the labeling reactions.

The primary source of failed aCGH experiments can be attributed to three main causes:

1. Poor DNA-sample quality—the best aCGH results are obtained with pure, intact test DNA samples. The best method for determining sample DNA quality is the spectrophotometric measurement of absorbance at 260 and 280 nm. The ratio of 260/280 should be 1.8 or higher, indicating the absence of significant contaminating protein. An aliquot should also be examined by agarose electrophoresis on a 1% native, ethidium-stained gel. The majority of the DNA should run as a large unresolved smear with an average size more than 20 kbp. Degraded DNA can adversely impact aCGH results by skewing sample labeling and degraded samples should be avoided.

2. Failed sample labeling—the random primer labeling reaction (*see* **Subheading 3.3.**) requires the use of components that are sensitive to their environmental conditions. The dye-labeled primers are light sensitive and should be stored at –20°C when not in use and protected from light during handling.

   The Klenow polymerase is temperature sensitive and should be stored at –20°C in a nonfrost free freezer. Maintain reaction components on ice during reaction assembly. Other components of the reaction are temperature sensitive and must be stored properly. For example, the prepared stock of 0.1 M DTT should be aliquoted and stored in O-ring sealed microcentrifuge tubes. Remove individual tubes for use and discard any unused portions to avoid repeated freeze/thaw cycles.

   Another factor that can adversely affect labeling reaction is a failure to denature the fragmented DNA before the addition of the polymerase (*see* **Subheading 3.3., step 4**). If the template is not denatured and snap-chilled, the reaction yield is greatly reduced.

3. Posthybridization loss of signal—a recent report suggests that cyanine dyes, particularly Cy5, are prone to degradation and a reduction in signal on exposure to even low levels (5–10 ppb) of atmospheric ozone for even brief periods of time *(14)*. Ozone levels vary seasonally with the highest levels observed in the summer months. Methods to control ozone include the creation of an ozone-free room by the installation of an activated carbon filtration system, an ozone scrubber, or the use of reagents to scavenge ozone before array drying. There is a useful discussion of the topic with links to resources on monitoring and controlling ozone available on the Internet *(15)*. The dyes are at highest risk in their dry state, before array scanning. The exposure is time dependent, so coordinating the final array-drying (*see* **Subheading 3.6., step 6**) with scanner access can reduce ozone-mediated degradation. Alternatively, arrays can be stored in an argon-filled chamber before to scanning. If the scanning steps require the use of substantially higher PMT settings than normal for the Cy5 channel than Cy3, this might indicate that the Cy5 signal is being degraded by ozone. It is worthwhile to monitor local levels of ozone and, if necessary, try the corrective measures described to minimize Cy signal loss.

The bioinformatics, microarray and molecular tools, and technologies described here represent a concerted effort to combine a variety of disciplines into a flexible and powerful platform for the genome-wide identification and mapping of copy-number changes. One advantage of the approach described here is that it is an "open platform" where array designs can be customized for specific genomes and subsections of genomes. Although the bioinformatics sections of this chapter might seem impenetrable to the cell biologist and the molecular techniques sections might be a foreign language to the bioinformatician, genome-scale technologies require the cooperation of multidisciplinary teams. We hope this chapter has proven valuable to genomicists of all stripes who have an interest in the advances in aCGH methodology and their application to a broad range of genomic research challenges.

References

1. Nuwaysir, E. F., Huang, W., Albert, T. J., et al. (2002) Gene expression analysis using oligonucleotide arrays produced by maskless photolithography. *Genome Res.* **12,** 1749–1755.
2. Lipshutz, R. J., Fodor, S. P., Gingeras, T. R., and Lockhart, D. J. (1999) High density synthetic oligonucleotide arrays. *Nat. Genet.* **21,** 20–24.
3. Barrett, M. T., Scheffer, A., Ben-Dor, A., et al. (2004) Comparative genomic hybridization using oligonucleotide microarrays and total genomic DNA. *Proc. Natl. Acad. Sci. USA* **101,** 17,765–17,770.
4. Sambrook, J., Fritsch, E. F., and Maniatis, T. (1989) *Molecular Cloning. A Laboratory Manual. 2nd ed.* Cold Spring Harbor Laboratory Press, Cold Spring Harbor, NY.
5. http://www.sanger.ac.uk/Software/analysis/SSAHA/.
6. http://www.nimblegen.com/products/software/arrayscribe/.
7. Workman, C., Jensen, L. J., Jarmer, H., et al. (2002) A new non-linear normalization method for reducing variability in DNA microarray experiments. *Genome Biol.* research0048.1–research0048.16.
8. http://genomebiology.com/2002/3/9/research/0048.
9. Petri, A., Fleckner, J., and Matthiessen, M. W. (2004) Array-A-Lizer: a serial DNA microarray quality analyzer. *BMC Bioinformatics* **5,** 12.
10. http://www.sanger.ac.uk/Software/formats/GFF/GFF_Spec.shtmL.
11. Olshen, A. B., Venkatraman, E. S., Lucito, R., and Wigler, M. (2004) Circular binary segmentation for the analysis of array-based DNA copy number data. *Biostatistics* **5,** 557–572.
12. http://www.bioconductor.org.
13. http://genome.ucsc.edu/.
14. Fare, T. L., Coffey, E. M., Dai, H., et al. (2003) Effects of atmospheric ozone on microarray data quality. *Anal. Chem.* **75,** 4672–4675.
15. http://genomics.princeton.edu/dunham/ozone.html.

10

Design and Fabrication of Spotted Long Oligonucleotide Microarrays for Gene Expression Analysis

Cheng-Chung Chou and Konan Peck

Summary

DNA microarray technology has advanced rapidly since the first use of cDNA microarrays almost a decade ago. For gene expression studies on organisms, for which the genomes have been sequenced, cDNA microarrays are being gradually replaced by gene-specific oligonucleotide microarrays. Although, cDNA microarrays give higher signal intensity than oligonucleotide microarrays, they cannot be used for the measurement of gene-specific expression, whereas, oligonucleotide microarrays can. To obtain both a high signal intensity and specificity in gene expression measurements, gene-specific oligonucleotide probes as long as 150-mers, designed using sequence databases and algorithms to identify unique sequences of genes, are used as microarray probes. In order to achieve a high signal intensity, specificity, and accurate measurement of expression, in addition to the length and sequence of the probes, it is necessary to optimize other parameters such as the surface chemistry of the microarray slides, the addition of spacers and linkers to the probes, and the composition of the hybridization solution.

Key Words: Gene-specific microarray; gene expression; microarray probe; oligonucleotide microarray; probe design; probe optimization.

1. Introduction

In recent years, DNA microarrays have made complex genome-wide gene expression analysis a relatively simple and routine process. The two main DNA microarray platforms are cDNA and oligonucleotide microarrays. cDNA microarrays *(1)* are made with long double-stranded DNA molecules generated by enzymatic reactions, such as PCR, whereas oligonucleotide microarrays employ oligonucleotides spotted by either robotic deposition or *in situ* synthesis *(2)* on a solid substrate. The probes generated by PCR (>500 bp) for cDNA microarrays are often too long to avoid nonspecific cross-hybridization. Therefore, gene-specific oligonucleotides are used to avoid this. However,

From: *Methods in Molecular Biology, vol. 381: Microarrays: Second Edition: Volume 1: Synthesis Methods*
Edited by: J. B. Rampal © Humana Press Inc., Totowa, NJ

several parameters must be carefully evaluated for the optimal use of oligonucleotides to replace cDNA microarrays for gene expression profiling. The immobilized DNA is referred to as the probe and the labeled samples for hybridization as the targets (*see* **Note 1**).

Discordant expression results have been reported using different microarray platforms *(3–5)*, most of these being for gene transcripts of low abundance and gene transcripts of ultrahigh abundance. Low abundance gene transcripts may not yield sufficient signal intensity for reliable detection and highly abundant gene transcripts may cross-hybridize to probes of low specificity. Because 99.8% of genes in human cells have transcript copy numbers <50 *(6)*, the detection sensitivity of the probes is a critical factor in obtaining reliable hybridization results. Of all the possible factors affecting oligonucleotide microarray performance, the sequence and length of the probes are most critical. Short oligonucleotide probe (25- to 30-mers) microarrays give great sequence-dependent hybridization variation *(7,8)* and poor hybridization efficiency, so multiple probes for a single gene *(1)* and RNA amplification by in vitro transcription (IVT) with T7 polymerase *(9)* have been used to obtain reliable quantitative information on gene expression using this system (e.g., Affymetrix GeneChip, which is based on light-directed chemical synthesis to generate thousands of probes on one chip *[2]*). Longer oligonucleotides (50- to 70-mers) provide significantly better detection sensitivity *(10,11)*. To reduce manufacturing cost, long oligonucleotide probe microarrays contain only one probe per gene. However, oligonucleotide hybridization is highly sequence-dependent *(8)* and each oligonucleotide probe should be tested empirically to ensure a good hybridization signal. For microarrays containing tens of thousands of spots, large-scale screening is time-consuming and costly. As shown in the previous study on the optimal length and number of probes *(8)*, a single probe per gene can be selected without the need for experimental validation if the length of the probe is ≥150-mer. However, 150-mer DNA probes are either too long or too expensive to be generated by chemical synthesis. An alternative approach for the production of gene-specific 150-mer probe microarrays were presented here. First, a computer algorithm has been developed to locate the unique sequences of genes in the last exon in the coding sequence regions or in 3′ untranslated regions (3′-UTRs) that do not have a splice junction. Second, to generate the primer pairs for the production of the microarray probes, the sequence-tagged site concept was used *(12)*, i.e., each gene is defined by a pair of PCR primers. This gene-specific primer pair is then used to generate the corresponding gene-specific long oligonucleotide probe (~150-mer) within the unique sequence region in the gene by PCR using human genomic DNA as the template.

Compared with 150-mer long oligonucleotide probes, oligonucleotide probes of 50- to 70-mers in lengths can be produced by straightforward chemical synthesis. The algorithm for designing the 150-mer probes can also be applied to designing probes of 50- to 70-mers in lengths. In order to use 50- to 70-mer long oligonucleotide probes, either the probes must be validated by experimental selection or multiple probes must be used for each gene. In the previous study on optimal probe design *(8)*, it was found that, when using 50- to 70-mer oligonucleotide probes, at least three probes per gene are needed to obtain a reliable measurement of expression. It is too expensive to synthesize tens of thousands of long oligonucleotide probes and test them all empirically (*see* **Note 2**).

2. Materials

1. TRIZOL reagent (Invitrogen, Carlsbad, CA).
2. QIAGEN Genomic-tips (QIAGEN, Chatsworth, CA).
3. GeneChip (Affymetrix, Santa Clara, CA).
4. 50 bp DNA molecular weight marker (Invitrogen).
5. QIAquick 96 PCR purification kit (cat. no. 28183, QIAGEN).
6. RNeasy MiniElute cleanup kit (cat. no. 74204, QIAGEN).
7. SuperScript indirect cDNA labeling system (cat. no. L1014-02, Invitrogen).
8. Cy3 mono-reactive dye pack (cat. no. PA 23001, Amersham, Biosciences, Iscataway, NJ).
9. Cy5 mono-reactive dye pack (cat. no. PA 25001, Amersham).
10. Amine spotting solution (cat. no. K2025, Genetix, Hampshire, UK).
11. Pronto! microarray hybridization kit (cat. no. 40026, Corning, Corning, NY).
12. UltraGAPS and epoxide slides (Corning).
13. Epoxide-coated slides (Corning).
14. Universal spotting solution for UltraGAPS slides and Pronto (Corning).
15. Expoxide spotting solution for Corning® Epoxide slides (Corning).
16. CodeLink activated slides (SurModics, Inc., Eden Prairie, MN).
17. 0.2 mM dNTP (Amersham).
18. 0.05 U/µL of Klentaq, 1% DMSO (Sigma, St. Louis, MO).
19. 10 mM Betaine (Sigma).
20. A 96-well plate (ABgene, Surrey, UK).
21. 96-Channel pipetting workstation (Hamilton Microlabs 4200, Hamilton, Reno, NV).
22. Pipetting workstation (Tecan Genesis Workstation 150).
23. PCR machine (PTC-225 DNA Engine Tetrad, MJ Research, Waltham, MA).
24. 384-Channel UV-VIS spectrophotometer (SPECTRAmax PLUS 384, Molecular Devices, Sunnyvale, CA).

3. Methods

3.1. Probe Design

3.1.1. Strategy for the Identification of Unique Sequences

To identify the unique sequences for each human gene transcript, human gene structures were analyzed to find the most sequence-divergent regions in nucleotide sequence databases available to the public, i.e., the expressed sequence tags (dbEST), mRNA (RefSeq), UTR, and human genome sequence databases. We collected 15,665 named full-length human gene transcript sequences from the UniGene database (build no. 153), 20,991 3′-UTR and 18,536 5′-UTR sequences of human genes from the UTRdb database (v15) *(13)*, and 235,914 human exon sequences from the Ensembl database (v7.29). Exon sequences were aligned with the 3′- and 5′-UTR sequences using the Basic Local Alignment Search Tool (BLAST) sequence alignment program to estimate the abundance of exon/intron splice junctions in UTR. The cross-hybridization probability was calculated by aligning each sequence with the 176,699 putative human transcripts archived in the human gene index created by The Institute of Genomic Research (TIGR) *(14)* (HGI v9) and 104,170 UniGene sets (build no. 153).

Table 1 shows that the 3′-UTRs have the smallest probability of cross-hybridization and are the most divergent regions in human genes. If the repetitive elements are masked using the RepeatMasker program (http://www. repeatmasker.org/), the cross-hybridization frequency in the 3′-UTR is reduced by one half to 11%, and can then be eliminated completely using the gene-specific sequence identification algorithm (UniProbe algorithm), which was described in **Subheading 3.1.2.** Using current reverse transcription and labeling methods, gene-specific probes biased to the 3′-ends are more preferable. However, 12.7% of 3′-UTRs contain splice junctions and are not suitable for probe design. For gene transcripts with splice junctions in the 3′-UTRs, single exons close to, or encompassing a portion of, the 3′-UTR sequence was used in the computation process.

3.1.2. Unique 150-mers Probe Design Algorithm

Based on the empirical data reported in the literature *(15)*, the percentage of local sequence homology, i.e., 75% sequence similarity within a 50 nucleotide (nt) window size was used to calculate potential cross-hybridization among human gene transcripts. The procedure for identifying the unique sequences of a gene transcript that can serve as probes is outlined next and depicted in **Fig. 1**.

1. Mask repetitive elements in the 3′-UTR sequences (no splice junction) of the named transcripts in the UniGene and the UTRdb databases with *X* using the RepeatMasker program.
2. Align the 3′-UTR sequences with all the sequences in the UniGene and TIGR human gene index databases (all against all) and mask the cross-hybridization regions between the query sequences and the target sequences with *N*.
3. Align the sequences masked with *X* and *N* in **steps 1** and **2** in each transcript and extract the unmasked sequence segments that are unique to each transcript in the databases used.

Table 1
Analysis of Human Gene Structure

	5′-UTR	Coding sequence	Exon	3′-UTR
Average length (nt)	244	1530	227	900
Standard deviation	357	1577	395	916
Maximum length (nt)	14,477	80,781	11,487	9450
Minimum length (nt)	13	5	1	14
Cross-hybridization (%)	27.6	42–57[a]	47	22.4
Percentage with at least one splice junction (%)	30	87	0	12.7
Percentage with at least one repeat element (%)	16.3	16.6	4.8	45.4

[a]The first value represents the cross-hybridization probability calculated using the UniGene database and the second that calculated using the TIGR human gene index.

4. Apply primer design to the unmasked sequence segments with a length ≥150 nt using the Primer 3 program (http://frodo.wi.mit.edu/cgi-bin/primer3/primer3_www.cgi). The parameters for PCR primer design are a probe length of 150–170 bp, a T_m of 57–60°C, and a GC content of 40–60%.
5. For unmasked sequence segments with a length less than 150 nt, repeat **step 1**, but using exons close to, or encompassing part of, the 3′-UTR as the query transcript sequences.
6. Use MUMmer 2 program *(16)* to exclude probes containing sequences of more than 15 contiguous nucleotides identical to those in other transcripts in the UniGene and TIGR human gene index databases.
7. For transcripts containing more than one probe obtained from **step 6**, select the probe that targets the sequence with the maximal Gibbs free energy of secondary structure, calculated using the M-fold program *(17)*.
8. Use the above transcript-specific primer pairs and their corresponding amplicon sizes as the input data for the e-PCR program *(18)* to check if they are unique in the entire human genome sequence. If not, repeat **step 4** to obtain appropriate primer pairs.

The specificity of 150-mer probes is demonstrated by our experimental results in a study of the keratin gene family. Among the members of the gene family, keratins 16–18 are highly homologous to each other (**Fig. 2**). Because the 3′-UTR of keratin 16 is shorter than 150 bp, the unique sequence was selected in exon 8, which covers all of the 3′-UTR (**Fig. 2A**). The unique sequence of keratin 17 was located within the 3′-UTR (**Fig. 2B**). For keratin 18 (**Fig. 2C**), the sequence of its 3′-UTR was found to highly similar to those of the other keratin transcripts, so the unique sequence was selected from exon 6, which is very close to the 3′-UTR. In **Fig. 2D**, PCR products of keratin 16 (AI367677), 17 (AA159573), and 18 (AA101381) were reverse transcribed

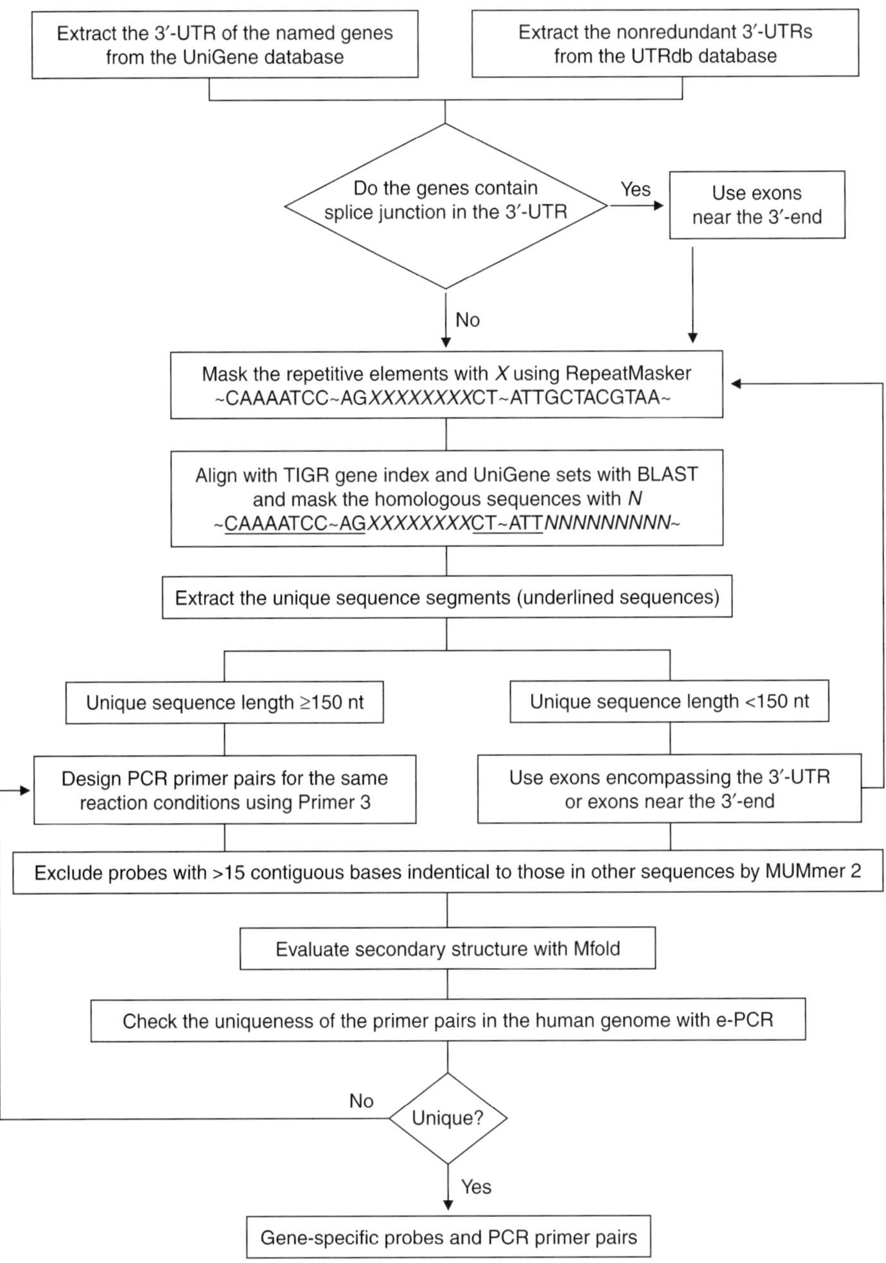

Fig. 1. Flowchart for identifying 150-mer gene-specific probes and PCR primer pairs.

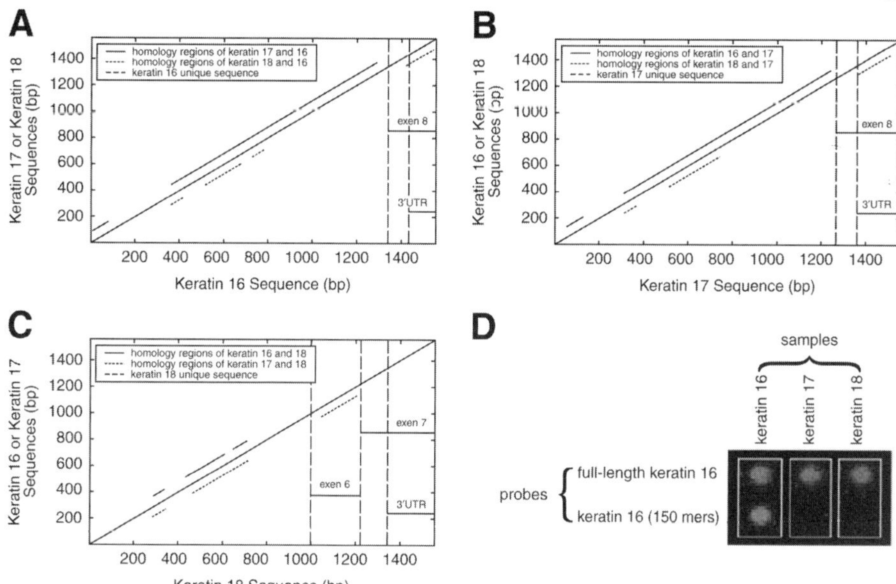

Fig. 2. Illustrations of 150-mer gene-specific probe design and the specificity of the probes. (**A–C**) Pairwise sequence alignment plots for keratin 16 (**A**), 17 (**B**), and 18 (**C**). (**D**) Specificity of the 150-mer keratin 16 gene-specific probe designed using the UniProbe algorithm. The keratins 17 and 18 samples do not give a signal with the keratin 16-specific 150-mer probes; in contrast, the long keratin 16 cDNA probe cannot discriminate between the three keratin gene transcripts.

using random hexamers, labeled with Cy3-dUTP using SuperScript III reverse transcriptase, and separately hybridized to microarrays containing two keratin 16 gene probes, one being the full-length cDNA and the other the specific 150-mer probe designed using the UniProbe algorithm.

The detailed labeling and hybridization reactions are described in the **Subheading 3.1.2.** As shown in **Fig. 2D**, the specific probe designed using the algorithm showed a good discrimination capability for the highly homologous sequences of this gene family. The results demonstrate that the UniProbe algorithm can pinpoint the regions in genes with the least sequence similarity.

3.1.3. Unique 30–70-mer Probe Design Algorithm

The same algorithm can also be applied to unique 30- to 70-mer probe design. The GC content was calculated for every oligonucleotide probe ranging from 30- to 70-mer selected within the unique sequences (the unmasked sequences described in *see* **Subheading 3.1.2.**, **step 3**) of the gene transcript by

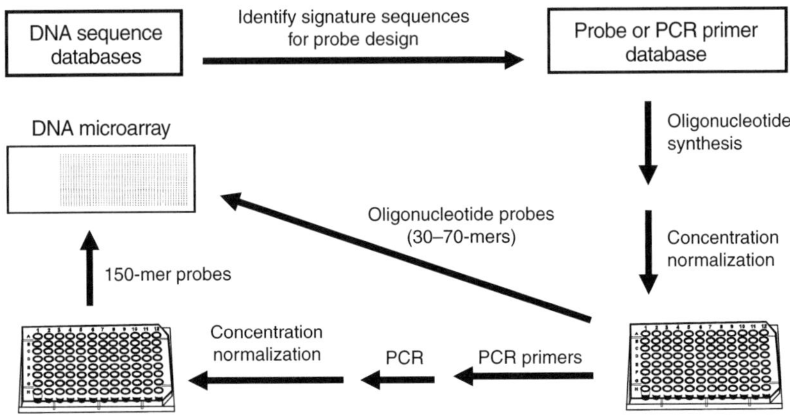

Fig. 3. Flowchart of gene-specific DNA microarray production.

a one-base-shift tiling method. Gene-specific probes with a GC contents 45–55% were selected, then the M-fold program was used to screen for probes that bind to the target sequences with the maximal Gibbs free energy.

3.2. Probe Preparation

A flowchart for the generation of these two types of probe is shown in **Fig. 3**. The detailed procedures are as follows.

3.2.1. Long Oligonucleotide Probes (150–160-mer)

1. Extract human genomic DNA from cell cultures or tissues using the Qiagen Genomic-tip System.
2. Fragment the genomic DNA by sonication and adjust the sonication parameters to ensure a DNA size distribution between 800 and 8000 bp. The size distribution is examined by gel electrophoresis.
3. The gene-specific primer pair sequences of human genes can be obtained from the UniProbe database available on the Internet (*see* **Note 3**). Forward primers should be modified with a 5′ amino-linker, which allows the covalent attachment of long oligonucleotide probes generated by PCR. The double-stranded probes can be rendered single-stranded by heating in boiling water to release anti-sense DNA strands after probe immobilization on array slides.
4. Prepare 100 μL of PCR reaction mixture containing 20 ng of sonicated genomic DNA, 1X PCR buffer (500 mM Tris-HCl, pH 9.1, 160 mM ammonium sulfate, 35 mM MgCl$_2$, 1.5 mg/mL bovine serum albumin), 0.2 mM dNTP, 0.3 μM each of the forward and reverse gene-specific primers, 0.05 U/μL of Klentaq, 1% DMSO, and 10 mM betaine (*see* **Note 4**).
5. Transfer 100 μL of the PCR reaction mixture to each well of a 96-well plate and start thermal cycler. The following PCR conditions are recommended: 94°C for 2 min; 32 cycles at 95°C for 50 s, 57°C for 30 s, and 72°C for 1 min; and 72°C for 10 min.

6. Run 2 µL of each PCR product on a 2% agarose gel together with a 50-bp DNA molecular weight marker for quality control and check the size of each PCR product.
7. Purify the amplified PCR products using a QIAquick 96 PCR purification kit. Measure the concentrations of the PCR products by the absorbance at 260 nm with a spectrophotomer and normalize to a final concentration of 1 μM in 5–10 µL of spotting buffer in 384-well plates using an automated liquid handling workstation.

3.2.2. Oligonucleotide Probes (30–70-mer)

1. Oligonucleotide probes modified with a 5′ or 3′ amino-linker are preferable, but not absolutely necessary (*see* **Note 5**).
2. Oligonucleotides generated by chemical synthesis must be desalted to ensure high immobilization efficiency. High-performance liquid chromatography or gel-purified oligonucleotides are preferred (*see* **Note 5**).
3. The addition of a spacer between the probe sequence and the amino-linker is strongly recommended for short oligonucleotide probes (*see* **Note 6**).
4. Appropriate amounts of oligonucleotides are transferred to 384-well plates and concentrated by vacuum centrifugation until nearly dry. The oligonucleotide pellets are then resuspended to a final concentration of 20–40 μM in 5–10 µL of spotting solution (*see* **Note 7**).

3.3. Probe Immobilization

1. Surface-modified microarray slides are readily available from several vendors and there are different microarray slides optimized for different lengths of oligonucleotide probes. Microarray slides from different vendors were tested and found that different surface coatings can result in signal-to-background ratio differences as great as 16-fold. UltraGAPS coated slides were regularly used and manufacturer's instruction was followed for spotting microarray probes (>50-mer single- or double-stranded DNA) without an amino-linker attached at the end. The probes are immobilized on the UltraGAPS slides by UV crosslinking.
2. For immobilization of amino-modified oligonucleotides (>30-mer), epoxide-coated slides or, CodeLink-activated slides are suitable.
3. By experimenting with various spotting solution formulations, it was found that Genetix Amine Spotting Solution is good for spotting UltraGAPS and CodeLink slides, giving uniform spots with no donut shape under relaxed printing conditions (i.e., a wide range of relative humidity, temperature, and arraying pin impact speed). Universal spotting solution for UltraGAPS slides and Pronto. Expoxide spotting solution for Corning Epoxide slides also work well.

3.4. RNA Extraction and Labeling

3.4.1. RNA Extraction

TRIZOL reagent for total RNA extraction and RNeasy MiniElute cleanup kits for purification are routinely used. This combination gives increased RNA purity and has been found to be necessary for guaranteeing the success of microarray experiments in our laboratory.

3.4.2. cDNA Synthesis and Labeling

In short oligonucleotide microarray experiments (e.g., GeneChip), RNA amplification by IVT is a necessary step before fluorescence labeling. IVT is not needed for long oligonucleotide probe or cDNA probe microarrays unless the amount of RNA derived from tissue specimens is low or a high sensitivity is required. For short oligonucleotide probe microarrays, the optimal protocol for achieving a high signal-to-noise ratio consists of RNA amplification, RNA fragmentation, and labeling with a large fluorescent protein, phycoerythrin. These procedures for short oligonucleotide microarrays are outside the scope of this chapter, so we limit our discussion to methods and protocols for long oligonucleotide (50- to 150-mer) probe microarrays.

A modified SuperScript Indirect cDNA Labeling System for cDNA synthesis and Cy3/Cy5 labeling is used. The kit synthesizes cDNA using SuperScript™ III reverse transcriptase at temperatures up to 55°C, providing increased specificity, a higher yield of cDNA, and a greater yield of full-length products. In addition, instead of the oligo-dT or random hexamers used in most protocols, both anchored oligo-dT (0.12 µg/µL) and amine-modified random hexamers (0.06 µg/µL) are used for cDNA synthesis (*see* **Note 8**). These modifications allow a smaller amount of total RNA to be used (2–5 µg) and more accurate hybridization results to be obtained.

4. Notes

1. Immobilized probes and the labeled samples for hybridization as the targets are suggested in **ref. *19***.
2. It is too expensive to synthesize tens of thousands of long oligonucleotide probes and test them all empirically. The *in situ* synthesized custom probe array production service provided by NimbleGen (Madison, WI) and CombiMatrix (Mukilteo, WA) is a feasible alternative approach for prescreening long oligonucleotide probes before synthesizing them for large scale array production.
3. An online web-based UniProbe database was constructed, where approx 14,000 experimentally validated human gene-specific DNA probes and PCR primer sequences were deposited, which were obtained by using the computational procedures described in **Subheading 3.1.2.** In addition, the UniProbe database provides external links to the National Center for Biotechnology Information LocusLink database for a more detailed description of gene function and to the Kyoto Encyclopedia of Genes and Genomes and BioCarta databases for information on cellular signaling pathways by UniGene ID, GenBank accession number, or gene symbol hyperlinks. The database is available online at http://genestamp. sinica.edu.tw/uniprobe/uniprobe.php.

4. The two PCR enhancing agents, DMSO and betaine, used in our PCR reaction mixture formula significantly improve the yield and specificity of difficult targets in PCR amplification of complex genomic DNA.

5. A significant proportion of long oligonucleotides are not full-length owing to the imperfect coupling efficiency in chemical synthesis. To obtain higher hybridization efficiency, it is preferable to purify the full-length oligonucleotides by high-performance liquid chromatography or gel electrophoresis. In addition, it is advantageous to add a spacer and linker to the 5′- or 3′- end of the oligonucleotides and, because only full-length oligonucleotides can be attached and have the linker, only the full-length oligonucleotides are covalently immobilized on the array slides.

6. Above the surface of a solid substrate, there is an unstirred layer as thick as 1 μm, which limits the diffusion of target molecules in solution to probes immobilized on the solid support. The addition of a spacer to short oligonucleotide probes lifts the probe sequence away from the surface, allowing better access to the target molecules and improving hybridization efficiency.

7. In the study *(8)* indicated that long DNA probes (e.g., 150-mer), which extend farther away from the slide surface than oligonucleotide probes (30- to 70-mers), are more accessible to free target molecules for hybridization and give a more intense hybridization signal at the same spotting concentration. Although, oligonucleotide probes are less accessible to target molecules in the hybridization solution, a high surface density resulting from a high spotting concentration largely improves their poor hybridization signal intensity. For example, 50- to 70-mers probes at a concentration of 20–40 μ*M* gave a similar hybridization signal to 150-mer or cDNA probes at a concentration of 1 μ*M (8)*.

8. Both amine-modified random and oligo-dT primers are used for reverse transcription to generate a mixture of target molecules with a size distribution between 100 and 1000 bases. The length of the target molecules is an important factor in hybridization efficiency. Shorter target molecules diffuse faster and contain less secondary structure, giving higher hybridization efficiency than longer target molecules. Furthermore, the addition of an amine-modified base to the ends of the random primers results in more dye incorporation into cDNA and increased signal intensity *(20)*.

Acknowledgments

We thank Hung-Hui Chen, Ping-Ann Chen, Li-Jen Huang, and Bi-Yun Lee for preparation of the oligonucleotides and the gene specific probes. We also thank Te-Tsui Lee for setting up the UniProbe WWW site. This project was supported by the National Research Program for Genome Medicine (NSC 92-2318-B-001-004-M51 and NSC 93-3112-B-001-013-Y) of the National Science Council.

References

1. Schena, M., Shalon, D., Davis, R. W., and Brown, P. O. (1995) Quantitative monitoring of gene expression patterns with a complementary DNA microarray. *Science* **270,** 467–470.

2. Lockhart, D. J., Dong, H., Byrne, M. C., et al. (1996) Expression monitoring by hybridization to high-density oligonucleotide arrays. *Nat. Biotechnol.* **14,** 1675–1680.

3. Holloway, A. J., Van Laar, R. K., Tothill, R. W., and Bowtell, D. D. (2002) Options available-from start to finish-for obtaining data from DNA microarrays II. *Nat. Genet.* **32,** 481–489.

4. Kuo, W. P., Jenssen, T. K., Butte, A. J., Ohno-Machado, L., and Kohane, I. S. (2002) Analysis of matched mRNA measurements from two different microarray technologies. *Bioinformatics* **18,** 405–412.

5. Zhang, L., Zhou, W., Velculescu, V. E., et al. (1997) Gene expression profiles in normal and cancer cells. *Science* **276,** 1268–1272.

6. Jordan, B. R. (2004) How consistent are expression chip platforms? *Bioassays* **26,** 1236–1242.

7. Selinger, D. W., Cheung, K. J., Mei, R., et al. (2000) RNA expression analysis using a 30 base pair resolution *Escherichia coli* genome array. *Nat. Biotechnol.* **18,** 1262–1268.

8. Chou, C. -C., Chen, C. -H., Lee, T. -T., and Peck, K. (2004) Optimization of probe length and the number of probes per gene for optimal microarray analysis of gene expression. *Nucleic Acids Res.* **32,** E99.

9. Wang, E., Miller, L. D., Ohnmacht, G. A., Liu, E. T., and Marincola, F. M. (2000) High-fidelity mRNA amplification for gene profiling. *Nat. Biotechnol.* **18,** 457–459.

10. Hughes, T. R., Mao, M., Jones, A. R., et al. (2001) Expression profiling using microarrays fabricated by an ink-jet oligonucleotide synthesizer. *Nat. Biotechnol.* **19,** 342–347.

11. Relogio, A., Schwager, C., Richter, A., Ansorge, W., and Valcarcel, J. (2002) Optimization of oligonucleotide-based DNA microarrays. *Nucleic Acids Res.* **30,** E51.

12. Olson, M., Hood, L., Cantor, C., and Botstein, D. (1989) A common language for physical mapping of the human genome. *Science* **245,** 1434–1435.

13. Pesole, G., Liuni, S., Grillo, G., et al. (2002) UTRdb and UTRsite: specialized databases of sequences and functional elements of 5′ and 3′ untranslated regions of eukaryotic mRNAs. Update 2002. *Nucleic Acids Res.* **30,** 335–340.

14. Lee, Y., Tsai, J., Sunkara, S., et al. (2005) The TIGR Gene Indices: clustering and assembling EST and known genes and integration with eukaryotic genomes. *Nucleic Acids Res.* **33,** D71–D74.

15. Kane, M. D., Jatkoe, T. A., Stumpf, C. R., Lu, J., Thomas, J. D., and Madore, S. J. (2000) Assessment of the sensitivity and specificity of oligonucleotide (50-mer) microarrays. *Nucleic Acids Res.* **28,** 4552–4557.

16. Delcher, A. L., Phillippy, A., Carlton, J., and Salzberg, S. L. (2002) Fast algorithms for large-scale genome alignment and comparison. *Nucleic Acids Res.* **30,** 2478–2483.

17. Walter, A. E., Turner, D. H., Kim, J., et al. (1994) Coaxial stacking of helixes enhances binding of oligoribonucleotides and improves predictions of RNA folding. *Proc. Natl. Acad. Sci. USA* **91,** 9218–9222.

18. Schuler, G. D. (1997) Sequence mapping by electronic PCR. *Genome Res.* **7,** 541–550.

19. Southern, E., Mir, K., and Shchepinov, M. (1999) Molecular interactions on microarrays. *Nat. Genet.* **21,** 5–9.

20. Xiang, C. C., Kozhich, O. A., Chen, M., et al. (2002) Amine-modified random primers to label probes for DNA microarrays. *Nat. Biotechnol.* **20,** 738–742.

11

Construction of *In Situ* Oligonucleotide Arrays on Plastic

Jang B. Rampal, Peter J. Coassin, and Robert S. Matson

Summary

The concept of DNA arrays was first introduced in the early 1980s, by Sir Edwin Southern. Since then, many research institutions and biotechnology companies have investigated the potential use of arrays in fields ranging from genetic diagnostics to forensics investigations. A 64-channel automated chemical delivery system, known as the Southern Array Maker, which synthesizes oligonucleotides directly onto an aminated polypropylene substrate has been constructed. Many different arrays have been synthesized for the purpose of detecting single point mutations, which might be either indicators of, or directly responsible for, many different types of genetic diseases and cancers. These include cystic fibrosis, *H-ras*, *K-ras*, and other mutations. In addition to the synthesized arrays, we are also looking into various alternative methods of producing both high- and low-density DNA arrays. This chapter is intended to demonstrate the synthesis of oligoarrays by *in situ* method using standard phosphoramidite chemistry. Phosphoramidate linkage to the aminated polypropylene is quite stable under oligo cleavage and deprotection conditions. Oligonucleotide density is approx 3 pmole or 10^{12} molecules/mm^2.

Key Words: DNA microarrays; oligoarrays; phosphoramidite chemistry; probe density; solid-phase synthesis; hybridization.

1. Introduction

In recent years, oligonucleotide arrays have emerged as promising tools in the analysis of genomic material. Sir Edwin Southern and others have described the synthesis and potential applications of oligonucleotide arrays in molecular biology and diagnostic fields. There are two well-established approaches of immobilizing oligonucleotides at specific sites on solid matrices. In the first approach known as *in situ* method, successive addition of activated phosphoramidite monomers is used to produce the desired oligonucleotide sequence.

From: *Methods in Molecular Biology, vol. 381: Microarrays: Second Edition: Volume 1: Synthesis Methods*
Edited by: J. B. Rampal © Humana Press Inc., Totowa, NJ

In the second approach, presynthesized oligonucleotides are attached to the solid support at a specific location or pattern.

Generally arrays are constructed on glass *(1,2)*, silicon *(3,4)*, membranes, plastics *(5,6)*, or polyacrylamide matrices *(7)*. Fodor et al. *(8–10)* have developed arrays system based on photolithography techniques. Other matrices including polypropylene ($-CH [CH_3] CH_2-)_n$ film were investigated for array synthesis and automation of the process *(11–14)*. Polypropylene was selected because it is inert to most chemical reagents. Also it has negligible fluorescent-background, low-nonspecific adsorption of proteins, nucleic acids, and other biomolecules, easy-to-handle, and very inexpensive. Native polypropylene is easily derivatized by a radiofrequency plasma discharge (RFPD) process *(15)*.

For array synthesis, a specially designed 64-channel automated chemical delivery system Southern Array Maker (SAM) is utilized for the synthesis of 64-lane array panels. A noncleavable phosphoramidate linkage to the polypropylene is quite stable under oligonucleotide cleavage and deprotection conditions *(11–13)*. Oligonucleotide probes are synthesized (*see* **Note 1**) onto amino polypropylene film using standard phosphoramidite chemistry. The one-dimensional arrays are then investigated by solid-phase hybridization. Similarly two-dimensional arrays could be constructed *(16)*. By introduction of a 3′ succinyl linkage between amino polypropylene and the growing oligonucleotide chain, a cleavable oligonucleotide can be obtained to evaluate both the quantity and quality of the product. It is also possible to incorporate other linkers such as alkyl groups or polyethers. High quality primers for PCR are obtained from such oligonucleotide arrays *(17)*.

In this chapter, we will discuss the *in situ* synthesis of DNA arrays using amino polypropylene film. A number of model DNA array panels related to cystic fibrosis (CF), K-ras, H-ras, and short tandem repeats were prepared. In array synthesis, each lane might contain any probe sequence specified by the user so that from 1 to 64 different sequences can be synthesized on the panel. Hybridization to oligonucleotide arrays created on polypropylene is based upon the reverse-blot format. The arrays are hybridized with nonisotopic-labeled synthetic oligonucleotide or PCR-amplified targets. Nonisotopic labels were used because they are easier to handle, and are stable for longer periods of time as compared with isotopic labels. A cooled CCD camera is used for detection of images. For detailed information on hybridization analysis *see* Chapter 14.

2. Materials

2.1. Reagents

1. Polypropylene: native biaxial polypropylene film (~1 mil in thickness) (Mobil Film Division, Catalina Plastics, Calabasa, CA).
2. Amino polypropylene: plasma amination of polypropylene by plasma generating apparatus (*see* **Note 2**).

3. Phosphoramidites: dA^{Bz}-, dC^{Bz}-, dG ^{iBu}-, T-phosphoramidities, and other DNA reagents (e.g., deblock, activator, capping A, capping B, oxidizer, CPG-nucleosides, 3′-succinylated nucleosides, and reverse phosphoramidites) for arrays and targets synthesis (Beckman Coulter, Fullerton, CA) (*see* **Note 3**). For drying activator reagent, add approx 1 g molecular sieves (4 Å) to it.
4. Cyanine dyes (Amersham Pharmacia, Piscataway, NJ).
5. Ammonia gas cylinder.
6. Ammonium hydroxide solution (J. T. Baker, Phillipsburg, NJ).
7. Amino-ON™-, fluorescein-ON™-, biotin-ON™-phosphoramidites (Clontech Lab, Palo Alto, CA).
8. Dimethoxytrityl chloride (Aldrich, Milwaukee, WI).
9. Tetra-N-butylammonium perchlorate (Aldrich).
10. 2,4,6-Collidine (Aldrich).
11. Sulfo-succinimidyl-4-O-(4,4′-dimethoxytrityl)-butyrate (S-SDTB) (Pierce, Rockford, IL).
12. 1-Ethyl-3-(3-dimethylaminopropyl)carbodiimide hydrochloride (EDAC) (Pierce).
13. 4-Dimethylamino pyridine (DMAP) (Aldrich).
14. Triethylamine (TEA) (Aldrich).
15. Anhydrous acetonitrile, 50-L drum (B & J or J. T. Baker).
16. Hybridization buffer: 2X–6 SSPE (sodium phosphate buffer), 0.01% sodiumdodecyl sulfate (SDS): dilute 20X SSPE to 2X or 6X SSPE (per liter: 3 M NaCl; 0.2 M NaH$_2$PO$_4$; 0.02 M ethylenediaminetetraacetic acid; adjust to pH 7.4 with 10 N NaOH), followed by addition of 0.01% SDS (v/v).
17. Streptavidin alkaline-phosphatase conjugate (PE Biosystems, Foster City, CA).
18. Streptavidin-FITC conjugate (PE Biosystems).
19. ELF reagent (Molecular Probes, Eugene, OR) (*see* **Note 4**).
20. Oxidizer solution (dilute) for oligonucleotide synthesis was for prepared by mixing the following reagents (4 L solution). All the reagents were purchased from Aldrich (*see* **Note 5**):
    a. 12 g Iodine.
    b. 200 mL Pyridine (spectrophotometric grade).
    c. 80 mL Double-distilled H$_2$O.
    d. 3720 mL Tetrahydrofuran.

2.2. Instrument/Apparatus

1. Ammonia plasma generator (Model PS0150E, Plasma Sciences, Foster City, CA) (*see* **Note 2**).
2. SAM research breadboard (Beckman Coulter).
3. Oligo1000 DNA synthesizer (Beckman Coulter).
4. Aluminum block for fabrication of 64-channel (Beckman Coulter).
5. P/ACE™ 2000: capillary electrophoresis system for the analysis of oligonucleotides with eCAP™ single-stranded (ss)DNA 100-R kit (Beckman Coulter, cat. no. 477480).
6. High-performance liquid chromatography (HPLC) system for the analysis of oligonucleotides: System Gold HPLC Programmable Solvent Module 126 equipped with a Diode Array Detector Module 168 (Beckman Coulter).

7. Electron spectroscopy for chemical analysis (ESCA) of the polypropylene.
8. Energy source: OmniPrint 1000 (Omnichrome, Chino, CA).
9. CCD camera from (Photometrics, Tucson, AZ; Model CH250).
10. PhosphorImager (Molecular Dynamics, Sunnyvale, CA).
11. VistraImager (Molecular Dynamics).
12. Microglass slides (Gold Seal, NH, cat. no. 3010).
13. Thermocycler.
14. LabVIEW software for running the SAM synthesizer (National Instruments, Austin, TX).
15. Oligo® software for sequences optimization and T_m analysis (NBI, Plymouth, MN).
16. NIH Image software for image analysis (National Institutes of Health, Bethesda, MD).

3. Methods

3.1. Preparation of Amino Polypropylene Film

Polypropylene is aminated in an ammonia plasma, generated by RFPD process *(15)*. Ammonia plasma consists of mixture of free radicals, ions, and other metastable highly reactive species. We use a radiofrequency generator set at 13.56 MHz to create an ammonia plasma confined within a vacuum chamber. Free radicals within the plasma collide with native polypropylene, abstracting a hydrogen atom from the carbon–carbon backbone to create surface free radicals. In the RFPD process, polypropylene sheets are placed in the plasma chamber. The chamber is evacuated and ammonia gas pumped into the chamber. The electrodes are turned on for approx 2 min to create the plasma. This plasma converts native polypropylene into amino polypropylene (*see* **Fig. 1**). The plasma reaction takes place within the first 10–1000 Å depth of the surface. The entire process takes about 10–15 min. The process produces 2–5% of amination of the surface.

3.2. Determination of Amine Density

Typically, the amine density is calculated by dimethoxytsityl chloride (DMT-Cl) or S-SDTB methods. Results give approx 5–6 pmoles/mm² as shown in **Table 1**. This amine density appears to be satisfactory for array synthesis. The following methods used for amine density determination.

3.2.1. DMT Analysis

Mix amino polypropylene films in an equimolar (0.5 *M*) solution DMT-Cl and tetra-n-butylammonium percholate in dichloromethane containing 1.5 equivalent of 2,4,6-collidine for 5 min *(18)*. Wash the film successively in dichloromethane and acetonitrile. Treat the DMT-film with 3% trichloroacetic

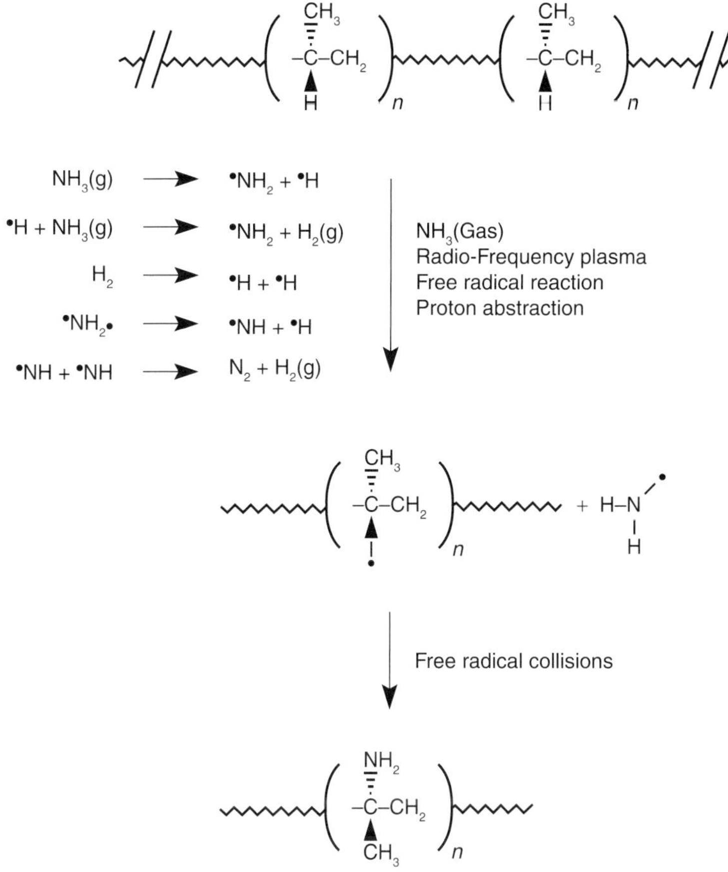

Fig. 1. Radiofrequency plasma discharge (RFPD) process.

acid for 30 s and measure the released orange color of DMT cation by visible ultraviolet (UV) spectrophotometer at wavelength λ_{498} nm.

3.2.2. S-SDTB Analysis

Amino polypropylene can be derivatized with S-SDTB reagent under aq conditions by using the protocol provided by manufacturer. The orange color, DMT cation, is measured at λ_{498} nm. Both procedures produce similar results.

3.3. ESCA Analysis

ESCA is a physical method to analyze the bond types and elemental composition of the material (*see* **Note 6**). Native polypropylene should have very little character since it largely consists of carbon–carbon bonds. On the other

Table 1
Comparison of Ligand Densities Among Various Substrates

Array substrate	Probe density	Bound target	Occupancy
PP-amine (1–20 mil)	5–6 pmoles/mm^2 or 3 × 10^{12} molecules/mm^2		
PP-oligo/cleaved	3 pmole/mm^2 or 1.8 × 10^{12} molecules/mm^2	0.52 fmole/mm^2 or 3.1 × 10^8 molecules/mm^2	0.017%
Glass-oligo	4 pmole/mm^2 or 2.4 × 10^{12} molecules/mm^2	1 fmole/mm^2 or 6 × 10^8 molecules/mm^2	0.025%
PAGE-oligo	5 pmole/mm^2 or 3 × 10^{12} molecules/mm^2	5 fmole/mm^2 or 3 × 10^9 molecules/mm^2	0.10%

hand, amino activated and or oligonucleotide attached polypropylene gives different composition of elements (*see* **Fig. 2** and **Table 2**).

3.4. Synthesis of DNA Arrays

3.4.1. Construction of 64-Channel Array Block

The 64-channel fabricated array block is made from aluminum material. The bottom part of the block has 64 parallel machined channels. Each channel is 500-µ wide, 500-µ deep, 1-mm center, and 6.5 cm in length (*see* **Fig. 3**). The top part of metal block has a ground flat surface. The top and bottom parts are held together with four screws and a clamp. Through the tiny tubing, inlet channels are connected to the research breadboard DNA array maker and the outlet channels attached to the waste bottle. The 8 × 8 cm^2 aminated polypropylene film is placed in between two metal blocks, the activated side down facing toward the channels. This activated film used for making arrays also functioned as a gasket during chemical delivery of reagents (*see* **Note 7**).

3.4.2. CF Probe Design

As an example for probe design, a portion of a CF array is discussed in which wild-type and mutant probes are shown in **Fig. 4** (*see* **Note 8**). For the purpose of optimizing probe-target interaction, probe sequences are varied to normalize the entire panel (16- to 21-mer) from an initial T_d ranging from 43 to 69°C using Nearest Neighbor thermodynamic matrix calculations. On the array panel, the wild-type (WT) and one mutant (ΔF508) appear in lanes 11, 13, respectively. The panel (16- to 21-mer) is identified as the T_d approx 55°C panel.

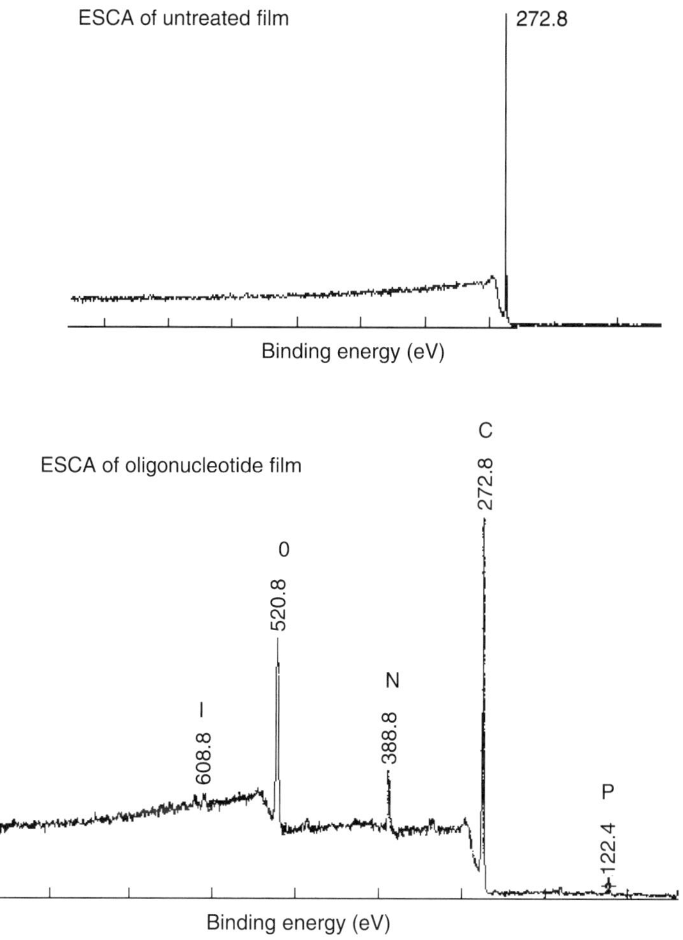

Fig. 2. Surface determination of plastic film using electron spectroscopy for chemical analysis. ESCA of native polypropylene film is shown in top graph of **Fig. 2** and with attached probes in the bottom graph of **Fig. 2**.

All the other probes represent other mutant of the CF gene. The wide T_d range CF panel requires that hybridization be conducted at an elevated temperature, followed by number of washing steps at various temperatures to achieve stringency. In order to further increase the discrimination between fully matched vs mismatched targets, allows hybridization to be performed near room temperature, the CF panel probe sequences are modified as shown in **Fig. 5**. Essentially, sequences are 3′ and/or 5′ truncated to normalize all T_d to a narrow range (37 + 1.5°C), using a Nearest Neighbor calculation. Regions of mismatch are placed near the middle of each sequence to achieve maximum discrimination *(16)*.

Table 2
Surface Modification of Polypropylene by Radiofrequency Plasma Discharge

Aminated polypropylene film (PPNH2)
Untreated film   98 ± 2.8% carbon
Treated film   89.1 ± 3.7% carbon
                               6.9 ± 2.9% nitrogen
                               0.8 ± 1% oxygen

Oligonucleotide film (line panel)

DNA side (%)		Blank side (%)
75	Carbon	84
7.2	Nitrogen	2.3
15	Oxygen	10
1.1	Phosphorus	0
0.2	Iodine	0.09
1.4 Ca, Na	–	3.4 Si

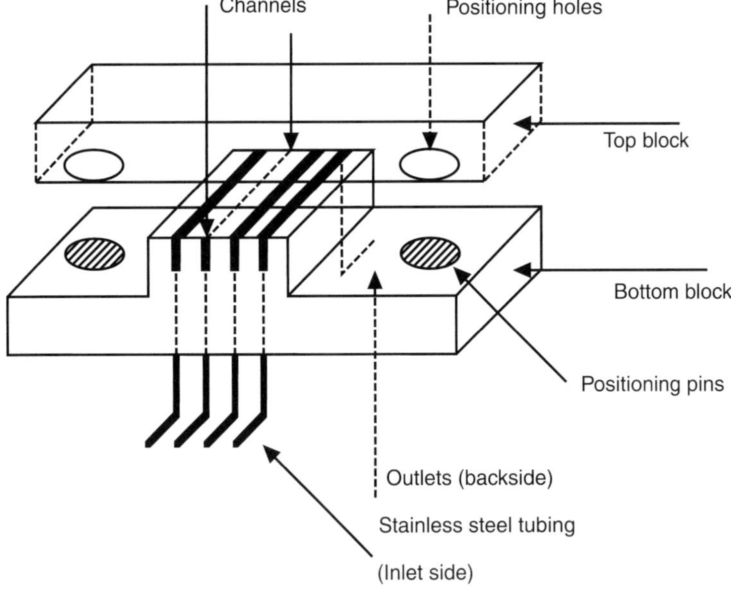

Fig. 3. 64-Channel plate device made from aluminum for DNA array synthesis. Channel dimensions: 0.5-mm width, 0.5-mm depth, 1 mm apart from centers. Parallel synthesis of 64 probes onto polypropylene surface.

Length	3'---->5'	ASO	Lane
19	CGCTATCTCGCAAGGAGGA	R117	1
19	CGCTATCTCACAAGGAGGA	R117H	2
19	ATACGGATCTATTTAGCGC	Y122	3
19	ATACGGATCAATTTAGCGC	Y122X	4
19	ACGTAAGGTTACACTACTT	I148	5
19	ACGTAAGGTCACACTACTT	I148T	6
19	CTTCATAATGGAAGAATAT	621/G	7
20	CCTTCATAATTGAAGAATAT	621/T	8
19	AAGACAAGAGTCAAAAGGA	Q493	9
19	AAGACAAGAATCAAAAGGA	Q493X	10
19	CTTTTATAGTAGAAACCAC	I507/F508	11
20	TTCTTTTATAGAAACCACAA	²I507	12
20	TTCTTTTATAGTAACCACAA	²F508	13
19	GAAACTACTGCGAAGACAT	V520	14
20	CGAAACTACTTCGAAGACAT	V520F	15
19	CCTCTACAGGATAATGGTT	1717/G	16
19	CCTCTACAGAATAATGGTT	1717/A	17
16	TCAAGAACCTCTTCCA	G542	18
16	TCAAGAAACTCTTCCA	G542X	19
16	GAGCAACTGGAGGTGA	G551	20
17	GAGCAACTAGAGGTGAG	G551D	21
16	TCACCTCCAGTTGCTC	R553	22
17	CTCCAGTTACTCGTTCT	R553X	23
19	GAAATCGTTCCACTTATTG	R560	24
19	GAAATCGTTGCACTTATTG	R560T	25
21	CACAGAATGAGCGGTAAAATA	3849/C	26
21	CACAGAATGAGTGGTAAAATA	3849/T	27
16	CGAAAGGAGGTGACAA	W1282	28
17	CCGAAAGGAAGTGACA	W1282X	29

Fig. 4. Sequence and design of CF panel: wide range T_d from 43 to 69°C.

This modified panel (12- to 16-mer) is identified as the T_d approx 37°C panel. Mismatch sequences are placed in the top half of each panel, whereas corresponding wild-type sequences are placed in the bottom half of the panel (probe sequences partially shown in **Fig. 5**). The new panel also contains a short tandem repeat sequence in every fifth lane as an internal marker for indexing purposes. It is possible to differentiate the WT signal from mutant signal (ΔF508) by optimizing the array panel design and hybridization conditions to a large extent (*see* **Note 9**).

Length	3′-->5′ Sequence	ASO	Lane
	Blank	Mutations	1
15	CGCGCGCGCGCGCGC	marker	2
15	CTATCTCACAAGGAG	R117H	3
14	ACGGATCAATTTAG	Y122X	4
16	GTAAGGTCACACTACT	1148T	5
16	TCATAATTGAAGAATA	621T	6
15	CTTTTGATTCTTGTC	Q493X	7
15	CGCGCGCGCGCGCGC	marker	8
15	TGTGGTTTCTATAAA	^{2}507	9
16	TTGTGGTTACTATAAA	^{2}508	10
15	ACTACTTCGAAGACA	V520F	11
16	TCTACAGAATAATGGT	1717A	12
14	GGAAGAGTTTCTTG	G542X	13
15	CGCGCGCGCGCGCGC	marker	14
14	AGCAACTAGAGGTG	G551D	15
14	GAACGAGTAACTGG	R553X	16
14	ATAAGTGCAACGAT	R560T	17
15	GAATGAGTGGTAAAA	3849T	18
14	GAAAGGAAGTGACA	W1282X	19
15	CGCGCGCGCGCGCGC	marker	20
15	TTTCTTGAGACTCTT	G134E	21
13	TGTTCTTGGGACT	1078^{2}T	22
15	AAAAGGTCTCCTACT	R334W	23
12	GGTACCCGTCTT	R347P	24
12	CGTTGGAGGTTG	A455E	25
15	CGCGCGCGCGCGCGC	marker	26
14	TTCTGAGTCGAGTG	R1162X	27
14	TGAACATCCAAATG	3659^{2}C	28
14	CTAGGTTGAAAAAA	N1303K	29

Fig. 5. Sequence design of CF panel: narrow range T_d 37 + 1.5°C.

3.4.3. LabVIEW Software

LabVIEW software is used for running SAM array maker. It is based on a graphical programming language, G, to create programs in block diagram format, and has complete libraries of functions and subroutines for most programming needs. **Figure 6** shows the use of graphic symbols instead of textual language. LabVIEW uses terminology, icons, and other common features that are familiar to most of the users. The top portion of **Fig. 6** illustrates the main control panel screen and the bottom portion describes the chemical delivery

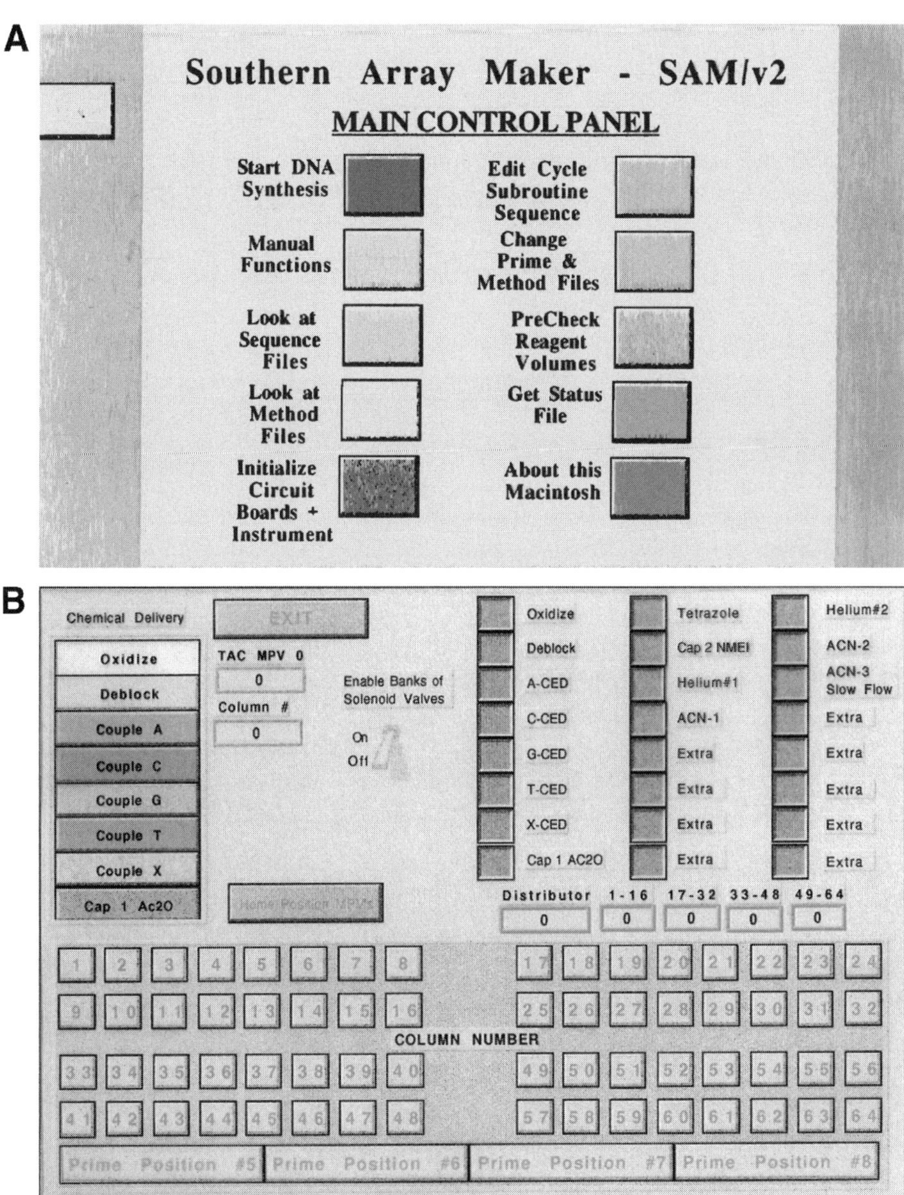

Fig. 6. (**A**) Picture of manual control panel using LabVIEW software, (**B**) picture of specific locations of reagents and channels.

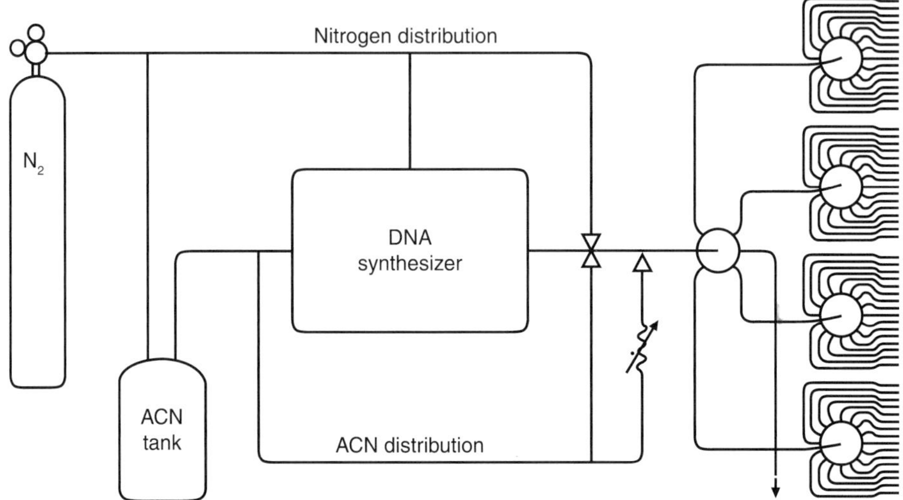

Fig. 7. Southern Array Maker schematic. SAM is based on a conventional DNA synthesizer. Its unique characteristic is a distribution manifold to direct reagents selectively to each of the 64 synthesis channels. Fast acetonitrile flush flow path 2 mL/min. For slow acetonitrile delivery path, 0.25 mL/min.

system and the position of array channel in use for synthesis. The software programs are also known virtual instruments (VIs) because their appearance and operation imitate actual instrument.

3.4.4. SAM Array Maker

In situ noncleavable arrays are synthesized using an automated SAM DNA synthesizer as shown in **Fig. 7**. The instrument is based on Oligo 1000 DNA Synthesizer (Beckman Coulter). In SAM, four additional valves, each having 16 ports is introduced in parallel for delivery of reagents to the 64 channels. Probe sequences (or allele specific oligonucleotide, ASO) that are to be synthesized are first designed based on T_m values and then are written into an Excel spreadsheet, which is subsequently converted into a binary run file. SAM array synthesizer is run using a Macintosh driven software program called LabVIEW. Standard phosphoramidite chemistry is utilized during array synthesis. The probes are synthesized from 3'- to 5'-end. The 3'-end of the nucleotide monomer is attached to the amino polypropylene film, for example, the first lane sequence is PP-3'-CGCTATCTCGCAAGGAGGA-5' and so on (*see* **Note 10**). PP stands for polypropylene. Steps involved in synthesis are (*see* **Fig. 8**):

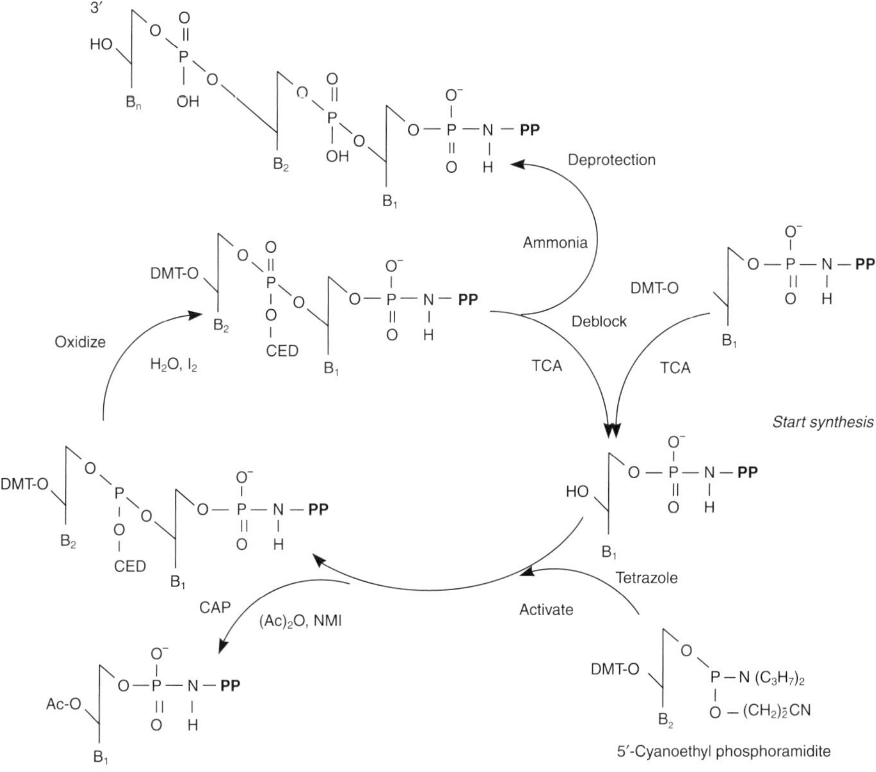

Fig. 8. SAM synthesis cycle: oligonucleotide array synthesis on aminated poly-propylene film. First base is attached by a highly stable phosphoramidate linkage directly to amino polypropylene film. Standard phosphoramidite solid phase chemistry is used for creating an array.

1. Deblocking.
2. Coupling of monomers.
3. Capping of failure sequences (*see* **Note 11**).
4. Oxidation step (*see* **Note 5**).
5. Deprotection.

Synthesized array is removed from the 64-channel block and is treated in ammonium hydroxide solution in a sealed vial at 70°C for 90 min. The cova-lently attached oligonucleotides or probes have an amidate linkage that is stable to cleavage conditions used in solid-phase DNA synthesis. A complete synthesis cycle running through to all 64 channels requires approx 1 h (*see* **Table 3**). The arrays are used as such for staining, or direct hybridization with complementary or mismatched targets. Most of the probe sequences syn-thesized are from 6 to 35 bases long.

Table 3
Steps Involved in Array Synthesis by SAM

Steps	Reagents/solvents	Total time (min) for 64 channels (one cycle)
Deblock	3% TCA in CH_2Cl_2	5
Wash	Dry CH_3CN (<30 ppm water)	5
Flush	Argon	2
Couple	Phosphoramidites (1 g/20 mL CH_3CN), 0.5M tetrazole in dry CH_3CN	17
Wash	Dry CH_3CN (<30 ppm water)	5
Flush	Argon	2
Capping[a]	Capping A (10% acetic anhydride in THF, pyridine, water); Capping capping B (17% N-methyl imidazole in THF)	5
Wash	Dry CH_3CN (<30 ppm water)	5
Flush	Argon	2
Oxidation	Composition, *see* **Subheading 2.4.** *(19)*	5
Wash	Dry CH_3CN (<30 ppm water)	5
Flush	Argon	2

[a]Capping optional.

3.5. Synthesis of Cleavable Oligonucleotide for Quality Control

3.5.1. Synthesis of Succinate Cleavable Linker

For the synthesis of cleavable oligonucleotide from the polypropylene material, the aminated polypropylene is coupled with active ester of 3′-succinyl-5′ DMT deoxynucleotide. The derivatized polypropylene is subsequently used for oligonucleotide synthesis using standard phosphoramidite solid phase chemistry. Procedure for coupling of 3′-succinylated C^{Bz} to amino polypropylene is (*see* **Note 3**):

1. Add 5–10 amino polypropylene films (8×8 cm^2), 0.733 g (1 mM) 3′-succinylated-C^{Bz}, 0.975 g (5 mM) EDAC, 30 mg 4-dimethylamino pyridine, and 160 μL triethylamine to 10 mL dry pyridine in a sealed vial.
2. Shake the reaction mixture under anhydrous conditions for 20 h at room temperature.
3. Wash polypropylene film with dry pyridine (5×10 mL) and dry acetonitrile (5×10 mL).
4. Add 5 mL capping reagent A and 5 mL of capping reagent B, shake reaction mixture for 1 h at room temperature to protect the unreacted amino groups.
5. Wash polypropylene film with dry acetonitrile (5×10 mL). Use this film for cleavable oligo synthesis and for determination of ligand density.

3.5.2. Synthesis of Cleaved Oligonucleotide

1. Synthesize the identical probe sequences in 10–12 channels with 5′DMT-ON in amino polypropylcnc film as produccd in **Subhcading 3.5.1.**, using SAM array maker in order to obtain enough material for analysis.
2. Treat polypropylene solid support in ammonium hydroxide solution for 1 h at room temperature for cleavage and deprotection.
3. Heat in a sealed vial at 70°C for 90 min.
4. Decant off the cleaved oligonucleotide solution.
5. Evaporate the solution to dryness.
6. Add 1 mL water.

3.5.3. Analysis of Cleaved Oligonucleotide

1. For concentration determination, run UV at wavelength λ_{260} nm.
2. RP-HPLC analysis: inject 25 µL solution into C_{18} Ultrasphere column (3 µm ODS, 4.6×7.5 cm^2) using a 70% gradient from 0.1 M ammonium acetate, pH 6.8, to acetonitrile at a flow rate of 1 mL/min for separation.
3. CGE analysis (*see* **Note 12**): evaporate oligonucleotide from **Subheading 3.5.2.**, **step 6**, having a 5′-DMT group under reduced pressure to dryness. Redissolve the dried material in 100 µL aq 80% acetic acid for 30 min at room temperature. Remove the acetic acid under reduced pressure. Prepare sample solution in HPLC grade water to a concentration of 0.2–0.5 OD/mL. A coated capillary 27 cm in total length (20 cm to detector window) from eCAP ssDNA 100-R kit, is used for oligonucleotide analysis. Inject sample by voltage injection of 7.5 kV/3 s. Separate at 300 V/cm. Detect the absorbance at λ_{260} nm (for electropherogram, *see* **Fig. 9**).
4. Coupling efficiency: based on UV-Visible spectroscopy and HPLC analysis data of the oligonucleotide, the step-wise yield is calculated, which is approx 97–99% **(19,20)**.

3.5.4. Probe Density

Oligonucleotide probe density is calculated as follows:

1. Channel width = 500 µ = 0.5 mm.
2. Channel length = 65 mm.
3. Total area per lane = 35.5 mm^2.
4. Cleaved oligonucleotide per lane = approx 0.1 nmole (based on UV analysis).
5. Ligand density = 0.1/32.5 = 0.003 nmole/mm^2 = 3 pmole/mm^2.
   Oligonucleotide density is approx 3 pmole/mm^2 or 1.8×10^{12} molecules/mm^2 (*see* **Table 1**). This probe density appears to be optimal for most hybridization.

3.6. Uniformity of Signals

Table 4 demonstrates representative lanes of strip that shows the uniformity and quality array synthesize on SAM. Example of 64-channel array synthesis,

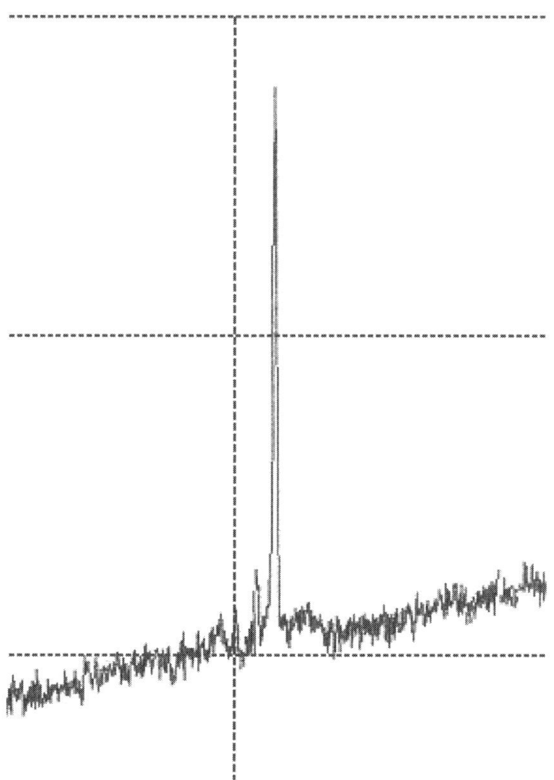

Fig. 9. CGE analysis of DMT-off cleaved product on P/ACE 2000: the largest peak correspondence to the peak of interest (main peak). Small peaks on the left side of the main peak show failure sequences.

where probes in lanes are running across the aminated polypropylene film. Full film 1×64 array is cut into strips of dimension 0.5×8 cm^2. Examine signal uniformity after hybridization with fluorescein-labeled target. Analyze strips using VistraImager. Coefficients of variation (CVs) of some of the bands are described.

3.7. Synthesis of Synthetic Targets and Primers for PCR

Synthesize labeled and unlabeled synthetic targets and primers for PCR using conventional phosphoramidite chemistry *(21)* on Oligo 1000 DNA synthesizer. The 5′-end of the oligonucleotides are attached with following labels according to the protocols provided by the manufacturer.

1. 5′-Biotin-ON.
2. 5′-Fluorescein-ON.
3. 5′-CY3 or CY5.

Table 4
Uniformity of Synthesis Bands Based on Standard Deviation Analysis

Standard Deviation		
Standard deviation within a strip no. 2	(14 bands)	10.49%
Standard deviation within a strip no. 7	(14 bands)	9.45%
Standard deviation within a strip no. 11	(14 bands)	10.44%

Labeled oligonucleotides (targets, primers, and so on) and PCR amplified products are purified by using Chroma Spin™ columns according to the protocol provided by the manufacturer. The purified samples are then used as such for array hybridization studies as described in Chapter 10.

4. Notes

1. Probe is defined as an oligonucleotide; a single stranded sequence of DNA, typically 6–30 bases that is attached to the solid surface. Probe can be attached through 3′- or 5′-end to the polypropylene surface. DNA sample (single- or double-stranded sequence) in solution is defined as a target.
2. Derivatized hydroxyl and carboxylated polypropylene materials can be obtained under appropriate conditions for array synthesis. Plasma derivatization of polypropylene is available from Science Plasma, Foster City, CA.
3. Synthesized reverse phosphoramidities (3′-DMT-deoxynucleotides), 3′- and 5′-succinate nucleosides in-house. Reverse phosphoramidites are commercially available from Glen Research, Sterling, VA.
4. Biotinylated hybrids conjugated with streptavidin alkaline-phosphatase and enzyme-labeled fluorescence (ELF) substrate (ex: λ_{365} nm; em: λ_{515} nm) for signal amplification. For more information *see* Chapter 14.
5. Sometime light staining of the film is noticed with commercially available oxidizer reagent. But the problem is eliminated by using dilute oxidizer solution.
6. Paul Dryden from Surface Analysis Lab, Dept. of Bioengineering, University of Utah, Salt Lake City, UT, performed ESCA.
7. Activated film itself acts as a gasket. Generally, no leakage of any fluid is observed. An additional layer of the polypropylene film can be used for extra precaution, if needed.

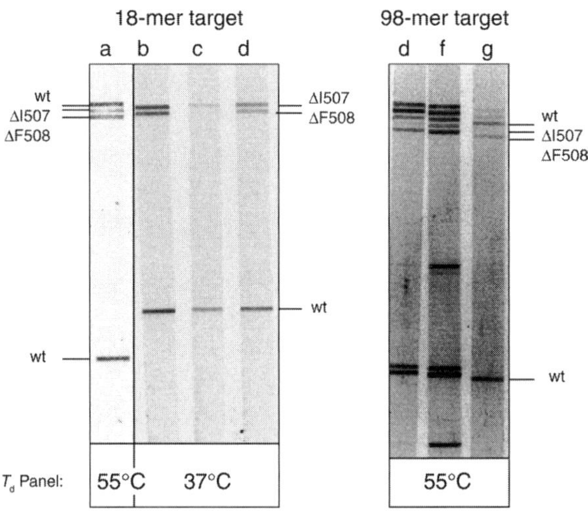

Fig. 10. The 18-mer and 98-mer 5′-fluorescinated oligonucleotide targets hybridized to wild-type I507/F508 probes, but failed to adequately discriminate between wild-type and mutant probes using the 55°C **A** in 6X SSPE, 0.02 *M* ethylenediaminetetraacetic acid, 0.01% SDS, pH 7.4 buffer for 1 h at 30°C. The use of dextran sulfate (10% solution in water) in **F** shows accelerated probe-target interaction, but further increased the level of cross-hybridization for the 98-mer target with the 55°C panel as compared with **E** in 6X SSPE buffer. This effect is less pronounced for the 18-mer target on the 37°C **D** as compared with hybridization in 6X SSPE buffer (*see* **B**). The use of 6X SSPE in combination with formamide (30%) shows best level of discrimination for both the 18-mer target on the 37°C **C**, as well as, the 98-mer target on the 55°C **G**.

8. CF probes originally designed for use in a Southern (forward, Type 1) blot application. Sue Richards and Thomas Caskey, Baylor College of Medicine, Houston, TX, provided the sequences.

9. Array panels described in **Figs. 4** and **5** have average T_d values approx 55 and 37°C, respectively. Hybridization to CF array panels is shown in **Fig. 10**. The 18-mer and 98-mer synthetic targets to wild-type I507/ΔF508 failed to adequately discriminate between wild-type and mutant probes using the 55°C **A** under a variety of salt stringencies. The use of dextran sulfate (**F**), which is known to accelerate probe-target interaction further increased the level of cross-hybridization for the 98-mer target with the 55°C panel. This effect is less evident for the 18-mer target on the 37°C **D** as compared with hybridization under high salt conditions observed in **B**. Although the use of high salt in combination with formamide appeared to achieve the best level of discrimination for both the 18-mer target on the 37°C **C**, as well as, the 98-mer target on the 55°C **G**, but these conditions

appeared to reduce the overall level of hybridization. So, probe design and optimization of hybridization conditions are important issues *(22)*.

10. 5'- to 3'-array can be constructed with reverse phosphoramidites, for example, PP-5'-CGCTATCTCGCAAGG-AGGA-3'. The array panel is used for hybridization, chain extension and for terminal transferase reactions.

11. Array synthesis without the capping step was performed and it was found that the quality of array is still satisfactory for hybridization studies.

12. Make sure during analysis instrument is on reverse polarity. Follow manufacturer's instructions as provided in the eCAP ssDNA 100-R kit (Beckman Coulter).

Acknowledgments

We thank Jim Osborne for his encouragement and supporting the project.

References

1. Southern, E. M., Maskos, U., and Elder, J. K. (1992) Analyzing and comparing nucleic acid sequences by hybridization to arrays of oliogonucleotide: evaluation using experimental models. *Genomics* **13,** 1008–1017.

2. Roger, Y. H., Baucom, P. J., Huang, Z. J., Bogdanov, V., Anderson, S., and Jacino, M. T. B. (1999) Immobilization of oligonucleotides onto glass support via disulfide bonds: a method for preparation of DNA microarrays. *Anal. Biochem.* **266,** 23–30.

3. Chrisey, L. A., Lee, G. U., and O'Ferrall, C. E. (1996) Covalent attachment of synthetic DNA to self-assembled monolayer films. *Nucleic Acids Res.* **24,** 3031–3039.

4. Lamture, J. B., Beattie, K. L., Burke, B. E., et al. (1994) Direct detection of nucleic acid hybridization on surface of a charge coupled device. *Nucleic Acid Res.* **22,** 2121–2125.

5. Nikiforov, T. T. and Rogers, Y. H. (1995) The use of 96-well polystyrene plates for DNA hybridization-based assays: an evaluation of different approaches to oligonucleotide immobilization. *Anal. Biochem.* **227,** 201–209.

6. Matson, R. S. and Rampal, J. B. (2003) DNA arrays: past, present, and future. *Am. Genomic/Proteomic Technol.* **April/May Issue,** 37–44.

7. Khrapko, K. R., Lysov, Y. P., Khorlin, A. A., et al. (1991) A method for DNA sequencing by hybridization with oligonucleotide matrix. *J. DNA Sequencing Mapping* **1,** 375–388.

8. Fodor, S. P. A., Read, J. L., Pirrung, M. C., Stryer, L., Lu, A. T., and Solas, D. (1991) Light-directed, spatially addressable parallel chemical synthesis. *Science* **251,** 767–773.

9. Pirrung, M. C., Fallon, L., and McGall, G. (1998) Proofing of photolithographic DNA synthesis and 3',5'-dimethoxybenzoinyloxycarbonyl-protected deoxynucleoside phosphoramidities. *J. Org. Chem.* **63,** 241–246.

10. Pease, A. C., Solas, D., Sullivan, E. J., Cronin, M. T., Holmes, C. P., and Fodor, S. P. A. (1994) Light-generated oligonucleotide arrays for rapid DNA sequence analysis. *Proc. Natl. Acad. Sci. USA* **91,** 5022–5026.

11. Matson, R. S., Rampal, J. B., and Coassin, P. J. (1994) Biopolymer synthesis on polypropylene supports. 1. Oligonucleotide. *Anal. Biochem.* **217,** 306–310.

12. Koester, H. and Coull, J. M. (1990) Membranes with bound oligonucleotides and peptides. US Patent No. 4,923,901.

13. Rampal, J. B. (2000) Covalent attachment of biomolecules to derivatized polypropylene supports. US Patent No. 6,013,789.

14. Wehnert, M. S., Matson, R. S., Rampal, J. B., Coassin, P. J., and Caskey, C. T. (1994) A rapid scanning strip for tri- and dinucleotide short tandem repeats. *Nucleic Acids Res.* **22,** 1701–1704.

15. Hollahan, J. R. and Stafford, B. B. (1969) Attachment of amino groups to polymer surfaces by radiofrequency plasmas. *J. Appl. Polymer Sci.* **13,** 807–816.

16. Matson, R. S., Rampal, J., Pentoney, S. L., Anderson, P. D., and Coassin, P. (1995) Biopolymer synthesis on polypropylene supports: oligonucleotide arrays. *Anal. Biochem.* **224,** 110–116.

17. Weiler, J. and Hoheisel, J. D. (1996) Combining the preparation of oligonucleotide arrays and synthesis of high-quality primers. *Anal. Biochem.* **243,** 218–227.

18. Reddy, M. P., Rampal, J. B., and Beaucage, S. L. (1987) An efficient procedure for the solid phase tritylation of nucleosides and nucleotides. *Tetrahedron Lett.* **28,** 23–26.

19. The evaluation and purification of synthetic oligonucleotides (1987) DNA Synthesis User Bulletin by Applied Biosystems no. 13.

20. Warren, W. J. and Vella, G. (1994) Analysis and purification of synthetic oligonucleotides by high-performance liquid chromatography, in *Protocols for Oligonucleotide Conjugates,* (Agrawal, S., ed.), Humana, Totowa, NJ, pp. 233–264.

21. Matteucci, M. D. and Caruthers, M. H. (1981) Synthesis of deoxyoligonucleotides on polymer support. *J. Am. Chem. Soc.* **103,** 3185–3191.

22. Rampal, J. B. (1999) Hybridization detection by pretreating bound single-stranded probes. US Patent No. 5,985,567.

12

Detecting Ligated Fragments on Oligonucleotide Microarrays

Optimizing Chip Design, Array Multiplex Ligation-Dependent Probe Amplification Modification, and Hybridization Parameters

Ian R. Berry, Carol A. Delaney, and Graham R. Taylor

Summary

Copy-number polymorphisms at specific genomic loci have been implicated in numerous human and animal disease phenotypes. Multiplex ligation-dependent probe amplification (MLPA) is a molecular genetic technique allowing targeted quantification of genomic copy-number changes (deletions and duplications), with potential for multiplexing up to 50 loci in one assay, and resolution down to the single nucleotide level. Modification of the MLPA technique to include Cy-labeled amplification primers permits parallel product detection by capillary electrophoresis and microarray hybridization. Detection and quantification of products by sequence-specific hybridization rather than size-specific capillary electrophoresis increases the potential for probe multiplexing possible in one assay and also allows for more flexible and efficient MLPA probe design. Protocols for the printing of synthetic oligonucleotide probe-sets for the detection of MLPA products, MLPA-probe amplification using array-compatible primers, and parallel product detection by quantitative capillary electrophoresis and microarray hybridization have been optimized.

Key Words: Copy number; Cy-labeled PCR; MLPA; oligonucleotide array; printing; quantitative capillary electrophoresis.

1. Introduction

Oligonucleotide microarrays offer a high density and highly flexible platform for interrogating mixes of nucleic acid fragments by means of hybridization (*1*). Within the field of molecular genetics, the use of synthetic oligonucleotides as detection probes offers numerous advantages over traditional alternatives, such as genomic insert clones (e.g., bacterial artifical chromosomes [BACs]), PCR

From: *Methods in Molecular Biology, vol. 381: Microarrays: Second Edition: Volume 1: Synthesis Methods*
Edited by: J. B. Rampal © Humana Press Inc., Totowa, NJ

products, or cDNAs. Synthetic oligos can be tailored to detect small fragments (such as PCR products) and microarrays are sufficiently sensitive in the optimal experimental conditions to be able to differentiate single-base changes by hybridization specificity *(2)*. Oligonucleotide arrays also possess great flexibility as they do not rely on time-consuming cloning and amplification techniques and specific sequences can be made to order rapidly and with a high degree of purity. Furthermore, oligonucleotide probe/target detection utilizes potentially small fragments, which might be incorporated as modifications of existing molecular genetic techniques. Incorporating a set of known "zip-code" tag oligosequences into PCR primers, molecular probes, and so on *(3)* can allow semiquantitative analysis of the results of different molecular techniques by hybridization to a universal array of antitags. This allows the potential analysis of a far greater number of sequences in multiplex than would be possible with traditional molecular genetic detection methods such as capillary electrophoresis and Southern blotting. Although, only two differentially labeled nucleic acid mixtures (traditionally, a "test" and "normal control" comparison) may be interrogated in one array using the established cyanine-derived dye-set (Cy3 and Cy5), the potential for additional dye-labels to generate triple and quadruple target arrays *(4)*, and the vast number of fragments (>10,000) that might be interrogated on one array make oligonucleotide microarrays a promising tool for increasing the scope of molecular genetic analysis.

Copy-number changes caused by genomic rearrangement, have been implicated as the sole or partial genetic cause of a large number of diseases *(5)*, and detection of such changes at a genetic level can be diagnostically important *(6)*. Multiplex ligation-dependent probe amplification (MLPA) is a revolutionary molecular genetic technique allowing determination of the copy number (and potentially single-nucleotide polymorphisms/mutations) at fewer than 50 probe-specific loci in a sample of genomic DNA *(7)*. Adjacent M13-derived probe pairs are designed complementary to the genomic region of interest and ligated across the junction by a thermostable ligase. Probes are not ligated, where the binding site has been deleted or where the nucleotides at the ligation junction are mismatched. The number of hybridized and ligated probe fragments is proportional to the copy number of the probe-binding site in the DNA sample. PCR amplification of the ligated fragments with primers (one of which is fluorescently labeled) for the common probe tails, therefore, generates a quantitative fluorescent signal for each probe. Incorporation of several probes for regions of normal genomic-dosage provides internal "normal" controls. Signals might be quantified by capillary electrophoresis, with the probe fragments separated at characteristically different sizes as a result of a variable length "stuffer" fragment incorporated into one of the M13 probes. MLPA reactions might be designed in a similar way using synthetic oligonucleotide probes *(8)*. These are technically easier to produce than the M13-derived fragments, but have a ceiling on their useful size at around 100 nucleotides because of the limitations of oligosynthesis, which limits their capacity for multiplexing *(7)*.

Adaptation of MLPA for detection on an oligonucleotide microarray platform (based on fragment sequence) rather than capillary electrophoresis (based on fragment size) might liberate MLPA-probe design from the requirement for the variable-sized "stuffer" fragment. This might allow potential for developing higher-order multiplexes (increasing the resolution of the technique). Furthermore, use of synthetic oligonucleotide probes rather than M13-derived fragments improves the ease with which new MLPA-based assays might be designed and optimized.

MLPA products might be detected on microarrays either by using a 45-mer oligonucleotide probe-set complementary to unique sections of each probe fragment in a specific reaction, or potentially by incorporating 25-mer "zip-code" sequences into the probes of proprietary assays, detected by a universal 25-mer array. MLPA reactions using M13-derived or synthetic oligoprobes might be modified to include Cy3- and Cy5-labeled primers. Oligonucleotide array probes are printed and immobilized on glass slides using the "acrydite" chemistry. MLPA products are hybridized to these probes in a reverse-labeled "test" vs "control" (known normal genotype) reaction. In optimized hybridization conditions, synthetic oligonucleotide array probes report differential ratios of copy number between Cy3- and Cy5-labeled targets as a comparable change in signal-intensity ratio, when compared with the probes of normal dosage. MLPA reactions incorporating differentially sized products might be detected in parallel by electrophoresis and microarray by incorporating an array-compatible dye on the forward primer and 6-carboxyfluorescein (6-FAM) on the reverse primer.

2. Materials

2.1. Printing Oligonucleotide Probes

1. Synthesized oligonucleotide capture probes: 25- or 45-mer synthetic oligos complementary to sequence within the forward strand (Cy-labeled strand) of the MLPA products. G/C content should be more than 40% and T_m of the probes (which might be checked using an automated T_m calculator *[9]*) should have as little variability as possible. Avoid using sequences with significant (25%+) complementarity to more than one MLPA product. If using a commercially supplied kit, refer to probe details to define suitable sequences. If desired, a defined 25–30 base insertion in the probe might be used as a universal "zip-code" detection fragment for designing MLPA probes. Probes should contain a 5′ acrylamide (acrydite) modification (Matrix Techcorp, Hudson, NH) (*see* **Note 1**).
2. Negative control DNA probes: these might be synthesized oligonucleotides of similar size to the capture probes, designed using the same parameters (including 5′ acrylamide/acrydite linker) (*see* **Note 2**).

3. "Landing Light" orientation probe. A 10-mer synthesized oligonucleotide fragment dual-labeled; 5′ acrylamide/acrydite and 3′ with a detectable fluorophore. In our hands, an oligo of sequence ATTCAGTCCT with a 3′ ROX (Sigma Genosys, The Woodlands, TX) or Cy3 label has been effective. ROX is generally detected by CCD and laser scanners in the Cy3-specific detection channel, and is preferable as a result of its photostability.
4. Sonicating bath (XB2, Grant, Fisons, Leicester, UK).
5. Pin wash solution: 0.1% (w/v) Tween-20.
6. Aqueous printing buffer: 1X EZ-Rays™ acrydite spotting buffer (Matrix Techcorp).
7. 15% (w/v) Glycerol, 0.005% Sarkosyl (Matrix Techcorp) (*see* **Note 3**).
8. MicroGrid MGII compact (Genomic Solutions, Ann Arbor, MI).
9. GeneScan software (Applied Biosystems, Foster City, CA).
10. EZ-Rays universal slides (Matrix Techcorp).
11. EZ-Rays activator solution (Matrix Techcorp) made up 14 g/L (or 0.014 g/mL) in water.
12. 1X EZ-Rays Quench Solution (Matrix Techcorp).
13. 0.5 *M* Sodium hydroxide.

2.2. Modifying the MLPA Reaction for Array Detection

1. MLPA kit (MRC-Holland, Amsterdam, The Netherlands): the kits contain all the reagents required for generating MLPA products with 6-FAM labeling, which might be detected by capillary electrophoresis. Kit should include: MLPA buffer, SALSA probes, Ligase 65 buffers A and B, Ligase 65, SALSA PCR buffer, SALSA PCR primers, enzyme dilution buffer, and SALSA polymerase. The modifications are optimized on the *BRCA1* gene kit (P002), but are applicable to any MLPA kit using the same amplification primers.
2. MLPA primer mix: a common reverse primer of sequence 5′-GTGCCAGCAAGATCCAATCTAGA-3′ is 5′ labeled with 6-FAM to enable capillary electrophoretic detection of the Cy5-labeled products. The forward primer, of sequence 5′-GGGTTCCCTAAGGGTTGGA-3′, is 5′ labeled with Cy3 or Cy5 to enable array detection. For each array experiment, the sample to be tested and the normal control may be differentially labeled with Cy3 and Cy5 by choosing either forward primer. Primers are routinely stored at −20°C in 50% glycerol (w/v) 1X TE (Tris-EDTA [ph 7.0]) solution, at a concentration of 100 pmol/µL. Primer mix is made up to 50 µL using 8 µL of the 6-FAM-labeled reverse primer stock, 8 µL of the Cy3 or Cy5 primer stock (depending on which label is required for the MLPA array products), 5 µL from a stock of 20 m*M* dNTPs (set of dATP, dGTP, dCTP, and dTTP all at 20 m*M* concentration) (Amersham Biosciences, Piscataway, NJ).

2.3. Detecting MLPA Products by Capillary Electrophoresis on the ABI 3100

1. ABI Rox-500 size standard (Applied Biosystems).
2. Deionized formamide ("Hi-Di" Formamide, Applied Biosystems).
3. 96-Well 0.2 reaction plates and 96-well septa (Applied Biosystems).

4. GenoTyper (Applied Biosystems).
5. POP4 polymer (Applied Biosystems).

2.4. Detecting MLPA Products by Microarray Hybridization

1. MinElute PCR purification columns (Qiagen, Valencia, CA).
2. Milli-Q purification systems (Millipore, Billerica, MA).
3. 5X EZ-Rays hybridization/wash buffer (Matrix Techcorp) working solution (*see* **Note 4**).
4. Micro-Array Hybrid Chambers (Camlab, Cambridge, UK).
5. 21 × 26 mm^2 Deckglaser glass cover slips. Larger cover slips might be used for larger arrays (Menzel-Glaser, Braunschweig, Germany).
6. 3MM filter paper (Whatmann, Maidstone, UK).
7. Vulcanising rubber solution (Weldtite, Barton-on-Humber, UK).
8. Shake 'n' Stack hybridization oven (Thermo Electron, Waltham, MA) or equivalent unit with temperature range 35 to 70°C and shaking/tilting agitation platform.
9. Slide array location template; sheet of transparent polythene with the outline of an array slide drawn on the underside, the array locations accurately marked (this can be achieved by printing a "dummy" array on a plain glass slide and tracing the slide and array outlines). The slide may be cleaned between uses by isopropanol wipes and DNA-Away (Molecular BioProducts, San Diego, CA).
10. Wash 1: 5X EZ-Rays hybridization/wash buffer.
11. Wash 2: 1X EZ-Rays hybridization/wash buffer.
12. Wash 3: 0.5X EZ-Rays hybridization/wash buffer.
13. 2X Tris-ethylenediaminetetraacetic acid (EDTA), pH 7.6.
14. Easy-dip slide washing stations and vertical rack (VWR International, West Chester, PA), or equivalent slide washing stations (requirements: slides can be fully immersed and agitated, container may be fully closed).

3. Methods

3.1. Printing Oligonucleotide Probes

Oligonucleotide probes are deposited by contact printing using a split-pin arrayer robot, such as the MicroGrid MGII Compact. The best printing results for oligonucleotides are achieved by maintaining a systematically low level of run-to-run variation in terms of the printing program, source solutions, and atmospheric conditions. Rigorous control of these variables is vital to minimize printing batch variations in spotting quality and deposition. The minimum possible experimental variation is introduced into the hybridization steps; this is a universal aim for all oligonucleotide printing procedures.

Split pins (otherwise known as "quill" pins) are preferable to solid pins as a means of source deposition. Split pins can generate multiple spots from material collected from just one source visit, delivering the source material through an internal reservoir capillary by kinetic force on contact with the slide.

If optimal conditions are maintained throughout the print run, split pins can generate spots with a low degree of size and morphology variation, and print more rapidly than solid pins from a smaller required source volume. Ceramic pins are the current "gold standard," offering a reported longer life span and less variable printing than metal alternatives, although in our hands tungsten pins (MicroSpot Quill, Genomic Solutions) have produced excellent results in small-scale assays.

The best printing results are obtained from microarray printers that have integral pin-washing stations (the option of static and circulating baths, including an automated pin drier, is preferable), internal controls for relative atmospheric humidity and (if possible) temperature, and a printing arm on an *xyz*-axis. Ability to control the velocity at which the pins make contact with the printing table is useful, though by no means necessary, to establish an optimal print run.

The following protocol has been extremely successful as a means of delivering nonvariable, high-quality oligonucleotide probes to glass array slides, although investigators would be encouraged to optimize aspects of the protocol for their own printing set up and requirements (*see* **Note 5**).

1. The required number of EZ-Rays slides are activated by incubating with gentle agitation for 45 min in Activator Solution and rinsing twice in water. Slides are dried with compressed N_2 (*see* **Note 6**) or centrifugation at full speed in a benchtop centrifuge in a slide rack.
2. A 384-well plate (consult arrayer manufacturer's guidelines) is loaded sequentially with all the MLPA detection probe oligonucleotides, negative control oligonucleotides, negative control buffer (printing buffer containing no nucleic acid material), and ROX-labeled landing light. All samples are aliquoted at a concentration of 20 μM in 1X aqueous printing buffer.
3. The arrayer is programed to deposit one array onto each slide (multiple arrays may be printed only if closed hybridization chambers are used on each array). Each array should contain at least three replicate spots from each sample, spatially separated in a manner leaving space between each replicate equivalent to at least half the diameter of the array. An example of a 384-well printing program suitable for this method is shown in **Fig. 1**. A more economical alternative is to print a different probe spot in each pin position, allowing 384 samples to be printed from one source plate but requiring replicates to all be positioned within the same quarter of the array, using the same pin.
4. Printing pins are sonicated in pin wash solution for 2 min before the print run, immersing the tips approx 1 cm under the surface.
5. Printing is performed at a relative external humidity <40%, at a temperature of 19–23°C. Humidity inside the print chamber is set at 50%. After each source visit, 15 "prespots" are deposited at full speed onto a plain glass slide to blot away any material on the outside of the split pin shafts. Pins descend at full speed to the target slides, with a "soft-touch" set for the last 2 mm of descent into the slide. Pin

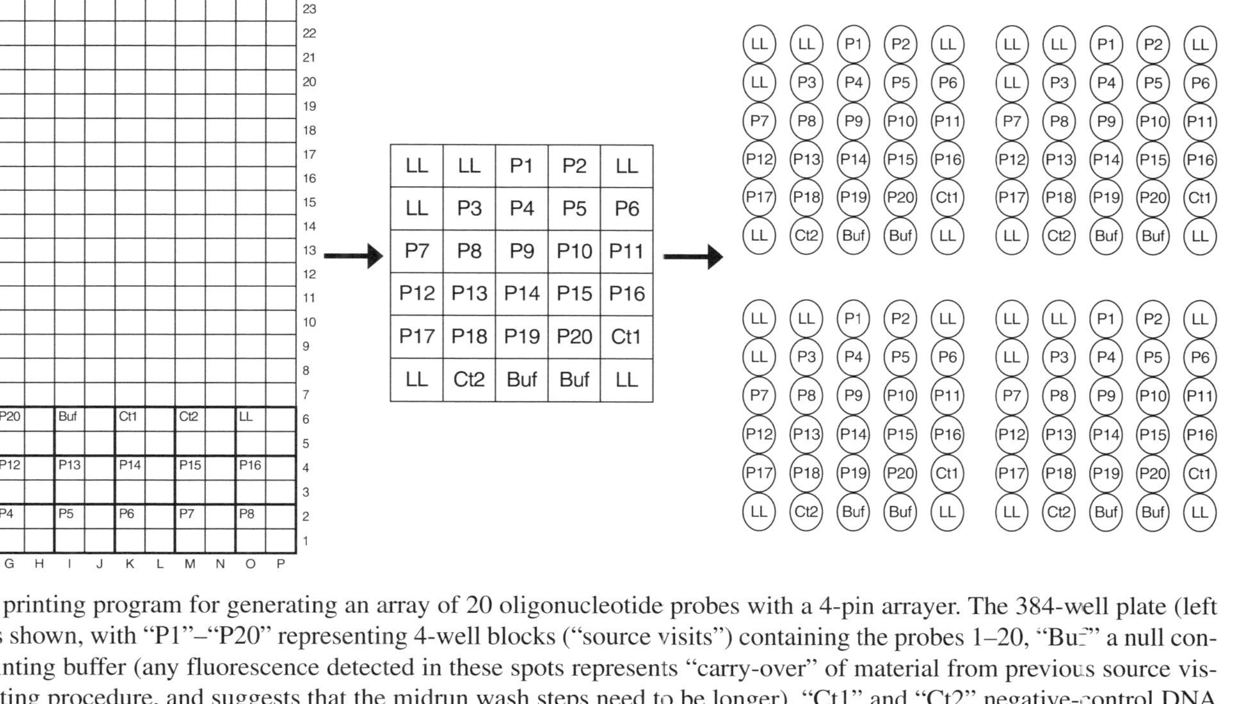

Fig. 1. Typical printing program for generating an array of 20 oligonucleotide probes with a 4-pin arrayer. The 384-well plate (left image) is set up as shown, with "P1"–"P20" representing 4-well blocks ("source visits") containing the probes 1–20, "Buf" a null control containing printing buffer (any fluorescence detected in these spots represents "carry-over" of material from previous source visits during the printing procedure, and suggests that the midrun wash steps need to be longer), "Ct1" and "Ct2" negative-control DNA spots (*see* protocol) and "LL" the ROX-labeled "landing light" probe. The pin array (center image) represents the pattern at which the pin arm is programed to deposit the spots, corresponding to the source visits in the 384-well plate. The printed array (right image) is laid down by the four pins as shown. Note that the presence of three landing light spots in the top-left-hand corner, and one in each other corner, of each pin array allows orientation of the array, and from a single action four replicate spots for each probe are deposited, with an intermediate distance of half the diameter of the array. Printing of all replicates on an array with a single pin contact allows multiple arrays (potentially several on one slide) to be generated very rapidly (although it does require enough source material to fill four wells in the source plate). Printing each probe with four different pin heads prevents a spotting problem with one pin eliminating an entire set of replicate probes. Ninety-six targets can be printed from each source plate using this method.

target depth is set at 0.8 mm below the level of the slide (*see* **Note 7**). Replicate spots are printed one spot at a time on each slide, rather than printing all replicates on one slide. This minimizes the possibility of all replicates on one slide being misprinted due to a midrun spotting issue.

6. Pin heads are washed twice prerun, twice postrun, and once before each new source visit. The wash procedure involves a 15 s dip in 0.5 *M* NaOH (*see* **Note 8**) with horizontal movement ("wiggle"), followed by 10 washes in the automated wash bath, each consisting of 1 s of water wash and 5 s of drying. If a large number of slides are to be printed, it is advisable wash and refills the pins after every 50–100 spots (exact figure might be optimized locally).

7. Slides are incubated overnight (or approx 12 h) at room temperature, 80% relative humidity. At this humidity the spots should not completely dry out and the covalent bonding of the active groups may progress to equilibrium. If spots are allowed to completely dry, "doughnutting," "halos" and other spot variability artifacts might occur. A diamond-tipped scribe might be used to make a small mark on the bottom right hand corner of each slide, to aid in orientation once the spots are washed off.

8. The processed slides are incubated with gentle agitation for 60 min in 1X Quench solution (which blocks the free active binding sites in the slide substrate) and briefly immersing twice in water. Slides are dried with compressed N_2 or centrifugation at full speed in a benchtop centrifuge in a slide rack.

3.2. Modifying the MLPA Reaction for Array Detection

MLPA kits supplied commercially will contain a set of reagent tubes allowing the steps of the reaction to be carried out quickly and efficiently. Proprietary MLPA protocols are based on the principles applied in the commercial kits but use locally derived reagents *(8)*. All MLPA buffers, enzymes, and reagents in this protocol are used as supplied in the MRC-Holland kits. Any alternative reagents and procedures (i.e., in a proprietary protocol) that have been previously optimized may be retained. The probe hybridization and ligation steps are not altered from the original protocols; the end-labeled primer modifications are introduced at the PCR step.

1. DNA-denaturation and hybridization of the MLPA probes: 0.2-mL microtubes are the most appropriate vessel for carrying out the MLPA reactions (*see* **Note 9**). Five microliters of genomic DNA at a concentration of 50 µg/mL in 0.8X Tris-EDTA (i.e., 50 ng of genomic DNA) is denatured by heating for 5 min at 98°C, then cooling to 25°C, in a Peltier-heating thermal cycler. 1.5 µL MLPA probe mix and 1.5 µL MLPA buffer are added to each sample (as a "master-mix") and mixed by pipetting. The mixture is then denatured at 95°C for 1 min, then incubated at 60°C for 16 h to allow sufficient time for the probes to hybridize. Investigation after this stage can identify any tubes with reduced sample volume (as a result of evaporation), which are liable to probe-to-probe variation in the result.

2. Ligation of specifically bound probes: for each sample, a mix of 3 µL Ligase-65 Buffer A, 3 µL Ligase-65 Buffer B, 25 µL water, and 1 µL Ligase-65 enzyme

(all prepared on ice) is added and mixed by pipetting while tubes are in the thermal cycler at 54°C. The mixture is incubated at 54°C for 15 min (facilitating the specific ligation of the probes by the thermostable ligase enzyme) then heated at 98°C for 5 min, denaturing the probe–genomic DNA complex and the ligase enzyme. The ligation products may be stored at 4°C following this stage of the process, and may retain quantitative ligated probe concentrations for several months.

3. PCR amplification of ligated fragments: for each sample, 2 µL MLPA primer mix (Cy3 or Cy5), 2 µL enzyme dilution buffer, 5.5 µL water and 0.5 µL SALSA polymerase is prepared on ice as a polymerase "master-mix" (*see* **Note 10**). Four microliters SALSA PCR buffer, 26 µL water, and 10 µL ligation mix (from **step 2**) are pipetted into fresh 0.2-mL reaction tubes. Ten microliters of the polymerase master mix is added to the reaction tubes while held at 60°C on the thermal cycler. A 5 min 95°C heating step is followed by 33 cycles of the following PCR amplification: 30 s at 95°C, 30 s at 60°C, and 60 s at 72°C. A final extension step of 20 min at 72°C is advisable. MLPA PCR products should be stored in the short-term in dark conditions at 4°C. Although representative electrophoresis and array results can be obtained after 1 yr of storage (unpublished data), it is good practice to analyze the products within 2 wk.

3.3. Detecting MLPA Products by Capillary Electrophoresis on the ABI 3100

The following protocol has been tested in our laboratory in a diagnostic setting on the 3100 Avant. It is adequate for the separation and detection of all MRC-Holland kits, modifications thereof (such as the Cy-labeled MLPA described in **Subheading 3.2.**) and, with minor optimization, including other commercial kits. MRC-Holland report that separation can be achieved using similar run settings on several commercial electrophoresis units, including the ABI 3100 (Avant, Applied Biosystems) *(10)*.

Spreadsheets for downstream analysis and quantitation of MLPA data are available for download from the World Wide Web *(11)*.

1. Using any compatible 0.2-reaction plate, 8 µL of deionized formamide and 1 µL ROX-500 size standard is dispensed into a number of wells equal to the number of MLPA products to be analyzed. One microliter of each MLPA product is added to each well, in a predefined order. Care must be taken to minimize incident light on the Cy-labeled products and capillary electrophoresis plate. The plate is vortexed to mix all samples, and briefly centrifuged to return the products to the bottom of the wells. A 96-well septum is applied to seal the plate. The plate is incubated at 98°C for 2 min to denature the products, and immediately transferred to a deep tray containing ice. Products might be stored in the dark for up to 10 h before electrophoresis, though best results will be achieved by running the electrophoresis immediately.

2. The order of samples in the plate is programmed into the electrophoresis station as per the manufacturer's instructions. The "red" channel is checked as the size standard, and the "no size standard" analysis module (*see* **Note 11**) should be selected to allow

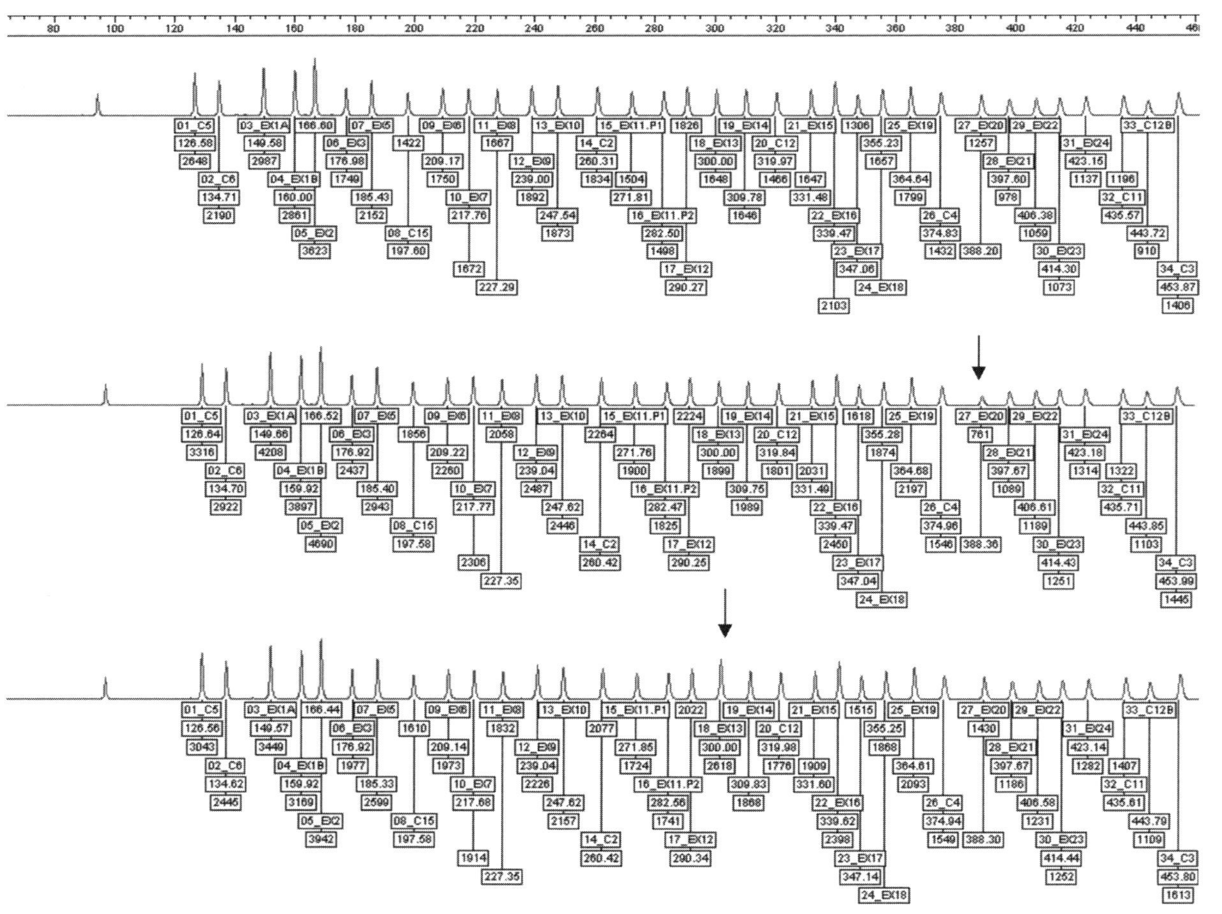

256

manual selection of sizing fragments after the run. The selected dye-set/filter must be able to detect the FAM-labeled MLPA products and ROX-labeled size standard in different detection channels with no significant "bleed-through" (contact manufacturer for current standard). The selected run module should inject at 1 kV for 15 s, running at a current of 100 µA and voltage of 15 kV for 25 min and at 60°C. Capillaries (32 cm) filled with POP4 polymer generate adequate results on the ABI 3100. For proprietary (and non-MRC-Holland) MLPA kits, these parameters might require optimization; manufacturer's guidelines should be followed.

3. The completed run files are exported to GeneScan software, or equivalent. A new project comprising all the MLPA products should be created. The ROX-500 size-standard peaks (red channel) are manually labeled (*see* manufacturer's product insert for correct sizes) and checked by eye for accuracy. The FAM-labeled SALSA and Cy5 products should be analyzed in the blue channel. In our experience the Cy3 is best detected *not* by analysing the FAM-labeled reverse strand products, but by analyzing the Cy3-labeled forward strand. The Cy3 dye tends to "bleed through" into the blue FAM channel but will usually be detectable specifically in the green or yellow channels (green, in the case of the ABI 3100 Avant run under the previously described parameters). For example, MLPA-electropherograms are shown in **Fig. 2**. The area containing the peaks is analyzed and exported to GenoTyper or equivalent software.

4. GenoTyper is used to label each peak with its identity, size in basepairs, and peak height (*see* **Note 12**). Peak heights for each labeled probe might be analyzed by eye or, more accurately, by exporting the data to a spreadsheet. In our experience, dosage results are best quantitated by peak height, rather than peak area. Examples of spreadsheet analysis of Cy-labeled MLPA dosage data is shown in **Fig. 3**.

Fig. 2. *(Opposite page)* Electropherograms of analyzed MLPA products using Cy3-labeled primers. The MRC-Holland *BRCA1* MLPA kit (P002) contains 34 probes, nine normal controls (from constitutive $n = 2$ regions of the different human chromosomes), and 25 exonic probes for the *BRCA1* gene. Two probes are present for BRCA1 exon 11, which is larger than all the other transcribed exons taken together. Deletions and duplications in this gene are implicated in the elevated risk of familial breast and ovarian cancer. The probes are labeled with an α-numeric tab representing the probe order and identity (e.g., 01_C5 being probe 1, chromosome 5 control, 03_EX1A being probe 3, exon 1A) approximate size in nucleotides (horizontal scale) and peak height in arbitrary units. The top image is a sample of normal genomic dosage (normal control) for all exonic probes. The middle image sample has a heterozygous exon 20 deletion (arrow), represented by an approx 50% decrease of the 27_EX20 probe ratio in comparison to the wild-type. The bottom image sample has a single exon 13 duplication (arrow) represented by an approx 50% increase of the 18_EX13 probe ratio in comparison to the wild-type. Cy3-labeled MLPA products were electrophoresed on the 3100 Avant and electropherogram traces analyzed and auto-labeled with size, peak height, and category (identity) using the GenoTyper software .

Category	height	chromo	height		Average	C5	C6	C15	C2	C12	C4	C11	C12B	C3	Max	Min	SD
C5	3316	C5	2689	C5	1.0168	1	0.9096	0.945	0.963	0.9979	1.088	1.0985	1.0036	1.145	1.1453	0.9096	0.0780064
C6	2922	C6	2155.3	C6	1.1178	1.0994	1	1.0389	1.058	1.097	1.196	1.2077	1.1033	1.259	1.2591	1	0.0857575
EX1A	4208	EX1A	3042	EX1A	1.1405	1.1217	1.0204	1.0601	1.08	1.1193	1.221	1.2323	1.1258	1.285	1.2847	1.0204	0.087503
EX1B	3897	EX1B	2845	EX1B	1.1294	1.1108	1.0104	1.0497	1.069	1.1084	1.209	1.2202	1.1148	1.272	1.2721	1.0104	0.0866472
EX2	4690	EX2	3608	EX2	1.0718	1.0541	0.9588	0.9962	1.015	1.0518	1.147	1.158	1.0579	1.207	1.2072	0.9588	0.0822267
EX3	2437	EX3	1728	EX3	1.1628	1.1436	1.0403	1.0808	1.101	1.1412	1.245	1.2563	1.1478	1.31	1.3098	1.0403	0.0892109
EX5	2943	EX5	2233.3	EX5	1.08	1.0099	0.972	1.0099	1.029	1.0663	1.163	1.1739	1.0724	1.224	1.2238	0.972	0.0871478
C15	1856	C15	1422.3	C15	1.0759	1.0582	0.9625	1	1.019	1.0559	1.152	1.1624	1.062	1.212	1.2119	0.9625	0.0825435
EX6	2260	EX6	1733	EX6	1.0752	1.0575	0.9619	0.9994	1.018	1.0552	1.151	1.1617	1.0613	1.211	1.2111	0.9619	0.0824928
EX7	2306	EX7	1688.3	EX7	1.1261	1.1076	1.0075	1.0467	1.066	1.1052	1.205	1.2167	1.1116	1.268	1.2685	1.0075	0.0863987
EX8	2058	EX8	1632.7	EX8	1.0393	1.0222	0.9298	0.966	0.984	1.02	1.112	1.1229	1.0259	1.171	1.1707	0.9298	0.0797359
EX9	2487	EX9	1936.3	EX9	1.059	1.0415	0.9474	0.9843	1.003	1.0393	1.133	1.1442	1.0453	1.193	1.1928	0.9474	0.081246
EX10	2446	EX10	1882	EX10	1.0716	1.0539	0.9587	0.996	1.015	1.0517	1.147	1.1578	1.0577	1.207	1.207	0.9587	0.0822135
C2	2264	C2	1767.3	C2	1.0562	1.0388	0.9449	0.9817	1	1.0366	1.13	1.1412	1.0426	1.19	1.1897	0.9449	0.0810334
EX11.P1	1900	EX11.P1	1507	EX11.P1	1.0395	1.0224	0.93	0.9662	0.984	1.0202	1.113	1.1231	1.0261	1.171	1.1709	0.93	0.0797529
EX11.P2	1825	EX11.P2	1489.7	EX11.P2	1.0101	0.9935	0.9037	0.9389	0.956	0.9913	1.081	1.0914	0.997	1.138	1.1378	0.9037	0.0774961
EX12	2224	EX12	1760	EX12	1.0419	1.0247	0.9321	0.9684	0.986	1.0225	1.115	1.1257	1.0284	1.174	1.1736	0.9321	0.0799334
EX13	1899	EX13	1624.3	EX13	0.9639	0.948	0.8624	0.8959	0.913	0.946	1.032	1.0415	0.9515	1.086	1.0858	0.8624	0.0739531
EX14	1989	EX14	1635.7	EX14	1.0026	0.9861	0.897	0.9319	0.949	0.984	1.073	1.0833	0.9896	1.129	1.1293	0.897	0.0769212
C12	1801	C12	1457.3	C12	1.0189	1.0021	0.9116	0.9471	0.965	1	1.091	1.1009	1.0058	1.148	1.1477	0.9116	0.0781738
EX15	2031	EX15	1645.3	EX15	1.0178	1.001	0.9105	0.946	0.964	0.9989	1.089	1.0996	1.0046	1.146	1.1464	0.9105	0.078084
EX16	2450	EX16	2013.7	EX16	1.0032	0.9866	0.8975	0.9324	0.95	0.9845	1.074	1.0839	0.9902	1.13	1.13	0.8975	0.0769635
EX17	1618	EX17	1322.3	EX17	1.0089	0.9922	0.9026	0.9377	0.955	0.9901	1.08	1.09	0.9958	1.136	1.1364	0.9026	0.0774005
EX18	1874	EX18	1582.3	EX18	0.9765	0.9604	0.8736	0.9076	0.925	0.9583	1.045	1.055	0.9639	1.1	1.0999	0.8736	0.0749166
EX19	2197	EX19	1774.3	EX19	1.0209	1.0041	0.9133	0.9489	0.967	1.0019	1.093	1.103	1.0077	1.15	1.15	0.9133	0.0783251
C4	1546	C4	1364.3	C4	0.9343	0.9189	0.8358	0.8684	0.885	0.9169	1	1.0094	0.9222	1.052	1.0524	0.8358	0.0716796
EX20	761	EX20	1219	EX20	0.5147	0.5062	0.4605	0.4784	0.487	0.5052	0.551	0.5561	0.5081	0.58	0.5798	0.4605	0.03949
EX21	1089	EX21	976.33	EX21	0.9197	0.9045	0.8227	0.8548	0.871	0.9026	0.984	0.9936	0.9078	1.036	1.0359	0.8227	0.0705563
EX22	1189	EX22	1018.3	EX22	0.9627	0.9468	0.8612	0.8948	0.911	0.9448	1.03	1.0401	0.9502	1.084	1.0844	0.8612	0.0738581
EX23	1251	EX23	1034	EX23	0.9919	0.9307	0.8924	0.9272	0.944	0.979	1.068	1.0778	0.9846	1.124	1.1236	0.8924	0.079661
EX24	1314	EX24	1068	EX24	1.0144	0.9977	0.9075	0.9429	0.96	0.9956	1.086	1.096	1.0013	1.143	1.1426	0.9075	0.077827
C11	1322	C11	1177.7	C11	0.9256	0.9103	0.828	0.8603	0.876	0.9084	0.991	1	0.9136	1.043	1.0425	0.828	0.0710093
C12B	1103	C12B	897.67	C12B	1.0131	0.9964	0.9063	0.9416	0.959	0.9943	1.084	1.0946	1	1.141	1.1412	0.9063	0.0777261
C3	1445	C3	1342	C3	0.8878	0.8732	0.7942	0.8252	0.841	0.8713	0.95	0.9592	0.8763	1	1	0.7942	0.0681117

3.4. Detecting MLPA Products by Microarray Hybridization

1. The MLPA products are purified using MinElute columns, one for each sample, following the manufacturer's protocol. Care must be taken to minimize the incidence of light during this procedure, preferably by wrapping the columns in aluminium foil and performing the purification in dark room conditions. The final elution of the purified product is performed by adding 20 µL hybridization buffer directly to the center of the membrane, allowing the column to stand for 3 min, and centrifuging into a clean 1.5-mL microtube at 10,000*g*.

2. The printed array slide(s), hybridization chambers (with a piece of moistened Whatmann paper in the internal well), cover slips and pipet tips are brought up to hybridization temperature in the oven (45°C for 45-mer probes, 40°C for 25-mer). Washing stations containing each wash solution are brought up to temperature. Two washing stations containing 2X Tris-EDTA are prepared.

3. Five microliters of each of the two MLPA products (a test sample and a "normal" control sample, differentially labeled with Cy3 and Cy5) are mixed by vortexing and denatured at 98°C for 2 min in a thermal cycler with heated lid. Tubes are immersed in a covered ice bath (minimizing incident light) for 2 min, and should be used immediately.

Fig. 3. *(Opposite page)*   Spreadsheet analysis of capillary electrophoresis results using Cy3-labeled MLPA products from the *BRCA1* kit (MRC-Holland). This data represents results taken from an exon 20 deletion sample (shown in **Fig. 2**) compared with the mean of three known normal controls. Data was exported from GenoTyper into an Excel (Microsoft) spreadsheet (available for download from the author's laboratory, *see* **ref. 11**). The "Category" column lists the probes from top to bottom in descending order of size (C5 being chromosome 5 normal control, EX1A being exon 1A), the first "height" column lists the peak heights for each probe in the tested sample, the second "height" column lists the mean heights of the normal control samples. Under the columns C5–C3, the ratio of the proportional peak height of each probe in the test sample is compared with the normalized mean of the normal controls. This is accomplished by dividing the ratio of each sample peak height to the appropriate sample control peak height (first "height" column) by the same ratio in the mean of the normal control samples. Thus, the relative ratio of each exonic peak in the test sample to all the normal control peaks is derived, and then divided by the same ratio in the control samples, allowing the height of each peak to be normalized against the internal normal control probes and external normal control samples. The resulting "dosage quotients" for each peak are indicated in the "Average" column. In this case, the exon 20 probe (shaded) shows dosage quotient of approx 0.5 (±0.15), corresponding to a heterozygous deletion of that exon. Single-exon deletion results should be validated by an alternative method to rule out possibility of an MLPA probe binding-site polymorphism. A result of 1.5 (±0.15) for any probe would indicate a duplication. The "SD" column lists the standard deviation for each probe's normalized ratios against the control probes. A score of <0.1 in these cells is desirable.

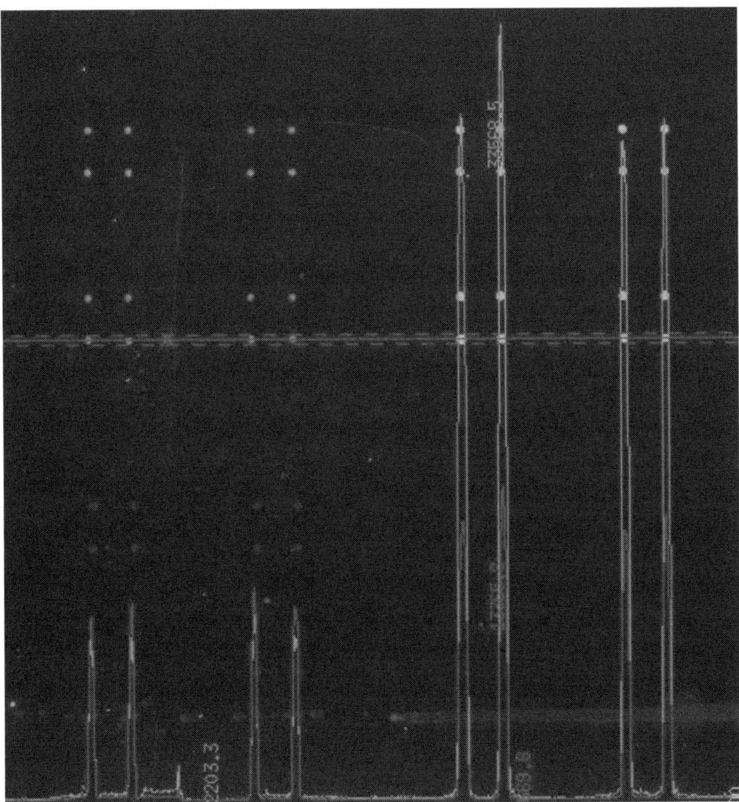

Fig. 4. Small array probeset hybridized to show a deletion result. Four array probes to detect the MRC-Holland *BRCA1* MLPA products are printed (16 replicates of each) in the four quarters of the array. The peak heights show signal intensity for Cy3 and Cy5 of the eight spots intersected by the dashed line. The four spots on the left are *BRCA1* exon 13 spots of normal dosage in both Cy3 and Cy5 channels, with peak heights almost identical for both channels (±0.2). The spots on the right show a deletion in the Cy3 channel for exon 3. The Cy3 peak is about half that of the Cy5 peak for these spots. When normalized against the other three probes on the array, the Cy3-labeled sample has a dosage quotient of 0.46 compared with the Cy5-labeled "normal control" sample, which would correspond to a heterozygous deletion of this genomic probe (0.5 ± 0.15). Signal intensities for the exon 3 probes are comparatively greater than for the exon 13 probes because of the smaller product size of the exon 3 products (this pattern of signal intensity diminishing as product size increases tends to be observed throughout the MLPA probe-set).

4. A cover slip is laid onto the array outlines on the slide template sheet. Eight microliters of the MLPA sample is pipetted evenly into the center of the cover slip, working rapidly and taking care not to introduce air bubbles. A slide is lowered array side downwards onto the cover slip, and pressed down. The sample should

spread evenly across the cover slip with no air bubbles and adhere to the slide. Any excess sample outside the cover slip is gently dabbed off with a piece of dry tissue. A vulcanising rubber seal is piped around the outside of the cover slip to seal the array. The slide(s) are immediately clipped, array side upward, into the hybridization chamber(s) and hybridized for 4–5 h in the oven.

5. Washes are then performed sequentially in slide staining containers. The slide(s) are removed from the hybridization chamber, rubber seal and cover slip removed by pealing off with a scalpel (the slide is orientated vertically and the scalpel blade cut down into the seal, pulling it away from the slide). The slide must *immediately* be immerzed in wash solution 1 in the staining rack (*see* **Note 13**).

6. Slides are incubated with gentle agitation in the staining rack at 45°C in wash buffer 1 for 5 min, then the rack is directly transferred to wash buffer 2 and incubated in the same conditions for a further 5 min. If experience demonstrates that this does not produce representative results after analysis with any probe-set, further 5 min incubation in wash buffer 3 may be applied (*see* **Note 14**).

7. The rack of slides is briefly (<5 s) dipped into the two Tris-EDTA washes in turn, to remove any salts and detergents from the previous wash steps, and the slide(s) are *immediately* dried by centrifugation at 10,000*g* in a benchtop centrifuge with slide rack, or by compressed N_2 or air (*see* **Note 13**).

8. Slides are scanned to detect intensity of the Cy3 and Cy5 signals using a CCD camera or confocal laser scanner. The test/control ratio for each probe is calculated by dividing the test signal by the control signal (Cy3/Cy5 or Cy5/Cy3) (*see* **Note 15**). The dosage quotient for the test probes can then be derived by comparing the test/control ratio for each probe to the mean test/control ratios for all the control probes. A dosage quotient of approx 0.5 corresponds to a deletion of test material, 1.5 to duplication. An example of a comparative deletion of target material hybridized to a small arrayed probe-set is shown in **Fig. 4**.

4. Notes

1. Alternative methods are available for attaching short oligonucleotides to glass slides, and many of these may be perfectly acceptable for the protocols described. Any 5′ attachment that covalently bonds the oligonucleotide to the slide (such as application of 5′ amine-modified oligonucleotides to an aldehyde [silylated] substrate) will effectively immobilize the array probes. However, it should be noted that methods relying on crosslinking of reactive groups within the DNA backbone to a reactive slide surface may be considerably less effective for oligonucleotide probes (<100 bases) than for larger cDNA or genomic fragments. Furthermore, printing buffer requirements, probe-to-slide attachment requirements, binding capacity of the slide substrate and hybridization behavior of the immobilized probes might be significantly different when other attachment chemistries are used. Preoptimization of printing and/or reference to supplier's guidelines would be essential in this case.

2. Negative control probes and landing lights should be based on nonrepetitive sequences, and should be BLASTed/checked for T_m complementarity with the *entire* length of the MLPA products to be hybridized to the array. Examples

include: GGAGCATTAAGACAGACAAATTCGAGTACATCTCTTACGAAG-
GAG and ATATGTTACAAACCGCACTGACTATTGAATCTTCCAATGCTCTTG
(45-mer), CAGCCTTACTGATCTCAAAGTTGAT and CGCTTATCAATGAG-
CATCTTAGAGC (25-mer), taken from the *Saccharomyces cerevisiae* YBL061C
chitin synthase III recruiter gene.

3. All buffers prepared for array work must be free from particulate matter and ionic
   salts, which can cause optical interference if deposited on slides. Double-distilled
   H_2O (resistivity >18 $M\Omega$-cm, filtered to remove all particles >0.22 μm) should be
   used. Milli-Q purification systems are suitable for laboratory water purification to
   these standards. Bottles and other solution containment vessels used must be
   clean, and should only be used specifically for array work. Many investigators
   purchase molecular grade water and use disposable plastic or Pyrex solution bottles to avoid having to wash and reuse solution containers, a process which may
   introduce optical contaminants. Unless otherwise specified, "water," when referred
   to in these protocols, confers to these standards.

4. 5X saline-sodium citrate (pH 7.0) (SSC), 0.1% sodium dodecyl sulfate (SDS)
   might be used as an alternative hybridization and washing buffer; with care this
   buffer yields slightly increased hybridization signals and lower background. SDS
   rapidly crystallizes as the solution dries and rapid slide handling is essential. The
   final SSC wash should not contain SDS.

5. Before printing onto expensive oligo-binding slides, and especially when using a
   new or nonoptimized array printer or printing procedure, it is essential to perform
   a "dummy" run on plain glass slides. The slides should be scrutinized to ensure all
   spots have been printed effectively. The print quality and morphology may be
   investigated by printing landing light spots with each pin which are diluted in spotting buffer containing a gradient of glycerol content (e.g., 0.5, 1, 2, 5, 10, 15, and
   20%). The landing light oligos should be bound to the slide according to the usual
   protocol. Scanning the slide and observing the quality and morphology of the
   spots can determine the appropriate composition of the printing buffer and any
   pin-specific spotting issues. An alternative experiment involves printing multiple
   test slides at different humidity levels, ranging from 40 to 65%.

6. Compressed N_2 or air may be used to dry the slides. However, compressors or gas
   cylinders which use oil as a lubricant must contain an effective oil and particulate
   trap to prevent deposition on the slides, such as the Quik-Dry air trigger (Matrix
   Techcorp).

7. Even in commercial arrayers designed for a high degree of accuracy, print platforms tend not to be completely level. Accurate and reproducible printing requires
   the software controlling the pin heads to be calibrated to the correct height of the
   slide platform each time the unit is assembled or moved to a new surface.

8. Sodium hydroxide is damaging to some pin types, always check with the manufacturer to ensure the pins are compatible.

9. The ratio of probe signals was found to be heavily dependent on the volume of the
   solution at the probe hybridization step of the MLPA protocol. Microtubes with a
   tendency to leak evaporated buffer during this 16 h incubation are therefore not

acceptable; performing a set of "dummy" hybridizations with water instead of reagents, and measuring the volume before and after the incubation, can demonstrate the efficacy of any brand of 0.2-mL microtube. The thermal cycler must also have a heated lid, to minimize evaporative deposition on the lids of the tubes. Experience has demonstrated that variation between thermal cyclers may cause some variation in the MLPA results, particularly when Cy-labeled primers are used, thus keeping using the same thermal block and same programme can improve reproducibility of results. We have found the 0.2-mL tubes supplied by Stratagene (La Jolla, CA) to be effective vessels for MLPA, and the range of Peltier thermal cycling units supplied by MJ Research, such as the PTC-200 (Genetic Research Instrumentation Ltd., Essex, UK), to generate reproducible results of diagnostic quality. Because of the requirement for adding reagents while tubes are in thermal cyclers, and the need to limit the incidence of light on the Cy-labeled primers, MLPA in this format can be technically demanding. Thus, attempting to process more than 15–20 samples at once without automated sample handling is not recommended.

10. Cy-labeled primers, probes and MLPA products are extremely sensitive to light. It is thus imperative to use opaque or foil-wrapped tubes for the storage of all Cy-labeled materials, and to take measures at each step of the MLPA and arraying process to minimize incident light. Performing the dilutions, MLPA PCR steps and array hybridizations in darkroom conditions is the ideal standard, but the use of opaque boxes for storing reagent tubes, covers for microplates/tubes, while working with them during the MLPA setup, and the use of opaque hybridization vessels and a low light environment for the array set up is a perfectly appropriate alternative.

11. If a "no size standard" module, or equivalent, is not programmed as a default option on your electrophoresis station, it is advisable to consult the manufacturer or supplier for advice. Alternatively, if ROX-500 size standard is commonly used on the unit, and an analysis module which correctly sizes the ROX-500 fragments is available, this might be used. It should be noted, however, that many capillary electrophoresis stations (including the 3100 Avant) will detect some "background" from the Cy3-labeled MLPA products in the "red" run channel, and thus manual selection of the correct sizing peaks is preferable. In cases where this "bleed-through" becomes an insoluble problem, further optimization of the run settings or use of a differently labeled size standard, which covers the same range of sizes (e.g., TAMRA-500, ABI) might be necessary.

12. Sizing of the MLPA probes by capillary electrophoresis rarely reports completely accurate results, with a typical standard error of 3–4 bp for each peak. Thus, when designing category labeling on GenoTyper or equivalent software, "binning" areas of 10 bp should be used to detect each probe, with the highest peak in the range ± 5 bp from the actual probe size flagged as the probe peak. The size of the binning area is obviously restricted in MLPA probe-sets, which contain consecutive probes with less than 10 bp difference in size. Statistically advanced methods for quantifying MLPA products are available *(12)*, but beyond the scope of this chapter.

13. Drying (even partially) of hybridization or wash buffers onto the slide surface will generate nonspecific background deposition and potentially ruin the slide. This is a major consideration when processing multiple slides, which should have their cover slips removed one by one and then be transferred to a rack fully immersed in a bath of wash buffer 1 at room temperature before continuing with the washing protocol. The same process can be repeated before transferring the slides to the Tris-EDTA rinse steps and drying step at the end of the slide washing procedure.

14. Optimization of hybridization wash steps is essential in producing representative signal intensities. Although this wash protocol has been effective in our hands for 45-mer probe-set, different probe-sets might respond better to other treatments. Optimization experiments for a probe-set should involve the following: performing the hybridization in forgiving conditions (e.g., 45°C in minimum 5X SSC or saline sodium phosphate EDTA salt solution, for short oligos), scanning this after rinsing off the labeled material with hybridization buffer, and then comparing the result to identical experiments performed in gradually more stringent washes. Increasing the hybridization temperate by 3–5°C, or adding deionized formamide (toxic) at an increased concentration of 5%, for each iteration may elucidate the best conditions. If further optimization is required for small probe-sets, using Cy3- and Cy5-labeled synthetic oligoreverse compliment targets to the arrayed probes might allow the hybridization behavior of individual probe-target pairs to be investigated. These can be used to demonstrate and overcome background effects caused by a single target species on mismatched probes, and to individually validate all probes. However, note that the behavior of single-stranded 25-mer oligo targets might not necessarily be totally representative of the behavior of longer double-stranded MLPA products in the same conditions.

15. Quality control should be applied to each printed spot of the hybridization. This can either be done by manually scanning the raw image and signal intensity data of all the replicate spots and discarding any apparent unusual or outlying results, which is time-consuming and potentially inaccurate, or by applying a software algorithm to automatically achieve the same thing. Many array analysis packages allow rejection of outlying replicate spots (or marking a comment, such as "rejected," in part of their data output) which lie outside certain parameters. It is good practice to reject any spots, where either the Cy3 or Cy5 signal intensity differs statistically significantly from the mean of the other replicates at greater than a 95% confidence interval. Automated "checking" algorithms which can reject misshapen, smudged, or high signal-to-noise ratio spots might be applied, where deemed appropriate.

Acknowledgments

The author would like to thank Jan Schouten for his help with MLPA methodology and modification, Mushtaq Mohammed for technical assistance, Fahd Al-Mulla, the coauthors, and all the staff at the DNA Lab, St. James's University Hospital, Leeds, for their advice and support during this work.

References

1. Lipshutz, R. J., Fodor, S. P., Gingeras, T. R., and Lockhart, D. J. (1999) High density synthetic oligonucleotide arrays. *Nat. Genet.* **21,** 20–24.
2. Ji, M., Hou, P., Li, S., He, N., and Lu, Z. (2004) Microarray-based method for genotyping of functional single nucleotide polymorphisms using dual-color fluorescence hybridization. *Mutat. Res.* **548,** 97–105.
3. Favis, R., Day, J. P., Gerry, N. P., Phelan, C., Narod, S., and Barany, F. (2000) Universal DNA array detection of small insertions and deletions in *BRCA1* and *BRCA2. Nat. Biotechnol.* **18,** 561–564.
4. Forster, T., Costa, Y., Roy, D., Cooke, H. J., and Maratou, K. (2004) Triple-target microarray experiments—a novel experimental strategy. *BMC Genomics* **5,** 13.
5. Lupski, J. R. (1998) Genomic disorders: structural features of the genome can lead to DNA rearrangements and human disease traits. *Trends Genet.* **10,** 417–422.
6. Taylor, C. F., Charlton, R. S., Burn, J., Sheridan, E., and Taylor, G. R. (2003) Genomic deletions in *MSH2* or *MLH1* are a frequent cause of hereditary nonpolyposis colorectal cancer: identification of novel and recurrent deletions by MLPA. *Hum. Mutat.* **22,** 428–433.
7. Schouten, J. P., McElgunn, C. J., Waaijer, R., Zwijnenburg, D., Diepvens, F., and Pals, G. (2002) Relative quantification of 40 nucleic acid sequences by multiplex ligation-dependent probe amplification. *Nucleic Acids Res.* **30,** E57.
8. White, S. J., Vink, G. R., Kriek, M., et al. (2004) Two-color multiplex ligation-dependent probe amplification: detecting genomic rearrangements in hereditary multiple exostoses. *Hum. Mutat.* **24,** 86–92.
9. Oligonucleotide Properties Calculator http://www.basic.northwestern.edu/biotools/oligocalc.html.
10. Generating Raw Data From Your Sequencer (ABi-3100, Beckman CEQ-8000) http://www.mlpa.com/pages/support_mlpa_aralysic_intopag.html.
11. Gene Dosage (including Regression-enhanced MLPA data analysis) http://Leedsdna.info/science/dosage.html.
12. Analysis of Raw MLPA Data by Statistical Methods. http://www.mlpa.com/pages/support_mlpa_analysis_overigpag.html.

13

Detection of Single-Nucleotide Polymorphisms in Cancer-Related Genes by Minisequencing on a Microelectronic DNA Chip

Alexandre Ho-Pun-Cheung, Hafid Abaibou, Philippe Cleuziat, and Evelyne Lopez-Crapez

Summary

The ability to realize simultaneous genotyping of multiple single-nucleotide polymorphisms or mutations is valuable in DNA samples from complex multigenic pathologies such as cancer. In this way, the complexity (number of hybridization units per chip) of the developed MICAM® DNA chip, and the orientation of the grafted pyrrole oligonucleotides, make it particularly well adapted to the analysis of single-nucleotide polymorphisms/mutations in multiple potential tumoral markers. The proposed genotyping methodology is based on solid-phase minisequencing, where oligonucleotides are designed to anneal immediately upstream of the polymorphism sites, and labeled dideoxynucleotides are used as substrates for polymerase extension. The developed assay was applied to the analysis of the *TP53* codon 72 polymorphism on DNA from cell lines and human colorectal samples.

Key Words: DNA chip; minisequencing; SNP detection; *TP53* codon 72; tumoral marker; electroaddressing; pyrrole residue.

1. Introduction

DNA microarrays are powerful tools to detect in parallel different genetic alterations. The ability to proceed to the simultaneous genotyping of multiple single-nucleotide polymorphisms (SNPs) or mutations is valuable for medical diagnosis purposes. Indeed, those DNA modifications may be associated to complex multigenic pathologies, such as diabetes *(1)*, cardiovascular diseases *(2)*, or cancer *(3)*.

Particularly in the oncology field, an increasing number of mutations and polymorphisms have been characterized in cancer related genes *(3–6)*. Detection of some of these alterations may contribute to the diagnosis of the pathology, as other alterations are linked with clinical drug response *(7)* and survival prognosis *(8)*.

From: *Methods in Molecular Biology, vol. 381: Microarrays: Second Edition: Volume 1: Synthesis Methods*
Edited by: J. B. Rampal © Humana Press Inc., Totowa, NJ

Although numerous studies that use high-throughput methods for expression analysis *(9)* or SNPs identification *(10)* have been published, less are the ones targeted to the genotyping of specific disease known mutations and/or polymorphisms.

The existing technologies applied to the detection of known mutations/-polymorphisms can be broadly classified as either hybridization- or enzyme-based methods. It has been shown that the use of enzymes brings an increased accuracy to the assay *(11)*. Several approaches such as cleavage (e.g., the Invader® assay Third Wave Technologies, Inc. *[12]*), ligation (e.g., oligonucleotide ligation assay [OLA] *[13]* and ligation detection reaction [LDR] *[14]*), or polymerase extension (e.g., pyrosequencing *[15]* or minisequencing *[16]*) have been developed. Minisequencing, also known as single nucleotide primer extension [SNuPE], genetic bit analysis [GBA], arrayed primer extension [APEX], and template-directed dye-terminator incorporation represents one of the simplest methods for known mutation/polymorphism detection. In this format, an oligonucleotide is designed to anneal immediately upstream of the polymorphism site, and labeled dideoxynucleotides are used as substrates for polymerase extension. The main advantages of primer extension are the power of discrimination between the homozygous and heterozygous genotypes, and the robustness of the assays.

Various methods were used to detect the incorporated dideoxynucleotide ranging from capillary electrophoresis and fluorescent homogeneous formats *(17–20)* to microarray-based assays using immobilized extension primers *(21,22)*. The use of the Microsystem for analysis in medicine (MICAM) polypyrrole DNA chip, was described recently, based on the electrocopolymerization of 5′-pyrrole-labeled oligonucleotides and pyrrole, to detect *K-ras* mutations in tumor samples *(23)*. Genotyping was made using allele-specific oligonucleotide hybridization, where labeled-target DNA was hybridized to the probes and fluorescently detected. This approach did not take advantage of the 3′-end of the probes, which remain accessible after the electropolymerization process *(24)*. Here, we report the ability to perform solid-phase minisequencing reactions directly onto the polypyrrole DNA chip thanks to the free 3′-ends of grafted probes. The developed assay was applied to the *TP53* codon 72 SNP genotyping of DNA from cell lines and human colorectal samples. This polymorphism was found to be associated to cellular apoptosis *(25)*. Moreover, it has been demonstrated that this SNP is an important factor for prediction of both treatment response and survival in clinical oncology *(26)*.

2. Materials

2.1. Electroaddressed Oligonucleotide Silicon Chip

The Apibio Company (Grenoble, France) designs and manufactures the MICAM® biochips based on probe coupling by electropolymerization of pyrrole (*see* **Table 1** and **Note 1**).

Table 1
Characteristics of the Grafted Probes

Probe Name	5′ → 3′ Sequence[a]	Function
P72S	GCTCCCAGAATGCCAGAGGCTGCTCCCC	*TP53* codon 72 genotyping probe
QC	CCTACCGTACCCCCTCCCGT	Hybridization/ extension control
SEP-A	[b]TTTAGCCTTAACGCCT**A**TGACGTCA	Extension control
SEP-C	TTTAGCCTTAACGCCT**C**TGACGTCA	Extension control
SEP-G	TTTAGCCTTAACGCCT**G**TGACGTCA	Extension control
SEP-T	TTTAGCCTTAACGCCT**T**TGACGTCA	Extension control
CP-Bio	[c]Bio-GCCTTGACGATACAGCTA	Detection control

[a]Each probe was 5′-labeled with a pyrrole residue and a $d(T)_{10}$ oligonucleotide linker.
[b]For each SEP, nucleotide in **bold** enabled the template-independent extension with the complementary chain terminator.
[c]Bio for biotin label.

2.2. Biological Material

1. Breast and pancreatic carcinoma cell lines with known *TP53* codon 72 genotype, SK-BR-3 (homozygous, CGC [Arg]) and MIA PaCa-2 (homozygous, CCC [Pro]), respectively, were purchased from the American-type culture collection and grown under standard conditions (*see* **Note 2**).
2. Tissue samples were obtained during surgery from patients with colorectal cancer (CRLC Val d'Aurelle, Montpellier, France) (*see* **Note 3**).

2.3. Generation of the Targets From Genomic Extracted DNA

1. HotStarTaq® DNA Polymerase and 10X PCR buffer (Qiagen, Courtaboeuf, France).
2. 100 m*M* dNTP solutions (Amersham Biosciences, Orsay, France) (*see* **Note 4**).
3. 100 µ*M* 5′-phosphorylated forward primer (5′P-GAAGACCCAGGTCCA-GATGA-3′) (*see* **Note 5**).
4. 100 µ*M* Reverse primer (5′-GGTGTAGGAGCTGCTGGTG-3′) (*see* **Note 5**).
5. RNase and DNase-free water (Sigma, St. Louis, MO).
6. 10 U/µL Exonuclease I (USB Corporation, Staufen, Germany).
7. 10 U/µL λ Exonuclease with specific 10X reaction buffer (USB Corporation).
8. DNA1000 kit, 2100 Bioanalyzer, chip priming station and adapted vortexer (Agilent Technologies).

2.4. Genotyping Process

1. MICAM biochip.
2. 50X Denhart's solution: dissolve 5 g of Ficoll (Type 400, Amersham Biosciences), 5 g of polyvinyl pyrrolidone, and 5 g of bovine serum albumin (Fraction V, Sigma) in 500 mL distilled water. Filter and store at –20°C.

3. 2X Buffer A: dissolve eight phosphate-buffered saline tablets (Sigma) and 38.2 g NaCl in 350 mL distilled water. Add 2.64 mL of 33% Tween-20, and make up to a final volume of 400 mL with distilled water. Filter and store at 4°C.

4. 4X Hybridization buffer: 36.64 mM phosphate, 9.89 mM KCl, 2 M NaCl, 0.2% Tween-20, 4X Denhardt, 40 µg/mL salmon sperm DNA. Mix 18.32 mL 2X buffer A, 1.6 mL 50X Denhardt's solution, and 80 µL salmon sperm DNA (10 µg/µL). Store at 4°C.

5. RNase and DNase-free water (Sigma).

6. 10 µM PolyA's solution.

7. 10 µM Solutions of the four quality control targets (QCT) (*see* **Note 6**).

8. 10 µM ddNTP solutions (Amersham Biosciences).

9. 10 µM Biotinylated ddNTP solutions (Amersham Biosciences).

10. Thermo Sequenase™ (32 U/µL), thermo sequenase reaction buffer, and thermo sequenase dilution buffer (Amersham Biosciences).

11. Streptavidin-R-phycoerythrin (1 mg/mL) (Molecular Probes, Paris, France).

12. 50% Formamide.

13. 15% Glycerol.

14. HybriSlip™ hybridization covers (Grace Bio-Labs, Bend, OR).

15. Combitips® plus 10 mL and Multipette® plus (Eppendorf, Le Pecq, France).

16. Hermetic reaction chamber.

17. Kimwipes® Lite (Kimberly-Clark, Saint-Cloud, France).

18. Apimager® (Apibio).

19. AnalySIS® software (Soft Imaging System).

3. Methods

A flowchart and overview for the genotyping of *TP*53 codon 72 SNP is presented in (**Fig. 1**).

3.1. DNA Extraction

High-molecular weight DNA from cell lines and patient's colorectal samples was extracted with a classic proteinase K digestion and phenol–chloroform protocol (*see* **Note 7**).

3.2. Generation of the Targets From Genomic Extracted DNA

1. Amplify by PCR a region (80 bp in the assay) containing the polymorphic site of interest. For this purpose, prepare a PCR master mix with dedicated PCR reagents, pipetman, and consumable supplies in the location within your laboratory dedicated to preparing PCR reactions. Prepare the mix in a 200 µL PCR reaction tube according to the following reaction volumes:

RNase and DNase-free H$_2$O	13.6 µL
10X PCR buffer	2.5 µL (1X final concentration)
5 mM dNTPs working solution	1 µL (200 mM final concentration)
10 µM Forward primer	1.2 µL (0.48 µM final concentration)
10 µM Reverse primer	1.2 µL (0.48 µM final concentration)
5 U/µL HotStarTaq DNA Polymerase	0.5 µL (2.5 U/reaction)

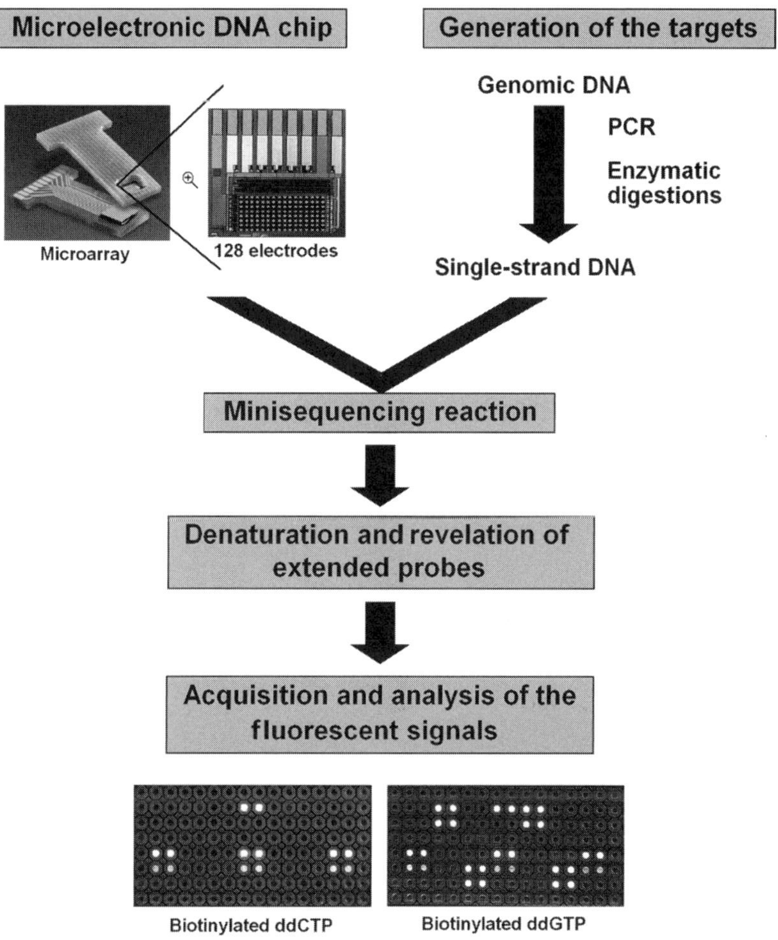

Fig.1. Minisequencing on microelectronic DNA chip: method overview.

2. Vortex the tubes.
3. Using pipetman dedicated to genomic DNA; add 5 µL containing 100 ng of extracted DNA (cell line, patient sample) to each PCR tube.
4. Place the tubes in a thermocycler. The enzyme is first activated during an initial denaturation step for 15 min at 95°C, followed by 17 cycles at 94°C for 30 s, 66°C for 30 s, 72°C for 90 s, with a touch down of –0.5°C for the annealing temperature each cycle. Then 14 normal cycles are carried out at 94°C for 30 s, 58°C for 30 s, and 72°C for 90 s. A final extension step of 10 min at 72°C achieves the reaction.
5. Perform a qualitative and quantitative analysis of amplicons by analyzing 1 µL of product with a DNA1000 kit and the 2100 Bioanalyzer (Agilent Technologies, Massy, France). Follow the protocol described in the user's manual.

6. Eliminate the excess of primer by digesting the remaining 49 μL of PCR product with 1 μL (10 U) of exonuclease I at 37°C for 30 min. Then inactivate the enzyme by incubating the mix at 80°C for 15 min. Both reactions are performed consecutively on a thermocycler.

7. Generate single-strand DNA by digesting 50 μL of amplicons with 1 μL (10 U) of λ exonuclease in the presence of 6 μL of the λ exonuclease 10X reaction buffer in a final volume of 60 μL. Perform the digestion for 45 min at 37°C on a thermocycler, and inactive the enzyme by a step at 80°C for 15 min.

3.3. Minisequencing Reactions

1. All minisequencing reactions were carried out in the microreaction chamber of the chip created by the packaging ($4 \times 5 \times 1$ mm^3). After the proper mixture is added, prehybridization, hybridization, and extension steps were performed by placing chips in a hermetic reaction chamber, which is then put in a temperature-controlled incubators.

2. Set the first incubator at 45°C, and the second at 60°C.

3. Prepare slides by cutting HybriSlip hybridization covers into 7-mm wide squares.

4. Prepare the washing buffer (0.5X Buffer A): 10 m*M* phosphate, 2.7 m*M* KCl, 545 m*M* NaCl, 0.0545% Tween-20. Prepare the washing buffer (10 mL/chip) by diluting 1 volume of 2X Buffer A with three volumes of RNAse and DNase-free water.

5. Hydrate the MICAM biochip at room temperature with 20 μL of RNase and DNase-free water for 5 min.

6. During **step 5**, prepare 1X prehybridization buffer using 4X hybridization buffer, RNase and DNase-free water and 10 μ*M* polyA's solution. 1X prehybridization buffer consists of 1X hybridization buffer, 0.2 μ*M* polyA.

7. Remove the water (*see* **Note 8**) from the chip's surface and perform the prehybridization step by adding 20 μL of 1X prehybridization buffer. Cover the microreaction chamber with a slide (*see* **Note 9**) and place the chip in a hermetic reaction chamber at 45°C for 15 min.

8. During **step 7**, prepare the hybridization mix in a 200-μL PCR tube: 5 μL 4X hybridization buffer, 500 fmol of the proper quality control oligonucleotide (*see* **Note 6**), 1 pmol of the single strand PCR product. Complete to 20 μL with RNase and DNase-free water, vortex briefly and heat the tube at 95°C for 5 min on a thermocycler.

9. After the prehybridization incubation, eliminate prehybridization buffer (*see* **Note 8**). Add the hybridization mix from **step 6** in the microreaction chamber and cover with a slide (*see* **Note 9**). Put the biochip in the hermetic reaction chamber and place it at 45°C for 1 h.

10. After the incubation, remove the chip from the chamber and wash it with two spurts of 1 mL washing buffer using a Combitip® plus 10 mL and Multipette plus. Incubate for 5 min at room temperature with 20 μL of washing buffer in the microreaction chamber, then wash again the chip with two spurts of 1 mL washing buffer. Eliminate the residual washing buffer (*see* **Note 8**).

11. Dilute the Thermo Sequenase™ to 4 U/µL in its dilution buffer. Then prepare the extension mix in a 200-µL PCR reaction tube according to the following volumes (for one reaction):
    a. 8 µL RNase and DNase-free H_2O.
    b. 2 µL of the previously diluted Thermo Sequenase (8 U).
    c. 2 µL of 10X Thermo Sequenase reaction buffer.
    d. 2 µL of the 10 µ*M* solution of the appropriate biotin-labeled ddNTP.
    e. 2 µL of each 10 µ*M* solution of the corresponding three other unlabeled ddNTPs.
    f. Vortex briefly, adds 20 µL of the mix in the microreaction chamber and cover with a slide (*see* **Note 9**). Incubate at 45°C for 5 min, then at 60°C for 15 min.

3.4. Denaturation and Revelation of Extended Probes

1. In order to stop the reaction and melt target/extended-probe complexes, wash the chip for 10 s with distilled water using a wash bottle. Dip the chip for 3 min in a 50% formamide bath heated at 60°C. Wash the chip for another 10 s with distilled water. Eliminate the residual water (*see* **Note 8**).
2. Prepare the revelation buffer by adding 1 vol of a 50 mg/L solution of streptavidin-R-phycoerythrin conjugate to 19 vol of RNase and DNase-free water. During the revelation step, it is necessary to prevent direct illumination of the streptavidin-R-phycoerythrin label.
3. Add 20 µL of the revelation buffer in the microreaction chamber, and incubate for 10 min at room temperature, in the dark.
4. Wash the microreaction chamber with four spurts of 1 mL washing buffer using a Combitip plus 10 mL and Multipette plus. Incubate for 5 min in the dark at room temperature with 20 µL of washing buffer in the microreaction chamber, then wash again the chip with four spurts of 1 mL washing buffer. Incubate for another 5 min in the dark at room temperature with 20 µL of washing buffer in the microreaction chamber. Wash again the chip with four spurts of 1 mL washing buffer. Eliminate the residual washing buffer (*see* **Note 8**).

3.5. Acquisition and Analysis of the Fluorescent Signals

Spread homogeneously 2 µL of 15% glycerol on the silicon surface, and analyze the chip on the Apimager®. Using the specific phycoerythrin filter and the ×10 objective, illuminate the biochip and accumulate photons for 10 ms. The phycoerythrin incorporated in the extended probes is excited, and fluorescent signals are measured by the instrument. The fluorescence levels are then allocated to each probe with the AnalySIS software. Typical results obtained are presented in **Table 2**.

Several control probes (*see* **Table 1**) have been implemented to the chip for the validation of the different steps. Thus, there is no need to conduct the analysis in

Table 2
Representative Results Obtained by Minisequencing on MICAM

Probe name	Description	Mean value of the fluorescent signal[a] (fluorescence units)
P72S[b]	*TP53* codon 72 genotyping probe	1434
QC	Quality control	1954
SEP[b]	Self elongating primer	852
CP-bio	Revelation control	3189
Au	Nonaddressed gold electrode	11

[a]The results represent the mean of 23 minisequencing reactions carried out with single strand DNA targets from cell lines or patients.

[b]Only the cases with specific incorporations were included.

duplicate. First, four self elongating probes (SEP) *(27)* are used to validate the extension step. The 3′-ends of these probes are self complementary and able to form dimers directly on the solid support. Each SEP, differing from another by only one nucleotide, allows the incorporation of a specific ddNTP in a template-independent extension reaction. Second, a biotinylated probe (CP-Bio), corresponding to a sequence from the exon 1 of the *K-ras* gene (outside potential mutated codon 12 and 13 regions), allows the control of the revelation step by the streptavidin-R-phycoerythrin conjugate. Third, a quality control probe (QC) is used to check the whole genotyping process. For each biotinylated ddNTP incorporated, a nonhuman, specific QCT (*see* **Note 6**) complementary to the QC probe will undergo the entire minisequencing reaction.

3.6. Result Interpretation

To allow comparison between different experiments, results interpretation is based on relative data rather than absolute values. For this purpose, the QC probe is used for normalization. On the one hand, the technological validation (TV) ratio (QC signal/Au signal), i.e., the signal-to-noise ratio, that reflected the quality of the minisequencing reaction, is calculated. The QC signal represents the fluorescence mean value for the QC probes, and the Au signal is the fluorescence mean value for the nonaddressed gold electrodes. For a TV ratio less than the threshold fixed at 20, the analysis is considered as invalid. Thus, a false diagnosis will be avoided.

On the other hand, the specific incorporation of one biotinylated ddNTP is determined with the ID percentage: (probe of interest [POI] signal/QC signal) ×100. POI signal represents the fluorescence mean value for the probe of interest, i.e., the P72S probes in this assay. For a selected biotinylated ddNTP, an ID >20%

Table 3
Result Interpretations

Patient number	Biotinylated ddNTP	TV[a]	ID[b] (%)	Genotype
00291	ddGTP	343	76	G/G
	ddCTP	164	11	
97B219	ddGTP	187	1	C/C
	ddCTP	33	128	
95B390	ddGTP	151	36	G/C
	ddCTP	178	77	

[a]The experience is considered valid if TV >20.
[b]There is a specific ddNTP incorporation to the probe of interest if ID >20%.

corresponds to a specific incorporation for the probe of interest. An example of interpretation is presented in **Table 3**.

It has been noticed that the different ddNTP do not have the same incorporation efficiency. This observation is not assay format dependent since it had yet been observed *(28)*. Indeed, for heterozygous samples, the incorporation of the biotinylated ddCTP is higher than the incorporation of the biotinylated ddGTP, and this variation is taken into account for the interpretation.

4. Notes

1. MICAM biochips are the fruit of microelectronic research and are made on a silicon support bearing 128 gold electrodes *(29)*. Probes to be grafted (*see* **Table 1**) are *ex situ* synthesized and purified. They are labeled at their 5′-end with a pyrrole residue using pyrrole-phosphoramidite building blocks and tailed with a $(T)_{10}$ spacer arm. The pyrrole group enables probes fixation on the gold electrodes using an electrocopolymerization process previously described *(30)*. Briefly, each of the 128 electrodes (50 µm) was successively functionalized by electrochemical polymerization (1 V/satured calomel electrode) in an aqueous 0.1 *M* $LiClO_4$ solution containing 20 m*M* pyrrole and 1 µ*M* of the pyrrole-oligonucleotide to be addressed.
2. The *TP53* codon 72 sequence for each cell line was checked by standard direct automatic fluorescent sequencing.
3. Surgical samples from colorectal carcinomas and corresponding normal mucosa were selected macroscopically by the anathomopathologist and then immediately frozen in liquid nitrogen before nucleic acid extraction.
4. Mix the dATP, dTTP, dCTP, and dGTP solutions and dilute with the proper volume of RNase and DNase-free water to obtain a 5 m*M* dNTPs working solution.
5. Prepare working solutions of amplification primers at a 10 µ*M* concentration.
6. The QCT (5′-GTAGTGACTTCATTA**X**ACGGGAGGGGGGTACGGTAGGTACG-GCTTGTTCTT-3′) with no homology with the human genome, are used to check the global quality of the different steps (hybridization, extension, revelation).

The nucleotide X is the template for the single extension of the QC probes. It is complementary to the biotinylated ddNTP used in the minisequencing reaction.

7. DNA extraction is performed as previously described in **ref.** *31*. All DNA samples were stored at –80°C. Determination of *TP53* codon 72 polymorphism was achieved through direct sequencing or PCR-RFLP as previously described in **refs.** *32* and *33*.

8. After the hydration step, it is important to prevent complete dessication of the silicon surface. Therefore, eliminations of the buffers are performed by dripping the biochip on Kimwipes Lite.

9. It is important to prevent the evaporation of the reaction mix during incubation at 45 or 60°C. To this purpose, cover the microreaction chamber with a 7-mm wide squared-slide cut from a HybriSlip hybridization cover. Take care to avoid bubble formation when deposing the slide.

References

1. Diamond, J. (2003) The double puzzle of diabetes. *Nature* **423,** 599–602.
2. Day, I. N. and Wilson, D. I. (2001) Science, medicine, and the future: genetics and cardiovascular risk. *BMJ* **323,** 1409–1412.
3. Fearon, E. R. and Vogelstein, B. (1990) A genetic model for colorectal tumorigenesis. *Cell* **61,** 759–767.
4. Nathanson, K. L., Wooster, R., and Weber, B. L. (2001) Breast cancer genetics: what we know and what we need. *Nat. Med.* **7,** 552–556.
5. Abate-Shen, C. and Shen, M. M. (2000) Molecular genetics of prostate cancer. *Genes Dev.* **14,** 2410–2434.
6. Buyru, N., Budak, M., Yazici, H., and Dalay, N. (2003) p53 gene mutations are rare in human papillomavirus-associated colon cancer. *Oncol. Rep.* **10,** 2089–2092.
7. Kleyn, P. W. and Vesell, E. S. (1998) Genetic variation as a guide to drug development. *Science* **281,** 1820–1821.
8. Andreyev, H. J., Norman, A. R., Cunningham, D., et al. (2001) Kirsten ras mutations in patients with colorectal cancer: the "RASCAL II" study. *Br. J. Cancer* **85,** 692–696.
9. Hippo, Y., Taniguchi, H., Tsutsumi, S., et al. (2002) Global gene expression analysis of gastric cancer by oligonucleotide microarrays. *Cancer Res.* **62,** 233–240.
10. Wang, D. G., Fan, J. B., Siao, C. J., et al. (1998) Large-scale identification, mapping, and genotyping of single-nucleotide polymorphisms in the human genome. *Science* **280,** 1077–1082.
11. Pastinen, T., Kurg, A., Metspalu, A., Peltonen, L., and Syvanen, A. C. (1997) Minisequencing: a specific tool for DNA analysis and diagnostics on oligonucleotide arrays. *Genome Res.* **7,** 606–614.
12. Kwiatkowski, R. W., Lyamichev, V., de Arruda, M., and Neri, B. (1999) Clinical, genetic, and pharmacogenetic applications of the Invader assay. *Mol. Diagn.* **4,** 353–364.
13. Landegren, U., Kaiser, R., Sanders, J., and Hood, L. (1988) A ligase-mediated gene detection technique. *Science* **241,** 1077–1080.

14. Barany, F. (1991) Genetic disease detection and DNA amplification using cloned thermostable ligase. *Proc. Natl. Acad. Sci. USA* **88,** 189–193.

15. Nordstrom, T., Nourizad, K., Ronaghi, M., and Nyren, P. (2000) Method enabling pyrosequencing on double-stranded DNA. *Anal. Biochem.* **282,** 186–193.

16. Syvanen, A. C., Aalto-Setala, K., Harju, L., Kontula, K., and Soderlund, H. (1990) A primer-guided nucleotide incorporation assay in the genotyping of apolipoprotein E. *Genomics* **8,** 684–692.

17. Chen, X. and Kwok, P. Y. (1997) Template-directed dye-terminator incorporation (TDI) assay: a homogeneous DNA diagnostic method based on fluorescence resonance energy transfer. *Nucleic Acids Res.* **25,** 347–353.

18. Pastinen, T., Raitio, M., Lindroos, K., Tainola, P., Peltonen, L., and Syvanen, A. C. (2000) A system for specific, high-throughput genotyping by allele-specific primer extension on microarrays. *Genome Res.* **10,** 1031–1042.

19. O'Meara, D., Ahmadian, A., Odeberg, J., and Lundeberg, J. (2002) SNP typing by a pyrase-mediated allele-specific primer extension on DNA microarrays. *Nucleic Acids Res.* **30,** E75.

20. Lopez-Crapez, E., Bazin, H., Chevalier, J., Trinquet, E., Grenier, J., and Mathis, G. (2005) A separation-free assay for the detection of mutations: combination of homogeneous time-resolved fluorescence and minisequencing. *Hum. Mutat.* **25,** 468–475.

21. Syvanen, A. C. (1999) From gels to chips: "minisequencing" primer extension for analysis of point mutations and single nucleotide polymorphisms. *Hum. Mutat.* **13,** 1–10.

22. Tonisson, N., Zernant, J., Kurg, A., et al. (2002) Evaluating the arrayed primer extension resequencing assay of TP53 tumor suppressor gene. *Proc. Natl. Acad. Sci. USA* **99,** 5503–5508.

23. Lopez-Crapez, E., Livache, T., Marchand, J., and Grenier, J. (2001) K-ras mutation detection by hybridization to a polypyrrole DNA chip. *Clin. Chem.* **47,** 186–194.

24. Livache, T., Roget, A., Dejean, E., Barthet, C., Bidan, G., and Teoule, R. (1994) Preparation of a DNA matrix via an electrochemically directed copolymerization of pyrrole and oligonucleotides bearing a pyrrole group. *Nucleic Acids Res.* **11,** 2915–2921.

25. Dumont, P., Leu, J. I., Della Pietra, A. C., George, D. L., and Murphy, M. (2003) The codon 72 polymorphic variants of p53 have markedly different apoptotic potential. *Nature Genet.* **33,** 357–365.

26. Sullivan, A., Syed, N., Gasco, M., et al. (2004) Polymorphism in wild-type TP53 modulates response to chemotherapy in vitro and in vivo. *Oncogene* **23,** 3328–3337.

27. Kurg, A., Tonisson, N., Georgiou, I., Shumaker, J., Tollett, J., and Metspalu, A. (2000) Arrayed primer extension: solid-phase four-color DNA resequencing and mutation detection technology. *Genet. Test.* **9,** 1–7.

28. Dubiley, S., Kirillov, E., and Mirzabekov, A. (1999) Polymorphism analysis and gene detection by minisequencing on an array of gel-immobilized primers. *Nucleic Acids Res.* **27,** E19.

29. Caillat, P., David, D., Belleville, M., et al. (1999) Biochips on CMOS: an active matrix address array for DNA analysis. *Sens. Actuator B* **160,** 154–162.
30. Livache, T., Roget, A., Dejean, E., Barthet, C., Bidan, G., and Teoule, R. (1994) Preparation of a DNA matrix via an electrochemically directed copolymerization of pyrrole and oligonucleotides bearing a pyrrole group. *Nucleic Acids Res.* **11,** 2915–2921.
31. Sambrook, J., Fritsch, E. F., and Maniatis, T. (1989) Analysis and cloning of eukaryotic genomic DNA, in *Molecular Cloning: A Laboratory Manual.* Cold Spring Harbor Laboratory Press, NY, pp. 9–16.
32. Thirion, A., Rouanet, P., Thezenas, S., Detournay, D., Grenier, J., Lopez-Crapez, E. (2002) Interest of investigating TP53 status in breast cancer by four different methods. *Oncol. Rep.* **9,** 1167–1172.
33. Merlo, G. R., Cropp, C. S., Callahan, R., and Takahashi, T. (1991) Detection of loss of heterozygosity in tumor DNA samples by PCR. *Biotechniques* **11,** 166–168.

14

Hybridization Analysis Using Oligonucleotide Probe Arrays

Robert S. Matson and Jang B. Rampal

Summary

This chapter describes methodology for the labeling, hybridization, and detection of amplicon target DNA to arrays of oligonucleotide probes attached to plastic substrates. A systematic approach to target discrimination based on both hybridization and wash stringency is provided.

Key Words: Array; biotin; chemiluminescent; discrimination; electrophoresis; ELF; fluorescence; hybridization; oligonucleotide; PCR; stringency; SYBR green 1.

1. Introduction

Hybridization of nucleic acids is a well-established methodology for genetic analysis *(1,2)*. Most information regarding the utility of these methods is based on the thermodynamic and kinetic studies of solution hybridization; much of which continues to be applied to solid-phase work involving, for example, the spotting of DNA on membranes commonly referred to as the Southern (or Northern if involving RNA) dot blots. Here target nucleic acid (genomic, PCR, amplicons, cDNA, RNA) is immobilized to a solid support such as a charged nylon membrane, which in turn is bathed in a solution of labeled probe DNA (e.g., radiolabeled allele specific oligonucleotide or ASO) to affect hybridization. A more recent approach has been to employ DNA arrays or reverse blots, where probes are attached to the support and labeled target is applied to the array *(3–5)*. Reverse blot arrays are constructed either by attachment of presynthesized probes to the support or by direct *(in situ)* synthesis on the support. Many different solid phases have been used for reverse blot supports including glass *(5)*, nylon *(6)*, silicon chips *(7)*, and plastics *(8,9)*. In this chapter, work from our laboratory on the hybridization of labeled targets to oligonucleotide

From: *Methods in Molecular Biology, vol. 381: Microarrays: Second Edition: Volume 1: Synthesis Methods*
Edited by: J. B. Rampal © Humana Press Inc., Totowa, NJ

probe arrays created by *in situ* synthesis on a polypropylene substrate will be described. This plastic substrate has some unique features such as very low intrinsic fluorescent backgrounds, chemical inertness, good mechanical strength, and low nonspecific adsorption of nucleic acids and proteins. In general, prehybridization and blocking protocols are not required; whereas, the use of linker chemistries to direct the probe away from the surface are also unnecessary *(8,10,11)*. The methods to be described in the use of the polypropylene oligonucleotide array include those for quality control (SYBR green staining of array panels; biotin-labeled oligonucleotide target hybridization); preparation of targets (PCR; electrophoresis; gel staining); estimations of target concentration and biotin incorporation; PCR sample hybridization; and detection methods (chemiluminescence; enzyme-labeled fluorescence). Finally, a description of the *in situ* synthesized array used in our studies is provided next.

The oligonucleotide array panel is made up of a clear polypropylene film or sheet to which oligonucleotides are synthesized *(in situ)* from the 3′- to 5′-end using standard phosphoramidite monomers and conventional DNA synthesis chemistries *(8,9)*. Lanes of ASO probes are created across the sheet. A modified oligonucleotide synthesizer is employed, where the synthesis column is replaced by a multiport valve fluidic delivery system, which directs reagents to a multichannel reaction block *(8)*. The 64-channel reaction block is designed such that the untreated side of the film serves as a seal, while the treated or reactive side is exposed to 64 independent synthesis channels that create the 64 parallel lanes of ASOs. Each lane may contain any DNA sequence specified by the user so that from 1 to 64 different sequences can be produced on the support. Usually, oligonucleotides in the size range from 12- to 24-mer are synthesized on the support depending upon the application (*see* Chapter 11). An 8×8-cm^2 sheet of activated film is used in the channel block. Upon completion of synthesis the sheet is cut into dipsticks. The dipsticks are approx 0.5×6.6 cm^2 and require about a 75-µL sample volume to cover the entire surface when used in a "cover slip" fashion over the sample. An example of the "dipstick" panel used to detect cystic fibrosis mutations is provided in **Fig. 1**.

The following discussion is aimed at presenting a general plan of attack for the discrimination of hybrids based on a reverse-hybridization format for *in situ* synthesized arrays. In this approach, temperature and salt conditions are varied for both hybridization and washing.

1.1. Hybridization Theory

For targets whose sequences show little homology (e.g., CF mutations), hybridization may be performed under low stringency or permissive conditions; while targets with stretches of homologous sequence (e.g., *K-ras* point mutations) more stringent conditions are required. A single hybridization condition may

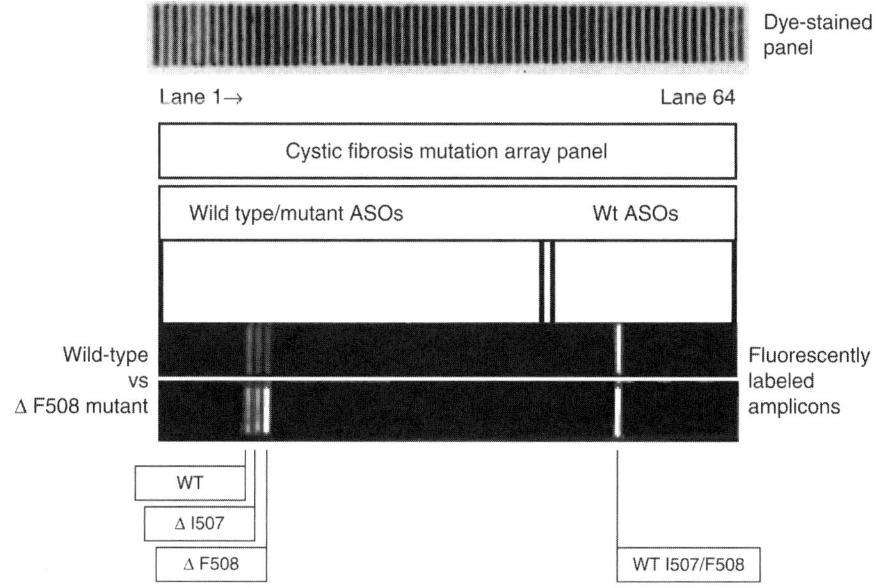

Cystic Fibrosis Mutation "Dipstick" Panel

Fig. 1. Cystic fibrosis mutation "Dipstick" panel. The panel consisted of 22 wild-type and 23 mutant probe sequences (16- to 21-mers) in situ synthesized in lanes 1–45. The first of 19 wild-type probes were repeated in lanes 46–64. The top panel has been developed with the invitrogen DNA stain. The bottom panels represent the result of hybridization of wild-type or Δ-F508 fluorescently labeled PCR products to the panels. In this example, the Δ wild-type and I 507 probes weakly cross-hybridize with the wild-type and F508 homozygous amplicons.

not be adequate to resolve similar sequences. Likewise, wash conditions need to be varied to discriminate among homologs. Hybridization depends upon the nucleation frequency, while washing depends upon the thermal stability (T_m) of the hybrids. Therefore, a stringent hybridization followed by a stringent wash is preferred for differentiation of closely related sequence *(1)*.

1.2. Hybridization Studies for Differentiation of Closely Related Sequences

With the exception of studies on short tandem repeats, most arrays used for genotyping will need to discriminate related targets; and involve the use of very closely related probes. For example, consider how to differentiate *K-ras* wild-type and mutations. These are one base mismatches surrounded by a significantly high G–C content that results in considerable hybrid stabilization; in fact the mutant target may form a more stable hybrid than the wild-type under

Table 1
Hybridization Protocol: Find Most Stringent Conditions at an Acceptable Signal

HYB temp →	T_d-n(°C)	T_d-25	T_d-20	T_d-15	T_d-10	T_d-5
X SSC ↓	NaCl conc. (mM)	–	–	–	–	–
6	1000	Very permissive	–	–	–	Stringent
2	330	–	–	–	–	–
0.3	50	–	–	–	–	–
0.06	10	–	–	–	–	–
0.02	3.3	–	–	–	–	–
0.006	1	Stringent	–	–	–	Very stringent

certain conditions. Thus, careful solid-phase probe design is essential to the outcome of the discrimination process especially in the above case where there is the possibility of interstrand solid-phase probe duplex formation. However, equally important is the impact on the approach to hybridization and an understanding regarding what factors are significant (*see* a list of such factors in the **Notes 1–6**). There are no set hybridization rules, just general guidelines that can be followed; and most of what is known is based upon solution-phase interactions. Each array system (i.e., *H-ras*, *K-ras*, CF, HLA, and so on) will most certainly require a different set of optimization conditions.

The following **Tables 1** and **2** are provided as an example of one approach, which has worked in laboratory. First, determine the best conditions for hybridization of both wild-type and mutant targets to the array. Given equivalent amounts of wt and Δ amplicons, determine how they hybridize to the probes. Can hybridization conditions be found that would promote hybrid formation of one amplicon over the other? Is the signal at an acceptable level? Use the table below as a starting point (*see* **Table 1**). The approx T_m of the mismatched oligonucleotide can be determined by:

$$T_m = T_d - 1.2 \times (\% \text{ mismatch}),$$

where T_d represents dissociation of target from the complementary solid-phase probe calculated using nearest neighbor approximations for basepair annealing. Software programs such as Oligo™ (Molecular Biology Insights, Inc., Cascade, CO) can be used to calculate this parameter.

Once a set(s) of hybridization conditions have been found proceed to determine wash condition that preferentially remove one hybrid (wt vs Δ) over the other (*see* **Table 2**). It may not be possible to completely remove a

Table 2
Wash Protocol: Find Most Stringent Conditions for Discrimination at an Acceptable Signal/Noise

WASH temp →	T_m-n(°C)	T_m-25	T_m-20	T_m-15	T_m-10	T_m-5
X SSC ↓	NaClconc. (mM)	–	–	–	–	–
6	1000	Very permissive	–	–	–	Stringent
2	330	–	–	–	–	–
0.3	50	–	–	–	–	–
0.06	10	–	–	–	–	–
0.02	3.3	–	–	–	–	–
0.006	1	Stringent	–	–	–	Very stringent

cross-hybridizing species, therefore, examine the discrimination ratio of the wt/Δ (or Δ/wt) signal and set threshold values, i.e., what level is acceptable, a ratio of 2:1 or 10:1 or 100:1, and so on?

2. Materials

2.1. SYBR GREEN 1 Array Panel Staining Protocol

1. SYBR GREEN™ 1 stain (Molecular Probes, Inc., cat. no. S-7563).
2. Tris-borate EDTA buffer (TBE) 10X stock (Digene, Inc., cat. no. 3400-1036).
3. Methanol.
4. Acetic acid (glacial).
5. UV lightbox, 254 nm excitation.

2.2. PCR Protocol

1. Taq DNA polymerase (Promega, cat. no. M2861).
2. 10X PCR buffer (Promega, cat. no. M1882).
3. dCTP-biotin (Tropix, cat. no. AU14B).
4. dNTPs unlabeled, 100 mM stocks (Promega, cat. no. U1240).
5. MgCl$_2$, 25 mM stock (Promega, cat. no. A3511).
6. Sample template.
7. Primers, 20 µM stocks each.
8. Mineral oil.
9. Thermocycler.

2.3. Electrophoresis Protocol

1. Gel loading buffer (Digene, cat. no. 3400-1046).
2. Electrophoresis apparatus (Novex Xcell II, cat. no. EI9001).

3. Precast 4–20% TBE gels (Novex, cat. no. EC62255).
4. TBE running buffer (Digene, cat. no. 3400-1036).
5. Molecular size ladder (Bio-Rad, cat. no. 170-8206).
6. Power supply (Stratagene, Feather Volt 250).

2.4. Gel Staining Protocol

1. SYBR GREEN™ 1 stain (Molecular Probes, Inc., cat. no. S-7563).
2. Tris-borate EDTA buffer (TBE) 10X stock (Digene, Inc., cat. no. 3400-1036).
3. CCD camera system w/520-nm lens optical filter.
4. Light source, UV lamp, 350 nm.

2.5. PCR Amplicon Concentration Estimate Protocol

1. Gel loading buffer (Digene, cat. no. 3400-1046).
2. DNA mass marker ladder (Bio-Rad, cat. no. 170-8206).
3. TBE running buffer (Digene, cat. no. 3400-1036).
4. Precast 4–20% TBE gels (Novex, cat. no. EC62255).
5. Electrophoresis apparatus (Novex Xcell II).
6. Power supply (Stratagene, Feather Volt 250).
7. CCD camera system w/520-nm lens optical filter (SpectraSource, Teleris 2 cooled CCD camera model).
8. Light source, UV lamp, 350 nm.
9. Image analysis software (NIH Image).
10. Graphing and mathematical analysis software (Microsoft Excel).

2.6. Biotin Label Quantification Protocol

1. TE buffer (1X stock) (Digene, cat. no. 3400-1039).
2. Vacuum Filter Plate apparatus (Bio-Rad, Bio-Dot SF, cat. no. 170-3938).
3. Nylon membranes (Zeta Probe Nylon, Bio-Rad, cat. no. 162-0153).
4. Denaturing solution: 1.5 M NaCl, 0.5 M NaOH (Digene, cat. no. 3400-1000).
5. Neutralizing solution: 3 M sodium acetate, pH 5.5.
6. Blocking solution (Denhardt's): 50X stock (Digene, cat. no. 3100-1017) containing 5X SSPE and 0.2% sodium dodecyl sulfate (SDS).
7. 2X Standard saline citrate (SSC) containing 0.01% SDS: dilutions from 20X stock (Digene, cat. no. 3400-1024).
8. Streptavidin alkaline phosphatase conjugate (Tropix, cat. no. APA10).
9. Chemiluminescent signal development reagent, CDP-Star™ (Tropix, cat. no. MS050R).
10. BCIP colorimetric development reagent (Sigma, cat. no. B-5655).

2.7. Quality Control Hybridization Protocol

1. Hybridization buffers (2X to 6X SSPE): prepared from dilutions of 20X SSPE (Digene, cat. no. 3400-1028). Per liter: 3 M NaCl, 0.2 M NaH$_2$PO$_4$, 0.02 M

EDTA, adjusted to pH 7.4 with 10 *N* NaOH; followed by addition of 0.01% SDS (v/v).
2. Streptavidin alkaline phosphatase conjugate (Tropix, cat. no. APA10).
3. ELF™ reagent (Molecular Probes, Inc., cat. no. E-6601).
4. CCD camera system or UV light box.
5. DNA stain—DNA Dipstick™ Kit (Invitrogen, cat. no. K5632-01).

2.8. Sample Hybridization Protocol

1. Denaturation solution, 0.5 *M* NaOH, 0.15 *M* NaCl.
2. Neutralization-hybridization solution, 2.4X SSC, 0.016% SDS, 0.28 *M* Tris-HCl, 0.028 *M* NaCl, pH 7.5.

2.9. CDP-Star™/Replica Blot Protocol

1. 2X SSPE.
2. Streptavidin alkaline phosphatase conjugate.
3. CDP-Star™ (Tropix, cat. no. MS050R).
4. Sapphire™ II enhancer (Tropix, cat. no. LAX250).
5. DEA buffer (Tropix, cat. no. AD120).
6. Nylon membrane.
7. CCD camera system.

2.10. ELF™ Signal Development Protocol

1. 2X SSC buffer containing 0.01% SDS.
2. Streptavidin alkaline phosphatase conjugate.
3. ELF reagent.
4. CCD camera system.

3. Methods
3.1. SYBR GREEN I Staining/Destaining

3.1.1. Staining

Time allocation: 30 min (*see* **Notes 7–15**).

1. Dilute the dye stock concentrate (1:100,000 v/v) in 1X TBE buffer, for example, take 1 µL dye concentrate and mix with 10 mL TBE, then remove 1 mL of this solution and dilute with 9 mL TBE. The resulting solution is approx 1/100,000 diluted dye.
2. Place the 1/100,000 diluted dye solution in a dark plastic staining tray and cover with a foil lid.
3. Place the strips DNA side down in the solution for 10 min.
4. Remove the strips and rinse briefly with TBE buffer.
5. Place the strips on a clean, untreated clear polypropylene film sheet for viewing on the 254-nm UV-lightbox. Turn the UV light on the minimum power setting and examine, shielding your eyes with UV filter eye goggles.
6. Discard dipsticks that do not have all lanes visible and complete.

3.1.2. Destaining

Time allocation: 80 min to overnight.

1. Prepare the destaining solution, methanol/H_2O/acetic acid (50:49:1 v/v by mixing for example, 1 mL of glacial acetic acid added to 49 mL of deionized water and mixed, followed by the addition of 50 mL methanol to a final volume of 100 mL in a graduated cylinder.
2. Place the stained strips in the destain solution and incubate for 1 h with agitation. Change the destain solution after the first 30 min with fresh solution.
3. After destaining rinse the strips in TBE buffer and recheck under the UV-lightbox. If significant stain remains repeat the destaining process until no further change is observed.
4. Soak the strips in distilled water for 10 min then blot dry. These may now be stored dry in a clean plastic or glass container or used directly for hybridization.

3.2. PCR

Time allocation: preparation time, 45–60 min; thermocycling time, approx 200 min (*see* **Notes 16** and **17**).

1. Templates, primer stock, Taq enzyme and buffers, $MgCl_2$, and dNTP are removed from $-20°C$ freezer and thawed. After thawing, each reagent is given a quick spin to gather the entire volume at the bottom of the tube. The reagents are then placed on ice.
2. Total reaction volume is typically 100 µL, of which the template consists of 10 µL. Because only template DNA varies among PCR samples, a master mix is made for all other reagents. The components of the master mix are: primers, Taq buffer, $MgCl_2$, and dNTPs. 10X PCR buffer is added to give a final $MgCl_2$ concentration of 1.5 mM. Additional $MgCl_2$ is added to raise the final concentration by an additional 0.5 mM. Each of the four dNTPs is present at a final concentration of 0.2 mM. The master mix is carefully mixed by pipetting up and down and then given a quick spin to gather the entire volume to the bottom of the tube.
3. The templates are received from a library and then diluted 1:100 in water before they are used. Dispense 10 µL of the template into each PCR tube. Add 89 µL of the master mix into the PCR tube and mix with the template. The PCR mix is now ready except for the addition of Taq DNA polymerase.
4. Prior to addition of Taq, each tube is given a 10 min 95°C hot start. After the hot start, each PCR tube is quick spun to gather all condensate, then 1 µL of Taq (5 activity units) is mixed into each tube.
5. A 50 µL mineral oil overlay is added to reduce evaporative loss.
6. PCR conditions are 80°C for 10 min, followed by 30 cycles of 1 min denaturation at 96°C, 1 min annealing at 56°C, and 1 min extension at 74°C. At the end of the thirty cycles is a 7 min final extension at 74°C.
7. The **Table 3** may be used to make from 1, 10, or 15 PCR reaction tubes.

Table 3
Flow Chart for Preparing PCR Reactions

Reagents	Stock concentration	Volume for 1 PCR (µL)	Master mix for 10 PCRs (µL)	Master mix for 15 PCRs (µL)
Forward primer	20 µM	5	50	75
Reverse primer	20 µM	5	50	75
dATP	10 mM	2	20	30
dGTP	10 mM	2	20	30
dTTP	10 mM	2	20	30
dCTP	10 mM	1	10	15
dCTP-biotin	0.4 mM	25	250	375
10X Taq buffer	15 mM (MgCl2)	10	100	150
$MgCl_2$	25 mM	2	20	30
Water	–	35	350	525

3.3. Electrophoresis

Time allocation: preparation time, 15–20 min; gel run time, approx 60 min (*see* **Notes 18–20**).

1. PCR products are quick spun to gather all condensates to the bottom of the tubes.
2. Extract 2 µL of the aqueous layer and mix with 2 µL of gel loading buffer.
3. The Novex XCell II electrophoresis apparatus is prepared by inserting the precast polyacrylamide gel and TBE running buffer. The gel is rinsed under a stream of deionized water as the comb and tape is removed prior to placement in the apparatus.
4. Each PCR product and the size ladder are loaded into separate wells on the gel.
5. The gel is run at 110 V for approx 60 min or until the dye front has gone three-fourths of the way down the gel. A Stratagene power supply is employed.
6. Following electrophoresis the gel is carefully removed from its retaining plate (cassette) and placed immediately in SYBR Green 1 staining solution.

3.4. Gel Staining

Time allocation: staining, 10 min (*see* **Notes 21–23**).

1. The gel is removed from its cassette and stained for 10 min in 40 mL of SYBR Green 1 diluted 1:10,000 in TBE buffer.
2. The staining try is covered with aluminum foil and placed on a shaker.
3. After staining, the solution is decanted into a waste container and the gel carefully laid onto a piece of polypropylene film.
4. The gel is photographed using a CCD camera. A 350-nm UV light is used to excite the SYBR Green 1 stained gel; and a 520-nm camera lens filter is used for capturing the fluorescence image.
5. Contaminated or failed amplifications are discarded at this point.

3.5. PCR Amplicon Concentration Estimate

Time allocation: preparation time, 30 min; gel run time, 60 min; staining time, 10 min; data analysis time, approx 60 min (*see* **Notes 24** and **25**).

1. Based on the intensities of the first gel, the volume of sample to be added for the second gel is estimated. Typically, 1–3 µL of sample is loaded into each lane. Each sample is mixed with 2 µL of gel loading buffer.
2. The mass marker is quickly spun and then mixed by repeated pipetting up and down. Serial dilutions of 0.5X, 0.25X, 0.125X, and 0.0625X are prepared in TBE running buffer. Mix 2.5 µL of each dilution with 2 µL of gel loading buffer.
3. The samples and the mass marker dilutions are loaded into individual wells of Novex TBE 4–20% gels and run at 110V for approx 60 min or until the dye front has run three-fourths the height of the gel.
4. The gel is removed from its cassette and stained in SYBR for 10 min in an aluminum foil covered tray on a shaker. After the staining, a fluorescent image is taken using the CCD camera with a 350-nm excitation and 520-nm emission filtering at the camera.
5. The intensity of each band is measured using Scion Image. Knowing the amount of mass marker loaded per lane and determination of the corresponding gel intensity band, a plot of intensity vs mass can be constructed. The resulting curve and equation of the curve is then used to estimate the mass of sample loaded into each lane by extrapolation.
6. Knowing the molecular weight, mass of sample loaded, and volume of sample loaded, the concentration of the sample is then calculated using a spreadsheet, for example, Microsoft Excel 97.

3.6. Biotin Label Quantification

Time allocation: stepwise times provided next. Total time approx 6.5 h (*see* **Notes 26** and **27**).

3.6.1. Preparation of the Dot Blot

Time allocation: total blotting time: 190 min.

1. Dilute the PCR product in 1X TE buffer, pH 7.4 according to extended dilution scheme (1:10–1:10,000) in **Table 4**.
2. Cut one Bio-Rad Zeta Probe Nylon Membrane to fit BIO-Dot filter paper. Assemble Bio-Rad BIO-Dot apparatus with membrane and three filter papers. Tighten screws and check that membrane and filter papers are held firmly in apparatus.
3. Premoisten the membrane with 500 µL DI water in each slot. Allow the water to soak in a few min. Then apply a vacuum to remove excess moisture (use aspirator to apply vacuum).
4. Apply 10-µL aliquots of PCR product into the appropriate slots. Allow the aliquot to soak in a few min. Then apply a vacuum to remove excess moisture.

Table 4
Preparation of Dot-Blot Solutions

	Dilution factor	Composition PCR product × TE buffer	Total volume (μL)
a.	1:10	2 μL Neat PCR × 18 μL TE	20
b.	1:50	6 μL Dilution a × 24 μL TE	30
c.	1:100	5 μL Dilution b × 15 μL TE	30
d.	1:200	15 μL Dilution c × 15 μL TE	30
e.	1:500	15 μL Dilution d × 22.5 μL TE	37.5
f.	1:1000	15 μL Dilution e × 15 μL TE	30
g.	1:2000	15 μL Dilution f × 15 μL TE	30
h.	1:4000	15 μL Dilution g × 15 μL TE	30
i.	1:8000	15 μL Dilution h × 15 μL TE	30
j.	1:10,000	16 μL Dilution i × 4 μL TE	20

5. Apply 100 μL denaturing solution (1.5 M NaCl, 0.5 M NaOH) to each slot that contains PCR product and also to adjoining rows and columns (to avoid edge effects). Let stand for 10 min. Then, apply a vacuum to remove excess moisture.
6. Apply 100 μL neutralizing solution (3 M sodium acetate, pH 5.5) to each slot as above. Let stand for 10 min. Then, apply a vacuum to remove excess moisture.
7. Remove membrane from apparatus and place membrane in a glass Petri dish. Air-dry membrane for 20–30 min. Then, bake the membrane in 80°C oven for 1 h (make sure the membrane is not sticking to the Petri dish).
8. Remove membrane from oven and place in a square plastic Petri dish or plastic container the just fits the membrane. Add 8–10 mL blocking solution (5X Denhardt's solution, 5X SSPE, 0.2% SDS). Place membrane on rotary shaker set at 3 for 30 min.
9. Wash membrane three times with 8–10 mL 2X SSC, 0.01% SDS. Shake for 1–2 min for each wash.

3.6.2. Chemiluminescent Detection

Time allocation: total blot development time: 100 min.

1. Prepare 2 mL 1:100 dilution of streptavidin alkaline phosphatase conjugate in 2X SSC, 0.01% SDS. Apply the conjugate solution to the membrane. Incubate the membrane on the rotary shaker for 30 min.
2. Wash membrane three times with 8–10 mL 2X SSC, 0.01% SDS. Shake for 1–2 min for each wash.
3. Place membrane in a shallow plastic.
4. Apply 2–3 mL Tropix CDP-Star™ ready to use reagent to the surface of the membrane. Monitor chemiluminescence under the CCD camera. Take 600 s (10 min) exposures corrected for dark field starting as soon as possible at 30-min intervals until maximum signal exposure is obtained (usually within 2 h). Keep membrane in high humidity chamber between exposures to prevent it from drying out.

3.6.3. Colorimetric Detection

Time allocation: BCIP detection time: 45 min.

1. After the conclusion of chemiluminescence exposure, wash off membrane with 2X SSC, 0.01% SDS.
2. Place the washed membrane it in a container with 10–15 mL BCIP reagent. Allow a visible color to develop within 20–30 min.
3. Wash membrane in deionized water to stop BCIP reaction before a background build up. Blot the membrane with filter paper to remove excess moisture and seal the membrane in a plastic bag.

3.7. Quality Control Hybridization

Time allocation: total procedure time: 6.5 h (*see* **Notes 28–30**).

1. Prepare biotinylated oligonucleotide targets in 6X SSPE, 0.01% SDS at a concentration of approx 1 μM at 25°C. Do not heat.
2. Use 100 μL per strip.
3. Pipet the above solution (from **step 1**) onto glass Petri dish or glass microscope slide.
4. Place dipstick with DNA side down on top of solution.
5. Hybridize, 1 h, 25°C, in a closed, humidified chamber.
6. Rinse strip in 3X 10 mL of 6X buffer.
7. Place strip in streptavidin–alkaline phosphatase conjugate prepared in 6X buffer solution. Use 100 μL of a 1:100 (v/v) dilution of the enzyme conjugate.
8. Incubate 1 h, ambient temperature.
9. Rinse with 2X buffer, at least 3X 10 mL per strip.
10. Prepare ELF signal development reagent. Use 100 μL per strip and apply as described in the previous steps.
11. Incubate, 30 min, ambient temperature.
12. Rinse briefly in deionized water.
13. Analyze by CCD or visual inspection on a UV light box.

3.8. Sample Hybridization

Time allocation: preparation time, 20 min; hybridization time, 60 min to overnight (*see* **Notes 31–33**).

3.8.1. PCR Amplicon Hybridization

3.8.1.1. Heat Denaturation

1. Heat PCR product to 95°C, 10 min, in a water bath, then place on ice for 10 min.
2. Remove 10 μL and dilute into 90 μL of 2X SSC, 0.01% SDS, and then return to water bath and heat for an additional 5 min.
3. Mix well and pipet solution down onto a glass microscope slide.
4. Immediately place the dipstick DNA side down on top of the puddle.

5. With flat forceps gently lift the strip from one end and with an up a down motion spread the solution and removes air bubbles from under the strip.
6. Place the slide in a humidified chamber and incubate from 1 h to overnight at 25°C.

3.8.1.2. CHEMICAL DENATURATION

1. Remove 10 µL PCR product and transfer into a polypropylene tube.
2. Add 15 µL of denaturation solution (0.5 *M* NaOH, 0.15 *M* NaCl) and mix by pipet dispensing of the solution several times.
3. Incubate for 10 min at ambient temperature.
4. Add 75 µL of neutralization–hybridization solution (2.4X SSC, 0.016% SDS, 0.28 *M* Tris-HCl, 0.028 *M* NaCl at pH 7.5) and mix.
5. Repeat **steps 3–6** (*see* **Subheading 3.8.1.1.**).

3.9. CDP-Star™/Replica Blot

Time allocation: 2.5 h (*see* **Notes 34** and **35**).
Starting with hybridized, biotinylated target on PP film strip:

1. Rinse with 2X SSC, 0.01% SDS (2X buffer).
2. Dilute 10 µL streptavidin–alkaline phosphatase with 990 µL 2X buffer (2.5 µg/mL); use 100 µL per strip.
3. Incubate for 1 h, ambient temp.
4. Rinse with 2X buffer, at least three rinses, 25 mL each.
5. Dilute 0.5-mL sapphire II with 4.5 mL DEA buffer or 0.2 mL /1.8 mL DEA buffer (sapphire 1:10).
6. Mix 20 µL CDP-star concentrate with 1200 µL solution (5) or 30 µL/1800 µL of solution.
7. Soak a charged nylon strip (standard microscope size) with 200 µL solution (6).
8. Apply film with DNA side down and cover with another slide taking care to remove any air bubbles under film.
9. Incubate for 1 h or monitor under CCD camera. At this point, the film can be removed from the nylon membrane and the replica blot image recorded.

3.10. ELF Signal Development

Time allocation: 100 min (*see* **Notes 36–39**).
Starting with hybridized, biotinylated target on PP film strip:

1. Incubate at 25°C with 200 µL 2X SSC, 0.01% SDS buffer 10 min then rinse with fresh buffer.
2. Incubate with 200 µL streptavidin–alkaline phosphatase conjugate (diluted 1:100 v/v in 2X SSC, 0.01% SDS) for 1 h in a humidified chamber at 25°C.
3. Rinse 3X 5 min with 200 µL 2X SSC, 0.01% SDS buffer to remove residual enzyme conjugate.
4. Incubate with 100 µL of ELF reagent for 30 min at ambient temperature without shaking.

5. After the ELF development, quickly rinse the strips in deionized water to remove residual ELF.
6. Fluorescent image of the ELF precipitate is captured by CCD camera with excitation at 350 and 520 nm emissions filtering at the camera. Typical exposure time 30 s.

3.11. In General, Factors Influencing Hybridization-Discrimination

The hybridization process is complex and many factors may interplay, which affect the discrimination power of the technology. Several important factors and conditions to consider in the design of hybridization-based assays (*see* **Note 11**).

4. Notes

1. Solid-phase support.
   a. Probe density (especially site variability; influences nucleation rate via steric hindrance crowding, probe–probe duplex formation, hairpins).
   b. Nonspecific adsorption sites (increased background; surface modification).
   c. Chemical stability (resistance to solvation; remains inert, void of reactive groups; resistant to destruction or uncontrolled surface modification or release of contaminants by reagents used for probe attachment, hybridization, wash, and signal generation).
   d. Thermal stability (resistance to melt, which may cause alterations in surface properties).
   e. Wettability.
   f. Autofluorescence.
   g. Contaminating substances for example, mold release, antistatic agents, fingerprints, and so on.
2. Array probe.
   a. Placement of mismatch.
   b. 3′ vs 5′ tethering to support.
   c. Secondary structure (duplex; hairpins).
   d. Size and spacer arm requirements.
   e. Strand (use of sense vs antisense strand).
   f. T_d magnitude, variability, and potential for normalization.
   g. Target specificity.
3. PCR target.
   a. Length and complexity (diffusion nucleation rate).
   b. Probe specificity (cross-hybridization).
   c. Secondary structure (duplex, hairpins reassociation kinetics).
   d. Concentration.
   e. Characterization (purity fidelity).
4. Hybridization.
   a. Time.
   b. Temperature ($T_m - 25°C$).
   c. Buffer composition (ionic strength, H-bond cosolvents, and accelerators).

    d. Agitation rate and process (gyration, vibration, vortex, and so on).

    e. Analyte matrix: target concentration, interfering substances (DNA-binding proteins cross-hybridization or competing targets buffer composition).

    f. Prehybridization or support pretreatment.

5. Wash.

    a. Time.

    b. Temperature.

    c. Buffer composition.

    d. Desorption process: continuous vs stepwise vs gradient desorption (thermal, electrical, chemical).

    e. Agitation.

6. Signal generation and detection system.

    a. Influence of reporter or label on hybrid stability.

    b. Nonspecific adsorption of reporter.

    c. Threshold or substrate plateau.

    d. Dynamic range.

    e. Kinetics: target concentration, interfering substances (DNA binding proteins; cross-hybridization or competing targets; buffer composition).

    f. Prehybridization or support pretreatment.

7. The buffered solution is stable for 1 d at room temperature. Cover the solution with aluminum foil since it is light sensitive. The 1/10,000 stock solution maybe stored under refrigeration for several weeks. Dye stock is supplied as a 10,000X concentrate in DMSO and should be stored frozen and properly capped. Splitting the stocks into aliquots is recommended for use.

8. Do not expose the strips for more than a few min to the strong UV light in order to avoid photodamage to the DNA. A CCD image may be captured at a 10 s setting with a 520-nm emission filter placed in front of the camera. Avoid direct contact of the dyed strips with other surfaces. We have found that the dye is easily adsorbed onto glass and painted surfaces.

9. Use approx 10 mL of destain solution per one to two strips and about 50 mL per 8×8 cm^2 sheet.

10. Overnight destaining does not appear to adversely affect strip performance as measured from small oligonucleotide hybridization analysis.

11. This is a qualitative method and should only be used to verify the quality of the dipstick panel. Not all probe sequences will stain with equal intensity. For example, polyA and probes whose sequences are rich in dA tend to stain lighter than others do.

12. Use appropriate rated UV filtering goggles for viewing of films under UV light sources.

13. If lanes are missing this is an indication that synthesis has failed in the lanes. If lanes are partially visible improper seating of the film on the channel block or partial blockage of a fluidic channel is suggested.

14. By our convention, the bottom of each strip is identified by the lot no (e.g., no. 313) and the last lane is, therefore, no. 64, while the top of each strip is identified by the strip no (e.g., 1–12) and the first lane is no. 1. These two numbers

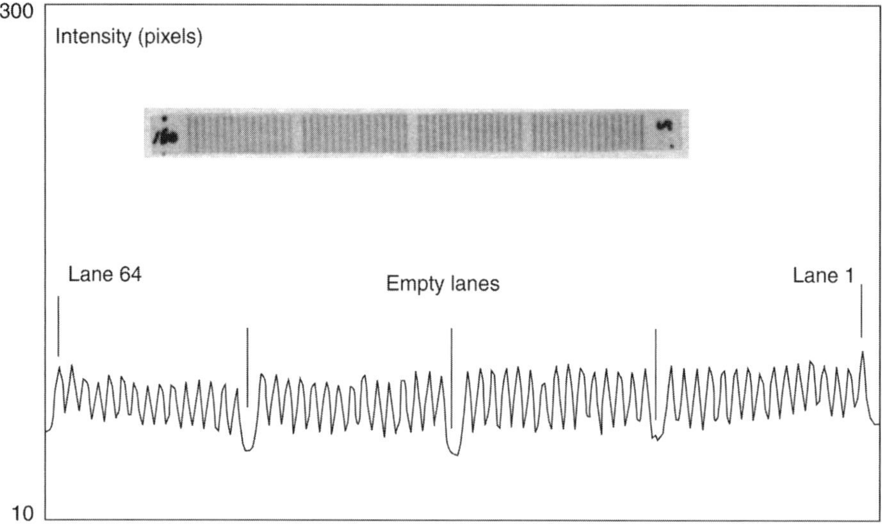

Fig. 2. Invitrogen DNA stained panel. A polypropylene dipstick panel was stained to reveal *in situ* synthesized probes. Empty lanes remain clear.

then can define each strip, for example: 313-1, 314-1, 316-3, and so on. The DNA side is opposite that of the label so if you can read the strip numbers, the DNA side is down. There is some difficulty encountered in reading these particular strips owing to a lack of visible reference lanes available for indexing. The Invitrogen DNA stain, DNA Dipstick™ Kit, was used, which will permanently stain each lane a light blue color (*see* **Fig. 2**). This stain, however, does not distinguish hybridized lanes from unhybridized lanes. These would have to be marked, for example, with a lab marker and the position determined after staining. Another approach is to insert polyA lanes as a ruler and develop these with a biotin labeled polyT. However, this will reduce the number of lanes available for ASO probes.

15. The Invitrogen DNA stain does not permit posthybridization because the probe oligonucleotide is chemically modified.

16. Vortexing is not used for mixing owing to the small volume of the reagent stocks. Instead, mixing is achieved by repeated pipetting up and down immediately before required volumes are aliquoted.

17. When the amplicon is to be biotin labeled, dCTP-biotin is substituted for 50% of the total dCTP. Various dCTP-biotins have been used but those that have the biotin moiety tethered off of the dCTP by a 10–14 atom linker are preferred for example, Tropix (AU14B). The pcr conditions described here are given as an example for internal incorporation of biotin using a biotin-dCTP analog. Although a typical thermocycling program is given in **Subheading 3.2. step 6** actual denaturation, annealing, and extension cycles may vary depending upon primer sequence, template, and desired amplicon product.

Table 5
Optimal Conditions for Hybridization of PCR Amplicons to a *K-ras* Probe Array

Hybridization Protocol: Find Most Stringent Conditions at an Acceptable Signal

HYBtemp→	T_d-n(°C)	T_d-35	T_d-25	T_d-20	T_d-15	T_d-10	T_d-5
K-ras Panel	54.3 ± 2.6°C	25°C	32°C	37°C	42°C	47°C	52°C
	330 m*M* salt 1 p*M* oligo (all probes) ΔG duplex –6.3 kcal/mol						
XSSC ↓	NaCl conc. [m*M*]	**Wild-type** **(T_m)**					
6	1000	Very Permissive					Stringent
2	330	**(55.8°C)**					
0.3	50	**(42.2°C)**					
0.06	10	**(30.6°C)**					
0.02	3.3						
0.006	1	**Stringent** **(14.0°C)**					**Very stringent**

*Shaded areas indicate probability for little or no hybridization signal development.
*Hatched area represents optimized hybridization conditions: 0.3 XSSC, 42°C.

18. The 4–20% polyacrylamide, TBE gels are used to obtain a higher resolution analysis of fragments, which allows not only an assessment of the quality of the PCR product but also information regarding the extent of primer-dimer affects on amplification efficiency.

19. Conventional agar-based submarine gels may also be used at this stage.

20. During the aqueous extraction, the pipet tip is dipped through the mineral oil layer first then the plunger is compressed to blow out any mineral oil, which might have entered the pipet tip. The outside of the pipet tip is also wiped using a Chemwipe tissue to remove residual mineral oil.

21. Prolonged staining is not recommended owing to increased backgrounds caused by excess dye accumulation in the gel.

22. After a gel has been run and visualized to determine whether or not the correct PCR amplicon was produced, those PCR tubes shown to contain the same amplicons are pooled together. Each PCR tube is again quick spun to gather all condensates to the bottom of the tube.

23. Avoid contact with both dye and polyacrylamide gel by wearing gloves.

24. This protocol is used for quantitative hybridization experiments in which case the target input concentration for hybridization or the applied mass need to be controlled.

25. Estimation of the amplicon concentration is required in **Subheading 3.6.**

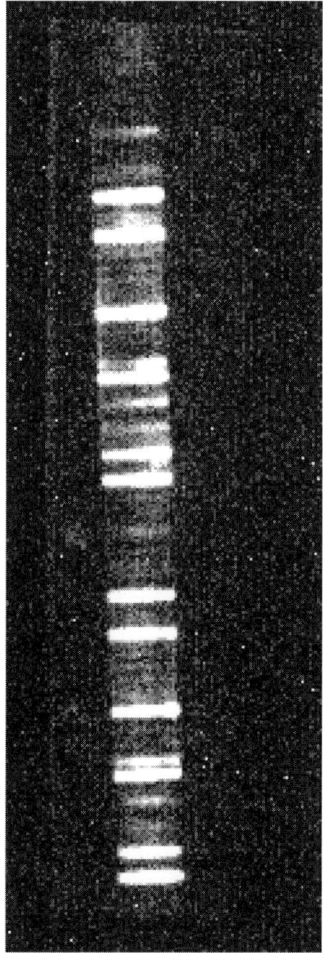

CCD camera image of a nylon replica blot of a dipstick array panel

Fig. 3. Example of a "dipstick" panel array with differentiation between the wild-type and mutation amplicons.

26. Biotin-dCTP at 50% of the total dCTP is furnished during pcr amplification. The degree of incorportion is dependent upon the number of available "C" sites within the amplicon, as well as, the amplification efficiency. The purpose of this protocol is to estimate the relative "signal" strength among targets used for hybridization. This is accomplished by titrating the streptavidin–alkaline phosphatase activity obtained for targets. Together with the estimate of the amplicon concentration, the relative "specific activity" of each labeled amplicon can be determined.
27. The BCIP staining is used to obtain a permanent record of the blot. Store dries, sealed in plastic in the lab notebook or other file away from strong light.

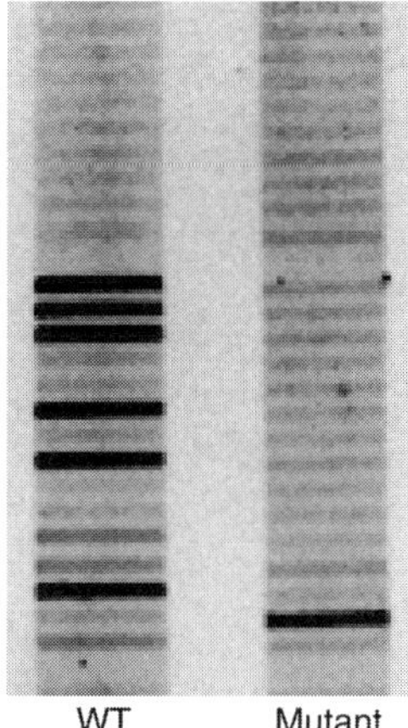

Example of a "dipstick" panel array with differentiation between
the wild-type and mutant amplicons

Fig. 4. CCD camera image of a nylon replica blot of a dipstick array panel.

28. This protocol allows for the rapid assessment of the panel using biotinylated oligonucleotide targets complementary to the *in situ* synthesized probes.

29. Stringency conditions for hybridization will require optimization and are dependent upon the sample. In some cases it may be necessary to perform an overnight incubation, especially if a weak signal is obtained after 3 h or a higher stringency is required for differentiation.

30. It is suggested that the specified temperature be controlled to within ±1°C for the reproducibility of hybridization experiments.

31. Both heat denaturation and chemical denaturation are equally effective.

32. The final hybridization solution contains: some amount of target, 1.8X SSC, 0.01% SDS, 0.21 *M* Tris-HCl, 0.06 *M* NaCl, pH 8.2.

33. This is a general hybridization protocol. For optimization refer to the tables provided in the **Subheading 1.**, for hybridization and washing conditions. An example is provided in **Table 5** for optimized hybridization conditions for a *K-ras* probe "dipstick" array. **Figure 3** shows the outcome for discrimination of a point mutation relative to the wild-type.

34. Although chemiluminescent reagents have been successfully employed with, e.g., charged nylon membranes, their use with plastics is problematic. This is largely owimg to the lack of a sufficiently charged surface that is needed to capture the light emitting intermediate. However, a "replica" blot of the dipstick panel can be easily produced by laying the plastic strip DNA-side down on top of a nylon membrane presoaked in chemiluminescent reagent (*see* **Fig. 4**).

35. The membrane should be fully saturated but only moist on the top surface with reagent. Decant off excess reagent before laying down the dipstick. It is advisable to weigh down the strip by covering it with a glass microscope slide so that the dipstick in firmly pressed down on the nylon membrane. Care should be taken in preparing the replica to avoid entrapment of air bubbles.

36. In **Subheading 3.10.**, alkaline phosphatase cleaves the ELF-phosphate into the insoluble and highly fluorescent ELF-alcohol. The excitation and emission wavelengths are broad with a large Stock's shift. The excitation max is centered at approx 350 nm with the emission max at 520 nm.
37. The ELF-alcohol precipitation process requires that the strip lay flat. Avoid agitation at this stage to prevent movement of the precipitate, which will reduce overall resolution of the signal image.
38. The developed dipstick can be stored dry in the dark for at least 1 yr. Some reduction in signal strength is experienced owing to photobleaching.
39. SSC buffer is used in the protocols employing alkaline phosphatase in order to avoid product inhibition from inorganic phosphate. Other buffers such as SSPE may be used during hybridization and then the strips rinsed and exchanged into SSC prior to the addition of the alkaline phosphatase conjugate.

References

1. Hames, B. D. and Higgins, S. J. (eds.) (1988) *Nucleic Acid Hybridization: A Practical Approach.* IRL, Oxford, UK.
2. Walker, J. M. (ed.) (1988) *Methods in Molecular Biology, vol 4, Nucleic Acid Techniques.* Humana Press, Totowa, NJ.
3. Dattagupta, N., Rae, P. M. M., Huguenel, E. D., et al. (1989) Rapid identification of microorganisms by nucleic acid hybridization after labeling the test sample. *Anal. Biochem.* **177,** 85–89.
4. Drmanac, R., Labat, I., Brukner, I., and Crkvenjakov, R. (1989) Sequencing of megabase plus DNA by hybridization: theory of the method. *Genomics* **4,** 114–128.
5. Southern, E. M., Maskos, U., and Elder, J. K. (1992) Analyzing and comparing nucleic acid sequences by hybridization to arrays of oligonucleotides: evaluation using experimental models. *Genomics* **13,** 1008–1017.
6. Saiki, R. K., Walsh, P. S., Levenson, C. H., and Erlich, H. A. (1989) Genetic analysis of amplified DNA with immobilized sequence-specific oligonucleotide probes. *Proc. Natl. Acad. Sci. USA* **86,** 6230–6234.
7. Pease, A. C., Solas, D., Sullivan, E. J., Cronin, M. T., Holmes, C. P., and Fodor, S. P. A. (1994) Light-generated oligonucleotide arrays for rapid DNA sequence analysis. *Proc. Natl. Acad. Sci. USA* **91,** 5022–5026.
8. Matson, R. S., Rampal, J. B., Pentoney, S. L., Anderson, P. D., and Coassin, P. (1995) Biopolymer synthesis on polypropylene support: oligonucleotide arrays. *Anal. Biochem.* **224,** 110–116.
9. Weiler, J. and Hoheisel, J. D. (1996) Combining the preparation of oligonucleotide arrays and synthesis of high-quality primers. *Anal. Biochem.* **243,** 218–227.
10. Matson, R. S., Rampal, J. B., and Coassin, P. J. (1994) Biopolymer synthesis on polypropylene supports, I. Oligonucleotides. *Anal. Biochem.* **217,** 306–310.
11. Wehnert, M. S., Matson, R. S., Rampal, J. B., Coassin, P. J., and Caskey, C. T. (1994) A rapid scanning strip for tri- and dinucleotide short tandem repeats. *Nucleic Acids Res.* **22,** 1701–1704.

15

In Situ Synthesis of Peptide Microarrays Using Ink-Jet Microdispensing

Bogdan V. Antohe and Patrick W. Cooley

Summary

The study of protein–protein and protein–DNA interactions is critical to understand biological processes. This article presents the methodology to create peptide microarrays *in situ* for the high-throughput screening of complex biomolecules. The *in situ* ink-jet peptide synthesis results in a conservation of costly reagent and amino acids, whereas providing a means to produce denser peptide arrays. A smaller amount of test sample is required to observe interaction when using these high-density peptide arrays.

Key Words: High-throughput screening; *in situ* synthesis; ink-jet dispensing; microdispensing; peptide synthesis; peptide microarray.

1. Introduction

Microarray chip technology has developed at an incredible pace with a multitude of companies fabricating microarray chips for numerous high-throughput screening applications. A similar evolution is anticipated for peptide microarrays largely driven by the range of applications they can address *(1,2)*. Functional genes and their products will continue to be identified with the rapid development occurring in genomics and proteomics. The elucidation of protein–protein and protein–DNA interactions is critical to understand biological processes and has tremendous value to the biotechnology and pharmaceutical industries. This task is daunting when faced with the complexity of the proteins and biomolecules involved. The approach of the microarray/chip technology has been used successfully for DNA sequencing, differential expression analysis, and other applications. Peptide microarrays have a similar potential that can be realized using a combinatorial approach. The synthetic peptides can be grouped in very

From: *Methods in Molecular Biology, vol. 381: Microarrays: Second Edition: Volume 1: Synthesis Methods*
Edited by: J. B. Rampal © Humana Press Inc., Totowa, NJ

large numbers in a single unit, a peptide library, thus allowing massively accelerated sample processing.

The potential applications of peptide microarrays are immense and many scientists advocate that peptide microarray technology should be emphasized as a part of the functional proteomics program *(3–5)*. Some applications of peptide microarray technology are briefly summarized next:

1. *Defining minimum protein–protein interaction domains:* overlapping peptides derived from a protein (e.g., a receptor) generated in a microarray format and the binding domain(s) of another protein (e.g., its ligand) can be quickly identified. Identification of protein–protein binding and their binding domain (epitope mapping) is extremely important in: (1) drug discovery process *(3–5)*, (2) phosphorylation studies for substrate specificity and mechanisms of reaction *(3)* and for the identification of new substrates for protein kinases *(4)*, (3) integrin binding studies with applications in cancer research and therapy *(5)*, platelet-related diseases *(6,7)*, and tissue engineering *(8)*, and (4) isolation of specific cell lines *(9)*.
2. *Key residues identification:* within an identified binding domain, multiple peptides containing different amino acids at one given position can be synthesized and assayed quickly. The data reveals the key residues for binding and provides critical information for computational modeling. This approach can be considered as a complete mutational assay *(10)*.
3. *General drug screening:* a peptide of a well-defined protein binding domain can be synthesized on the membrane and then incubated with both the drugs to be screened and its target protein *(11–13)*.
4. *Protein conformation probe:* when multiple binding domains are identified, one of the valuable findings for drug design is whether these domains bind to the protein in the same area. Soluble peptides representing the domains can be preincubated with the protein to check if such binding will inhibit protein binding to another binding domain on the microarray *(11)*.
5. *Interaction between protein and other molecules:* a synthesized peptide microarray could be used to study protein–DNA *(12)*, protein–polysaccharide, protein–cell interactions *(13)*, and metal-binding studies *(12)*.

2. Materials

2.1. Substrate Preparation

1. Amino-PEG500-UC cellulose sheets with typical loading 3–4 nmol/mm^2 (Intavis LLC, San Marcos, CA).
2. *O*-benzotriazole-N,N,N′,N′-tetramethyl-uronium hexafluorophosphate (HBTU) (Chem-Impex International, Inc., Wood Dale, IL).
3. P-[(R,S)-α-1-(9H-fluren-9-yl)-methoxy formamido]-2,4-dimethoxybenzylphenoxyacetic acid (Rink Amide Linker—RAL) (Chem-Impex International, Inc.).
4. N,N-diisopropylethylamine peptide synthesis grade (Applied Biosystems, Foster City, CA).

5. N,N-dimethylformamide (DMF) biotech—low water/amine free (Pharmco Products, Brookfield, CT).
6. Pipperidine biotech grade (Sigma-Aldrich, St. Louis, MO).
7. Acetic anhydride (Sigma-Aldrich).
8. Methanol high-performance liquid chromatography (HPLC) grade (VWR International, Batavia, IL).
9. Bromophenol blue (Sigma-Aldrich).
10. Chromatography paper (VWR International, West Chester, PA).
11. Fume exhaust hood.
12. Tweezers.
13. Trough for washing (Sigma-Genosys, The Woodlands, TX).
14. Glass disposable pipets (10 or 20 mL).
15. Pipet bulbs.
16. Micropipetor.
17. Microbalance.
18. Vortex mixer.
19. Rotary shaker for washing.
20. 2-mL Polystyrene vials.

2.2. Printing System and Piezoelectric Microdispenser Set Up

1. Printing system (MicroFab Technologies, Inc., Plano, TX) (*see* **Note 1**).
2. Glass capillary piezoelectric microjet device(s) (MicroFab Technologies) (*see* **Note 2** and **Fig. 1**).
3. Isopropyl alcohol (Sigma-Aldrich).

2.3. Synthesis

1. Fmoc-L-amino acid active esters (Sigma-Genosys).
2. 1-Methyl-2-pyrrolidinone (NMP) for peptide synthesis (Sigma-Aldrich).
3. 2-mL Polystyrene vials.

2.4. Antibody Probing Using ELISA

1. Trifluoroacetic acid (TFA; Sigma-Aldrich).
2. Triisobutyl silane (Sigma-Aldrich).
3. Nitro-blue tetrazolium (NBT; Mallinckrodt Baker, Inc., Phillipsburg, NJ).
4. 5-Bromo-4-chloro-3-indoyl phosphate, toludinium salt (BCIP; Calbiochem, San Diego, CA).
5. Mouse antienkephalin IgG (Chemicon International, Temecula, CA).
6. Bovine serum albumin (BSA; Sigma-Aldrich).
7. Goat alkaline phosphatase (AP)-conjugated goat antimouse IgG + IgM (H + L) (Biomeda, Foster City, CA).
8. AP detection reagent kit (Novagen, Madison, WI).
9. 1X AP buffer: 100 mM Tris-HCl, pH 9.5, 100 mM NaCl, 1 mM MgCl$_2$.

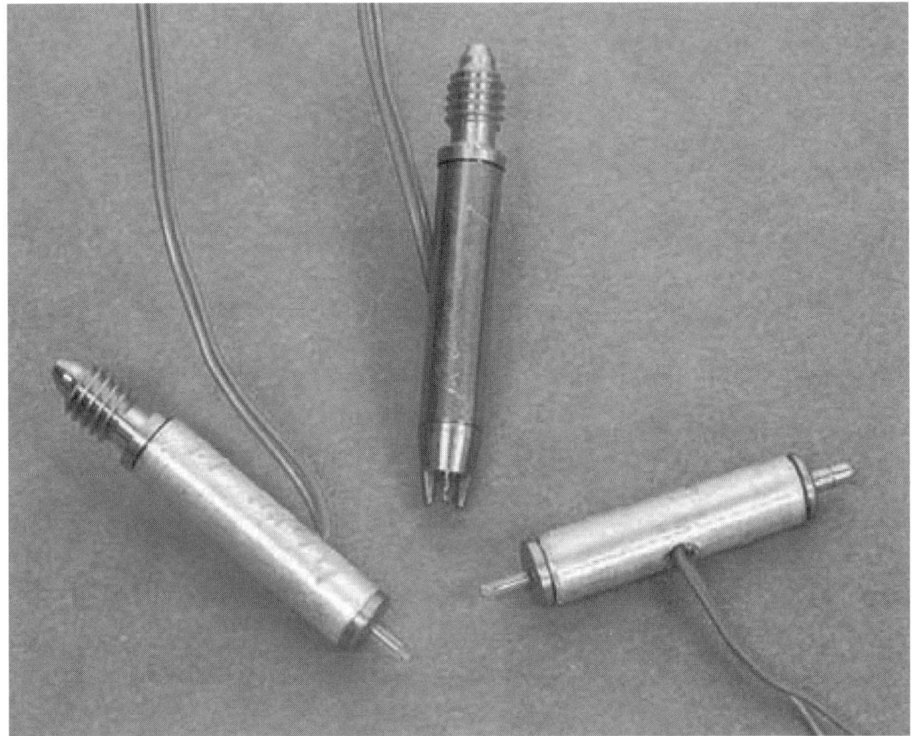

Fig. 1. Various configurations of piezoelectric microdispensing devices manufactured by MicroFab Technologies, Inc.

10. Tris buffered saline (TBS): 150 m*M* NaCl, 10 m*M* Tris-HCl, pH 7.5.
11. Tris Tween buffered saline (TTBS): 0.3 *M* NaCl, 20 m*M* Tris-HCl, pH 7.8, 0.1% (v/v) Tween-20, and 0.01% sodium azide.

2.5. Peptide Cleavage

1. HPLC water (Sigma-Aldrich).
2. Trifluoroacetic acid (TFA; Sigma-Aldrich).
3. Triisopropylsilane (Sigma-Aldrich).

3. Methods

This synthesis process follows the solid-phase synthesis procedure on a cellulose membrane *(14)*, but uses microdispensers to synthesize smaller quantities of peptides to permit denser microarrays with a larger number of peptides. The polyethylene glycol (PEG) linker allows for more efficient coupling to the membrane. A modification to the procedure is the substrate derivatization using a cleavable RAL that allows the peptides to be cleaved from the substrates. If the peptides are to be used for antibody analysis, no RAL derivatization is performed.

3.1. Substrate Preparation

The entire surface of the substrate is PEG derivatized. The RAL is also attached to the entire substrate to permit synthesis locations at any desired pitch. The RAL step is not required if the peptides do not need to be cleaved from the substrate.

1. Cut the PEG derivatized cellulose membrane to the desired size (*see* **Note 3**).
2. Using a pencil, mark two crosses toward the ends of the substrate. These will serve as alignment marks when replacing the substrate on the dispensing system.
3. Check the DMF solution for free amide groups (*see* **Notes 4** and **5**).
4. Prepare the RAL/HBTU solution by first weighing 0.081 g RAL and 0.057 HBTU in a 2-mL vial. Add 0.3 mL DMF to the vial and vortex.
5. Add 45 µL of N,N-diisopropylethylamine to the same vial and vortex. Leave for 2 min to activate. The solution color will be light brown.
6. Pipet the solution over the entire membrane and leave it for 15 min to couple.
7. Add 10 mL of DMF to the trough and set it on the rotary shaker at 8–10 rpm for 2 min. Empty the trough and repeat this step two more times. This step and all washing steps have to be performed in the fume exhaust hood.
8. Repeat the **steps 4–7** once again. Do not throw away the last DMF wash.
9. Add 250 µL of piperidine to the DMF solution in the trough. Continue the washing on the rocker for 5 min.
10. Empty the trough and add 10 mL of DMF. Wash for 2 min. Empty and repeat four more times.
11. Premix 100 µL of the bromophenol blue solution (*see* **Note 4**), add to the trough and rock for 2 min. If derivatization has taken place the membrane should be entirely blue. Empty trough.
12. Add 10 mL of methanol and discard after 2 min. Repeat two more times.
13. Dry the membrane with either a cool hair drier or by blotting between two pieces of chromatography paper.

3.2. Printing System and Piezoelectric Microdispenser Set Up

The processes required to prepare the piezoelectric microdispensers on MicroFab's printing system (jetlab®, MicroFab technologies) for the dispensing of the solutions for the *in situ* synthesis of the peptides are described next.

1. Load the piezoelectric microdispenser fluid reservoir with 500 µL NMP for microdispenser verification.
2. Prime the microdispenser by applying pressure to fill the glass capillary to the orifice with fluid. Adjust the vacuum control to maintain the fluid within the orifice.
3. Input the electronic drive parameters for operation of the microdispenser. Initially, set the rise and fall time to 3 µs, dwell time 30 µs, dwell voltage to 25 V and the frequency to 240 Hz (*see* **Note 6**).
4. Increase the dwell voltage in 3- to 5-V increments if no dispensing is observed (*see* **Notes 6** and **7**).

5. An unstable drop or drop with satellites can often be corrected by adjustment of the dwell time in 1- to 3-μs increments (*see* **Note 8**).
6. After verifying the dispenser operation purge the NMP out of the reservoir and purge with 1 mL DMF followed by 1 mL methanol.
7. Program the recipe(s) specifying the number of drops of the amino acid to be dispensed and the coordinates where to be dispensed (*see* **Note 9**).
8. Mount the membrane on the system holder.
9. Using the pencil markings on the derivatized substrate determine the offset and the angle between the substrate and the *x*-axis of motion (*see* **Note 10**).

3.3. Synthesis

The amino acid solutions are prepared and then loaded and printed one at a time. The substrate is removed and undergoes washing, blocking, and deprotection steps after completing the printing of all amino acids in a layer.

Check the NMP and DMF solutions for the presence of free amine groups (*see* **Note 5**).

1. Prepare a vial of 0.3 *M* of each of the amino acids to be used in the synthesis (*see* **Note 11**) in 200-μL NMP using the molecular weights as shown in **Table 1**.
2. Load the vial containing the first amino acid onto the printing system and follow the **steps 2–5** in **Subheading 3.2**.
3. Print the recipe corresponding to the first amino acid.
4. Wait 15 min for coupling.
5. Repeat **steps 3** and **4** for double coupling (*see* **Note 12**).
6. Apply vacuum to the top of the vial to draw the amino acid back into the vial. Remove the reservoir and seal tightly until next use.
7. Flush the microdispensing device with 1 mL DMF and then 1 mL methanol.
8. Repeat the **steps 3–8** until all amino acids that are used in the current layer are dispensed.
9. Remove the membrane from the system and place it in the washing trough. If this is the last amino acid in the synthesis sequence, skip forward to **step 26**. The following steps should be done in the fume exhaust hood.
10. Add 10 mL of DMF to the trough and set it on the rotary shaker at 8–10 rpm for 2 min. Empty the trough and repeat this step two more times. Do not throw away the last DMF wash.
11. Add 400 μL of acetic anhydride to the last DMF wash for blocking.
12. Leave it on rotary shaker at 8–10 rpm for 15 min.
13. If membrane is still blue add 150 μL acetic anhydride and shake for 15 more min (*see* **Note 13**).
14. Empty the trough then add 10 mL of DMF and set on the rotary shaker at 8–10 rpm for 2 min. Empty the trough and repeat this step two more times. Empty the trough.
15. Add 10 mL of 20% piperidine in DMF to the trough for deprotection.

Table 1
Molecular Weight of the Preactivated Fmoc Amino Acid Esters

Amino acid	Molecular weight
Alanine (Ala)	477.4
Arginine (Arg)	774.7
Asparagine (Asn)	520.4
Aspartic acid (Asp)	577.5
Cysteine (Cys)	751.8
Glutamic acid (Gln)	534.4
Glutamine (Glu)	591.5
Glycine (Gly)	463.4
Histidine (His)	643.6
Isoleucine (Ile)	519.5
Leucine (Leu)	519.5
Lysine (Lys)	634.6
Methionine (Met)	537.5
Phenylalanine (Phe)	553.5
Proline (Pro)	503.4
Serine (Ser)	528.6
Threonine (Thr)	542.6
Tryptophan (Trp)	592.5
Tyrosine (Tyr)	625.6
Valine (Val)	504.4

As described in **Subheading 3.3**.

16. Set the trough on the rotary shaker at 8–10 rpm for 5 min. Empty the trough.
17. Add 20 mL of DMF and set the rotary shaker at 8–10 rpm for 2 min. Empty the trough. Repeat this step four more times.
18. Premix 100 µL of 1% bromophenol blue solution to 10 mL DMF and add to the trough. This is the bromophenol blue staining step to verify peptide synthesis has occurred. This verification can be repeated for each amino acid layer.
19. Set the trough on the rotary shaker at 8–10 rpm for 5 min. Only the synthesis locations should become blue (**Fig. 2**). The membrane outside the synthesis location may become a light shade of blue. If all membrane and the solution become blue repeat **steps 18–21** once more. If the membrane still becomes blue the synthesis has been compromised (*see* **Note 13**).
20. Empty the trough.
21. Add 10 mL of methanol to the trough. Set the trough on the rotary shaker at 8–10 rpm for 2 min then empty. Repeat this step two more times.
22. Dry the membrane using a cool hair dryer or by blotting between two pieces of chromatography paper (*see* **Notes 14** and **15**).
23. Mount the membrane on the printing system holder.

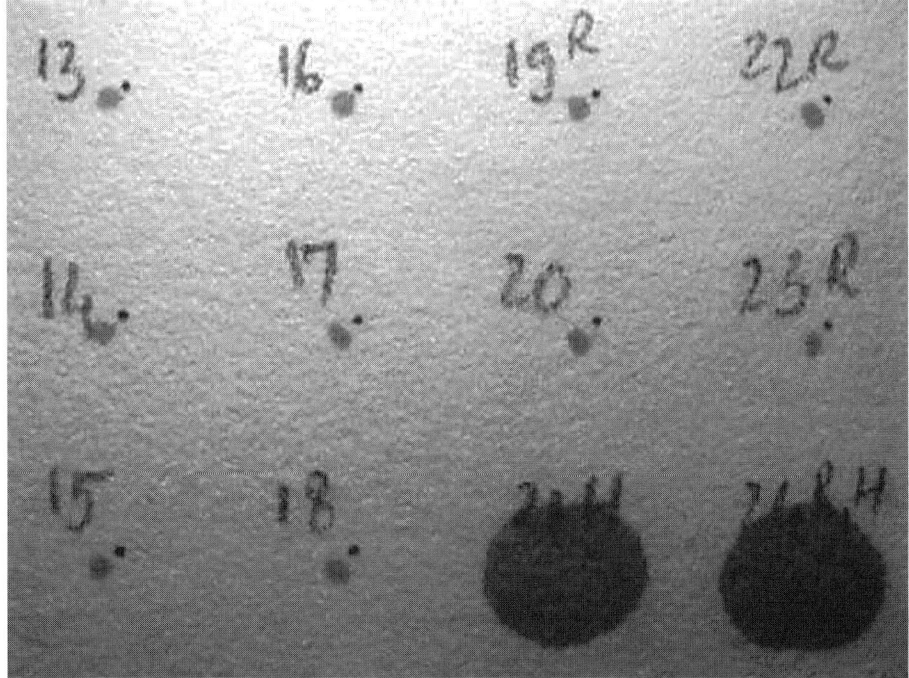

Fig. 2. Membrane after bromophenol blue staining. Enkephalin is synthesized at all locations. The spots marked with R were locally derivatized with RAL. At the large spots (21 and 24) the solution dispensing was done by hand pipetting.

24. Use the pencil markings on the derivatized substrate to determine the offset and the angle between the substrate and the *x*-axis of motion (*see* **Note 10**).
25. Repeat **steps 3–24** for all, but the last amino acid. For the last amino acid repeat **steps 3–9** only and continue with **step 26**. All subsequent steps have to be performed in the fume exhaust hood.
26. Add 10 mL of DMF to the trough and set it on the rotary shaker at 8–10 rpm for 2 min. Empty the trough and repeat this step two more times.
27. Add 10 mL of 20% piperidine in DMF to the trough for deprotection.
28. Set the trough on the rotary shaker at 8–10 rpm for 5 min. Empty the trough.
29. Add 20 mL of DMF and set the rotary shaker at 8–10 rpm for 2 min. Empty the trough. Repeat this step four more times.
30. Perform the bromophenol blue staining outlined in **steps 19–21**.
31. Use a pencil to mark the synthesis locations for future reference of synthesis spot location. This will be useful for antibody probing or when cutting the synthesis spots from the membrane.
32. Add 200 µL acetic anhydride in 10 mL DMF to the trough and sct on the rotary shaker at 8–10 rpm for 15 min for blocking.
33. If spots are still blue repeat **step 32**.

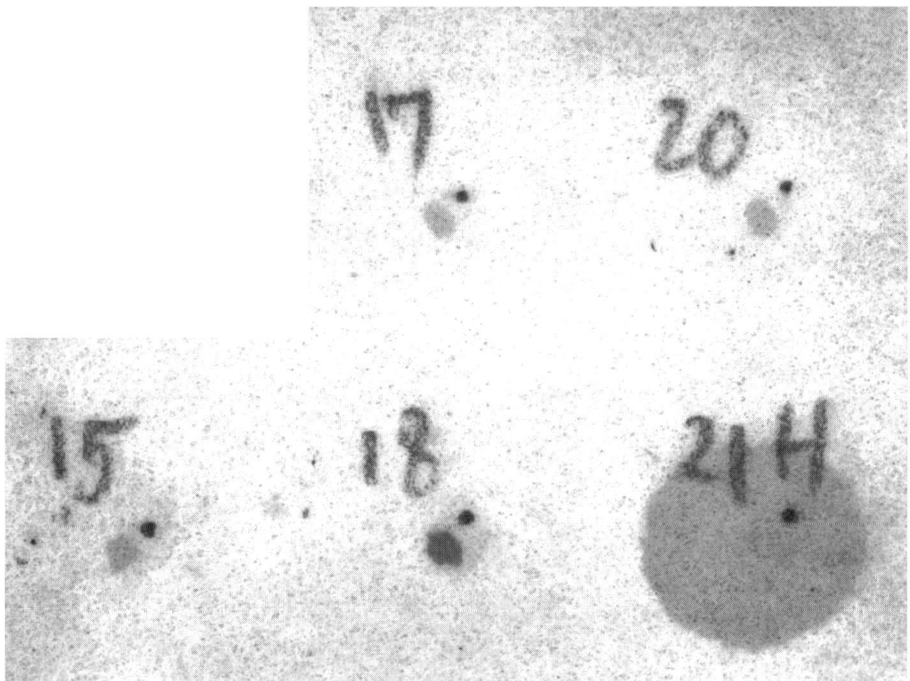

Fig. 3. Colorimetric result after ELISA analysis of the synthesized enkephalin shown in **Fig. 2**. The antienkephalin antibody was applied as a 1:20 at location 18, 1:100 at location 21, and 1:1000 at location 20. No antibody was added at location 15 and a non-specific antibody was added (1:20) was added at location 17. The results indicate specific binding at 1:20 dilution.

34. Add 10 mL of DMF to the trough and set it on the rotary shaker at 8–10 rpm for 2 min. Empty the trough and repeat this step two more times.
35. Add 10 mL of methanol to the trough. Set the trough on the rotary shaker at 8–10 rpm for 2 min then empty. Repeat this step two more times.
36. Dry membrane using a cool hair dryer or by blotting between two pieces of chromatography paper.
37. Store the membrane in a sealed plastic bag at –20°C.

3.4. Antibody Probing Using ELISA

The following describes the antibody probing of enkephalin peptide synthesized on the cellulose membrane (*see* **Note 16**). The colorimetric results of the probing with antienkephalin are shown in **Fig. 3**.

Typically, the peptides are synthesized to determine the interaction with a certain antibody or other biological compounds. The analysis conditions and materials may vary from the ones presented here.

1. Prepare 50% TFA and 0.025% triisobutylsilane in DMF.
2. Pipet enough of the above solution in the trough to cover the entire membrane surface (~20 mL).
3. Leave the membrane at room temperature for 1 h for peptide deprotection.
4. Wash the membrane three times for 5 min each in methanol using a rotary shaker at 8–10 rpm.
5. Wash the membrane three times for 5 min each in TBS buffer (pH 7.5) using the rotary shaker at 8–10 rpm.
6. Block the membrane overnight with 1% bovine serum albumin (BSA) in TTBS buffer at 4°C.
7. Wash the membrane three times for 5 min each in TTBS buffer using the rotary shaker at 8–10 rpm.
8. Dilute the antienkephalin antibody (test antibody) in 20 mL of 1% BSA. (*see* **Note 17**).
9. Add the diluted antibody to the membrane and shake 3–4 h and then empty the trough.
10. Add the TTBS buffer and shake the membrane for 10 min using the rotary shaker at 8–10 rpm. Repeat this twice to remove the unbound antibody.
11. Prepare a dilution of 1:3000 of the AP-conjugated goat antimouse IgG + IgM (or AP conjugated secondary antibody of choice) in 20 mL of the 1% BSA in TTBS buffer.
12. Incubate the membrane in the 20 mL of the 1:3000 dilution of AP-conjugated secondary antibody for 1 h.
13. Wash the membrane five times for 2 min using the TTBS buffer.
14. Based on a 10×10-cm^2 blot, prepare the colorimetric developing solution by combining 60-μL NBT (83 mg/mL NBT in 70% [v/v] DMF) and 60-μL BCIP (42 mg/mL BCIP in 100% DMF) with 15 mL of 1X AP buffer.
15. Place the membrane protein side up in a clean tray and add the colorimetric developing solution. Incubate the membrane at room temperature until purple color develops. A strong purple signal should appear within 2–10 min.
16. To stop the reaction, wash the blot thoroughly in deionized water and allow to air-dry. Store dry blot at room temperature wrapped in plastic.

3.5. Peptide Cleavage

Cleavage of the synthetic peptides from the membrane is not normally performed, because antibody probing is the most common use of the peptides synthesized on the cellulose membranes. The peptides are cleaved only when further analysis (HPLC or MS) is required. For this type of analysis the same peptide should be synthesized at more than one location (3–5) to produce enough material. **Figure 4** shows the result of the HPLC analysis of enkephalin synthesized using the described protocol. The peak observed at 9.858 min corresponds to enkephalin.

1. Prepare the cleavage cocktail by mixing together 450 μL TFA, 50 μL HPLC water, and 10 μL triisopropylsilane in a 1.5-mL centrifuge vial.
2. Prepare the cleavage buffer by mixing 50 μL of TFA with 50 mL of HPLC water.

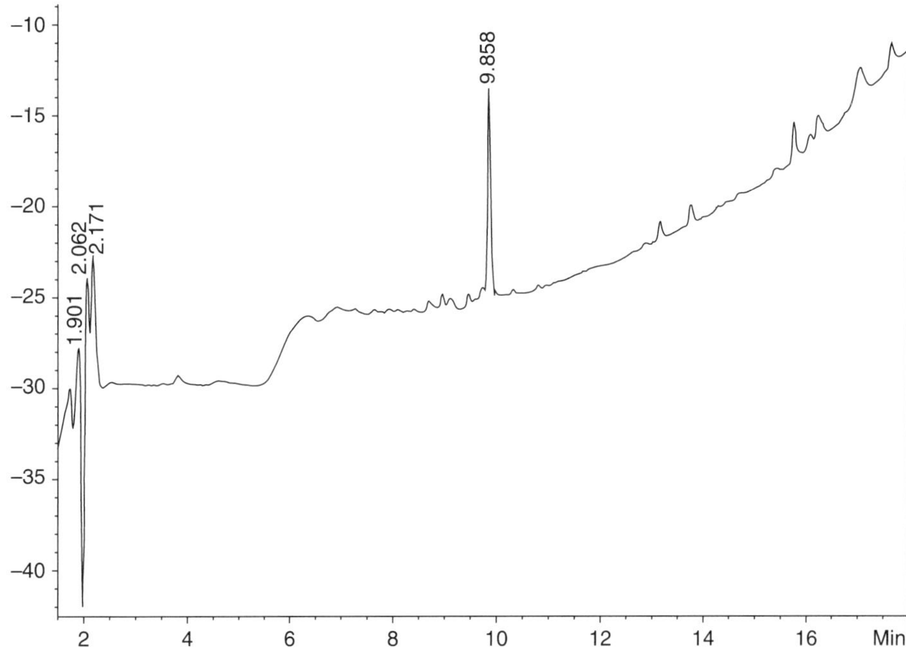

Fig. 4. HPLC result of the enkephalin synthesized at microscale (location 19 in **Fig. 2**). The peak corresponds to enkephalin with the Ac and NH_2 groups at the C and N terminals, respectively.

3. Take the bag with the membrane out of the freezer and let it equilibrate to room temperature before removing the membrane.
4. Use the pencil marks made in the last layer to determine the synthesis locations. Punch or cut out three to five spots of the same peptide.
5. Place the cut-out spots together into a 1.5-mL centrifuge vial.
6. Add 100 mL of the cleavage cocktail. Be sure that the cut-outs are covered with the solution.
7. Incubate at room temperature for 2 h.
8. Add 100 mL of the cleavage buffer into the cleavage tube and vortex.
9. Transfer quantitatively the crude peptide solution into a different vial. This solution can be used for HPLC or MS analysis.

4. Notes

1. The minimum equipment required for this application is an *xy* motion system. The substrate can then be mounted and moved by computer control to a desired location for dispensing the amino acid solutions. This system should provide for the programming of various printing recipes to designate the dispensing locations and volume.
2. A single piezoelectric dispensing device can be used to dispense all of the amino acid solutions. In this configuration the washing has to be performed when switching

between amino acids dispenses. A system incorporating a piezoelectric dispenser for each different amino acid solution can be also employed.

3. The derivatized membrane must be stored in a bag at –20°C. After removing the bag from the freezer wait for the membrane to get to room temperature before opening, otherwise condensation will occur on the membrane. Always handle the membrane with gloves or tweezers. The cellulose membrane is precut to a 96-well size. Cut to a smaller size if the desired number of peptide requires less membrane area. A 3-mm distance between locations (defined by the spreading of the dispensed solutions) was found acceptable when using 100 drops per spot of amino acid per coupling.

4. Weigh 20 mg of bromophenol blue in a 2-mL vial. Add 2 mL of DMF and vortex to dissolve. The resulting solution should be a yellow/orange color. If the solution has a different color (green/yellow or green/blue) the DMF contains free amides and requires purification (*see* **Note 5**). If the solution is acceptable, close the vial tightly and store at room temperature in the fume exhaust hood. Otherwise, prepare new bromophenol blue solution after DMF purification.

5. The DMF and NMP solutions should be tested for free amine groups. The test consists of adding 100 µL of the 1% bromophenol blue solution to 1 mL of NMP or DMF in a 1.5-mL centrifuge vial. Vortex and wait for 5 min. If the color is yellow the solutions are amine free. If the solution has a green/yellow or green/blue color they are not acceptable and need to be purified to eliminate the free amine groups *(15)*.

6. The surface tension and viscosity properties of each amino acid solution determine the electronic drive parameters required to obtain dispensing. Thus, each solution will require different electronic drive parameters. It is useful to record the dwell time and dwell voltage parameters required for each solution for future reference.

7. Particulate contamination can block the orifice resulting in drops that dispense intermittently or at an angle. Carefully wipe the microdispenser tip to remove such contamination. If this fails the particulate material can be removed by immersing the tip of the microdispenser into methanol and applying vacuum to backflush the device.

8. Check the following if the microdispenser is not dispensing:
   a. Does the fluid reach the orifice of the microdispenser? Adjust the pneumatics for positive pressure if the fluid does not reach the orifice.
   b. Are there air bubbles present in the glass tip of the microdispenser? Remove air bubbles by manipulating the pneumatics.
   c. Is there sufficient fluid in the reservoir? Check to confirm there is a sufficient amount of fluid in the reservoir.
   d. Are the wires of the microdispenser securely connected? Check to confirm the wires are connected.

9. The conventional way of writing a peptide amino acid sequence is from the amino (N)-terminal to the carboxy (C)-terminal. The attachment of the amino acid sequence to the synthesis substrate begins at the C-terminal. Thus, the peptides are synthesized in layers starting at the C-terminal. A printing layer consists of

dispensing the amino acid for that specific layer (e.g., third amino acid from the C-terminal) at the desired synthesis locations and the third amino acid can differ between the synthesis locations. If each amino acid solution for each layer can be loaded in separate individual dispensers, printing a layer can be programmed in a single recipe. If a single dispenser is used, the layer recipe has to be written for each amino acid in each layer. Thus, for each layer manual loading, washing, and unloading of the devices, when switching from one amino acid to the next one is required. Successful peptide synthesis was obtained when using 100 drops per location with double coupling (~20 nL of 0.3 M amino acid solutions).

10. MicroFab's jetlab software has built in capability to determine the offset and rotation of the substrate by surveying alignment marks on the substrate. The alternative is to place the device tip above the first marking and record the x- and y-coordinates. After doing the same for the second marker the two sets of coordinates can be used to determine the offset and the rotation.

11. Arginine is unstable and the arginine amino acid solution has to be prepared fresh immediately prior to dispensing the solution in each layer. A smaller amount (50 µL) can be prepared each time before the coupling. If the peptide synthesis extends beyond 1 d fresh amino acid solutions have to be prepared for the second day.

12. If coupling has occurred a discoloration of the blue membrane will be observed at the locations of the synthesis. The discoloration will vary between the various amino acids that are coupled to the membrane.

13. If the membrane is still blue after the second round of acetic anhydride, the synthesis or the bromophenol blue is compromised. Prepare new bromophenol blue solution and recheck the DMF and NMP.

14. The membrane must be dry to allow for the absorption of the next amino acid solution that is deposited. Otherwise, the amino acid solution will not flow completely into the membrane fibers and will spread out on the surface of the membrane.

15. A longer synthesis might take more than 1 d. This is the place to stop. Store the dyed membrane overnight in a bag in a –20°C freezer.

16. A common requirement of antibody probing is to avoid letting the membrane dry out during the antibody probing procedures. If drying occurs, a high background signal may result.

17. We have used dilutions of 1:20, 1:100, and 1:1000 based on a 1-mg/mL solution. For the specific results presented here antibody binding to a specific synthetic peptide was demonstrated using the 1:20 level and was found appropriate for the colorimetric detection. The detection level will depend on the specific peptides synthesized and antibodies used.

Acknowledgments

This work was supported by the NIH grant 5R44RR014397 and by the Drug Discovery Unit at Procter & Gamble.

References

1. Min, D. H. and Mrksich, M. (2004) Peptide arrays: towards routine implementation. *Curr. Opin. Chem. Biol.* **8,** 554–558.
2. Panicker, R. C., Huang, X., and Shao, Q. Y. (2004) Recent advances in peptide-based microarray technologies. *Comb. Chem. High Throughput Screen.* **7,** 547–556.
3. Fujii, K., Zhu, G., Liu, Y., et al. (2005) Kinase peptide specificity: improved determination and relevance to protein phosphorylation. *Proc. Natl. Acad. Sci. USA* **101,** 13,744–13,749.
4. Sun, H., Low, K. E., Woo, S., et al. (2005) Real-time protein kinase assay. *Anal. Chem.* **77,** 2043–2049.
5. Falsey, J. R., Renil, M., Park, S., Li, S., and Lam, K. S. (2001) Peptide and small molecule microarray for high throughput cell adhesion and functional assays. *Bioconjug. Chem.* **12,** 346–353.
6. Thorpe, D. S., Yeoman, H., Chan, A. W., Krchnak, V., Lebl, M., and Felder, S. (1999) Combinatorial chemistry reveals a new motif that binds the platelet fibrinogen receptor, gpIIbIIIa. *Biochem. Biophys. Res. Commun.* **256,** 537–541.
7. Bowditch, R. D., Tani, P., Foong, K. C., and McMilan, R. (1996) Characterization of autoantigenic epitopes on platelet glycoprotein Iib/IIIa using random peptide libraries, *Blood* **15,** 4579–4584.
8. Swan, E. E., Popat, K. C., and Desai, T. A. (2005) Peptide-immobilized nanoporous alumina membranes for enhanced osteoblast adhesion. *Biomaterials* **26,** 1969–1976.
9. Spear, M. A., Breakefield, X. O., Beltzer, J., et al. (2001) Isolation, characterization, and recovery of small peptide phage display epitopes selected against viable malignant glioma cells. *Cancer Gene Ther.* **8,** 506–511.
10. Houghten, R. A. (1985) General method for the rapid solid-phase synthesis of large numbers of peptides: specificity of antigen–antibody interaction at the level of individual amino acids, *Proc. Natl. Acad. Sci. USA* **82,** 5131–5135.
11. Reid, J., Betney, R., Watt, K., and McEwan, I. J. (2003) The androgen receptor transactivation domain: the interplay between protein conformation and protein–protein interaction. *Biochem. Soc.* **31,** 1042–1046.
12. Huang, X., Pieczko, M. E., and Long, E. C. (1999) Combinatorial optimization of the DNA cleaving Ni(II) x Xaa-Xaa-His metallotripeptide domain. *Biochemistry* **38,** 2160–2166.
13. Pennington, M., Lam, K., and Cress, A. E. (1996) The use of combinatorial library method to isolate human tumor cell adhesion peptides. *Mol. Diversity* **2,** 19–28.
14. Frank, R. (1992) Spot-Synthesis: an easy technique for the positionally addressable, parallel chemical synthesis on a membrane support. *Tetrahedron* **48,** 9217–9232.
15. (1999) Sigma-Genosys custom SPOT's technical manual, ver 3, Sigma-Genosys, The Woodlands, TX, p. 51.

16

Intein-Mediated Peptide Arrays for Epitope Mapping and Kinase/Phosphatase Assays

Ming-Qun Xu, Inca Ghosh, Samvel Kochinyan, and Luo Sun

Summary

Synthetic peptides are widely used for production and analysis of antibodies as well as in the study of protein modification enzymes. To circumvent the technical challenges of the existing techniques regarding peptide quantization and normalization, a new method of producing peptide arrays has been developed. This approach utilizes intein-mediated protein ligation that involves linkage of a carrier protein possessing a reactive carboxyl-terminal thioester to a peptide with an amino-terminal cysteine through a native peptide bond. Ligated protein substrates or enzyme-treated samples are arrayed on nitrocellulose membranes with a standard dot-blot apparatus and analyzed by immunoassay. This technique has improved sensitivity and reproducibility, and is suitable for various peptide-based applications. In this report, several experimental procedures including epitope mapping and the study of protein modifications were described.

Key Words: Intein-mediated protein ligation; kinase; nitrocellulose; peptide array; phosphatase; protein substrate.

1. Introduction

In the postgenomic era synthetic peptides are increasingly utilized for the investigation of posttranslational modifications, such as phosphorylation and dephosphorylation, owing to the ease of chemical synthesis and incorporation of various chemical groups *(1–4)*. In addition, synthetic peptides are widely used for the generation, purification and analysis of antibodies *(5–8)*. Enzyme-linked immunosorbent assays (ELISA) and arrays are commonly used for peptide-based assays *(4,5,9)*. One of the limitations of ELISA is that different peptides do not bind to polystyrene uniformly. SPOT is an array technique that involves the direct synthesis of peptides on cellulose membranes, often in a quantity that can hinder quantitative analysis as saturation is easily reached

From: *Methods in Molecular Biology, vol. 381: Microarrays: Second Edition: Volume 1: Synthesis Methods*
Edited by: J. B. Rampal © Humana Press Inc., Totowa, NJ

(8,10,11). Array analysis using peptide on low-cost nitrocellulose is hampered by ineffective and variable-binding efficiency that often results in low sensitivity and false-positives. Furthermore, peptides are usually not applicable to Western blot analysis owing to their small molecular mass. A new approach was presented and termed as intein-mediated peptide arrays (IPA), which enables the use of peptides for arrays as well as for enzymatic assays and Western blots *(12–14)*. This method takes advantage of the intein-mediated protein ligation (IPL) technique in which a peptide possessing an N-terminal cysteine can be efficiently linked to the carboxyl terminus of a reactive carrier protein through a peptide bond *(15–17)*. This linear carrier protein–peptide substrate can then be utilized in multiple applications. As in ELISA, array production is quick and simple, the number of peptides in the array can be adjusted as needed, peptide synthesis is kept separate from arraying, and small amount of peptide is arrayed so it is economical to use each array only once. In addition, the amount of peptide in each array feature is normalized by using a carrier protein when the array is produced. For IPA, labor-intensive peptide purification and accurate quantification are unnecessary; because the carrier protein dominates the binding event, the retention of peptide substrates on the matrix is less variable and the sensitivity is increased up to 10^4-fold *(12)*. In addition, because each carrier protein molecule has precisely one reactive site, the amount of peptide arrayed onto the membranes can be effectively normalized. Here, several examples to demonstrate the benefits of this IPA strategy are described.

1.1. Intein-Mediated Peptide Array

This new technique was built on the hypothesis that using an intein-generated carrier protein for arraying synthetic peptides will result in efficient binding of peptides to a membrane and an improved signal-to-noise ratio *(12)*. The IPL technique takes advantage of the catalytic activity of an engineered intein to form a thioester at the N-terminal splice junction (between the C-terminus of a target protein and the N-terminus of the intein) *(18)*. A reactive thioester can be readily generated at the C-terminus of the target protein (carrier protein) by the treatment of the target protein–intein fusion precursors with an appropriate thiol compound, such as 2-mercaptoethanesulfonic acid (MESNA), resulting in a stable carboxyl-terminal thioester group *(15)* (**Fig. 1**). Peptides possessing an N-terminal cysteine are linked to the carrier protein by the use of IPL *(16)*, also termed as expressed protein ligation *(17)*. The peptides are synthesized with an amino-terminal cysteine and ligated to the carboxyl-terminus of a carrier protein through a peptide bond. Ligation is conducted with at least 25-fold molar excess of synthetic peptide so peptide purity does not significantly influence the final yield of the carrier protein–peptide ligation product. Unlike conventional chemical methods for conjugating synthetic peptides to carrier

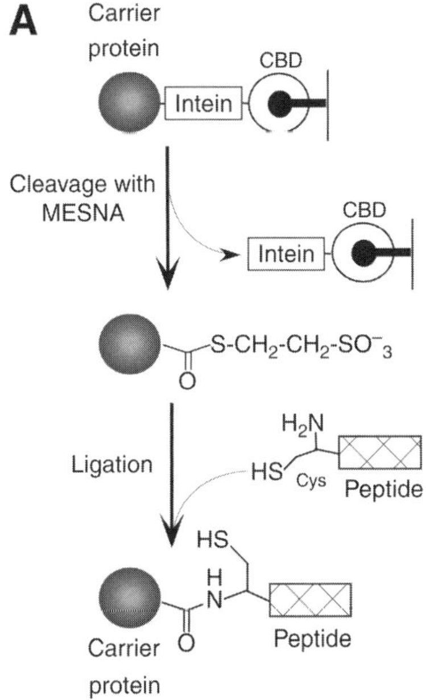

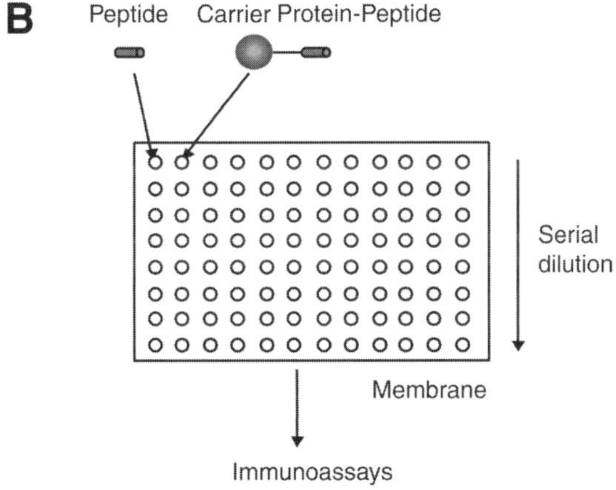

Fig. 1. A schematic of intein-mediated peptide array. (**A**) Diagram showing the production of a carrier protein-synthetic peptide product by intein-mediated protein ligation. The carrier protein is fused to the N-terminus of an intein, followed by a chitin binding domain. The carrier protein, possessing an active C-terminal thioester, is generated by incubation of the chitin bound fusion protein in the presence of 2-mercaptoethane-sulfonic acid. A synthetic peptide possessing an N-terminal cysteine is ligated to the C-terminus of the carrier protein through a native peptide bond. (**B**) Flowchart for arraying synthetic peptides after intein-mediated protein ligation. The ligation products and unligated peptides are serially diluted and arrayed onto a nitrocellulose membrane in a grid, as shown, and then reacted with an antibody.

proteins, the stoichiometry of an IPL reaction is precisely one-to-one, therefore, the amount of peptide in each array feature is determined by the amount of the carrier protein, instead of the absolute peptide amount or purity. The ligation efficiency can be readily evaluated by detecting a mobility shift on a sodium dodecyl sulfate polyacrylamide gel (SDS-PAGE) stained with Coomassie blue. Furthermore, the linear carrier protein-peptide product migrates as a single sharp band on SDS-PAGE so that the same set of substrates or assay samples can be examined by Western blot *(13,14)*. The IPL technique has also been utilized for site-specific biotinylation and immobilization of proteins in microarray production *(19,20)*.

1.2. Epitope Mapping by IPA

Synthetic peptide arrays are powerful tools for characterizing antibodies raised against peptide antigens. An epitope mapping experiment is intended to define the antigenic determinant recognized by a monoclonal antibody. This can be accomplished by probing an array that is generated from a library of short, overlapping peptides that span the antigenic protein sequence. Usually an epitope can be defined precisely by constructing a peptide library in which each epitope residue is substituted with either an alanine (alanine-scanning) or another amino acid to assess the contribution of each residue to antibody binding and to determine which substitutions affect antibody recognition (mutational analysis). The application of IPA in epitope mapping was demonstrated by designing a peptide library to define the epitopes of two commercially available monoclonal antibodies against hemagglutinin (HA) corresponding to residues 98–106 (YPYDVPDYA) of HA protein *(12)*. The data exhibited the enhanced sensitivity of carrier protein-HA peptide substrates as compared to unligated peptides in epitope mapping (**Fig. 2**).

1.3. Analysis of Antibody Specificity

Highly specific antibodies are crucial tools for investigation of gene function that involves the determination of protein expression patterns and post-translational modifications. An important application of IPA is to aid in the evaluation of antibody specificity. This approach was employed to demonstrate the specificity of a general phospho-tyrosine antibody *(12)*. Four groups of peptides containing a phospho-tyrosine, phospho-serine, phospho-threonine, or no phosphorylated residue were ligated to a carrier protein. Serial dilution samples of the ligation products were arrayed onto nitrocellulose and incubated with a phospho-tyrosine antibody that presumably recognizes degenerate peptide sequences containing phospho-tyrosine residue. Only one of the five unligated phospho-tyrosine peptides generated significant immunoreactivity

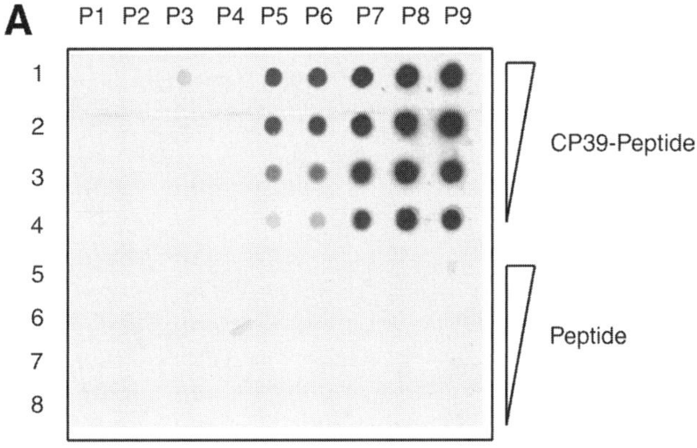

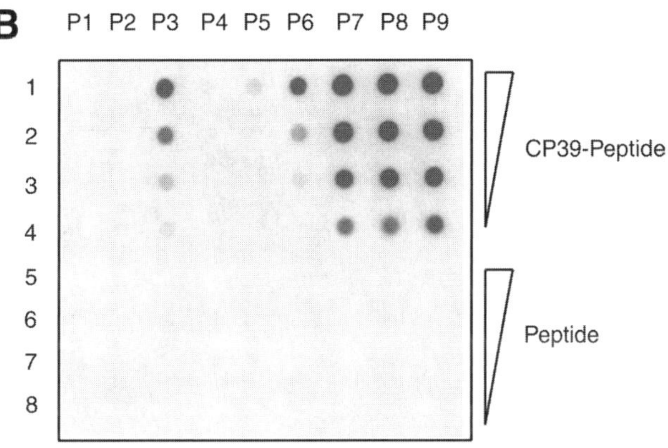

Fig. 2. Epitope mapping by dot-blot alanine-scanning assays. An alanine-scanning hemagglutinin (HA) peptide library (**Table 1**) was ligated to the carrier protein, CP39. The ligation products and peptides were diluted and arrayed onto two 0.45-μm nitrocellulose membranes (threefold serially diluted in rows 1–4), along with unligated synthetic peptides (threefold serially diluted in rows 5–8). One membrane was probed with a monoclonal antibody against HA from Zymed Laboratories (**A**), and the other membrane was probed with anti-HA monoclonal antibody from Cell Signaling Technology, Inc. (**B**).

(**Fig. 3**). The phospho-tyrosine antibody recognized all five phospho-tyrosine peptides that were ligated to the carrier protein, confirming its use as a phospho-tyrosine motif-specific antibody. The data also suggest that diluting the peptide substrates in array analysis is crucial for a successful quantitative evaluation of the affinity of this motif-specific antibody for different phospho-peptides.

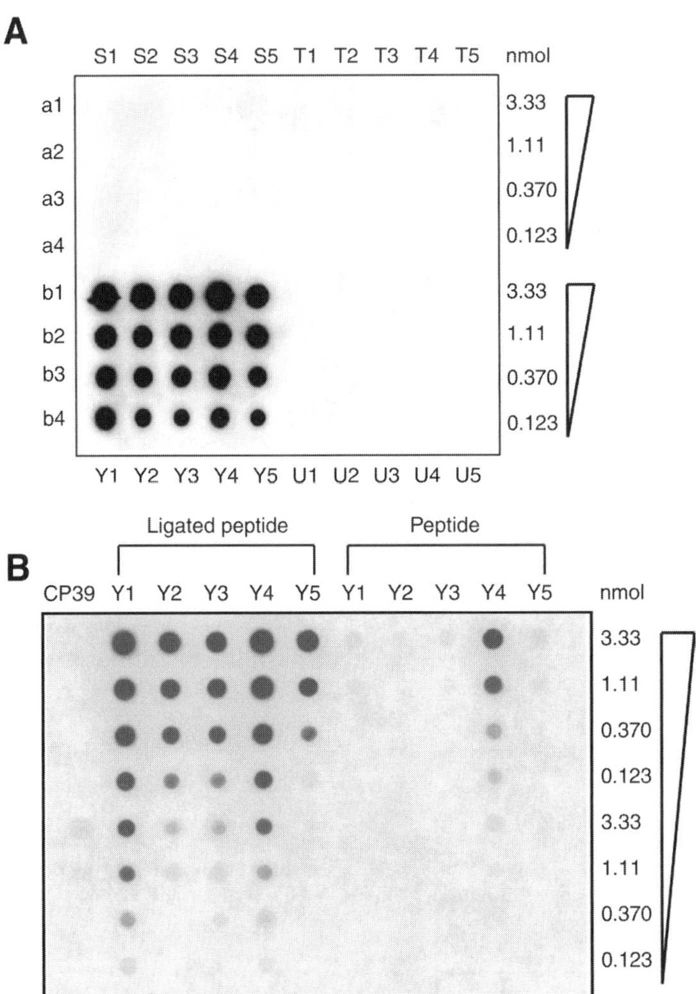

Fig. 3. Defining the specificity of a general phospho-tyrosine antibody by dot-blot assay. Four groups of peptides containing phospho-serine (columns S1–S5), phospho-threonine (columns T1–T5), phospho-tyrosine (columns Y1–Y5), or no phosphorylation (columns U1–U5) (**Table 2**), were ligated to CP39. (**A**) The ligation products were serially diluted threefold and blotted onto a 0.45-μm nitrocellulose membrane (rows a1–a4 for phospho-serine and phospho-threonine peptides and rows b1–b4 for phospho-tyrosine and nonphosphorylated peptides). The membrane was reacted with a monoclonal antibody against phospho-tyrosine. (**B**) Comparison of immunoreactivity of the ligated and unligated peptides. The five synthetic peptides containing a phospho-tyrosine were ligated to CP39. The ligation samples and unligated peptides were arrayed onto a 0.45-μm nitrocellulose membrane. Immunoblotting was performed with a phospho-tyrosine antibody. The amount of peptide is indicated on the right side.

1.4. Kinase Assays

Protein phosphorylation and dephosphorylation are predominant events in signal transduction pathways. Determination of the specificities of kinases and phosphatases facilitates the understanding of the molecular mechanisms and the development of drug targets. The use of synthetic peptide substrates has become a powerful tool to determine kinase specificities and dissect kinase signaling pathways, allowing mutational analysis and investigation into optimal phosphorylation sites *(3)*. Detection of kinase activity is often performed by the measurement of the incorporation of radiolabeled γ-phosphate from ATP. An alternate approach is to determine the immunoreactivity of a phosphorylated protein to a phospho-specific antibody. Three common methods for the determination of immunoreactivity are ELISA, SPOT technique, and Western blot analysis. Western blot is generally not applicable for the small synthetic peptides. Peptide synthesis and arraying are linked in the SPOT approach *(10)*. The SPOT membranes containing target peptide substrates are commercially available and can be used for screening kinase activity present in a cell extract or a purified sample. The amount of peptide in each SPOT array element is about 25 μg. When a saturating amount of peptide is used, the assays become less sensitive in distinguishing variations. In addition, SPOT membranes are relatively expensive and in order to reuse the membranes they must be stripped. To use IPA, a peptide possessing a phosphorylation site of interest is first synthesized with an amino-terminal cysteine residue. The peptide is then ligated to the cysteine-reactive carboxyl terminus of a carrier protein through a peptide bond to form a kinase substrate **(Fig. 4)**. Simple protocols have been developed for producing arrays of kinase substrates on commonly used membranes, such as low-cost nitrocellulose. The on-membrane approach is to prepare a membrane containing the ligated kinase substrates and subsequently, to treat the membrane with a cell extract or kinase. Alternatively, the substrates are first treated with kinase and then arrayed onto a membrane. Both approaches allow for arraying the substrates or assay samples and their serial dilutions onto nitrocellulose for analysis with phospho-specific antibodies. The second approach permits small-volume reactions (in 10–20 μL) thereby consuming less enzyme sample, and treatment of the substrates with different kinase samples. In addition, the untreated samples can be utilized as the controls for quantitative analysis. If a kinase exhibits autophosphorylation activity, this may cause a significant background signal when substrates are treated with the kinase prior to the protein blotting step. This limitation can be overcome by using the on-membrane assay approach or by conducting Western blot analysis to distinguish the signal of a kinase from that of a phosphorylated substrate *(14)*.

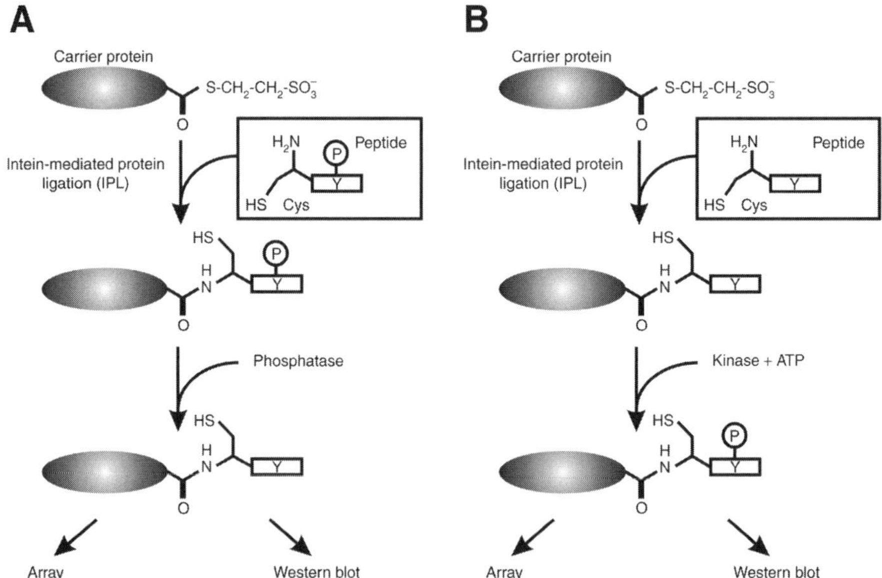

Fig. 4. Generation of substrates for kinase and phosphatase assays. **(A)** Diagram showing the production of carrier protein–peptide ligation product by intein-mediated protein ligation. A synthetic peptide possessing an N-terminal cysteine is ligated to the C-terminus of a carrier protein via a native peptide bond. The ligated substrate is treated with a kinase and subjected to array or Western blot analysis with a phospho-tyrosine antibody. **(B)** Diagram showing ligation of a phosphorylated peptide to a carrier protein using intein-mediated protein ligation. A peptide possessing a phospho-tyrosine is indicated. The ligated phospho-peptide is used as a substrate for phosphatase assays. The assay samples are serially diluted and arrayed onto a nitrocellulose membrane for analysis with a phospho-specific antibody (**Fig. 5A**). The same assay samples can also be used in Western blot analysis (**Fig. 5B**). Alternatively, the substrates can be arrayed onto nitrocellulose for on-membrane enzymatic assays and immunoassays (**Figs. 6** and **7**).

1.5. Phosphatase Assays

Use of synthetic peptides allows for the production of pure substrates containing phosphorylated tyrosine, serine, or threonine for phosphatase assays. The IPA approach for the investigation and screening of substrate specificity of various protein phosphatases were employed (**Figs. 4** and **5**). Similar to the protocols described earlier in kinase assays, phosphatase treatment can be performed before or after the protein blotting process, depending on the purpose of the experiment. Performing phosphatase assays prior to the blotting step consumes less enzyme and permits treatment of the substrates with different phosphatases. The measurement of phosphatase activity also relies on analysis with phospho-specific antibodies. A phosphorylated substrate without phosphatase treatment should be included as a control. In the IPA approach, the peptide synthesis and substrate preparation steps are kept separate from arraying, offering researchers flexibility in array production. Therefore, this technique provides a simple platform to produce peptide arrays in a laboratory.

2. Materials

2.1. Intein-Mediated Protein Ligation

1. Peptide with N-terminal cysteine. Dissolve the peptide in water or dimethyl sulfoxide (DMSO) to a concentration of 1–10 mM (*see* **Notes 1** and **2**).
2. Carrier protein 39 (CP39; 1 mg/mL; New England BioLabs, Inc., Ipswich, MA; cat. no. E6603S) or Carrier Protein 27 (CP27; 1 mg/mL, NEB, cat. no. E6607S) (*see* **Notes 3–6**).
3. 10X CP Ligation buffer: 1 M Tris-HCl, pH 8.5, 100 mM MESNA (Sigma) (*see* **Note 7**).

2.2. Sodium Dodecyl Sulfate Polyacrylamide Gel Electrophoresis

1. 3X SDS Sample buffer: 187.5 mM Tris-HCl, pH 6.8 at 25°C, 6% SDS, 0.03% bromophenol blue, 30% glycerol. Dithiothreitol should be added to the 3X SDS sample buffer to a final concentration of 40 mM dithiothreitol.
2. 12 or 10–20% Tris-glycine polyacrylamide gels (Invitrogen, Carlsbad, CA; cat. no. EC60052 or EC61352).
3. 1X Tris-glycine running buffer: 0.1% SDS, 190 mM glycine, 25 mM Tris-base.
4. Protein marker, broad range (2–212 kDa) (NEB, cat. no. P7702S).
5. Coomassie blue stain: 0.3% Brilliant blue R in destain solution (Sigma, cat. no. B-0149).
6. Destain solution: 25% methanol (v/v), 10% acetic acid (v/v).

2.3. Dot Blotting and Immunoassays

1. Bio-Dot® Microfiltration Apparatus (Bio-Rad, Hercules, CA; cat. no. 170-6545).
2. Nitrocellulose membrane (Protran, Schleicher and Schuell, Keene, NH; cat. no. BA85).
3. Zeta Probe nylon membrane (Bio-Rad, cat. no. 162-0153).
4. Phosphate-buffered saline (PBS): 58 mM sodium phosphate dibasic, 17 mM sodium phosphate monobasic, 68 mM NaCl, pH 7.4.
5. Phosphate-buffered saline with Tween-20 (PBST): PBS, 0.05% Tween-20.
6. Blocking solution: 5% nonfat dry milk in PBST.
7. Tris-buffered saline with Tween-20 and Triton X-100 (TBSTT): 20 mM Tris-HCl, pH 7.5, 150 mM NaCl, 0.2% Tween-20, 0.05% Triton X-100.
8. Blocking buffer: 5% nonfat dried skim milk in TBSTT.
9. Anti-HA monoclonal antibody (Cell Signaling Technology [CST], Beverly, MA; cat. no. 2362).
10. Mouse anti-HA monoclonal antibody (Zymed, San Francisco, CA; cat. no. 32-6700).
11. Monoclonal antibody against the HA epitope from either Zymed Laboratories or CST.

12. Phospho-Tyrosine monoclonal antibody (P-Tyr-100, CST, cat. no. 9411).
13. Anti-mouse immunoglobulin G (IgG), horse radish peroxidase (HRP)-linked antibody (CST, cat. no. 7076).
    a. Anti-rabbit IgG, HRP linked antibody (CST, cat. no. 7074); anti-mouse IgG, HRP-linked antibody (CST, cat. no. 7076).
    b. Saran wrap.
    c. Whatman 3M paper (Whatman Inc. Florham Park, NJ; cat. no 3030-6185).
    d. LumiGLO reagent and Peroxide (CST, cat. no. 7003).
    e. Hyperfilm™ ECL (Amersham Biosciences, Buckinghamshire, UK; cat. no. RPN3114K).

2.4. Kinase/Phosphatase Assay

1. Src Protein Tyrosine Kinase (Lake Placid, NY; cat. no. 14-326).
2. Abl Protein Tyrosine Kinase (NEB, cat. no. P6050S).
3. Alkaline phosphatase, Calf Intestinal (NEB, cat. no. M0290S).
4. Protein phosphatase 1 (NEB, cat. no. P0754S).
5. T-cell protein tyrosine phosphatase (NEB, cat. no. P0752S).
6. LAR protein tyrosine phosphatase (NEB, cat. no. P0750S).
7. 10X Kinase/phosphatase buffer (from enzyme's manufacturer).
8. Adenosine 5′-triphosphate (Sigma, cat. no. A-2383). Dissolve ATP in water and adjust the pH to 7.0 with 0.1 N sodium hydroxide. Store at –80°C.
9. 0.025-µm Membrane filter disc (Millipore, Bedford, MA; cat. no. VSWP02500).
10. Water.

3. Methods

3.1. Preparation of Samples

The IPA method utilizes a standard IPL protocol in which a peptide possessing an N-terminal cysteine is linked to the carboxyl terminus of a reactive carrier protein via a peptide bond (**Fig. 1**). The procedure starts with the design of peptide substrates possessing an N-terminal cysteine. A linker region can be incorporated to improve the peptide presentation or folding (*see* **Note 1**). In addition, this cysteine-containing peptide can be used for the generation of an immunogen and an affinity column for antibody production and purification *(13,21)*.

1. Peptides are synthesized with an additional cysteine residue at their N-terminus. Further high-performance liquid chromatography purification of peptides will improve their ligation efficiency to a carrier protein (*see* **Note 9**). Peptides are dissolved in water to prepare a 2- to10- mM stock solution. For insoluble peptides, it may be necessary to dissolve the peptide in DMSO before adding water (*see* **Note 10**).
2. CP39 and CP27 are commercially available at 1 mg/mL (*see* **Note 11**) and 0.2 mg/mL concentration from (New England BioLabs). Thioester-tagged proteins can be kept for more than 9 mo at −80°C without a significant loss of their ligation activity (*see* **Note 6**).

3. Bio-Dot microfiltration apparatus is used to transfer samples onto membranes (*see* **Note 12**). The apparatus is rinsed and dried before an experiment to prevent contamination. Ninety-six samples are arrayed onto a membrane following the manufacturer's recommended protocol.

3.2. Peptide Ligation to the Carrier Protein

The IPL procedure involves mixing the peptide with the carrier protein and incubating at 4°C for 4–16 h (*see* **Note 13**). Although either a crude or purified peptide can be used for ligation, use of a purified peptide generally results in higher ligation efficiency (50–95%) and sensitivity (*see* **Note 9**). The ligation products are then used as substrates for Western blot analysis (*see* **Note 14**), array analysis and enzyme assays (*see* **Note 11**). The ligation products are stable and can be stored at –80°C for later use (*see* **Note 15**). The ligation efficiency can be readily evaluated by observing a gel shift on a 12% SDS-PAGE stained with Coomassie brilliant blue (*see* **Note 16**).

1. Dissolve the peptide in water to make a peptide stock solution of 2 mM.
2. Thaw carrier protein (1 mg/mL) at room temperature. Set up the following reaction immediately.
   Mix the following samples in order:

2X Peptide solution (2 mM)	12.5 μL
10X Carrier protein reaction buffer	2.5 μL
Carrier protein (1.0 mg/mL)	10.0 μL
Total volume	25.0 μL

3. Incubate at 4°C overnight.
4. Store the reaction at –20°C if not used immediately.

3.3. Transferring Carrier Protein-Peptide Onto Membranes

Protein blotting is performed according to a standard protocol suggested by the manufacturer (Bio-Rad). The ligated substrates or assays are diluted in a 96-well microtiter plates before arraying onto a nitrocellulose membrane. It is important to change tips at each step to achieve accurate data.

1. Clean and dry the Bio-Dot apparatus consisting of the membrane support plate, sample template, and gasket prior to the assembly.
2. Apply a light vacuum to the Bio-Dot apparatus through an outlet port on the vacuum manifold. Attach tubing to the outlet that runs from the vacuum manifold through the three-way flow valve to a trap for collection of the waste liquid.
3. Place the membrane support plate into position in the vacuum manifold. (There is only one way to slide the plate into the manifold.)
4. Place the sealing gasket on top of the membrane support plate. The guide pins on the vacuum manifold help align the 96 holes in the gasket over the 96 holes

in the support plate. Visually inspect the gasket to make sure the holes are properly aligned. If the gasket is not centered, pull lightly at the corners until it is aligned.

5. Prewet the nitrocellulose or nylon membrane by slowly sliding it at a 45° angle into PBS solution. Wet the nitrocellulose in PBS for 10 min; soak the membrane completely. Remove the membrane from the solution; let the excess liquid drain from the membrane. Lay the membrane on the gasket in the apparatus so that it covers all the holes. The membrane should not extend beyond the edge of the gasket after the Bio-Dot apparatus is assembled. Always use forceps or wear gloves when handling membranes. Remove any air bubbles trapped between the membrane and the gasket.

6. Place the sample template on top of the membrane. The guide pins insure that the template will be properly aligned. Tighten the four screws. When tightening the screws, use a diagonal crossing pattern to ensure even application of pressure on the membrane surface.

7. Attach a vacuum source to the flow valve, with a waste trap set up and positioned between the vacuum outlet and the flow valve. Turn on the vacuum lightly and set the three-way valve to apply vacuum to the apparatus.

8. With the vacuum applied, repeat the tightening process using the diagonal crossing pattern. Tightening the screws while the vacuum is applied ensures a tight seal, preventing cross-contamination between slots. Failure to tighten the screws may lead to leaking of the samples between the wells.

9. Dilute the ligation product and controls (carrier protein and unligated peptide) in PBS in a 96-well microtiter plate; add 10 µL of each ligation sample into 140 µL PBS in the first row of the plate and subsequently add 100 µL of PBS to the rest of the wells in the plate.

10. Perform a threefold serial dilution in PBS by transferring 50 µL from the first sample row into the following row containing 100 µL PBS. Repeat this process by transferring the diluted samples into the next row until the last row of the plate. Discard the 50 µL of diluted sample from the last row of the plate. Using a 12-multichannel pipet will greatly facilitate this dilution process and ensure consistency.

11. Transfer the samples (100 µL) from each well of the microtiter plate to the corresponding wells on the Bio-Dot microfiltration apparatus.

12. Gently remove the solution from the wells by applying a light vacuum. Watch the sample wells; as soon as the buffer solution drains from all the wells, adjust the flow valve so that the unit is exposed to atmospheric pressure and disconnect the vacuum.

13. Each well of the apparatus is washed three times with 100 µL PBS; gently remove the PBS from the wells by light vacuum.

14. The Bio-Dot Microfiltration Apparatus is disassembled, the membrane is removed and rinsed with 40 mL PBS (the volume used depends on the size of the container). Do not allow the membrane to dry out.

15. A membrane arrayed with samples is ready for immunoblotting and enzymatic assays.

3.4. Determination of the Ligation Efficiency

As a general guideline, ligation efficiency can be assessed by electrophoresis on 12% SDS-PAGE stained with Coomassie blue. Ligation samples containing approx 0.5 µg of carrier protein are loaded in each lane and a sample of unligated carrier protein is used as a control. Ligation of peptide to a carrier protein results in a mobility shift on a SDS-PAGE gel (*see* **Note 5**).

1. These instructions assume the use of 12% Tris-glycine polyacrylamide gel.
2. Assemble the gel apparatus according to the manufacturer's suggested protocol. Fill the tank with Tris-glycine running buffer.
3. Wash each well with Tris-glycine running buffer using Pasteur pipet or glass pipet.
4. Add 12.5 µL of 3X SDS sample buffer to 25 µL ligation mix and boil the samples for 3 min (give a quick centrifugation if necessary).
5. Load 2–5 µL of boiled samples onto a gel along with 15 µL of a broad range protein marker.
6. Attach lid and connect electrical leads to the gel apparatus.
7. Turn-on the power supply with select mode to V (voltage) position.
8. Electrophoresis is done at a constant 100 V to separate protein samples.
9. Once the protein loading dye reaches the bottom of the gel, turn off the power supply.
10. Detach lid and electrical wires and remove the gel from the gel tank.
11. Stain the gel with 0.3% Coomassie brilliant blue solution for 25 min at room temperature.
12. Transfer the gel into destaining solution and change the solution a few times until the proteins are visible without a significant background blue color.

3.5. Immunoblotting

1. The membrane is blocked by incubating the membrane in 50 mL of 5% dry milk in TBSTT for 1 h at room temperature with shaking and then washed three times for 15 min each with 50 mL TBSTT.
2. For antibody incubation, the membranes arrayed with samples, are placed in trays containing the diluted primary antibody in 50 mL 2% dry milk in TBSTT overnight at 4°C.
3. The primary antibody solution is discarded and the membranes are washed with 50 mL of TBSTT three times for 15 min each.
4. The solution containing diluted HRP-linked secondary antibody in TBSTT with 2% dry milk is added to the membrane and incubated at room temperature for 1 h.
5. The solution is discarded and the membrane is washed three times for 15 min each with 50 mL of TBSTT.
6. Antibody binding is then detected by chemiluminescence using the LumiGLO solution. Add 1 mL of LumiGLO reagent A and 1 mL peroxide reagent B to 18 mL water. Add the mixed LumiGLO mix solution to the membrane and incubate for 1 min at room temperature with shaking.

Table 1
Peptides for Alanine Scanning of HA Epitope

		A1	A2	A3	A4	A5	A6	A7	A8	A9
P1	CAGAG	A	P	Y	D	V	P	D	Y	A
P2	CAGAG	Y	A	Y	D	V	P	D	Y	A
P3	CAGAG	Y	P	A	D	V	P	D	Y	A
P4	CAGAG	Y	P	Y	A	V	P	D	Y	A
P5	CAGAG	Y	P	Y	D	A	P	D	Y	A
P6	CAGAG	Y	P	Y	D	V	A	D	Y	A
P7	CAGAG	Y	P	Y	D	V	P	A	Y	A
P8	CAGAG	Y	P	Y	D	V	P	D	A	A
P9	CAGAG	Y	P	Y	D	V	P	D	Y	A

This library of nine peptides (P1–P9) corresponds to residues 98–106 (YPYDVPDYA) of the hemagglutinin protein as in **ref. *12***. P9 contains the wild-type sequence and each of the other eight peptides carries a single substitution with an alanine residue.

7. Remove the membrane from the solution and blot with Whatman 3M paper to remove excess solution from the membrane and wrap in Saran wrap (*see* **Subheading 2.3.**). Try to avoid bubbles or creases when the membrane is wrapped with Saran wrap.
8. The wrapped membrane is then placed in a film cassette and exposed to the film in the dark for an exposure time varying from 30 s to 5 min.

3.6. Epitope Mapping by Alanine-Scanning

In the following example, an alanine-scanning HA peptide library was designed and synthesized (**Table 1**). The HA peptide library consisted of a CAGAG peptide tag at the amino terminus of the epitope consisting of nine residues, 98–106, of the HA protein (YPYDVPDYA). The amino-terminal peptide tag allowed IPL to carrier proteins and served as a spacer for antibody binding. The library consisted of nine components, P1 through P9, one for each of the eight HA peptide residues substituted with Ala and P9 encoding the unaltered HA peptide sequence. The IPA method was used to ligate these peptides to a carrier protein and to generate arrays to map and differentiate the epitopes of two commercially available monoclonal antibodies against HA.

1. Each synthetic HA peptide was dissolved in water to make a 2-m*M* stock solution. Ligation to CP39 was set up according to the ligation protocol described earlier (*see* **Subheading 3.2.**).

2. The ligation products were arrayed in duplicate on 0.45-μm nitrocellulose membranes along with unligated synthetic peptides (*see* **Subheading 3.3.**).
3. Each membrane was then probed with a monoclonal antibody against the HA epitope from either Zymed Laboratories (**Fig. 2A**) or CST (**Fig. 2B**).

This array showed the similarities and differences between the two HA monoclonal antibodies. Both antibodies tolerated alanine substitutions at residues 7 and 8. Alanine residues at positions 1, 2, and 4 reduced binding to nearly undetectable levels, indicating those positions are essential amino acids of the epitope for the binding of both antibodies. However, the two antibodies recognized slightly different epitopes; the antibody from Zymed Laboratories is less sensitive to replacement at residues 5 and 6 than the antibody from CST, and conversely the CST antibody tolerated an alanine mutation at position 3 more readily than the Zymed Laboratories' antibody. Importantly, the unligated synthetic peptides showed no detectable signal, precluding this kind of detailed epitope analysis.

3.7. Substrate Profiling for Phospho-Specific Antibody

The following example demonstrates the characterization of substrate specificity of a phospho-specific monoclonal antibody. We constructed an array of unrelated synthetic peptides to determine the specificity of a general phospho-tyrosine antibody (P-Tyr-100, Cell Signaling Technology).

1. Four groups of peptides containing phospho-serine (S1–S5), phospho-threonine (T1–T5), phospho-tyrosine (Y1–Y5), or no phosphorylated residue (U1–U5) were designed and synthesized with an N-terminal cysteine (**Table 2**).
2. Each peptide was ligated to CP39 following the ligation protocol described earlier (*see* **Subheading 3.2.**).
3. The ligation products were serially diluted in a 96-well microtiter plate and transferred onto nitrocellulose following the steps described earlier (*see* **Subheading 3.3.**).
4. The blot was incubated with a phospho-tyrosine antibody (P-Tyr-100) that presumably recognizes peptides containing a phospho-tyrosine regardless of the surrounding peptide sequence, following the protocol described earlier (*see* **Subheading 3.5.**).

The phospho-tyrosine antibody (P-Tyr-100) specifically recognized only the phospho-tyrosine containing substrates (*see* **Fig. 3A**). Only one of the five unligated phospho-tyrosine peptides (Y4) yielded significant immunoreactivity, albeit at a level 27-fold lower than the same peptide ligated to the carrier protein (*see* **Fig. 3B**). This phospho-tyrosine antibody did recognize all five phospho-tyrosine peptides, but with different affinities, peptides Y1 and Y4 produced the strongest signals, followed by Y2 and Y3, and then by Y5. This antibody exhibited different affinity to various substrates when the substrates were

Table 2
Peptides for Characterization of Phospho-Specific Antibody

Peptide	Protein	Sequence
S1	hRad 17(Ser645)-P	CASELPApSQPQPFSA
S2	BAD-Ser128-P	CGMEEELpSPFRGRS
S3	PLD1(Ser561)-P	CRKFSKFpSLYKQLH
S4	PFK-2(Ser466)-P	CVRMRRNpSFTPLSS
S5	RYR2(S2808)-P	CNRTRRIpSQTSQV
T1	GCN2-Thr745-P	CSDPSGHLpTGMVGTA
T2	PERK-Thr980-P(mouse)	CAYATHpTGQVGTKL
T3	IRAK(Thr100)-P	CRARDIIpTAWHPPA
T4	DNA-PK(Thr2609)-P	CTPMFVEpTQASQGT
T5	B-Raf(Thr439)-P	CDRNRMKpTLGRRD
Y1	FcgammaRIIB(Tyr292)-P	CGAENTITpYSLLMHP
Y2	CSF1R(Tyr723)-P	CSQGVDTpYVEMRPV
Y3	PKC delta(Tyr64)-P	CFDAHIpYEGRVIQ
Y4	PTEN(Tyr240)-P	CREDKFpYMFEFPQP
Y5	C-Kit(Tyr719)-P	CSDSTNEpYMDMKPG
U1	SKP2(73–88)	CHPESPPRKRLKSKGSDKD
U2	Axl(Tyr866)	CAAEVHPAGRYVL
U3	Filamin A(266–278)	CKAKLKPGAPLRPK
U4	Smc1(1221–1233)	CTKYPDANPNPNEQ
U5	ARK5(Thr211)	CNLYQKDKFLQTF

Four groups of peptides containing phospho-serine (S1–S5), phospho-threonine (T1–T5), phospho-tyrosine (Y1–Y5), or no phosphorylation (U1–U5), were ligated to CP39.

The names of the proteins from which the peptides were derived are shown in the middle column.

A lower case p indicates a phosphorylated residue (in the column on the right side).

diluted at least 27-fold (i.e., for array features that contain 123 pmol peptide or less).

3.8. Phosphatase Assay and Its Detection on Membrane

This section describes phosphatase assays that use synthetic phospho-peptides ligated to a carrier protein (*see* **Fig. 4**). The ligated substrates are treated with phosphatases in a small volume (10–20 µL per reaction) before the samples are subjected to protein blotting and analysis with phospho-specific antibodies. The data presented in **Fig. 5** demonstrates the effectiveness of this phosphatase assay system.

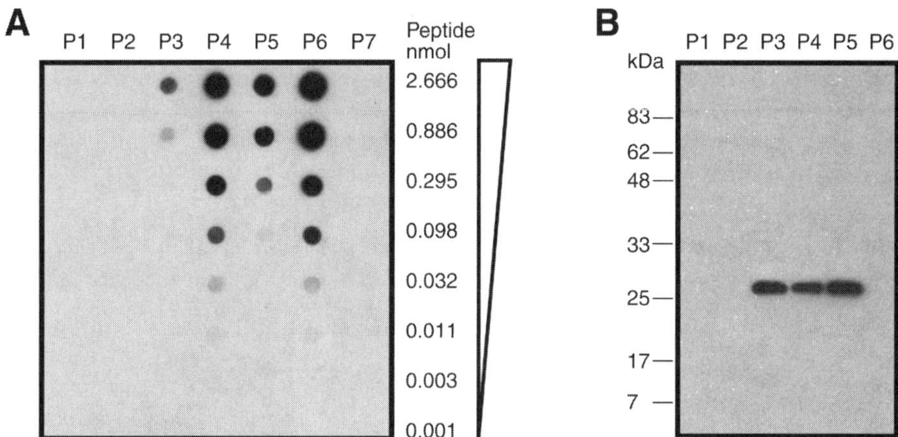

Fig. 5. Dot-blot and Western blot analysis of phosphatase assays using a ligated carrier protein–peptide substrate. **(A)** Dot-blot analysis of phosphatase activity. A phosphorylated peptide, CIGEGTpYGVVYK, corresponding to I^{10} to K^{20} of human cyclin-dependent kinase (Cdc2) was synthesized with an additional N-terminal cysteine residue. A nonphosphorylated peptide, Cdc2 (CIGEGTYGVVYK), was used as a control. Phospho-Cdc2 was ligated to CP27 and the ligation mixture was treated with Calf Intestinal Phosphatase (CIP), Protein Phosphatase 1 (PP1) or TC PTP. P1, CP27 alone; P2, CP27 ligated to Cdc2; P3, phospho-Cdc2 peptide alone; P4, CP27 ligated to phospho-Cdc2 (untreated); P5, CP27 ligated to phospho-Cdc2, treated with CIP; P6, CP27 ligated to phospho-Cdc2, treated with PP1; P7, CP27 ligated to phospho-Cdc2, treated with TC PTP. **(B)** Western blot analysis of phosphatase assays described in **(A)**. P1, CP27 alone; P2, CP27 ligated to Cdc2; P3, CP27 ligated to phospho-Cdc2 (untreated); P4, CP27 ligated to phospho-Cdc2, treated with CIP; P5, CP27 ligated to phospho-Cdc2, treated with PP1; P6, CP27 ligated to phospho-Cdc2, treated with TC PTP. The blots were probed with a phospho-tyrosine antibody. The results from dot-blot and Western blot analysis are consistent, with TC PTP exhibiting the highest phosphatase activity on the ligated phospho-Cdc2 substrate.

3.8.1. Substrate Preparation

3.8.1.1. SYNTHESIS OF PHOSPHORYLATED PEPTIDES

Phosphorylated peptides can be designed based on either a native protein sequence or consensus sequence for a dephosphorylation site. An extra cysteine residue or a linker (e.g., Cys-Gly) is included at the N-terminus of each peptide sequence (*see* **Note 1**). Nonphosphorylated peptides may be used as controls. Use of purified peptides increases ligation efficiency and sensitivity in immunoassays.

3.8.1.2. Ligation Reaction

Ligation is carried out at 4°C for 4–16 h in the presence of 0.5–1 mM peptide and 0.5 mg/mL of CP39 or CP27 in 100 mM Tris-HCl, pH 8.5 and 10 mM MESNA. Following is a typical ligation protocol for the generation of substrates for multiple reactions.

Carrier protein (1 mg/mL)	100 µL
Peptide solution (5 mM)	40 µL
Ligation buffer (10X)	20 µL
Water	40 µL
Total volume	200 µL

3.8.1.3. Assessment of Ligation Efficiency

Ligation efficiency can be evaluated by SDS-PAGE (*see* **Subheading 3.4.**).

3.8.2. Phosphatase Assay

Typically, a mixture of carrier protein–peptide ligation reaction (containing 1 mM peptide and 0.5 mg/mL CP39 or CP27) or unligated synthetic peptides (at 1 mM final concentration) are treated with phosphatase at 30°C for 1 h in 1X reaction buffer. A control assay is performed in the absence of phosphatase. The assays can be immediately subjected to protein blotting or stored at −20°C for later use. Commercially available enzymes are provided in a wide range of specific activity and defined by different substrates and methods. In an initial screening assay, it is convenient to use 1 µL of enzyme (in the range of 5000–400,000 U/mL) in a 20 µL reaction volume (*see* the next protocol). Enzyme titration can be subsequently carried out to compare the substrate preference and specificities of different enzymes.

Recommended assay protocol:

Substrate	8 µL
Reaction buffer (10X)	1 µL
Phosphatase	1 µL
Total volume	10 µL

3.8.3. Protein Blotting

Five microliters of each assay sample is diluted into 145 µL of PBS in a 96-well microtiter plate. Serial threefold dilutions are then made into PBS by mixing 50-µL of sample and 100 µL of PBS. The diluted sample (100 µL) in each well is applied to a nitrocellulose membrane using a dot-blot apparatus (Bio-Dot Microfiltration Apparatus). Each dilution series starts with a sample containing 2.66 nmol of peptide. The diluted samples are drawn through the membrane by applying a light vacuum and each well of the apparatus is rinsed three times with 100 µL of PBS. The entire apparatus is disassembled, and the membrane is rinsed with 40 mL of PBS.

3.8.4. Immunoassay

Immunoassay is carried out with a phospho-specific antibody (*see* **Subheading 3.5.**). The assay sample without phosphatase treatment is used as a positive control for antibody detection and as a standard for quantitation. Dephosphorylation of the substrates results in a decreased binding of phospho-specific antibody. The assay samples were also investigated by Western blot analysis and the data were consistent with the dot-blot analysis (*see* **Fig. 5**).

3.9. On-Membrane Phosphatase Assay

In the "on-membrane assay" approach, ligated carrier protein-peptide substrates are arrayed onto nitrocellulose for treatment with a kinase or a phosphatase. Ligated peptides are superior to free peptides when blotted onto a membrane, with significant improvement in the signal-to-noise ratio, as well as increased sensitivity that will ensure consistent data. Next is an example of using one set of phospho-tyrosine peptides (Y1–Y5, *see* **Table 2**) ligated to a carrier protein as substrates to test the specificity of T-Cell protein tyrosine phosphatase and LAR protein tyrosine phosphatase. Following is the experimental protocol:

1. The phospho-peptides were ligated to CP39 according to the method described earlier (*see* **Subheading 3.2.**).
2. The ligated products were serially diluted and arrayed onto nitrocellulose (*see* **Subheading 3.3.**). Each threefold dilution series starts with a sample containing 3.33 nmol of peptide.
3. The blots were rinsed with PBS and incubated with T-cell protein phosphatase (*see* **Fig. 6A**) or LAR protein tyrosine phosphatase (*see* **Fig. 6B**) for 1 h at room temperature. The blots (each 5×5 cm^2) were incubated in 5 mL 1X phosphatase reaction buffer and 100 U of phosphatase.
4. The reaction solution was discarded and blots were washed with TBSTT three times for 15 min each at room temperature.
5. The membranes were first blocked with 5% dry milk in TBSTT for 1 h at room temperature followed by three TBSTT washes of 15 min each.
6. The blots were incubated with a phospho-specific antibody following the protocol previously described (*see* **Subheading 3.5.**).
7. After washing the blots with 50 mL TBSTT three times for 15 min each, the blots were incubated with HRP-conjugated antibody in 2% dry milk in TBSTT for 1 h at room temperature.
8. The solution containing the secondary antibody was discarded and the blots washed with TBSTT three times for 15 min each.
9. The blots were developed with the LumiGLO solution and subsequently exposed to film.

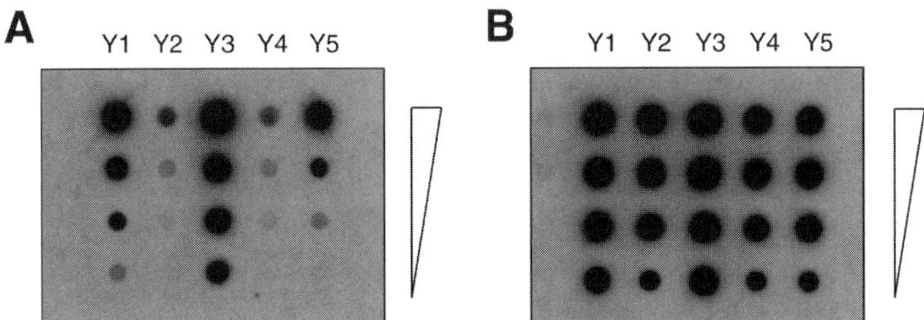

Fig. 6. On-membrane phosphatase assays. Five phospho-tyrosine containing peptides (Y1–Y5, **Table 2**) were ligated to a carrier protein, CP39. Threefold serial dilutions were prepared in PBS in 96-well microtiter plates before arraying onto duplicate nitrocellulose membranes. Each dilution series starts with a sample containing 3.33 nmol of peptide. The membranes were incubated with T-cell protein tyrosine phosphatase (**A**) or LAR protein tyrosine phosphatase (**B**) and then incubated with phospho-tyrosine monoclonal antibody.

The data showed that T-cell protein tyrosine phosphatase (TCPTP) phosphatase exhibited the most activity on ligated phospho-peptides Y2 and Y4. LAR phosphatase had little activity on these ligated phospho-peptide substrates, when an equal number of units of enzyme were used (*see* **Fig. 6**).

3.10. On-Membrane Kinase Assay

In this experiment, a group of peptides was designed for an on-membrane assay with Abl protein tyrosine kinase (**Table 3**). The ligated carrier protein–peptide substrates were arrayed onto a nitrocellulose membrane before subjecting the membrane to kinase treatment.

1. Peptides Abls, Ablsm, and Ablsp were ligated to CP39 (*see* **Subheading 3.2.**).
2. The ligated products were diluted in a 96-well microtiter plate and transferred onto a nitrocellulose membrane (4 × 4 cm²) (*see* **Subheading 3.3.**). Each threefold dilution series starts with a sample containing 3.33 nmol of peptide.
3. The blot was rinsed once with PBS and incubated with 200 U of Abl Protein Tyrosine Kinase for 1 h at room temperature in 5 mL of 1X Abl kinase reaction buffer and 200 μM ATP.
4. The reaction solution was discarded and the blot was washed with TBSTT three times for 15 min each.
5. The blot was incubated in 5% milk in TBSTT for 1 h at room temperature followed by three TBSTT washes of 15 min each at room temperature.
6. The membrane was incubated with a phospho-tyrosine antibody (P-Tyr-100) in TBSTT containing 2% nonfat dry milk for 1 h at room temperature.

Table 3
Peptide Substrates for Abl Kinase Assay

Peptide	Sequence
Abls	CGSNEAI YAAPFAKKK
Ablsm	CGSNEAI AAAPFAKKK
Ablsp	CGSNEAIpYAAPFAKKK

Abls contains a putative phosphorylation site, a tyrosine residue for Abl kinase as in **ref. 3**; Ablsm is the Abls peptide with the tyrosine residue substituted by alanine; Ablp possesses a phosphorylated tyrosine residue, thereby serving as a positive control for antibody detection.

7. After washing in TBSTT three times of 15 min each, the blots were incubated with HRP conjugated antimouse antibody in 2% dry milk in TBSTT for 1 h at room temperature.
8. The solution containing secondary antibody was discarded and the blots were washed three times for 15 min each.
9. The blots were developed with the LumiGLO solution and subsequently exposed to film.

A strong signal was detected from ligated CP39-Abls after incubation with Abl protein tyrosine kinase, indicating that the ligated Abls peptide was phosphorylated by Abl kinase, whereas, the unligated Abls peptide generated no signal (**Fig. 7**). The data indicated that use of the ligated peptide substrates had a significant advantage over the unligated peptides in membrane-based kinase assays, leading to improved sensitivity. In control samples, Ablsm, a peptide with the tyrosine residue substituted by alanine, showed no signal, whereas a phospho-peptide Ablsp, the positive control, had the greatest signal.

3.11. Kinase Assay and Its Detection on the Membrane

In the following protocol, phospho-peptides are ligated to a carrier protein, treated with kinase(s) and arrayed onto a nitrocellulose membrane before subjecting to detection with a phospho-specific antibody.

1. The peptide substrates are ligated to a carrier protein according to the protocol previously described.
2. The ligated samples are dialyzed against 5 mM Tris-HCl, pH 7.5, 50 mM NaCl to remove the unligated peptide using a 0.025-μm filter disc.
3. The kinase assays are carried out at 30°C for 60 min in a 10 μL reaction with each sample containing the following: 10 μM of the carrier protein, 1–10 U of kinase, 1X kinase buffer and 100 μM ATP.
4. The reaction products are arrayed on a 0.45-μm nitrocellulose membranes along with unligated synthetic peptides (*see* **Subheading 3.3.**).

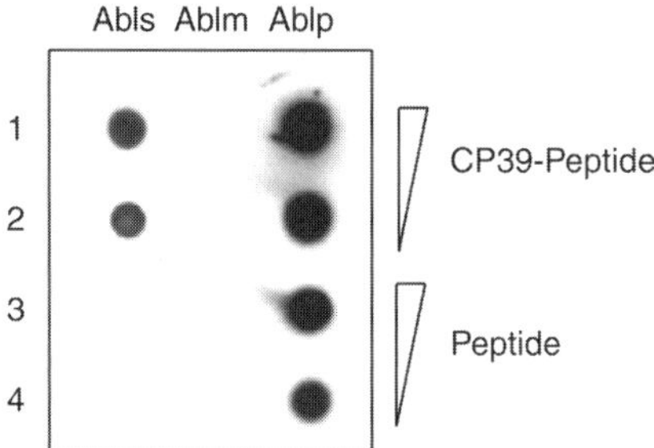

Fig. 7. On-membrane kinase assay. Abl kinase peptide substrate, Abls and its variants Ablsm and Ablsp (**Table 2**), were ligated to CP39. The ligated samples (rows 1 and 2) and unligated peptides (rows 3 and 4) were prepared by threefold dilution in PBS in 96-well microtiter plates before arraying onto the nitrocellulose membrane. Each series starts with a sample containing 3.33 nmol of peptide. The membrane was incubated with Abl protein tyrosine kinase followed by analysis with a monoclonal antibody against phospho-tyrosine.

5. The blots are blocked in 5% milk in TBSTT for 1 h at room temperature followed by three TBSTT washes for 15 min each at room temperature.
6. The membranes are incubated with a phospho-specific antibody in 2% nonfat dry milk TBSTT solution.
7. After three washes in TBSTT for 15 min each, blots are incubated with HRP-conjugated antibody in TBSTT with 2% nonfat dry milk for 1 h at room temperature.
8. Discard the solution containing the secondary antibody and wash the blots with TBSTT three times for 15 min each.
9. The blots are developed with LumiGLO solution and subsequently exposed to film.

4. Notes

1. A cysteine residue must be present or added to the N-terminus of a target peptide sequence. Inclusion of a short linker (e.g., CG or CAGAG) may increase the flexibility of a target peptide upon its attachment to a carrier protein.
2. If the molecular mass of the peptide is 2862 Da then the calculation for dissolving the peptide in water is as follows:

$$1\ M = M_r\ \text{gm/L}$$
$$1\ M = 2862\ \text{gm/L}$$
$$1\ \text{m}M = 2.862\ \text{mg/mL}$$

If a peptide is not soluble, you may first dissolve the peptide in 20 µL DMSO and then add water to make a 2- to 10-mM stock solution. The peptide solution should be stored at –80°C.

3. CP39 is a 39-kDa recombinant protein from *Haemophilus haemolyticus*, while CP27 is a 27-kDa recombinant protein from *Dirofilaria immitis*. Both are expressed

with a reactive C-terminal thioester and have low cross reactivity with mouse and rabbit sera *(13)*. Various dilutions of antibodies should be tested. We have tested crude mouse and crude rabbit sera and observed a single distinct band of the ligated product. In addition, the use of a ligated peptide (CP39-peptide or CP27-peptide) significantly increased the sensitivity of certain peptide substrates in ELISA's.

4. Ligation of CP27 to a peptide (e.g., small peptides of 10–15 amino acids) can be readily detected by a 12% Coomassie blue stained SDS-PAGE. The ligated CP39-peptide product can be detected by Western blot using an antipeptide antibody; however, its detection by Coomassie blue staining of a SDS-PAGE is difficult when the size of the peptide is smaller than 1.5 kDa. To achieve high ligation efficiency we recommend that ligation to CP27 or CP39 be conducted for at least 4 h and the peptide concentration is 1–2 mM. However, for production of positive controls for Western blot analysis this adjustment is not necessary.

5. Use of CP27 allows for the generation of an additional carrier protein–peptide fusion with a different size than the CP39-peptide fusion product. CP27 is useful when it is needed to determine the ligation efficiency for a small peptide (10–15 amino acids). The mobility shift of CP27 after ligation is more easily detected on a 12% Coomassie blue-stained SDS-PAGE, as compared with the mobility shift with CP39. It has been observed that in some cases, CP27 is a preferred carrier protein for generating an optimal peptide substrate for kinase assay, arrays, and so on *(14)*. The optimal carrier protein for a particular assay may be determined by testing the different carrier protein–peptide substrates.

6. The shelf life of the thioester tagged carrier protein is at least nine mo. Greater than 50% ligation was observed when the carrier protein was stored for 12 mo at −80°C.

7. An internal or a C-terminal cysteine in a peptide could affect the ligation efficiency of the peptide to a carrier protein. The inclusion of 10 mM MESNA in the ligation reaction can reduce the effect of disulfide bond formation of cysteine-containing peptides. In addition, an internal or C-terminal cysteine may undergo transthioesterification with the carrier protein, but this side reaction will readily reverse as a result of the presence of MESNA.

8. In addition to ligation with an intein-generated carrier for array and Western blot analysis, synthetic peptides containing an N-terminal cysteine can also be used as antigens, for conjugation with keyhole limpet hemacyanin, bovine serum albumin, or ovalbumin *(13)*.

9. Ligation was found to work with both crude and purified peptides was found. Ligation with crude peptides at 4°C overnight can result in a wide range of ligation efficiencies, usually in the range of 10 to 50%, whereas, ligation with purified peptides usually yields 75–90% ligation. Ligation of 10% efficiency can still yield a strong signal, though consistent results are obtained when peptide purity is more than 30% *(12)*. To ensure the greatest amount of ligated product we recommend using a purified peptide. We purify the peptide on a Vydac semipreparative C18 column for reverse-phase purification using a 180-min linear gradient, 10–100% B with a flow rate of 2 mL per min. buffer A is 0.1% trifluoroacetic acid (TFA) H_2O (V/V), and buffer B is 0.1% TFA/60% CH_3CN/40% H_2O (V/V/V).

10. Even if the peptide is not completely soluble, the amount of ligation product may be sufficient for detection by array or Western blot analysis. It was found that ligation of a carrier protein with a partially soluble peptide overnight at 4°C can result in 10–50% ligation efficiency and generate positive signals in immunoassays. If necessary, dissolve the peptide first in DMSO, add water, and use in ligation. Usually 15–25 µL of DMSO is added for up to 2 mg of peptide, and then add water to make a 2- to 10-mM stock solution. Also, add the solid peptide to the ligation reaction.

11. CP39 (Array, NEB, cat. no. E6603S) and CP27 (Array, NEB, cat. no. E6607S) contain a higher concentration of protein (1 mg/mL), which is necessary for increased sensitivity in array analysis (*12*). The use of a lower concentration of carrier protein in array analysis resulted in decreased signal. CP39 (Western, NEB, cat. no. E6602S) and CP27 (Western, NEB, cat. no. E6606S) contain protein at a lower concentration (0.2 mg/mL); but after ligation to a peptide, they routinely give a strong, sharp band on Western blot analysis (*13*).

12. The dot-blot apparatus used in this article limits the number of array features to 96. With this dot-blot apparatus, it is recommended to array at least 5 pmol ligation product (123 pmol of peptide). Under these conditions, 1 mg of peptide is ligated to a carrier protein in a volume of 1 mL, which after dilution for arraying produces 2700 array features, each 2 mm in diameter. For highly sensitive assays, arraying 2–4 nmols of peptide and a threefold serial dilution is recommended. However, ligation products can be arrayed with higher density devices, including microarray printers, onto nitrocellulose-coated glass slides. In practice, the number of features is·limited by the number of peptides available for arraying. SPOT membranes in the 384-well microtiter plate format typically have 20 nmol peptide in each feature; about 160-fold more than the amount was recommend for this new IPA method.

13. The possibilities for ligation problems are:
    a. Peptide does not possess an N-terminal cysteine or the sulfhydryl group is oxidized.
    b. The peptide solution may be very acidic and cause the pH of the reaction to drop significantly. Check the pH of your peptide solution. If the pH is <6.0 dissolve the peptide in 1 M Tris-HCl, pH 9.0.
    c. Peptide preparation contains impurities. Purify the peptide, using a Vydac semipreparative C18 column (*see* **Note 9**).
    d. Concentration of peptide is incorrect.
    e. Peptide or ligation product is insoluble (*see* **Notes 2** and **10**).
    f. The carrier protein has lost its ligation capabilities owing to repeated freeze–thaw cycles or long-term storage at −20°C. Use the control peptide, PB1 (NEB, cat. no. E6608S), to test ligation.

14. In Western blot analysis, even a small amount of ligated product (27 ng of CP39 or CP27 per well) yielded a positive signal (*13,14*).

15. Because peptides are ligated to carrier proteins through covalent bonds, the ligation products are stable and can be stored at −20°C for up to 1 mo or −80°C for assays at a later date.

16. Ligation efficiency can be assessed using SDS-PAGE gel stained with Coomassie blue *(13–16)*. To check the ligation run 0.2–1 µg of carrier protein and include a control of carrier protein (without peptide). A shift of the ligation product might not be detected on SDS-PAGE if a large carrier (e.g., CP39, 39 kDa) is ligated to a short peptide. Use a 20- to 30-kDa carrier, like CP27, for easy detection of ligation by 12% SDS-PAGE (*see* **Note 4**).

References

1. Houseman, B. T., Huh, J. H., Kron, S. J., and Mrksich, M. (2002) Peptide chips for the quantitative evaluation of protein kinase activity. *Nat. Biotechnol.* **20,** 270–274.
2. Lesaicherre, M. L., Uttamchandani, M., Chen, G. Y., and Yao, S. Q. (2002) Antibody-based fluorescence detection of kinase activity on a peptide array. *Bioorg. Med. Chem. Lett.* **12,** 2085–2088.
3. Songyang, Z., Carraway, K. L., Eck, M. J., et al. (1995) Catalytic specificity of protein-tyrosine kinases is critical for selective signaling. *Nature* **373,** 536–539.
4. Falsey, J. R., Renil, M., Park, S., Li, S., and Lam, K. S. (2001) Peptide and small molecule microarray for high throuput cell adhesion and functional assays. *Bioconjug. Chem.* **12,** 346–353.
5. Harlow, E. and Lane, D. (1988) Antibodies, in *Immunoassays*, (Harlow, E. and Lane, D., eds.). Cold Spring Harbor Laboratory, New York. pp. 553–612.
6. Martens, W., Greiser-Wilke, I., Harder, T. C., et al. (1995) Spot synthesis of overlapping peptides on paper membrane supports enables the identification of linear monoclonal antibody binding determinants on morbillivirus phosphoproteins. *Vet. Microbiol.* **44,** 289–298.
7. Reineke, U., Ivascu, C., Schlief, M., et al. (2002) Identification of distinct antibody epitopes and mimotopes from a peptide array of 5520 randomly generated sequences. *J. Immunol. Methods* **267,** 37–51.
8. Frank, R. and Overwin, H. (1996) SPOT synthesis. Epitope analysis with arrays of synthetic peptides prepared on cellulose membranes. *Methods Mol. Biol.* **66,** 149–169.
9. Reimer, U., Reineke, U., and Schneider-Mergener, J. (2002) Peptide arrays: from macro to micro. *Curr. Opin. Biotechnol.* **13,** 315–320.
10. Frank, R. (2002) The SPOT-synthesis technique. Synthetic peptide arrays on membrane supports—principles and applications. *J. Immunol. Methods* **267,** 13–26.
11. Reineke, U., Volkmer-Engert, R., and Schneider-Mergener, J. (2001) Applications of peptide arrays prepared by the SPOT-technology. *Curr. Opin. Biotechnol.* **12,** 59–64.
12. Sun, L., Rush, J., Ghosh, I., Maunus, J. R., and Xu, M. -Q. (2004) Producing peptide arrays for epitope mapping by intein-mediated protein ligation. *Biotechniques* **37,** 430–443.
13. Ghosh, I., Sun, L., Evans, T. C., Jr., and Xu, M. -Q. (2004) An improved method for utilization of peptide substrates for antibody characterization and enzymatic assays. *J. Immunol. Methods* **293,** 85–95.

14. Xu, J., Sun, L., Ghosh, I., and Xu, M. Q. (2004) Western blot analysis of Src kinase assays using peptide substrates ligated to a carrier protein. *Biotechniques* **36,** 976–981.
15. Evans, T. C., Jr., Benner, J., and Xu, M. Q. (1998) Semisynthesis of cytotoxic proteins using a modified protein splicing element. *Protein Sci.* **7,** 2256–2264.
16. Evans, T. C., Jr. and Xu, M. Q. (1999) Intein-mediated protein ligation: harnessing nature's escape artists. *Biopolymers* **51,** 333–342.
17. Muir, T. W., Sondhi, D., and Cole, P. A (1998) Expressed protein ligation: a general method for protein engineering. *Proc. Natl. Acad. Sci. USA* **95,** 6705–6710.
18. Chong, S., Mersha, F. B., Comb, D. G., et al. (1997) Single-column purification of free recombinant proteins using a self-cleavable affinity tag derived from a protein splicing element. *Gene* **192,** 271–281.
19. Lesaicherre, M. L., Lue, R. Y., Chen, G. Y., Zhu, Q., and Yao, S. Q. (2002) Intein-mediated biotinylation of proteins and its application in a protein microarray. *J. Am. Chem. Soc.* **124,** 8768–8769.
20. Tan, L. P., Lue, R. Y., Chen, G. Y., and Yao, S. Q. (2004) Improving the intein-mediated, site-specific protein biotinylation strategies both in vitro and in vivo. *Bioorg. Med. Chem. Lett.* **14,** 6067–6070.
21. Sun, L., Ghosh, I., and Xu, M. Q. (2003) Generation of an affinity column for antibody purification by intein-mediated protein ligation. *J. Immunol. Methods* **282,** 45–52.

17

Printing Low Density Protein Arrays in Microplates

Robert S. Matson, Raymond C. Milton, Michael C. Cress, Tom S. Chan, and Jang B. Rampal

Summary

Here, we provide methods for the creation of protein microarrays in microplates. The microplate consists of 96 wells with each well capable of holding a protein microarray at a spot density of up to 400 (20×20) individual elements. Arrays of capture monoclonal antibodies, corresponding to specific interleukins, were printed onto the bottom of the wells which had been surface activated for covalent attachment. A Biomek® 2000 laboratory automation workstation (Beckman Coulter, Inc., Fullerton, CA) equipped with a high-density replicating tool was used for printing low density 3×3 to 5×5 arrays. For higher density arrays, a microarrayer system (Cartesian PS7200, Genomic Solutions, Inc., Ann Arbor, MI) was employed. Multiple antigens were simultaneously analyzed without detectable cross-reactivity associated with capture antibody or secondary antibody interactions. Detection of interleukin antigens, spiked into cell culture media containing 10% fetal calf serum, was specific and sensitive.

Key Words: Antibodies; Biomek; cytokines; immunoassay; microarray; microplate.

1. Introduction

The microarray was developed originally as a hybridization-based platform tool for genomic analysis *(1)*. The successful use of the gene expression microarray has led others to adopt the format for proteomic research *(2)*. Printed slide microarrays have demonstrated its use as an antibody or antigen microarray *(3)*. Likewise, protein microarray slide experiments that attest to the resolving power of this approach in detecting a single protein is described in **ref. *4***. Therefore, over the past several years it has been a relatively straightforward adaptation of those processes and tools used for construction and analysis of the gene expression slide microarray for work with proteins. In particular, the antibody-based enzyme-linked immunosorbent assay (ELISA) has been rather easily miniaturized in the form of the antibody microarray. This is not

From: *Methods in Molecular Biology, vol. 381: Microarrays: Second Edition: Volume 1: Synthesis Methods*
Edited by: J. B. Rampal © Humana Press Inc., Totowa, NJ

very surprising. Indeed, is this not the rebirth of Ekin's "microspot" technology of the late 1980s *(5)*. Albeit, global assessment of the proteome by microarray analysis is only remotely related than the use of the ultrasensitive microspot immunoassay for diagnostics. Nevertheless, the fundamental processes of spotting and detecting an array of proteins on a substrate remain the same.

The glass slide is the most popular substrate for spotting down nucleic acids microarrays. As we have previously discussed the early protein microarray work was also based on the glass slide microarray. For the assessment of large numbers of different proteins from a single sample, the slide microarray is a convenient format. However, in the case of the micro-ELISAs in which only a few related proteins are to be analyzed from many samples the microplate format is better suited, especially for automated assays. Thus, there has been the emergence of the familiar 96-well microtiter plate footprint for these kinds of applications. This microarray design is sometime referred to as the "array of arrays" format. For example, protein microarrays created by printing antibodies onto a glass plate that had been subdivided into 96 ovals formed from a mask and coated with Teflon *(6)*. Thus, a hydrophobic material surrounded each of the arrays allowing the sample to be applied in each oval well while serving as a physical barrier to restrict liquid flow in order to prevent cross-talk between adjacent wells. However, this approach only permits the application of small volumes of reagents (e.g., 25–50 µL/well) that have to be carefully dispensed and subsequently aspirated from the plate surface. Without sophisticated automation for such precise liquid handling tasks this can be a tedious and time-consuming process.

For these reasons our laboratory first sought to adapt the conventional microplate for use with protein arrays *(7)*. The microplate is generally manufactured to be compliant for automated liquid handling workstations and thereby represents an integral tool for lab automation. It has been widely adopted for high throughput assays. However, in the mid-1990s microplates were not designed for use with microarrays. Very few flat-bottom plates having activated surfaces for immobilizing proteins were available. Printing into the bottom of plates using conventional microarrayers had also not been described. And, no plate reader was commercially available that would read microplate arrays. It was therefore necessary for our laboratory to develop our own plates, surface chemistries, printing processes, and reader *(7)*.

We found the Biomek high-density replicating tool (HDRT) to be a particularly useful tool for printing small arrays in the bottom of microplates. The HDRT is essentially a plate of nails. It was originally designed for replicate blotting of yeast or YAC colonies onto filter paper. We adapted it for use in making from 3 × 3 to 5 × 5 arrays in the bottom of microplates using the standard offsets provided in the Bioworks software for replica gridding. It was possible to grid 1–10 microplates (96 wells) with microarrays on the Biomek deck. The Biomek HDRT system was used in our laboratory in the initial stages during the development of an assay and will be described for that purpose. Others *(8)* have reported on the use of the Biomek 2000 workstation equipped with a custom print head that could be used for printing slide microarrays. The Biomek FX workstation equipped with a commercial fixture for slot pins (V&P Scientific, San Diego, CA) has also been used to prepare microarrays in microplates *(9)*.

For more precise spotting of proteins of larger arrays and smaller spot diameters we have utilized conventional microarrayers whose decks and software programs were modified specifically for use in printing microplate arrays. A Cartesian PS7200 microarrayer system with a customized deck from Genomic Solutions, Inc., (Ann Arbor, MI) has been routinely used here. The PS7200 model enabled us to print by either noncontact "dispense" or by replacing the dispense head with conventional print heads (TeleChem International Sunnyvale, CA or Majer Precision, Tempe, AZ) contact printing using quill pins is achieved.

Using either the Biomek workstation or the Cartesian PS7200 we are able to print reagents into commercial standard microtiter plates (flat-bottom), and also microplates of our own design. For example, a flexible microplate is described that is vacuum formed from plastic sheets. In most instances when arraying new proteins, preliminary printing experiments and micro-ELISA analysis using the Biomek workstation are first performed. Once printing conditions are optimized and assays developed, we can then transfer the process to a microarrayer system. However, this is not always necessary because with careful pin matching the Biomek workstation can achieve similar levels of performance for low density arraying of proteins.

2. Materials

1. Acyl fluoride activated microplates of a flexible polypropylene film (microarray substrates) were prepared from the reaction of (diethylamino)sulfur trifluoride (DAST) *(10)* with succinylated polypropylene *(11)* (*see* **Note 1**).
2. Microplate microarrays were created by contact printing using the Biomek 2000 workstation equipped with a 384-pin HDRT (high-density replicating tool) with 96 pins spaced to fit into a conventional microplate well (Beckman Coulter, Inc., Fullerton, CA).
3. Enzyme-labeled fluorescence (ELF) reagent, a fluorescent precipitating substrate for alkaline phosphatase was used for signal development (ELF-97 Endogenous Phosphatase Detection Kit; Molecular Probes, Inc.).
4. CCD camera system for digital images (Teleris 2, SpectraSource, Inc., Westlake Village, CA) (*see* **Note 2**).
5. Excitation light at 350 nm was generated using a UV mineral light with signal emission collected at 520 nm using a 10-nm band-pass-lens filter.
6. 16-Bit images were analyzed using ImaGene software (BioDiscovery, Inc., EL Sequndo, CA) then exported as eight-bit values into an Excel spreadsheet (Microsoft Redmond, WA) for calculation and graphic display.

7. PS7200 model for printing arrays (Genomic Solutions).
8. Microtiter plates (*see* **Subheading 3.1.**).
   a. Ten activated MTPs, for example, DAST-surface activated, flat-bottom, polypropylene 96-well plates.
   b. Source plates (384 well) for each anti-interleukin (IL) mouse monoclonal antibody (Ms MAb).
   c. Equipment.
      i. Biomek workstation programmed with method "HDRT 10 Multiplate."
      ii. Method "HDRT 10 Multiplate" has the HDRT at A1 and the "Thermal exchanger/probe plate" at B6. The method prints to "HDR Multiplates" at the other 10 positions.
      iii. Ten printing stages (high density replicator [HDR] Multiplates): Omnitray, single microtiter plates with lids (Nalge Nunc, cat. no. 242811) preregistered on the Biomek deck created by marking from a previous dummy printing run for positioning the wells.
      iv. The MTPs are affixed to the printing stages with either scotch removable double coated tape (3M, St. Paul, MN; no. 667) or scotch poster tape (3M, no. 109).
9. Packaging of printed and quenched MTP plates.
   a. Kapak® 1 pint pouches (Ampack Packaging, LLC, Cincinnati, OH; stock no. 402) (other sizes also available).
   b. Argon compressed gas (Air Liquide, Houston, TX; cat. no. UN 1006).
   c. Kapak Scotchpak® pouch sealer.

2.1. Preparation of Reagents

2.1.1. Concentrations of Proteins Used for Preparing Antibody Arrays

Protein	Molecular wt. (Daltons)	Stock (mg/mL)	Print conc. (mg/mL [M])	Moles/nL (moles/spot)$_{max}$
MAb (mouse) capture anti-IL	150,000	1	1 [6.7E-06]	6.67E-15 (67 fmoles)
Human immunoglobulin (Ig)G (polyclonal) Landing lights	150,000	28	1[6.7E-06]	6.67E-15 (67 fmoles)

2.2. Reagents Used at Each Stage of the HDRT Printing Process

2.2.1. Printing Buffer Preparation

2.2.1.1. Preparation of 1 M Carbonate Buffer, pH 9.0

1. Chemicals.
   a. Sodium hydrogen carbonate ($NaHCO_3$) (Aldrich, cat. no. 23,652-7).
   b. 1 N NaOH.
2. Equipment.
   a. 2-L Beaker with mark calibrated for 1 L.
   b. Stirrer and stir bar.
   c. pH meter (Beckman Coulter, Inc.).

3. Procedure.
   a. Weigh out 84.01 g sodium hydrogen carbonate and place it in the beaker with the stir bar.
   b. Add about 950 mL deionized (18 MΩ) water and stir to dissolve (*see* **Note 3**).
   c. Adjust the pH (from about 8.0) to 9.0 with 1 *N* NaOH solution.
   d. Make up the volume to 1 L.

2.2.1.2. PREPARATION OF 16% SODIUM SULFATE

1. Chemical.
   a. Sodium sulfate (Na_2SO_4) (Sigma, cat. no. S-6264).
2. Procedure.
   a. Weigh out 16 g sodium sulfate and place it in the bottle with the stir bar.
   b. Dissolve the sodium sulfate in deionized water and make up to 100 mL.

2.2.2. Preparation of the Printing Ink

1. Equipment.
   a. 15-mL Conical tubes.
   b. 0.2-μ*M* Syringe filter.
2. Procedure.
   a. Add 500 μL 1 *M* carbonate buffer, pH 9.0 to 2.5 mL 16% sodium sulfate and 7 mL of deionized water together in a 15-mL tube.
   b. Mix well and filter through 0.2-μm syringe filter.

2.2.3. Preparation of Antibodies in Printing Ink Buffer

1. Ms anti-Hu IL-4 MAb (Sigma, cat. no. I-7651 or R&D Systems, cat. MAB604).
2. Ms anti-Hu IL-8 MAb (Sigma, cat. no. I-1645 or R&D Systems, cat. no. MAB208).
3. Ms anti-Hu IL-10 Mab (Sigma, cat. no. I-9276 or R&D Systems, cat. no. MAB217) (*see* **Note 4**).

Add 500 μL 50 m*M* carbonate, 4% sodium sulfate, pH 9.0 (printing ink buffer) to one 0.5 mg vial of Ms anti-Hu IL MAb and mix well. Allow the reconstituted antibody to rehydrate for at least 12 h. This will be the 1 mg/mL antibody stock solution. Store at 4°C.

2.2.4. Preparation of Registration Markers (Landing Lights)

Ink component	Volume (μL)
Human IgG, 28 mg/mL (Pierce, cat. no. 31877)	36
1 *M* Carbonate buffer, pH 9.0	50
16% Sodium sulfate	250
Food dye (Schilling food colors, McCormick & Co.)[a]	250
Deionized water	414
Total volume	1000

[a]Prepare a landing light ink in each color: blue, red, and green.

2.2.5. Preparation of Source Plates

Prepare three printing (source) plates for printing with the Biomek workstation:

1. Mouse anti-Hu IL-4 MAb (blue plate):
   a. Combine two vials of reconstituted mouse anti-Hu IL-4 MAb and mix well. This is the anti-IL-4 printing ink. The volume is 1 mL and the concentration is 1 mg/mL.
   b. Apply 20-µL aliquots of the anti-IL-4 printing ink to the wells of a 384-well microtiter plate according to the printing plate map (*see* **Note 5**).
   c. Apply 20-µL aliquots blue landing light ink to the four corner wells (A1, A23, O1, and O23) as indicated in the plate map (*see* **Note 5**).
   d. Force inks to the bottoms of the well by centrifugation: place plate with counterweight plate in centrifuge (e.g., Hermle model Z320). Turn on centrifuge according to its procedure. When centrifuge speed reaches 2000 rpm (3000*g*), turn off centrifuge.
2. Mouse anti-Hu IL-8 MAb (red plate):
   a. Add 500 µL printing ink buffer to a vial of reconstituted mouse anti-Hu IL-8 MAb and mix well. This is the anti-IL-8 printing ink. The volume is 1 mL and the concentration is 0.5 mg/mL.
   b. Apply 20-µL aliquots of the anti-IL-8 printing ink to the wells of a 384-well microtiter plate according to the printing plate map (*see* **Note 5**).
   c. Apply 20-µL aliquots red landing light ink to the four corner wells (A1, A23, O1, and O23) as indicated in the plate map above.
   d. Force inks to the bottoms of the well by centrifugation: place plate with counterweight plate in centrifuge. Turn on centrifuge according to its procedure. When centrifuge speed reaches 2000 rpm (3000*g*), turn off centrifuge.
3. Mouse anti-Hu IL-10 MAb (green plate).
   a. Combine two vials of reconstituted mouse anti-Hu IL-10 MAb and mix well. This is the anti-IL-10 printing ink. The volume is 1 mL and the concentration is 1 mg/mL.
   b. Apply 20-µL aliquots of the anti-IL-10 printing ink to the wells of a 384-well microtiter plate according to the printing plate map (*see* **Note 5**).
   c. Apply 20-µL aliquots blue landing light ink to the four corner wells (A1, A23, O1, and O23) as indicated in the plate map.
   d. Force inks to the bottoms of the well by centrifugation: place plate with counterweight plate in centrifuge. Turn on centrifuge according to its procedure. When centrifuge speed reaches 2000 rpm (3000*g*), turn off centrifuge.

2.2.6. HDRT Pre/Postcleaning

1. Reagents.
   a. TBS with 0.05% SDS (Soapy Solution).
   b. Deionized water.

    c. Methanol (anhydrous) (VWR, cat. no. VW4300-3).
    d. Chromic acid.
2. Equipment.
    a. Rotary shaker.
    b. Containers sized to just fit HDRT.
    c. Latex gloves.
3. Procedure.
    a. Place about 100 mL of the soapy solution in a container. The container should be chosen such that the container length is about the same of the HDRT and slightly wider than the HDRT. When the HDRT is place in the container, the solution should cover the pins up to the pin plate. The liquid in this and subsequent steps should not come higher than the plate holding the pins.
    b. Put the container and HDRT on the shaker for 10 min. Set the shaker for gentle motion; the liquid should not be allowed to slosh over the black plate of the HDRT.
    c. Rinse the HDRT four times in deionized water: add about 100 mL deionized water to a container, place the HDRT in the container and shake on the rotary shaker for 5–10 min each time.
    d. Rinse the HDRT two times in methanol: add about 100 mL methanol to a container, place the HDRT in the container and shake on the rotary shaker for 5–10 min each time.
    e. Allow the HDRT to dry on its stand (*see* **Notes 6** and **7**).

2.2.7. Postprint Processing of MTP Plates

2.2.7.1. PREPARATION OF QUENCHING REAGENT

1 mg/mL casein in 50 mM carbonate, 0.15 M sodium chloride, pH 9.0.

1. Chemicals.
    a. Casein, Hammersten (Amersham Life Sciences [USB], GE Healthcare Piscataway, NJ; cat. no. US 12840).
    b. Sodium chloride (J. T. Baker, Mallinckrodt Baker, Inc., Phillipsburg, NJ; cat. no. 4058-1).
    c. 1 M Carbonate buffer, pH 9.0.
    d. Deionized water.
2. Equipment.
    a. 1-L Bottle.
    b. Stirrer and stir bar.
3. Procedure.
    a. Weigh 8.8 g sodium chloride and add to the bottle with the stir bar.
    b. Add 950 mL deionized water and 50 mL 1 M carbonate buffer, pH 9.0.
    c. Weigh 1 g casein and slowly add while stirring to the buffer solution in the bottle.
    d. Continue stirring until the casein has gone into solution. This will take several hours and the final solution will still be turbid. Casein solutions must be stored at 4–8°C (*see* **Note 8**).

2.2.8. Other Materials Required

1. Biomek 2000 Laboratory Automation Workstation, equipped with HDRT 384 with 96 pins (Beckman Coulter, Inc., Fullerton, CA).
2. Biomek Thermal Exchange Plate Unit (P/N no. 148276) equipped with a Cooling System (VWR Heated/Refrigerated Circulator, MDL no. 1160S).
3. Source plates (384 well, polypropylene, Greiner Bio-One no. 781280).
4. Convection oven or incubator.
5. Humidity chamber.

3. Methods

3.1. Setting Up the Biomek HDRT for Printing

1. Procedure.
   a. Remove the bags of DAST-activated MTPs from the freezer and allow them to come to room temperature before opening bags (*see* **Subheading 2., steps 8 and 9**).
   b. Affix the plates to the printing stages. Align the wells so that the print areas marked on the stage lid fall within the bottom of the wells. Wear gloves during this procedure so as not to leave marks on the plates. Make sure that plates are firmly fixed to the printing stages and that they lie flat on the surface. (It might be necessary to tape down edges of the plates if they are warped and do not lie flat.)
   c. Place the printing stages with plates at positions A2-A6 and B1-B5 on the Biomek worksurface. Make sure that the stages are properly placed and that the lids are pushed fully to the back and to the right.
   d. Turn on the cooler for the thermal exchanger.

3.1.1. Biomek HDRT Printing

3.1.1.1. Mouse Anti-Hu IL-10 MAb (Row C)

1. Place the Mouse anti-Hu IL-10 MAb source plate (green plate) on the thermal exchanger.
2. Edit the method "HDRT 10 Multiplate" for printing the third row of the 3×3 array:
   a. Edit the line begin loop: count = 45.
   b. Edit labware "HDR Multiplate": matrix rows = 3.
3. Begin the method "HDRT 10 Multiplate" without the HDRT in place. Monitor movement of the Biomek arm.
4. After printing Row B Column 3 (count = 6), stop the Biomek run program. Raise the head and manually place HDRT on it. Continue printing with HDRT in place for the three columns of Row C. Be very careful the first time the HDRT comes to the printing plate and make sure that the printing plate is properly aligned.
5. After printing Row C Column 3 (count = 9), stop the Biomek run program. Raise the head and manually remove the HDRT. Continue printing without the HDRT for Rows A and B.

6. Repeat **steps 4** and **5** (**step 4** at count = 15, 24, 33, and 42 and **step 5** at count = 18, 27, and 36) until the end of the run.
7. Clean and dry the HDRT according to the procedure (*see* **Subheading 2.2.6.**).

3.1.1.2. MOUSE ANTI-HU IL-8 MAB (ROW B)

1. Place the mouse anti-Hu IL-8 MAb source plate (red plate) on the thermal exchanger.
2. Edit the method "HDRT 10 Multiplate" for printing the second row of the 3×3 array:
   a. Edit the line begin loop: count = 30.
   b. Edit labware "HDR Multiplate": matrix rows = 2.
3. Begin the method "HDRT 10 Multiplate" without the HDRT in place. Monitor movement of the Biomek arm.
4. After printing Row A Column 3 (count = 3), stop the Biomek run program. Raise the head and manually place HDRT on it. Continue printing with HDRT in place for the three columns of Row B. Be very careful the first time the HDRT comes to the printing plate and make sure that the printing plate is properly aligned.
5. After printing Row B Column 3 (count = 6), stop the Biomek run program. Raise the head and manually remove the HDRT. Continue printing without the HDRT for Row A.
6. Repeat **steps 4** and **5** (**step 4** at count = 9, 15, 21, and 27 and **step 5** at count = 12, 18, and 24) until the end of the run.
7. Clean and dry the HDRT according to the procedure (*see* **Subheading 2.2.6.**).

3.1.1.3. MOUSE ANTI-HU IL-4 MAB (ROW A)

1. Place the mouse anti-Hu IL-4 MAb source plate (blue plate) on the thermal exchanger.
2. Edit the method "HDRT 10 Multiplate" for printing the first row of the 3×3 array:
   a. Edit the line begin loop: count = 15.
   b. Edit labware "HDR Multiplate": matrix rows = 1.
3. Begin the method "HDRT 10 Multiplate" with the HDRT in place. Be very careful the first time the HDRT comes to the printing plate and make sure that the printing plate is properly aligned. This run can proceed as an ordinary Biomek workstation run.
4. Clean and dry the HDRT according to the procedure (*see* **Subheading 2.2.6.**).

3.1.2. Postprint Processing of MTP Plates

1. Remove plates from the printing stages and examine them under the magnifier light.
2. Record any miss-registrations or missing spots observed.
3. Allow the plates to dry for 1 h after the end of printing.

3.1.3. Quenching of the MTP Plates

1. Procedure.
   a. Apply 80-μL aliquots of the quenching reagent to the wells of the printed MTPs.

b. Stack the plates in a box or a container with lid. Incubate the plates at room temperature for at least 1 h.
c. Shake out the quenching reagent and blot the plate on a Kimwipe.
d. Rinse the plates twice: dip the plate briefly into deionized water and shake out the water each time. Blot the plate on a Kimwipe each time.
e. Allow the plates to dry on Kimwipes (*see* **Note 9**).

3.1.4. Packaging of the Printed MTP Plates

1. Procedure.
   a. Put an MTP plate in a pouch and flush the pouch with argon gas.
   b. Seal the pouch in the sealer: hold the heat seal down for 5–6 s.
   c. Store printed plates at 4–8°C.

3.2. Microarray Printing

3.2.1. Quill Pin Printing

The Biomek HDRT method allows printing of low density arrays 3 × 3 up to 5 × 5 in a microplate well (**Fig. 1**). For higher density printing a microarrayer system is preferred. In our experiments, we utilized a Cartesian PS 7200 Dispenser equipped with a standard print head from Majer Precision for use with their Microquill split (quill-type) pins. Alternatively, the system may be adapted for use with other quill pins and holders such as for the CMP or SMP pins that are offered by Telechem, International. Because quill pins permit sub-nL delivery less material is required and spot features of 100 μ diameter or lower are possible. Printing on 200 μ, center to center spacing permits up to 20 × 20 features per well. A single plate of 3456 spots requires approx 45 min to print. In contrast, the Biomek 2000 workstation with the HDRT-96 solid pins (2–3 nL delivery) set on 500 μ, center to center spacing, prints 5 × 5 features per well for 10 plates (15,360 spots) within about 30 min.

3.2.2. Aspirate and Dispenser-Based Printing

An alternative to contact printing is the dispensing of microdroplets (pL volumes) to the surface. In this case, a nozzle is used to sip (aspirate) up a volume of reagent and then dispense droplets at defined locations into the bottom of the microplate well. Usually multiple droplets are streamed into the location to form a single spot. A Cartesian PixSys (SynQuad) Dispenser system was utilized for this purpose either from a single nozzle or eight-channel nozzle system depending on the printing task. The spotting rate is approximately that of the quill pin system previously described. In this case, the droplet dispense must be done close to the bottom of the well. As a result, the nozzle must go in an out of each well requiring additional run time. The advantage of the system is that of flexibility in the design of the array and printing setup. Also, source plate

Biomek HDRT printing of 5X5 Human IgG Array

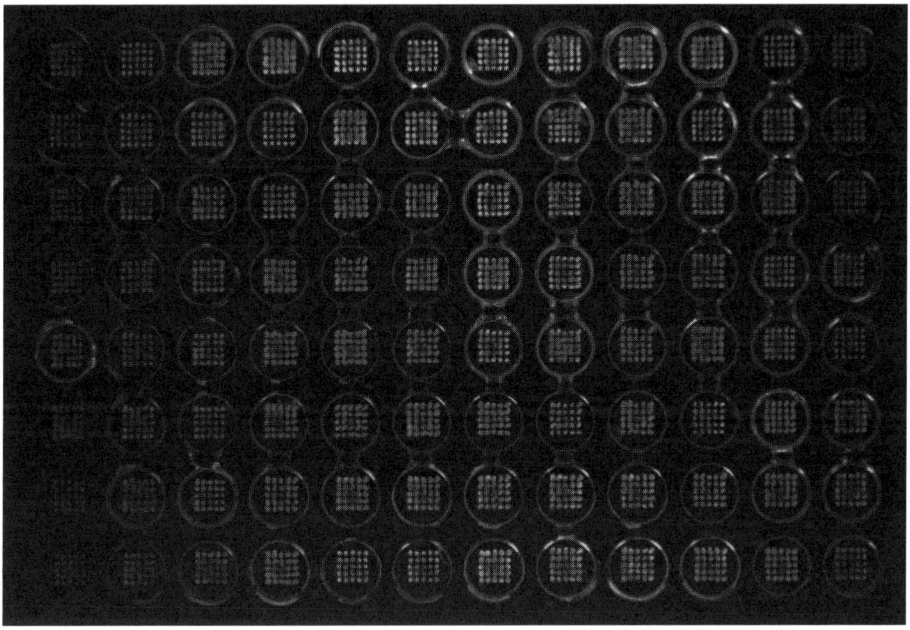

Fig. 1. Biomek HDRT printing into a microplate. Human Immunoglobulin G was printed down in a 5×5 pattern in each well of an acyl fluoride-activated polypropylene microplate using an HDRT-384 tool equipped with 96 solid pins. Mouse antihuman IgG biotin conjugate was used for detection followed sequentially by the addition of streptavidin-alkaline phosphatase and ELF reagent for signal development. The developed microarray microplate image was taken using a CCD camera.

preparation is simplified because software permits return aspirate from the same wells. This reduces the number of source plate well required for printing.

3.3. Demonstration Experiments

3.3.1. Printing With the Biomek Workstation

3.3.1.1. SOURCE PLATE PREPARATION

The mouse antihuman-IL antibodies at 1 mg/mL in printing ink were prepared in a 384-well source plate as described in the **Subheading 3.** The HDRT-384 is set up to print using 96 solid pins that match a 96-well plate. The Biomek workstation is programmed to permit the pins to touch the bottom of the source plate wells. To prevent displacement of the solutions from the wells or unwanted adsorption of antibody on the pin shaft, a minimal volume (5–10 µL) of printing ink is placed in each well. In order to prevent evaporation, the source plate is placed on a cooling block positioned on the Biomek deck. The outermost corner wells of the source plate are filled with human IgG to serve as Landing Lights (LL) for plate registration during the reading of the plate using a CCD camera system.

3.3.1.2. PRINTING

The Biomek deck is set up with the HDRT holder in place. The other positions on the deck include 1–10 activated MTPs and the source plate with holder (**Fig. 2**). In order to print using the HDRT it is necessary to use special labware icons. In this case select the HDR 384 gel plate icon from the Labware list and modify the settings so that the HDRT will print in the bottom of a 96-well, flat-bottom plate. This is accomplished by changing the grid spacing, well depth, and top height settings of the HDR 384 gel plate as shown in **Fig. 3**. The Biomek workstation is first calibrated for position accuracy and motor axis home. The printing program is then initiated as described in the Methods section. The HDRT moves from its holder to the source plate and then to the MTPs where it touches down onto the well bottom to deliver the first spot of each array simultaneously to all 96 wells. This is repeated for replicate spotting. After completion of the first set of replicate spots the Biomek workstation is paused and the HDRT removed for cleaning. It is then returned cleaned and air-dried. The program is resumed in order to print a second set of different replicates. This process is repeated until the array is complete. The arrayed plates are then removed from the deck and placed in an oven for incubation at 37°C for 30 min. Following this the wells were blocked in casein solution for 1 h then briefly rinsed in water and allowed to air-dry. The plates were then sealed in bags and placed under refrigeration until needed.

3.3.1.3. ASSAY

An assay was setup as shown in **Fig. 4** for the assessment of unknown amounts of cytokine standards spiked into RPMI + FCS (fetal calf serum) media. A standard cytokine mix was prepared containing IL-4 (10 ng/mL) + IL-8 (1 ng/mL) + IL-10 (100 ng/mL) in a diluent of RPMI with 10% FCS. In addition, for a blind study three different samples (D, E, and F) were furnished for determination. These were placed in wells as three replicates as indicated. The standards were likewise placed into wells in serial dilution as shown. The samples were incubated for 1 h at 25°C in a humidified shaker water bath.

Following the incubation the contents of the wells were removed and each well rinsed three times in TBST. A mixture of secondary detection antibodies (biotinylated) was prepared in TBS containing casein (1 mg/mL) at 1:1000 v/v dilution and aliquoted into each well except for well containing LLs. To these wells were added biotinylated mouse antihuman IgG at 1:1000 v/v dilution in diluent. The plate was incubated an additional 1 h under the previous conditions.

Biomek worksurface deck setup for printing array plates

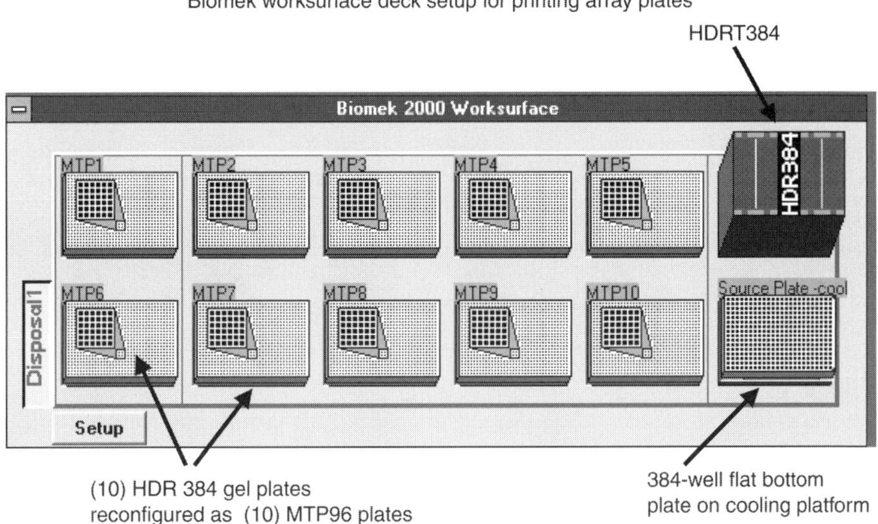

(10) HDR 384 gel plates
reconfigured as (10) MTP96 plates

384-well flat bottom
plate on cooling platform

Fig. 2. Biomek deck setup for microplate printing. The deck was loaded with up to 10 activated microplates for printing of arrays into wells. A standard 384-well microtiter plate was used as the source plate and contained the antibodies for printing. A separate plate was used for a different antibody of the array. The source plate was cooled on the deck using a special cooling platform.

Convert the HDR Gel 384 Labware for printing into a 96-well flat bottom plate

MTP 96-well plate setup

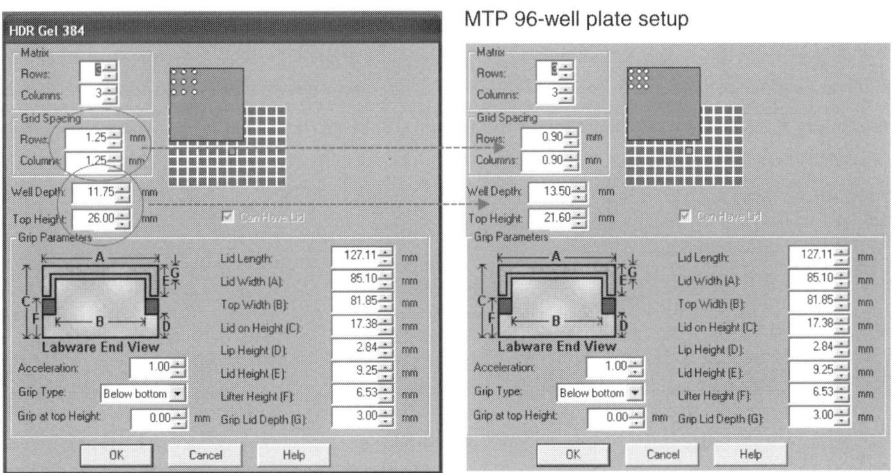

Change the grid spacing, well depth and top height to accommodate a 96-well plate

Fig. 3. HDR Gel 384 labware conversion for plate printing. In order to print into microplates using the Biomek the HDR tool must recognize that the appropriate HDR labware is installed on the deck. However, the labware dimensions require adjustment in order for the HDRT to print at the bottom of the wells of the microplate.

Multiplexed micro-ELISA MTP setup for analysis of cytokines

Standard Curves in RPMI/FCS												IL-4 Std. Curve ng/mL	IL-8 Std. Curve ng/mL	IL-10 Std. Curve ng/mL
LL	Landing Lights: ~1.3 mg/mL Hu IgG													
IL-n	Ms α-Hu Interleukins printed at 1.0 or 0.5 mg/mL													
1	2	3	4	5	6	7	8	9	10	11	12			
A: LL	IL-n	IL-n	IL-n	IL-n	IL-n	IL-n	IL-n	IL-n	IL-n	IL-n	LL	10	1.0	100
B: IL-n	IL-n	IL-n (E)	IL-n	IL-n	IL-n	IL-n (F)	IL-n	IL-n (D)	IL-n	IL-n	IL-n	5	0.5	50
C: IL-n	IL-n	IL-n (D)	IL-n	IL-n	IL-n	IL-n	IL-n	IL-n	IL-n	IL-n	IL-n	2.5	0.25	25
D: IL-n	IL-n	IL-n	IL-n	IL-n (D)	IL-n	IL-n	IL-n	IL-n	IL-n	IL-n	IL-n	1.25	0.125	12.5
E: IL-n	IL-n	IL-n	IL-n	IL-n	IL-n	IL-n	IL-n	IL-n	IL-n	IL-n (E)	IL-n	0.62	0.063	6.3
F: IL-n	IL-n	IL-n	IL-n	IL-n	IL-n	IL-n (F)	IL-n	IL-n	IL-n	IL-n	IL-n	0.31	0.031	3.1
G: IL-n	IL-n	IL-n	IL-n	IL-n (F)	IL-n	IL-n	IL-n	IL-n (E)	IL-n	IL-n	IL-n	0.16	0.016	1.6
H: LL	IL-n	IL-n	IL-n	IL-n	IL-n	IL-n	IL-n	IL-n	IL-n	IL-n	LL	0	0	0
Antigen Incubation (in Serial Dilutions):												Time:		
Diluent = RPMI + 10% Fetal Calf Serum												Sample D	Diluent	
Samples = Three replicates of three unknowns												Sample E	Sample F	
Diluent	IL Mix Std Curve	Diluent or Sample	IL Mix Std Curve	Diluent or Sample	IL Mix Std Curve	Diluent or Sample	IL Mix Std Curve	Diluent or Sample	IL Mix Std Curve	Diluent or Sample	Diluent	1 hr		

Detection (biotinylated) Ab: anti-IL Mix: all three bt-Ab's at 1:1000 in 1 mg/mL Casein in TBS													bt-Ms α-Hu IgG at 1:1000 dilution in Landing Lights
bt-Ms α-Hu IgG	anti-IL Mix	anti-IL Mix	anti-IL Mix	anti-IL Mix	anti-IL Mix	anti-IL Mix	anti-IL Mix	anti-IL Mix	anti-IL Mix	anti-IL Mix	bt-Ms α-Hu IgG	1 hr	
SA-AP diluted 1:1000 in 1 mg/ml Cascein in TBS													SA-AP in Landing Lights
SA-AP	SA-AP	SA-AP	SA-AP	SA-AP	SA-AP	SA-AP	SA-AP	SA-AP	SA-AP	SA-AP	SA-AP	30 min	
ELF	ELF	ELF	ELF	ELF	ELF	ELF	ELF	ELF	ELF	ELF	ELF	1 hr‡	
‡ Read ELF without rinse and leave overnight.													

Fig. 4. Microplate map for performing multiplexed micro-ELISA. For the determination of the amounts of cytokines in samples a standard curve for each analyte must be established. This is accomplished by serial dilution of cytokine standards in diluent using a column of wells. The cytokine standards are premixed and the ELISA performed simultaneously for each well. Likewise, samples and other controls are randomly placed in wells of the microplate for performance of the assay. This map shows each stage and reagent addition for the micro-ELISA.

After removal of the secondary antibody solutions the wells were again rinsed in Tris buffer saline Tween-20 (TBST). Next, streptavidin–alkaline phosphatase conjugate (SA-AP) was added at 1:1000 v/v dilution in TBS-casein. The plate was incubated for 30 min then rinsed in TBST. ELF reagent was then added and the plate incubated at room temperature without shaking for 1 h. The wells were then briefly rinsed in DDI water and the plate imaged using a CCD camera system.

3.3.1.4. RESULTS

Standard curves for each of the cytokines were prepared as shown in **Fig. 5** from the mean intensities ($n = 5$ serial dilution sets). The unknown cytokines were first identified as either IL-4, -8, or -10 and their amounts within the sample estimated from the respective ELF intensity (rfu) vs standard curve. Based on estimates in the linear portion of the standard curves the following determinations were made:

IL-4	measured = 90.3 pg/well	actual = 100 pg/well	recovery = 90%
IL-8	measured = 10 pg/well	actual = 15 pg/well	recovery = 67%
IL-10	measured = 231.5 pg/well	actual = 250 pg/well	recovery = 93%

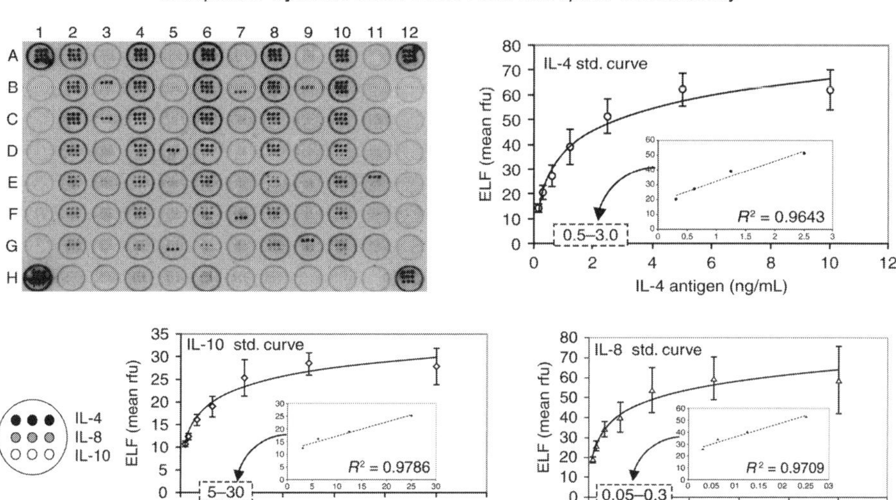

Fig. 5. Results from a microplate microarray ELISA for three cytokines. Using the assay map represented in **Fig. 4** the standard curves are established for each cytokine. Using these curves the concentration of the cytokine analyte in the various samples was estimated.

The lower recovery for IL-8 might be related to the higher affinity of the capture antibody. In other experiments this antibody appeared to saturate at lower antigen concentration thereby reducing the linear working range. Attempts to work in the linear range required significant dilution of IL-8 antigen which might have led to some denaturation as well.

3.3.1.5. BIOMEK 2000 WORKSTATION PRINTING PERFORMANCE

Although the Biomek workstation was not intended to be used for precision arraying it is possible to achieve reproducible arrays of low density using the HDRT system. For example, the well to well imprecision was measured for six plates that had been printed with human IgG homoarrays and found to be at a mean coefficient of variation (CV) = 5.5% (*see* **Fig. 6**).

However, it is critical that the HDRT and the 96 pins are properly cleaned after use. Salt or debris buildup on the pin shaft is the most obvious detriment to performance. Dirty pins will not freely slide up and down in the holder and thus touchdown on the microplate will be erratic leading to the deposit of different amounts of protein in wells. It is also important to maintain proper alignment of the HDRT with that of the source plate. The pins should not touch the walls of the source plate and minimal amounts of fluid should travel up the

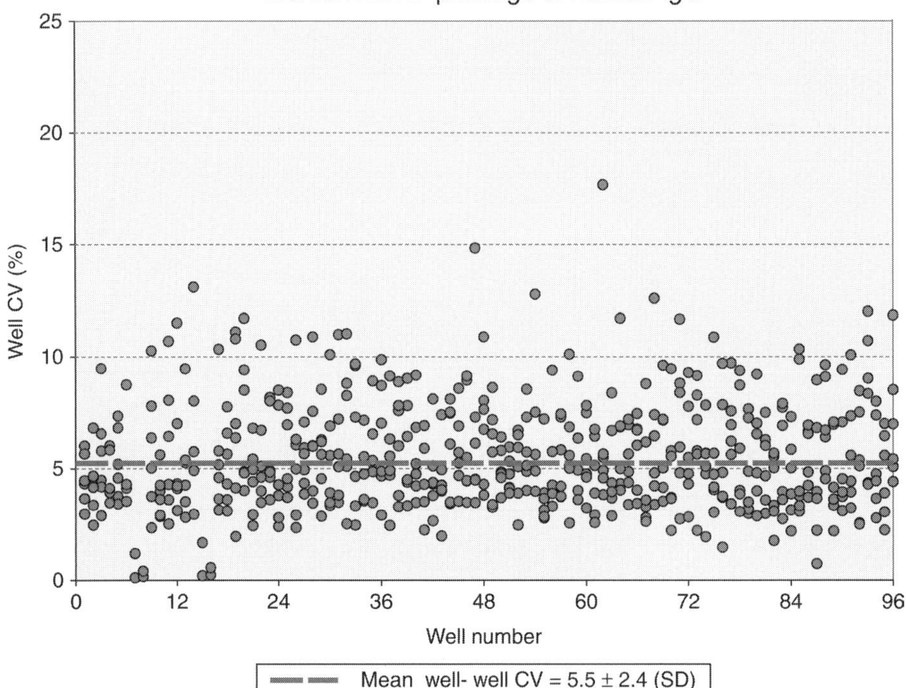

Fig. 6. Well to well imprecision for six microarray microplates. Imprecision (CV%) from each 5 × 5 array in each well (intrawell) across 96 wells (interwell) from six different microarray plates is plotted.

pin's shaft. Contact of the pins with the walls can cause uneven wetting of the pin and on occasion can lead to tip damage or even binding of the pin shaft. When operating the Biomek workstation it is also important to reduce the amount of vibration associated with movement of the HDRT. This is particularly important during the time when the pins are about to strike the bottom of the microplate to create the array. The net result will be the formation of uneven arrays or differing amounts of material being deposited. To avoid such problems it is advised that the HDRT be deaccelerated during touchdown (*see* **Note 10**).

3.3.2. Cartesian Pin Printing

As previously discussed quill pin printing using a conventional microarrayer system is capable of achieving higher density features than with the Biomek HDRT. However, the use of high salt content with quill pins is

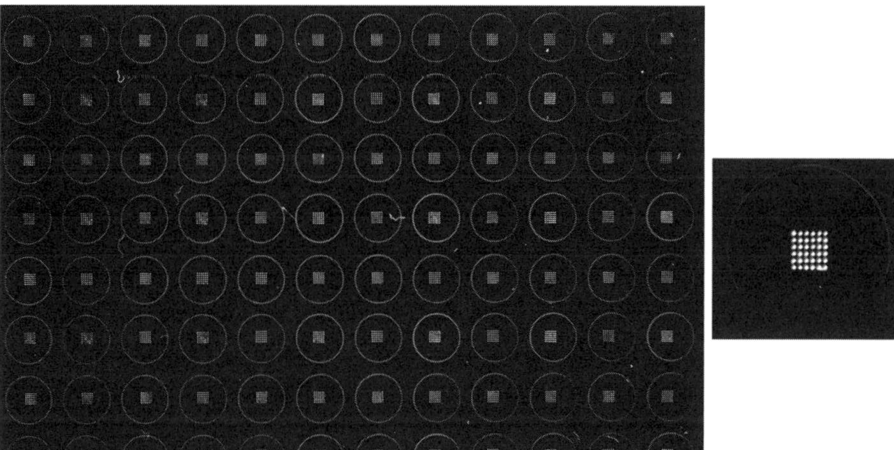

Fig. 7. Microplate microarray printing results using Quill Pins. A Human IgG microwell microarray was printed using a Cartesian PS 7200 printer equipped with Majer Precision Split (Quills) Pins.

problematic since salts will rapidly deposit in the capillary thereby obstructing fluid delivery. For that reason the use of 4% sodium sulfate is not recommended nor required as a means to provide uniform spotting. In the case of the Biomek HDRT which prints much larger spots the sodium sulfate served to produce more uniform spots with homogeneous surface morphology. On the other hand, quill pins are designed to produce smaller spots and these are usually of good spot morphology. As a result, the printing ink can be simplified. We have found that preparing proteins in 50 mM carbonate buffer, pH 9.0 at about 1 mg/mL final concentration is sufficient. The printing ink is prepared in 1X MSS (microarray spotting solution, TeleChem Int'l, Sunnyvale, CA) which uniformly wets out the quill capillary thereby improving spotting. It does not denature antibodies.

An example of a 6 × 6 array of human IgG printed on the bottom of the flexible microplate is provided in **Fig. 7**. This array was printed using the Cartesian PS 7200 equipped with 4 Majer Precision Microquills (Tempe, AZ) arranged in a 2 × 2 pattern in the holder. In this case, a fluorescent dye was added to visualize the printed spots (*see* close-up image). In other experiments, total imprecision of the array (including printing and assay) was determined. For example, using a single pin to print a 96-well plate with monoclonal antibody probes for IL-4 and -8 antigens a well to well CV of about 7% was achieved (*see* **Fig. 8**).

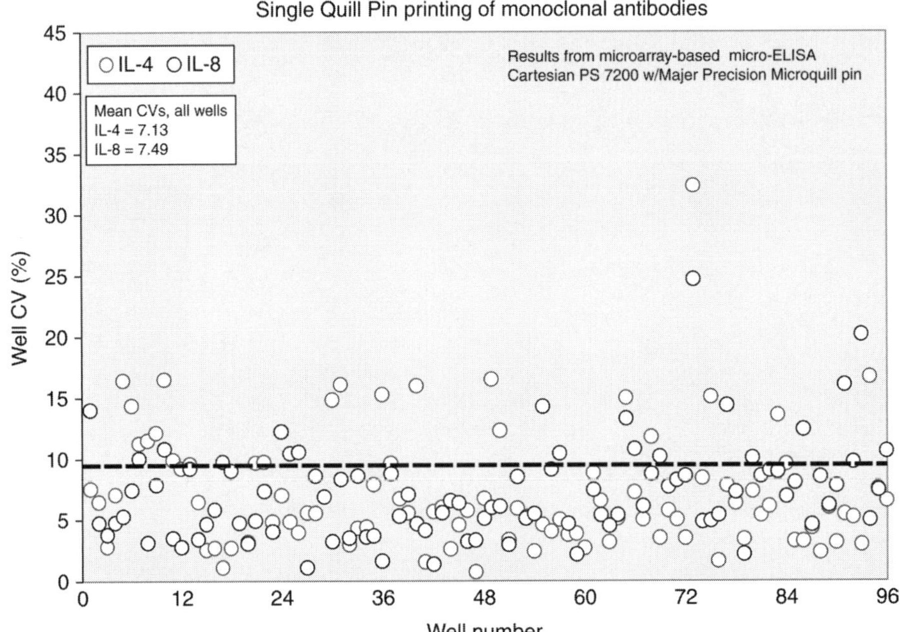

Fig. 8. Well to well imprecision for a Single Quill Pin printing into microplates. A comparison of individual cytokine signal variation for two cytokines derived from assays run across the same 96 wells of the microplate.

3.3.3. Cartesian Noncontact Dispenser Printing

We have employed a PixSys 4200 (Genomic Solutions, Inc.) with eight SynQuad (Genomic Solutions, Inc.) dispensers to print proteins as well. This system relies on volume metering using a solenoid valve and syringe displacement mechanism to deliver metered volumes between 10 and 20 nL droplets to the surface. The resulting spots are approximately the same diameter as achieved using the Biomek HDRT. The precision dispense permits close packed and highly ordered arrays. Source plate set up is straightforward and can be prepared using the Biomek 2000 workstation to dispense probes from stock solutions prepared in microtubes into a 384-well plate. In this case, 18 print cycles were used to prepare the 6 × 6 array pattern (*see* **Fig. 9**).

The PixSys allows for independent and "on the fly" dispensing for each dispense head to create the desired pattern based on the inputed software designed array. In **Fig. 10** the anti-IL capture antibody array (IL-4, -8, -10) is prepared using the PixSys 4200 to print into the flexible plate. The array design included a fifth LL (internal to the capture antibody array) spot for orientation. Other features included casein spots for assessment of nonspecific binding; and a final column containing

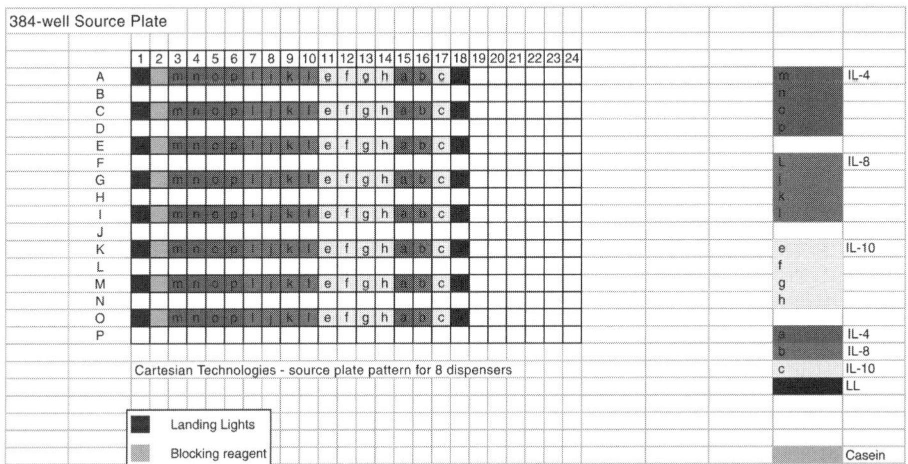

Fig. 9. Source plate well map for printing into microplates using dispense heads. Monoclonal anticytokine antibodies were prepared in a 384-well source plate for printing using an eight dispenser head configuration for the PixSys 4200 printer (Genomic Solutions, Inc.).

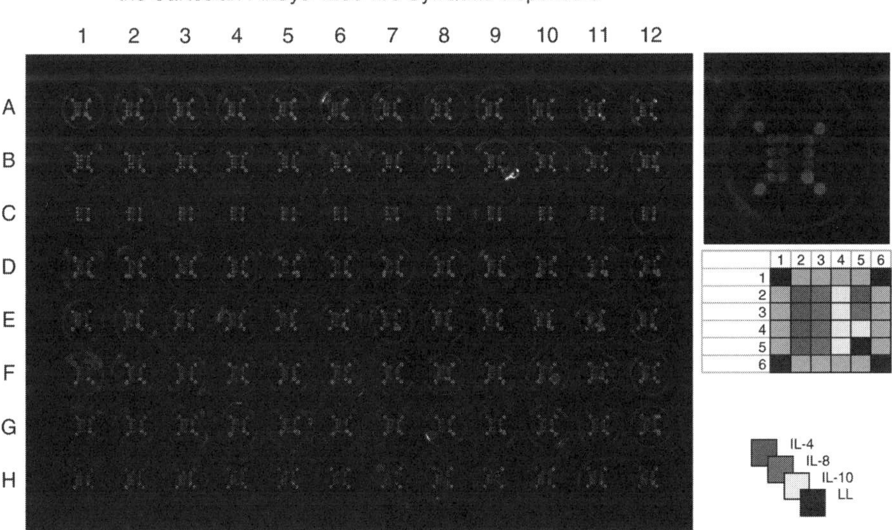

Fig. 10. Results from printing of antibodies into microplates using dispensers. The results of the eight-dispenser printing are shown. The inserted well map colors corresponds to those representing individual anticytokine capture antibodies of the source plate map from **Fig. 9**.

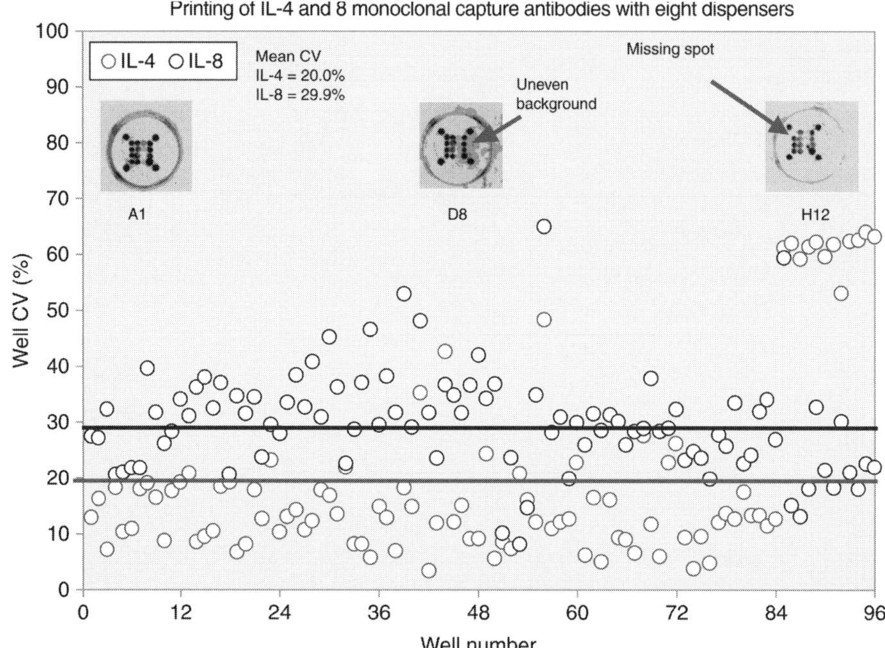

Fig. 11. Well to well imprecision for eight-dispenser head printing into microplates. Results from micro-ELISA for two cytokines of the microplate well microarray are compared across all wells. Spot morphology from wells having higher levels of imprecision are compared for wells A1, D8, and H12.

singlets of each of the capture antibodies to access cross-contamination from the nozzles that might arise from inadequate rinsing before aspirate.

Note that dispenser C failed to print the outer LL registration markers in the initial print cycle but otherwise performed adequately in completing the array. It is possible that the nozzle was blocked by an air-bubble that prevented either the aspirate or dispense to occur. For that reason it is very important that air bubbles are cleared from the source plate. The most effective method of accomplishing this is to centrifuge the plate in order to displace the air from the well.

This particular printing was problematic leading to higher well to well CVs. In **Fig. 11** is shown the overall (print + assay) well to well CVs for IL-4 and -8. Analysis of each dispenser head performance revealed that column 8 of the array in each case exhibited lower signal and greater variation. Examination of individual wells showed uneven background (*see* **Fig. 11**, A1 vs D8 inserts). This is most likely owing to post-print processing and may be due to incomplete blocking of these wells. For IL-4 wells at the end of the print run had missing spots (*see* **Fig. 11**, H12 insert) leading to a higher CV.

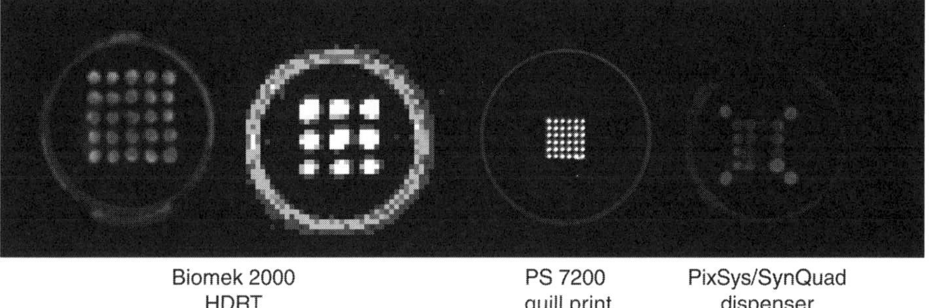

Fig. 12. Comparison of antibody arrays printed using different printer technologies. Spot diameters and morphologies for antibody arrays printed into microplate wells using the Biomek HDRT with solid pins, a Cartesian PS 7200 (Genomic Solutions, Inc.) equipped with split pins (Quills) and the Cartesian PixSys system (Genomic Solutions, Inc.) equipped with SynQuad dispenser heads (Genomic Solutions, Inc.) are compared.

3.4. Remarks

We have evaluated several printing systems for the low density arraying of antibodies in microplates **(Fig. 12)**. The Biomek HDRT provides a convenient method for producing such arrays for initial assay development. It was possible to achieve similar levels of printing and assay performance using either the Biomek workstation or a microarrayer. However for the construction of higher density arraying at smaller spot diameters it is necessary to move to a conventional microarrayer system. Both contact (quill pin) and noncontact (aspirate–dispense) arrayers are available for the printing of proteins. Under proper conditions it is possible to obtain comparable levels of performance.

4. Notes

1. The DAST activation chemistry was developed in our laboratory and is protected by patent. Other surface modification chemistries may be used. Suppliers of commercial plates offer amine, carboxyl, and aldehyde surfaces, which can be used to immobilize proteins.
2. Other reporter and detection systems can be used. Biotinylated secondary (reporter) antibodies are a convenient approach since a number of streptavidin-conjugates of enzymes and dyes are commercially available. A CCD camera system is employed in order to image wells of the microplate. There are commercial reader systems now available that can resolve the array of arrays format in microplates. Alpha Innotech (San Leandro, CA) offers the NovaRay Detection Platform and UVP (Upland, CA) offers the BioSpectrumAC Imaging System.
3. Unless otherwise noted deionized water refers to 18 MΩ quality.
4. It is highly recommended that matched pairs of primary (capture) and secondary (reporter) monoclonal antibodies are used whenever possible. R&D Systems

provides several sets including the recombinant antigen used to generate the antibodies.

5. The printing plate map is essentially using alternating wells of a 384-well plate to place the 96 aliquots of the printing ink. Verify that the HDRT with 96 pins registers with the selected wells. Each antibody requires a separate source plate.

6. When storing plates in a refrigerator for an extended time, they must be carefully sealed with Parafilm or a plate sealer; otherwise, the plates will dry out. Long term storage is not recommended.

7. If printing quality deteriorates remove pins from HDRT tool and soak in chromic acid solution for 10 min. Rinse pins extensively in deionized water. Air-dry before placing back in tool holder.

8. The casein solution might be sonicated to speed up solubilization of the casein.

9. The plates might be quenched overnight or longer in a refrigerator. In which case, additional quenching reagent should be introduced into the box or container to prevent drying out in the wells.

10. From time to time the HDRT solid pins might need to be replaced resulting from the damage of the tip or shaft. We have found it beneficial to characterize each of the 96 pins for delivery volume and spot quality then match a set of pins. Delivery volume of each pin can be determined by touchdown of a fluorescent marker such as fluorescein into a plate then diluting each well with a fixed volume of buffer. The plate can then be read in a fluorometer and the amount of fluorescein delivered for each pin calculated from a standard curve. Likewise, the fluorescent spot can be imaged under a CCD camera and checked for spot diameter and morphology. Pins can then be match based on these results.

References

1. Schena, M., Shalon, D., Davis, R. W., and Brown, P. O. (1995) Quantitative monitoring of gene expression pattern with a complementary DNA microarray. *Science* **270,** 467–470.

2. Matson, R. S. (2005) *Applying Genomic and Proteomic Microarray Technology in Drug Discovery.* CRC Press, Boca Raton, FL.

3. Haab, B. B., Dunham, M. J., and Brown, P. O. (2001) Protein microarrays for highly parallel detection and quantitation of specific proteins and antibodies in complex solutions. *Genome Biol.* **2,** research0004.1–0004.13.

4. MacBeath, G. and Schreiber, S. L. (2000) Printing proteins as microarrays for high-throughput function determination. *Science* **289,** 1760–1763.

5. Ekins, R., Chu, F., and Biggart, E. (1990) Fluorescence spectroscopy and its application to a new generation of high sensitivity, multi-microspot, multianalyte, immunoassay. *Clin. Chim. Acta* **194,** 91–114.

6. Mendoza, L. G., McQuary, P., Mongan, A., Gangadharan, A., Brignac, S., and Eggers, M. (1999) High-throughput microarray-based enzyme-linked immunosorbent assay (ELISA). *BioTechniques* **27,** 778–788.

7. Matson, R. S., Milton, R. C., Cress, M. C., and Rampal, J. B. (2001) Microarray-based cytokine immunosorbent assay. Oak Ridge Conference, Poster no. 20.

8. Macas, J., Nouzova, M., and Galbraith, D. W. (1998) Adapting the Biomek®2000 Laboratory Automated Workstation for printing DNA microarrays. *BioTechiques* **25,** 106–109.
9. Cleveland, P. H. and Koutz, P. J. (2005) Nanoliter dispensing of μHTS using pin tools. *Assay Drug Dev. Tech.* **3,** 213–225.
10. Milton, R. C. deL. (2000) Polymeric reagents for immobilizing biopolymers. US Patent no. 6,110,669.
11. Matson, R. S., Rampal, J. B., and Coassin, P. J. (1994) Biopolymer synthesis on polypropylene supports, I. Oligonucleotides. *Anal. Biochem.* **217,** 306–310.

18

Forward-Phase and Reverse-Phase Protein Microarray

Yaping Zong, Shanshan Zhang, Huang-Tsu Chen, Yunfei Zong, and Yaxian Shi

Summary

Protein microarray is a powerful tool for identifying disease biomarkers and therapeutical targets, and for systematically studying biological pathways with high efficiency. Although the protein microarray platform has been adopted by proteomic research and discovery, what remains problematic is how to maintain the activities and structures of printed proteins on slide surface. With the recent accomplishments in the R & D laboratory, now scientists around the world are able to preserve the biological functions of spotted proteins for high throughput analysis. Full Moon BioSystems (FMB) has developed general guidelines, which helps scientists efficiently prepare forward as well as reverse-phase protein microarray using FMB's proprietary polymer-coated slides, to obtain reliable and accurate array data with FMB's unique detection and analysis technology.

Key Words: Calibration slide; capture antibody; cell lysates; forward-phase protein arrays; protein slides; reserve-phase protein arrays; signal-to-noise ratio (SNR).

1. Introduction

With the completion of the human genome project a few years ago, scientists now focus on the determination of genes' functions and the interplay of genes' product(s). A high throughput platform allowing the interpretation of genome (genomic) information will be absolutely necessary in this regard. Microarray is a highly robust tool for screening and analyzing thousands of analytes in one experiment with high efficiency and accuracy. The technology was originally developed and has mainly been used for gene expression profiling and for structure analysis *(1,2)*. With the increasing availability of protein information, protein array has become a popular and mainstream platform for simultaneous identification, quantification, and functional analysis of proteins in basic and

From: *Methods in Molecular Biology, vol. 381: Microarrays: Second Edition: Volume 1: Synthesis Methods*
Edited by: J. B. Rampal © Humana Press Inc., Totowa, NJ

applied proteome research *(3,4)*. Recently, a powerful array-based sera protein profiling analysis has successfully been used to identify proteins associated with ovarian cancer *(5)*.

The two most important protein microarray formats, namely forward-phase protein microarray (as in **ref. 6**, for review and references therein) and reverse-phase protein microarray *(7)*, are commonly used in proteome discovery and in diagnostics *(8)*. The basic format of forward-phase protein microarray is similar to that of sandwich enzyme-linked immunosorbent assay. Briefly, in forward-phase protein microarray, the capture antibody is first immobilized on the slide surface. The immobilized antibody can then be used to capture the antigens it recognizes in a test sample. The test sample can be cells, cell lysates, blood, or other biological specimen. The captured analyte is then detected directly with a fluorescent dye-conjugated detection antibody, or detected indirectly with the detection antibody followed by a fluorescent dye-conjugated second antibody. Thus, forward-phase protein microarray offers a high degree of detection specificity as a result of pairing of capture and detection antibodies for an interested analyte. Nonetheless, identifying paired antibodies for detection can be time-consuming. Reserve-phase protein microarray provides a solution to this problem. In contrast to the forward format, the reverse-phase protein microarray allows test samples to be printed on the slide directly. Hundreds of thousands of biological samples can be processed in a single array. The array is then detected with dye-conjugated detection protein, such as antibody. Whereas, reverse-phase format offers the power and throughput for protein array, the specificity might be compromised to some degree owing to the use of single detection antibodies.

Although protein microarray has been successfully implemented in basic and applied proteomic research and discovery, the accuracy and efficiency for evaluating the significance of a biological sample or specimen relies solely on the data generated by the arrays. The selection of substrates or slides used for printing protein arrays, and the performance of detection scanner are the two key factors determining the reliability of array data. Not only does the slide provide the physical support for proteins, it is also responsible for retaining the three-dimensional (3D) structure and functions of the interested proteins. Additionally, the slide that produces excellent printing morphology, spotting consistency, and low detection background is also essential to data accuracy. For DNA and protein microarrays, expression profiling results are based on the analysis of data generated by microarray scanners. The performance of a microarray scanner also plays an important role in determining array data quality. A scanner's channel-to-channel cross-talk undermines the dependability of its readouts. Moreover, the "functional" dynamic detection range and sensitivity can directly affect the accuracy and reliability of array data. Thus, periodic validation or calibration of array scanners are absolutely necessary to ensure that the scanner produces consistent data. Consequently, the selection of right slides for protein printing and the use of adequate scanner calibration and validation tools together will assure reliable and accurate data from the protein microarray platform.

In this chapter, general guidelines for preparing forward- and reverse-phase protein microarray using Full Moon BioSystem's (FMB's) propriety protein slides are illustrated. The combination of FMB's detection technology and array data analysis tools enables the scientists to simultaneously identify, quantify, and functionally analyze proteins of interest. This robust assay platform has the potential to replace singleplex analysis systems.

2. Materials

1. Protein slides (Full Moon BioSystems, Inc., Sunnyvale, CA).
2. Scanner calibration slide (Full Moon BioSystems).
3. Protein Printing buffer (Full Moon BioSystems).
4. Dulbecco's phosphate buffered saline (Sigma, St. Louis, MO).
5. Tween-20, Sigma Ultra (Sigma).
6. Mouse antihuman interlukin (IL)-13 capture antibody (R & D Systems Inc., MN).
7. Anti-IL-6 capture antibody (R & D Systems).
8. Anti-tumor necrosis factor (TNF)-α antibody (R & D Systems).
9. Goat anti-human-IL13 (R & D Systems).
10. Rabbit anti-human Her2, IL-6, and TNF-α (Epitomics Inc., Burlingame, CA).
11. Cy3 monofunctional reactive dye (Amersham Biosciences, NJ).

3. Methods

3.1. Slide Selection

A 3D structure provides an adequate environment for the immobilized proteins to remain functional after printing. FMB has developed a proprietary polymer coating technology to fulfill this need. Under microscopic examination, the porous 3D structure is clearly visible on the surface of a polymer-coated slide (**Figs. 1** and **2**). Slides coated with FMB proprietary polymers contain multifunctional and high-affinity porous bonding sites for protein molecules. These features allow the printed proteins to nest in between the bonding sites on the slide, and also ensure the attached protein to retain its structure. Thus, the binding specificity of the protein to its biological counterpart is adequately retained. Moreover, proteins printed on FMB's slides can remain stable for a long period of time when handled properly. FMB's Protein Slides are well recognized in the protein microarray community. The slides offer many other advantages for proteins arrays, including excellence spot morphology and spotting consistency, and low detection background.

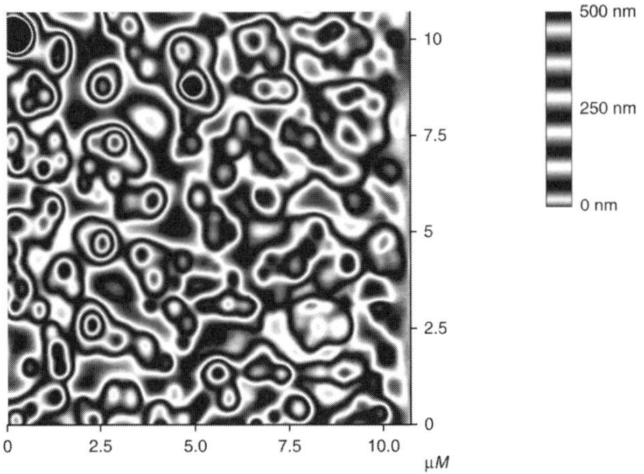

Fig. 1. The surface morphology of FMB protein slides. The image is obtained by atomic force microscope.

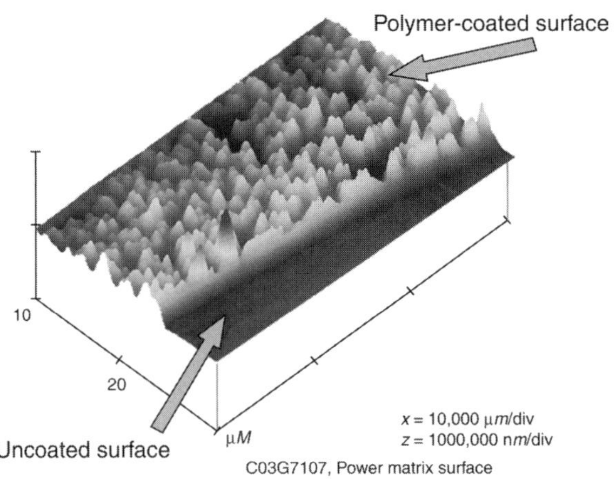

Fig. 2. The structures of coated and uncoated surface of a FMB protein slide. It clearly demonstrates the 3D porous structure on the slide surface, through which proteins can be immobilized with high efficiency, and activities are retained.

3.2. Scanner Calibration

For any DNA or protein microarray, the final results and conclusions derived are based on the readouts of a microarray scanner. The detection range, sensitivity and channel-to-channel cross-talk of the scanner are important factors in determining the accuracy and reliability of array data. FMB's microarray scanner calibration and validation slides are designed to evaluate and calibrate

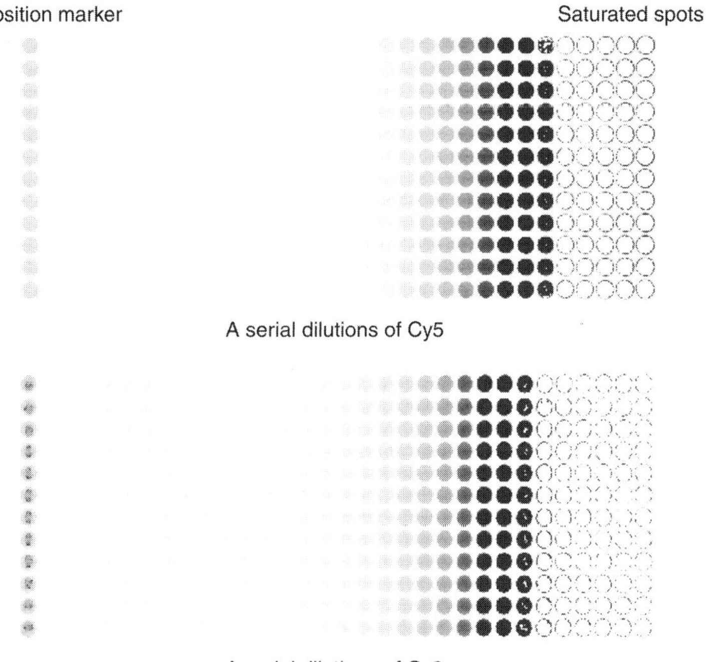

Fig. 3. Image of a FMB Scanner calibration slide scanned on Axon GenePix 4000 array scanner.

scanners' performance to ensure high quality array data. The scanner calibration slide from FMB contains two blocks of arrays. The first block contains a series of Cy5 dilutions with 12 repeats covering eight orders of dynamic range, and the second block contains a series of Cy3 dilutions also covering eight orders of dynamic range. An image of FMB scanner calibration slide is shown in **Fig. 3**.

R & D scientists use the protocol described later to evaluate the linear dynamic range and sensitivity of a microarray scanner. This protocol is easy to follow and also to modify, whenever required. After scanning, the image can be used to calculate the relative intensity using the accompanying software provided by the scanner's manufacturer.

1. Load a FMB scanner calibration slide on the scanner's slide holder.
2. Perform a quick scan to locate the arrays.
3. Based the initial image obtained from **step 2**, optimize PMT settings and make sure that the spots in last four to five columns on the far right side of the image with high dye concentrations produce saturated signals.
4. Scan the slide with both red and green lasers.

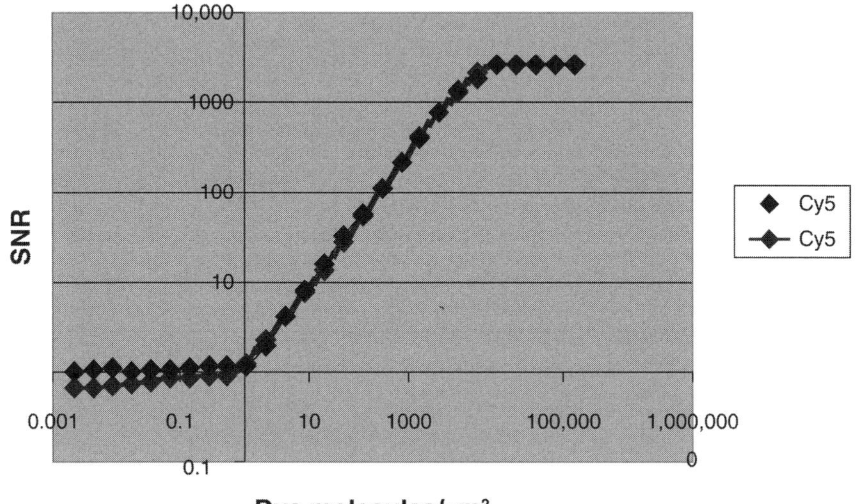

Fig. 4. Correlations between signal-to-noise ratios of Cy3 and Cy5 and the number of Cy3 and Cy5 dye molecules per micron square.

5. Analyze the image and calculate signal-to-noise ratio (SNR) based on the following equation:

$$SNR = \frac{(\text{Average signal intensity} - \text{Average background intensity})}{\text{Standard deviations of background signals}}$$

6. Plot SNR against dye molecules/μm^2 (reference data are provided along with the product).
7. Determine the linear dynamic response range and limit of detection based on SNR > 3.

An example plot is shown in **Fig. 4**. From this graph, the scanner detection limit and linear dynamic range are determined. The linear dynamic range in this experiment is about 3.5 orders for fluorescent dyes on the glass surface. After determining the linear dynamic range, the spotting concentration of sample proteins or antibodies can be optimized.

3.3. Procedure of Forward Phase Protein Microarray

Depending on the array format, proteins to be printed can be antibodies, antigens, or various biological samples. For reverse-phase protein microarray, spotted materials can be denatured or nondenatured proteins from a variety of sources. Ideally, samples should be printed in serial dilutions to generate an internal standard working curve by which direct quantitative measurements can be obtained. While the power of this type of protein arrays on analyzing signal transduction pathways using cells from microdissected clinical trial specimen has been demonstrated *(9)*, the scientists at Full Moon BioSystems have successfully developed and optimized easy to follow protein array protocols to be used with its Protein Slides. Presented next are the two practical procedures routinely used to prepare and process forward- and reserve-phase protein arrays in laboratories.

3.3.1. Printing Target Preparation

1. Prepare capture antibody solution in 1X PBS with 0.05% BSA (pH: ~7.4) at room temperature to a final concentration of about 100 µg/mL.
2. Transfer 20 µL of the antibody solution to a 96- or 384-spotting plate.
3. Quickly spin the plate to bring down the liquid to the bottom of wells.
4. Setup array spotter and spot antibodies on FMB's protein slides.

3.3.2. Humidity Treatment

After spotting, incubate the printed slides in a chamber with relative humidity of 65–75% for 4–6 h (*see* **Notes 1** and **4**). After incubation, allow the slides to dry by placing the slides in a low humidity environment (35% or less) at room temperature for 30 min.

3.3.3. Pretreatment

1. Prepare blocking solution: 1X PBS/1% BSA.
2. Incubate the slides in blocking solution for 30 min at room temperature on an orbital shaker at low speed.
3. Wash the slides briefly with TBS/0.1% Tween-20 at room temperature for 5 min.
4. Remove the slides and rinse them extensively with Milli-Q water and dry with a gentle stream of nitrogen blow.
5. Cover the spotted area with 30 µL of sample mixture prepared in blocking solution. Place a cover slip on top (*see* **Note 1**). Avoid any bubbles under the cover slip. Carefully place the slides into a humidified chamber with 100% humidity.
6. Incubate for 1 h.
7. After incubation, wash the slides three times with TBS/0.1%Tween-20 at room temperature for 5 min each time.
8. Dip the slides thoroughly in a staining jar containing Milli-Q H_2O at room temperature and then dry slides with a gentle stream of nitrogen immediately.

3.3.4. Detection With Cy Dye-Conjugated Antibody or Dye Conjugated Secondary Antibody

1. Cover the arrays with 30 µL of detection (conjugated or unconjugated) antibody mixture prepared with the blocking solution, and place a clean cover slip on top of the arrays. Avoid any bubbles under the cover slip. The concentration of antibodies may vary depending on the makeup of the experiment and the complexity and affinity of the labeled samples.
2. Carefully place the slides into a humidified chamber with 100% humidity.

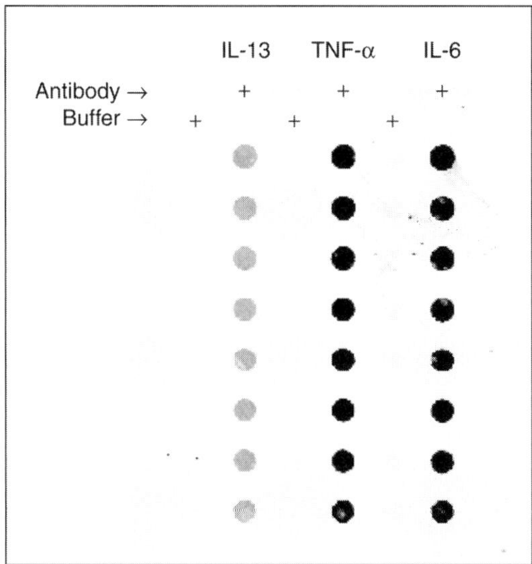

Fig. 5. An image of a forward-phase protein micrarray detected with Cy-3-conjugated detection antibodies. The slide was first printed with capture antibodies for IL-13, TNF-α, and IL-6 in eight repeats on FMB's protein slides. Printing buffer was used as negative controls. Serum sample from human was applied to the printed slide. After 1 h of incubation, the slide was washed and then detected with paired Cy-3-conjugated detection antibodies for IL-13, TNF-α, and IL-6.

3. Incubate for 1 h.
4. Wash with TBS/0.1% Tween-20 at room temperature for 5 min each time. Repeat twice.
5. Remove the slides and rinse extensively with Milli-Q H_2O and dry with a gentle stream of nitrogen blow.
6. Slides are ready for scanning, if dye-conjugated detection antibodies were used.
7. If dye-conjugated secondary antibodies were used for detection, covers the slide with 30 μL of secondary detection antibody mixture prepared with the blocking solution (*see* **Note 2**) and place a clean cover slip. Avoid any bubbles under the cover slip.
8. Carefully place the slides into a humidified chamber with 100% humidity.
9. Incubate for 1 h.
10. Wash with TBS/0.1% Tween-20 at room temperature for 5 min. Repeat twice. Remove the slides and rinse them extensively with Milli-Q H_2O and dry with a gentle stream of nitrogen.
11. Scan the slides.
12. A typical scanned forward-phase detection image is shown in **Fig. 5**. In this experiment, capture antibodies for several human proteins were first spotted on FMB's protein slides. After incubation with a serum sample, coupling Cy-3-conjugated

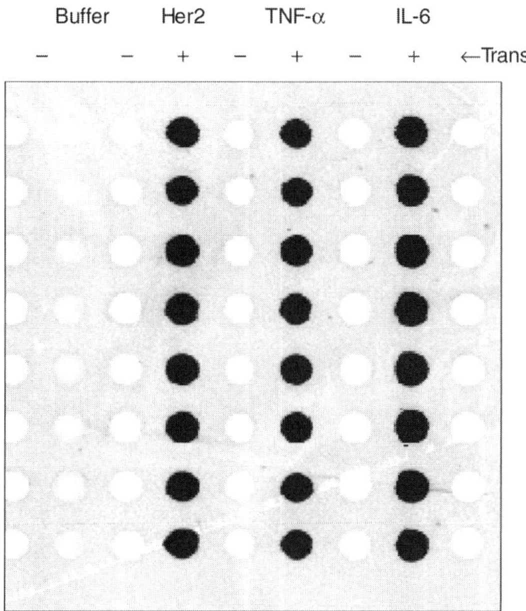

Fig. 6. Image of a reverse-phase protein microarray detected with Cy3-conjugated antibodies. 293T embryonic kidney cells were transfected with and without expression construct for Her2, TNF-α, or IL-6. Cell lysates were prepared after 48-h incubation at 37°C, and used to prepare reverse-phase protein micrarray on FMB's protein slide in eight repeats. Cy-3-conjugated anti-Her2, anti-TNF-α, and anti-IL-6 antibodies were used for detection.

antibodies for antigens were used for detection. Forward-phase protein microarray allows simultaneously detection of multiple proteins at physiological concentrations.

3.4. Procedure of Reverse-Phase Protein Microarray

3.4.1. Preparation of Targets for Printing

1. Prepare cell lysates or other biological samples with 1X PBS or FMB Protein Printing Buffer at the concentration of 0.2 mg/mL.
2. Transfer 20 µL of the printing sample to a 96- or 384-spotting plate.
3. Quickly spin the plate to bring the liquid down to the bottom of wells.
4. Set up array spotter and spot samples to FMB's Protein Slides.

3.4.2. Humidity Treatment

After printing, incubate the slides in an environment with 65–75% humidity for 2–4 h, and then air-dry (or low humidity) for 30 min.

3.4.3. Pretreatment

1. Block the slides with FMB blocking reagent or 1X PBS/1 BSA for 30 min at room temperature.
2. Wash the slides with FMB Protein Wash Buffer II or with TBS/0.1% Tween-20.
3. Rinse with deionized H_2O and dry with N_2.

3.4.4. Detection With Cy Dye-Conjugated Antibody

1. Apply 30 µL (0.5 µg/mL) dye-conjugated detection antibody to the slides and cover with cover slips. Avoid any bubbles under the cover slip.
2. Carefully place the slides into a humidified chamber with 100% humidity.
3. Incubate for 1 h.
4. Wash the slides with TBS/0.1%Tween-20 for 5 min/wash. Repeat twice.
5. Rinse the slides with deionized water and dry with N_2.
6. Slides are ready for scanning.
7. A typical scanned reverse-phase array image is shown in **Fig. 6**. Transfected and nontransfected cell lysates were spotted on FMB's protein slides. Mixed dye-conjugated antibodies were used to detect the expressed antigens. Reverse-phase protein microarray provides a means to detect multiple proteins simultaneously.

4. Notes

1. The need for humidity treatment may vary depending upon the specific properties of the printed samples. Experimentation and optimization may be needed to determine suitable conditions.
2. To prepare clean cover slips, wash cover slip with 70% ethanol, and then dry with nitrogen.
3. Optimal secondary antibody concentration may need to be determined.
4. A laboratory humidity chamber with 65–75% humidity can be put together using material readily available in laboratories. To do so, place saturated NaCl solution, this consists of 100 g of NaCl solids in 40–50 mL Milli-Q H_2O in a regular chamber with tight seals.

References

1. DeRisi, J., Penland, L., Brown, P. O., et al. (1996) Use of cDNA microarray to analysis gene expression patterns in human cancer. *Nat. Genet.* **14,** 457–460.
2. Lockhart, D. J. and Winzelerw, E. A. (2000) Genomics, gene expression and DNA arrays. *Nature* **405,** 827–836.
3. Grubb, R. L., Calvert, V. S., Wulkunle, J. D., et al. (2003) Signal pathway profiling of prostate cancer using reverse-phase protein arrays. *Proteomics* **3,** 2142–2146.
4. MacBeath, G. and Schreiber, S. L. (2000) Printing proteins as microarray for high-throughput function determination. *Science* **289,** 1760–1763.
5. Petricoin, E. F., Ardekani, A. M., Haitt, B. A., et al. (2002) Use of proteomic patterns in serum to identify ovarian cancer. *Lancet* **359,** 572–577.

6. Albala, J. (2001) Array-based proteomics: the latest chip challenge. *Expert Rev. Mol. Diagn.* **1,** 145–152.
7. Nishizuka, S., Charboneau, L., Young, L., et al. (2003) Proteomic profiling of the NCI-60 cancer cell lines using new high-density reverse lysate microarrays. *Proc. Nalt. Acad. Sci. USA* **100,** 14,229–14,234.
8. Paweletz, C. P., Charboneau, L., Bichsel, V. E., et al. (2000) Rapid protein display profiling of cancer progression directly from human tissue using protein biochip. *Drug Dev. Res.* **49,** 34–42.
9. Paweletz, C. P., Gillespie, J. W., Ornstein, D. K., et al. (2001) Reverse-phase protein microarrays pro-survival pathways at the cancer invasion front. *Oncogene* **20,** 1981–1989.

19

Cell Microarray for Functional Exploration of Genomes

David Castel, Marie-Anne Debily, Amandine Pitaval, and Xavier Gidrol

Summary

As more genomes are sequenced, challenge of rapidly unraveling the functions of genes was faced. To that end, cell microarrays have been recently described that permit transfection of thousands of nucleic acids in parallel and enable the analysis of phenotypic consequences of such perturbations. As many parameters can influence the efficiency of transfection and consequently protein expression or extinction, some important features in manufacturing cell microarrays for functional exploration of genomes were described.

Key Words: Cell microarray; functional genomics; high-throughput screening; reverse transfection; RNA interference; siRNA microarray.

1. Introduction

With the complete sequencing of the human genome, research priorities have shifted from the identification of genes to the elucidation of their functions. To expedite the functional exploration of the human genome, one needs technological developments to transfect thousands of nucleic acids in parallel and to simultaneously analyze thousands of resulting phenotypes.

Recently, it has been proposed a cheap and flexible cell-based microarray system for high-throughput analysis of gene overexpression *(1)*. Based on this reverse transfection format, a cell microarray was developed and optimized for simultaneous transfection of thousands of different nucleic acid molecules, expression vectors, and small interfering RNA (siRNA) *(2)*. Several groups have now adopted this technology for massively parallel transfection and phenotypic analyses *(3–6)*. The entire process can be divided in four distinct steps that will be described in details (*see* **Fig. 1**): (1) nucleic acid preparation, (2) microarray manufacturing, (3) cell reverse transfection, (4) data acquisition.

Although still in its infancy, this technology seems very promising to help characterize physiological roles of genes of yet unknown functions.

From: *Methods in Molecular Biology, vol. 381: Microarrays: Second Edition: Volume 1: Synthesis Methods*
Edited by: J. B. Rampal © Humana Press Inc., Totowa, NJ

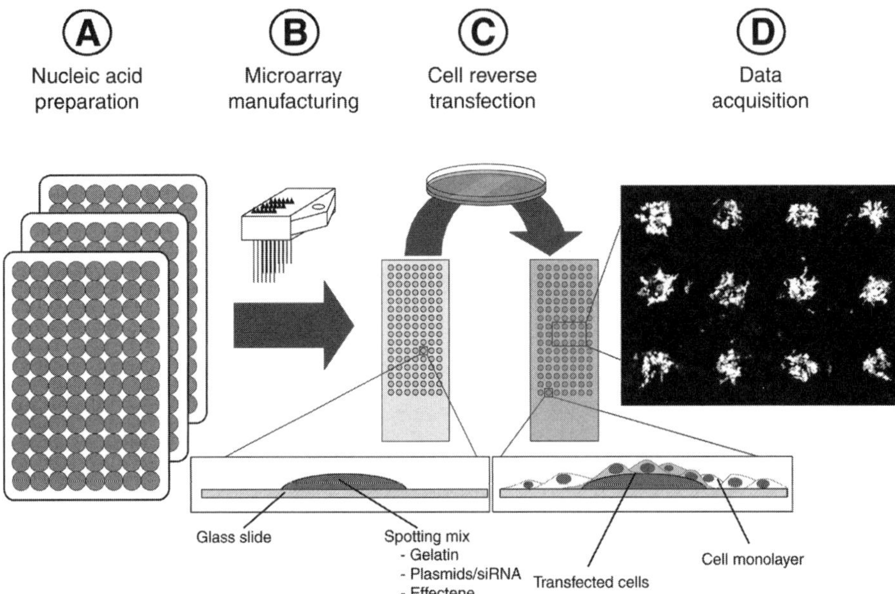

Fig. 1. Cell microarray workflow. (**A**) The genomic resources used, ideally pangenomic collections, can be double stranded RNA (small interfering RNA), or DNA (complementary DNA or open reading frame, cloned in expression vectors). (**B**) Nucleic acids are mixed with polymer and transfection reagents and then printed onto microscope slides using a microarrayer. Each slide can thus contain several thousands features. (**C**) Cells are then seeded onto the slide. A proportion of cells growing in direct contact with each spot are transfected. (**D**) The phenotype of each cluster of transfected cells can then be analyzed. Cells transfected with an expression vector encoding green fluorescent protein appear in white in the picture.

2. Materials

2.1. Nucleic Acid Preparation

2.1.1. Plasmids

1. pEGFP-C1 plasmid (Clontech, Palo Alto, CA), plasmid expressing enhanced green fluorescent protein (EGFP) (*see* **Note 1**).
2. Qiagen Plasmid Midi Kit (Qiagen, Hilden, Germany).
3. TE buffer: 10 mM Tris-HCl, 1 mM ethylenediamine tetra acetic acid (EDTA), pH 8.0. Store at room temperature.

2.1.2. Small Interfering RNA

1. Lyophilized synthetic siRNA specific to Lamin A/C (sense CUGGACUUCCAGA AGAACAdTdT, antisense UGUUCUUCUGGAAGUCCAGdTdT) 5 nM, HPP grade (purity >90%) (Qiagen). Store at –20°C.

2. Synthetic siRNA specific to EGFP, labeled at their 3′-end with rhodamine (sense GCAAGCUGACCCUGAAGUUCAU, antisense GAACUUCAGGGUCAGCU-UGCCG) 5 nM, HPP grade (purity > 90 %) (Qiagen). Store at –20°C.
3. Resuspension buffer: 30 mM HEPES-KOH pH 7.4, 100 mM potassium acetate, 2 mM magnesium acetate.

2.2. Microarraying

Poly-(lysine)-coated microscope slides (Polysine™; Menzel-Gläser, Germany) ($25 \times 75 \times 1$ mm^3). Store at room temperature until use (*see* **Note 2**).

1. Microgrid II arrayer equipped with MicroSpot 2500 pins (Biorobotics, Cambridge, UK).
2. Vacuum desiccators (Nalgene, Neerijse, Belgium) with indicating desiccant (contents: 97% CaS0$_4$, 3% CoCl$_2$) (Drierite, Xenia, OH).
3. Sucrose solution: 1.5 M sucrose (Sigma-Aldrich, St. Louis, MO) dissolved at 60°C in deionized water. Store at 4°C (*see* **Note 3**).
4. 0.5% (w/v) Gelatin solution: 2% (w/v) type B gelatin solution (Sigma-Aldrich, cat. no. G-1393) is diluted to 0.5% (w/v) with sterile deionized water. Aliquote and store the solution at 4°C (*see* **Notes 4** and **5**).
5. 1% (w/v) Gelatin solution: 2% (w/v) type B gelatin solution (Sigma-Aldrich, cat. no. G-1393) is diluted to 1% (w/v) in sterile deionized water. The solution is aliquoted and store at 4°C.
6. Effectene transfection kit (Qiagen) containing EC-Buffer, Enhancer and Effectene (*see* **Note 6**).

2.3. Cell Culture

1. Human embryonic kidney cell line HEK293T (CRL-11268; ATCC, Manassas, VA).
2. Dulbecco's supplemented medium: Dulbecco's modified Eagle's medium from Gibco™ (Invitrogen, San Diego, CA) containing 4.5 g/L glucose supplemented with 10% fetal calf serum (Hyclone, Logan, UT), 100,000 U/L penicillin, 50 mg/L streptomycin, and 200 mM glutamine. Store at 4°C.
3. Solution of trypsin (0.25%) and EDTA (1 mM) (Gibco). Store at –20°C.
4. 100-mm Cell culture dish (TPP, Trasadingen, Switzerland).

2.4. Data Acquisition

1. 4% (w/v) Paraformaldehyde solution: dissolve 40 g of paraformaldehyde (Sigma-Aldrich) in 1X phosphate-buffered saline solution (PBS) by heating at 70°C. Adjust pH between 8.0 and 9.0. Store 15-mL aliquots at –20°C until use (*see* **Note 7**).
2. Microscope cover slips ($24 \times 50 \times 0.15$ mm^3) from Menzel-Gläser (Braunschweig, Germany).
3. Antifade mounting medium: Vectashield® (Vector Laboratories, Burlingame, CA) containing 1.5 µg/mL DAPI (4,6-diamidino-2-phenylindole) for nuclear staining. Store at 4°C.

4. Scanarray® 5000 confocal microarray scanner (Packard Biochip Technologies, Billerica, MA) set up with an external laser for excitation at 490 nm and detection at 508 nm (for EGFP visualization).
5. Upright or inverted fluorescence microscope with a CCD (charge coupled device) camera, Axiocam MRc (Zeiss, Germany) and filter-sets for DAPI (excitation at 345 nm, emission at 455 nm), EGFP (excitation at 489 nm, emission at 508 nm) and rhodamine (excitation at 550 nm, emission at 580 nm) detection.

3. Methods

Plasmid DNA or siRNA is incubated with transfection reagents to promote their transfer into cells. The solution is then mixed with gelatin in order to spatially constrain nucleic acids transfection and to obtain clusters of transfected cells. The mixture is printed onto the slides using the microarrayer.

The spotted array is then covered with a monolayer of cells. Cells in direct contact with the spotted nucleic acid are reverse transfected. About 24–72 h after transfection, phenotypic effects of either gene overexpression or knockdown can be observed on the array. In the present study, EGFP is used to monitor plasmid and siRNA transfection efficiencies. EGFP fluorescence can be readily detected using either fluorescence microscope or standard scanners, and gene extinction is then quantified.

3.1. Nucleic Acid Preparation

3.1.1. Plasmid Preparation

1. The plasmids are purified using Qiagen Plasmid Midi Kit or any protocol allowing purification of high quality plasmidic DNA (*see* **Note 8**).
2. Plasmid concentration is monitored by ultraviolet absorbance; with the optical density of 280/260 nm ratio must be at least 1:8.

3.1.2. Small Interfering RNA Preparation

1. Add 250 μL resuspension buffer to the 5 n*M* siRNA and mix thoroughly. Solubilize siRNAs by heating at 37°C for 1 h. The siRNA solution obtained has a final concentration of 20 μ*M* (~0.3 μg/μL) and is ready for transfection.
2. Aliquot the solution and store frozen until use.

3.2. Microarray Manufacturing

3.2.1. Preparation of Solutions

3.2.1.1. PLASMID SOLUTION

1. Allow an aliquot of gelatin slowly warm up to room temperature (*see* **Note 9**).
2. Dilute 5 μL of pEGFP-C1 at 0.1 μg/μL with 6.5 μL of EC buffer. Mix thoroughly by vortexing or pipetting up and down (*see* **Note 10**).
3. Add 2 μL of Enhancer reagent and mix by pipetting (avoid vortexing). Incubate 5 min at room temperature.

4. Add 2 μL of Effectene reagent to the mixture. Mix gently by pipetting. Incubate 15 min at room temperature.
5. Add 12 μL of gelatin 0.5% and 1.2 μL of 1.5 *M* sucrose solution. Mix by pipetting and transfer to 96-well microplate for microarray printing.

3.2.1.2. PLASMID AND siRNA SOLUTION

Regarding siRNA microarrays, the general procedure is the same except for slight modifications of plasmids, siRNAs, and lipid complex formation.

1. Allow an aliquot of gelatin slowly warm up to room temperature (*see* **Note 9**).
2. Mix 1 μL of pEGFP-C1 at 0.6 μg/μL with 0.5 μL to 2 μL siRNA solution (corresponding to 0.15 to 0.6 μg), specific for EGFP or lamin A/C (*see* **Note 11**).
3. Dilute the mix in 11 μL of EC buffer.
4. Add 3.3 μL of Enhancer reagent and mix by pipetting (avoid vortexing). Incubate 5 min at room temperature.
5. Add 3.3 μL of Effectene reagent to the mixture. Mix gently by pipetting. Incubate 15 min at room temperature.
6. Add 6 μL of gelatin 1% and 2 μL of 1.5 *M* sucrose solution. Mix gently by pipetting and transfer to 96-well microplate for microarray printing.
7. Add 6 μL of a 1% gelatin solution.

3.2.2. Cell Array Printing

The printing is performed with a Biorobotics Microgrid II microarrayer equipped with MicroSpot 2500 pins (*see* **Note 12**).

1. Printing process is conducted in a strictly controlled atmosphere at 19°C and 65% humidity.
2. Average spot diameter obtained with MicroSpot 2500 pins is approx 250 μm.
3. The pitch (i.e., space between centers of two neighboring spots) must be at least twice the expected spot diameter to avoid spot fusion during printing and overlapping of transfected clusters during cell culture (*see* **Note 13**). In our study, 600 features were printed on the array with a pitch of 550 μm (*see* **Note 14**).
4. Before each spotting, pins are cleaned in a heated (60°C) sonication water bath filled with deionized water for approx 15 min.
5. Define a spotting area with a blank zone near edges of the slides, as spotting too close to the edge may be problematic (*see* **Note 15**).
6. During the printing process, when changing of source wells, pins are washed in two successive water baths and then dried in the wash station, before returning to the source plate.
7. We use a pin contact time of approx 50 ms and only a single hit at the same location (*see* **Note 16**).
8. Once the printing is completed, slides are labeled and transferred into a vacuum dessiccator for at least 48 h before use (*see* **Notes 17** and **18**).
9. Array quality control (missing spots, fused spots, regularity of features) can only be performed using a microarray scanner (*see* **Note 19**). Autofluorescence of the

printed mix allows easy visualization of spots, as it can be detected with high laser power and PMT sensitivity.

10. Slides can be stored for weeks in the vacuum desiccator before use (*see* **Note 20**).

3.3. Cell Transfection

3.3.1. Cell Culture and Reverse Transfection

1. Grow human embryonic kidney cell line (HEK293T) in Dulbecco's supplemented medium at 37°C in 5% CO_2 humidified atmosphere. Avoid reaching confluency 24 h before transfection, trypsinize and plate the cells at around 100,000 cells/cm^2 (*see* **Note 21**).
2. Dilute nine million cells in 15 mL of Dulbecco's modified Eagle's medium on the day of transfection, trypsinization and count cells were done.
3. Put the slide face-up in a 100-mm Petri dish and add the cell suspension by pouring the content of the tube onto the array to obtain final density of about 150,000 cells/cm^2 (*see* **Note 22**). (Avoid pouring the cell suspension directly on the slide.)
4. Cells are grown on the array for 24–72 h at 37°C and 5% CO_2 before the analysis of the subsequent phenotypes.

3.3.2. Cell Fixation

1. Remove the slide from the culture Petri dish.
2. Wash the slide once with 1X PBS.
3. Fix the cells on the slide by immersing it in a 4% paraformaldehyde solution for 15 min at room temperature (perform under a hood).
4. Wash the slide once with 1X PBS.
5. Microarrays can be stored at 4°C in 1X PBS for several weeks.

3.4. Data Analysis

Different methods can be used for data acquisition depending on the subsequent analysis envisioned. Common microarray scanner can be used when fast image capture and convenient quantitation of fluorescence is needed. Unfortunately, resolution is often limited to 5 μm per pixel, which might sometimes be too low. When a higher image resolution is essential (e.g., in subcellular phenotype analysis), images can be captured and analyzed either with an upright or inverted fluorescence microscope with a digital camera.

3.4.1. Data Acquisition

3.4.1.1. Scanner Image Acquisition

1. Fix cells on the slide (*see* **Subheading 3.3.2.**).
2. Rinse the slide with deionized water to eliminate remaining PBS.
3. Dry the slide at room temperature and load it into the scanner without any cover slip.

4. Scan the array to detect green fluorescent protein (GFP) with excitation at 488 nm and detection at 508 nm.
5. Fine-tune the scanner to minimize background and avoid saturated signal. Use the maximal resolution available (5 μm in this study).
6. Use any software provided with the scanner for data analysis (*see* **Subheadings 3.4.2.** and **3.4.3.**).

3.4.1.2. MICROSCOPE IMAGE ACQUISITION

1. Fix cells on the slide (*see* **Subheading 3.3.2.** and **Note 23**).
2. Put a drop of Vectashield containing DAPI on the slide.
3. Mount the slide with a cover slip.
4. Visualize stained nuclei using an appropriate filter and capture the image using a CCD camera.
5. Repeat the procedure for EGFP and Rhodamine fluorescence using their respective filters.
6. Overlap images for subsequent analysis (*see* **Subheadings 3.4.2.** and **3.4.3.**).

3.4.2. Assessment of Transfection

The utilization of pEGFP-C1 vector encoding EGFP allows for easy monitoring of transfection efficiency. Indeed, transfected cells exhibit a detectable green fluorescence. In case the images are acquired with a microscope, the ratio of transfected cells to the total number of cells on a spot can be calculated. As each cell nucleus is stained by DAPI coloration, overlapping the image with green fluorescence enables the estimation of percentage of transfected cells.

Alternatively, a microarray scanner can be used. In that case, a visual evaluation of image fluorescence can be first performed by the user. Then, more quantitative fluorescence measurement can be achieved with microarray analysis software.

3.4.3. Assessment of Knockdown

First, to evaluate siRNA transfection, rhodamine fluorescence (green excitation and red emission) of siRNAs is visualized in the cytoplasm of cells using a fluorescence microscope. Then, knockdown effect of siRNAs is assessed with a microarray scanner (as in the case of plasmid transfection, *see* **Subheading 3.4.2.**). For positive control, cells are cotransfected with EGFP specific siRNA and the EGFP coding vector. No green fluorescence should be measured in spots. For negative control, cotransfection is performed with siRNA targeting LaminA/C and the EGFP coding vector. In this case, fluorescence level similar to that observed in cells transfected with EGFP alone should be observed. This is a proof that no unspecific interference with EGFP expression occurred (*see* **Fig. 2**).

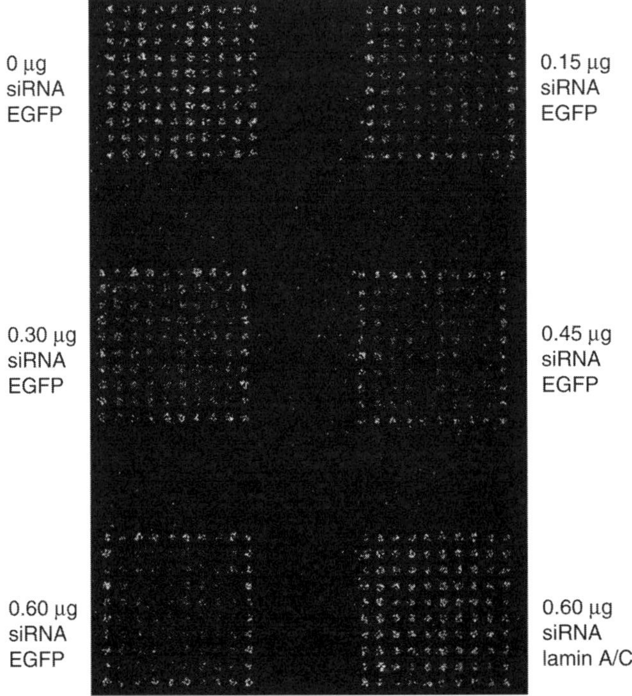

Fig. 2. Highly parallel gene silencing with enhanced green fluorescent protein (EGFP) targeting small interfering RNA (siRNA) cotransfected with the pEGFP-C1 plasmid into HEK293T (*see* **ref. 2**).

Six blocks of pEGFP-C1 (0.6 µg) ± EGFP (or lamin A/C) targeting siRNA were printed on a cell microarray and fluorescence was quantified after 48 h. Each block is made of a 10 × 10 surrounding square of control sample (pEGFP-C1 alone) and a 8 × 8 filled square of spots per sample (pEGFP-C1 [0.60 µg] cotransfected with, from left to right and top to bottom, 0, 0.15, 0.30, 0.45, and 0.60 µg EGFP, siRNA, and 0.60 µg lamin A/C siRNA for the bottom right square). The slide was fixed and scanned after 48 h incubation.

4. Notes

1. Any plasmid encoding a fluorescent protein under control of a strong mammalian promoter (Cytomegalovirus or equivalent) can be used to promote an efficient production of fluorescent proteins in the cells.
2. Slide coating (Superfrost Plus, polylysine, γ-amino propyl silane) used is critical as it interferes with cells' adhesion and viability during culture on the array. The choice of a particular coating depends on the cell type. The advantage of Polylysine treatment is to provide very fast adhesion of a large number of cell types.
3. Sucrose stabilizes DNA–lipid complexes during desiccation. Transfection efficiency is significantly decreased without sucrose.

4. Various gelatin concentrations should be tested to determine the optimum that yields both efficient transfection and spatial restriction.

5. Transfection efficiency is dramatically influenced by the type (types A or B) and forms (powder, ready-to-use solution) of gelatin used. We noticed that gelatin powder was not satisfying.

6. Alternative transfection reagents can be used, but best results are obtained with Effectene on HEK293T. In fact, transfection reagents should be optimized for each cell type as in **ref. *4***.

7. Alternative cell fixation protocol may be considered. However, when using fluorescent protein expression, ice-methanol/ethanol should be avoided as it dissolves fluorescent proteins.

8. Plasmid DNA (as well as siRNA) purity is critical for cell microarray as it is the limiting factor for efficient transfection. A minimum optical density 260/280 nm ratio of 1.8 is required to ensure high transfection efficiency on the array.

9. Heating of the gelatin is not recommended before spotting, in order to get spatial restriction in clusters of transfected cells.

10. Spotting solution has been optimized for HEK293T cells. One has to do so for each cell type. Moreover, the quantity of nucleic acids, as well as the type and amount of transfection reagent should be tested.

11. For cotransfection experiments (plasmidic DNA and siRNAs), it is essential to mix nucleic acids thoroughly before adding the Enhancer and Effectene in order to maximize rates of simultaneous transfection in cells. Otherwise, various populations of cells are obtained in each spots (those up taking one of the plasmids or both of them) resulting in biased interpretation.

12. Pin size and type should be carefully chosen as the quality and size of the features rely on it. Quill pins allow repetitive spots printing without refill, and spots seem to be more even (no donuts). On the other hand, quill pins are subject to clogging and need extensive washing.

13. First, the viscosity of the arrayed mixture leads to larger spots than indicated in pin manufacturer specifications. Second, transfection occurs on a wider area than physical spot size due to plasmid diffusion during cell culture. Because of these two phenomena, a minimal pitch of twice the theoretically expected spot size is recommended.

14. Nevertheless, slides can easily contain up to 5000 spots printed onto a single array. Each spot can be replicated several times, to get multiple measurements for each nucleic acid and allow for statistical analysis.

15. Cell adhesion and growth is altered within 3 mm of slide edges. Concerning the area of the slide near the frosted section, despite cell growth is normal; transfection appears to be compromised and is not restricted to spots.

16. The contact time of printing needle is important as it determines the volume of the mix spotted onto the array. It can be increased if transfection efficiency is low, but this may induce some "leaking" of plasmids outside the spots.

17. Avoid using stickers for slides identification, as this interferes with the cell culture. Also, do not use a permanent marker on slides before fixation; even if no toxicity is observed, it causes unwanted fluorescent background.

18. Arrayed slides should be extensively desiccated before transfection, otherwise transfection will not be restricted to the surface of the spots and cell clusters might overlap, compromising the subsequent analysis.
19. As salt concentration of the arrayed solution is low, dried spots are difficult to visualize, which prevents any quality control of the array.
20. Slides can be stored for several weeks in a vacuum desiccator. Use control spots with GFP to assess slide conservation. If no GFP-expressing cells are seen on the array after transfection, the batch should be discarded.
21. To increase transfection efficiency, pass the cells 24 h before transfection in order to stimulate their proliferation (transfection efficiency is higher with cycling cells).
22. The seeding density mentioned is for HEK293T cells. It should be adapted to each cell type in order to reach confluence after 24–72 h of culture. When cells are seeded too sparsely, the cell monolayer is not dense enough and the cell number is too low on each spot. To the contrary, a highly confluent seeding lowers transfection efficiency. An optimum should be determined regarding cell size as well as doubling time and transfection efficiency.
23. Transfection efficiency can also be visualized before cell fixation using an inverted microscope. This allows to check whether cell density and time of cell culture are satisfactory, and to study nucleic acid transfer kinetics.

References

1. Ziauddin, J. and Sabatini, D. M. (2001) Microarrays of cells expressing defined cDNAs. *Nature* **411,** 107–110.
2. Baghdoyan, S., Roupioz, Y., Pitaval, A., et al. (2004) Quantitative analysis of highly parallel transfection in cell microarrays. *Nucleic Acids Res.* **32,** E77.
3. Kumar, R., Conklin, D. S., and Mittal, V. (2003) High-throughput selection of effective RNAi probes for gene silencing. *Genome Res.* **13,** 2333–2340.
4. Yoshikawa, T., Uchimura, E., Kishi, M., Funeriu, D. P., Miyake, M., and Miyake, J. (2004) Transfection microarray of human mesenchymal stem cells and on-chip siRNA gene knockdown. *J. Control Release* **96,** 227–232.
5. Silva, J. M., Mizuno, H., Brady, A., Lucito, R., and Hannon, G. J. (2004) RNA interference microarrays: high-throughput loss-of-function genetics in mammalian cells. *Proc. Natl. Acad. Sci. USA* **101,** 6548–6552.
6. Mousses, S., Caplen, N. J., Cornelison, R., et al. (2003) RNAi microarray analysis in cultured mammalian cells. *Genome Res.* **13,** 2341–2347.

20

Quantification of Mixed-Phase Hybridization on Polymer Microparticles by Europium(III) Ion Fluorescence

Kaisa Ketomäki and Harri Lönnberg

Summary

A protocol for quantification of oligonucleotide hybridization on polymer microparticles by europium(III) ion fluorescence is described. The procedure involves modification of commercially available amino-functionalized microparticles in such a manner that oligonucleotide probes may be assembled *in situ* on these particles or, alternatively, they may be immobilized postsynthetically. The oligonucleotide-coated particles obtained are then used as the solid phase in a mixed-phase hybridization assay. The efficiency of hybridization is quantified with the aid of oligonucleotides tagged with a europium(III) chelate. Either, the fluorescently tagged probe is hybridized directly to a complementary particle-anchored oligonucleotide, or a sandwich-type assay set up, where a third oligonucleotide complementary both to the tagged and particle-bound probe mediates the attachment to the particles, is exploited. The number of europium(III) ions attached to the solid-phase is determined by the DELFIA® protocol, involving release of the europium(III) ions in solution and development of the fluorescence by addition of an enhancement solution. Alternatively, the fluorescence intensity of the photoluminescent chelate may be measured directly from a single particle.

Key Words: DELFIA®; immobilization; microparticles; mixed-phase hybridization; oligonucleotides; time resolved fluorometry.

1. Introduction

Solid-supported oligonucleotide arrays belong among the most rapidly developing experimental tools for systems biology. Numerous applications, including sequencing, quantification of gene expression, detection of single-nucleotide polymorphism and evaluation of accessibility of RNA to hybridization (*1–8*), have gained interest. Accordingly, understanding of the factors that affect the efficiency and selectivity of hybridization at the interface of the solution and

From: *Methods in Molecular Biology, vol. 381: Microarrays: Second Edition: Volume 1: Synthesis Methods*
Edited by: J. B. Rampal © Humana Press Inc., Totowa, NJ

solid phase is of considerable importance. Oligonucleotide-coated microscopic polymer particles offer a convenient technique to study the underlying principles of mixed-phase hybridization *(9–14)*. In addition, suspension arrays of categorized microparticles in principle allow multiplexed hybridization assays similar to those carried out on oligonucleotide slides *(15,16)*. Oligonucleotide-coated microparticles are rather straightforward to prepare. The desired probe may be assembled by machine-assisted oligonucleotide synthesis on an appropriately derivatized particle *(11)*, or prefabricated probes bearing a suitable functionality may be immobilized postsynthetically *(10,14)*.

Oligonucleotides tagged with either an organic fluorophore or a lanthanide ion chelate may be used for quantification of the hybridization event *(9,17)*. Lanthanide ion chelates, however, exhibit some major advantages over organic prompt fluorophores. Time-resolved fluorometry may be applied, which allows elimination of the short-lived background fluorescence *(17)*. In addition, owing to large Stokes shift, the lanthanide ion chelates does not suffer from the inner filter effect and, hence, the fluorescence emission remains proportional to the concentration over several orders of magnitude *(10–12)*.

The present article describes a protocol for quantification of oligonucleotide hybridization on polymer microparticles by europium(III) ion fluorescence. For this purpose, conversion of commercially available amino-functionalized microparticles to hydroxyl- and mercapto-derivatized particles are first described. Then preparation of oligonucleotide-coated particles is outlined. Finally, preparation of fluorescently tagged oligonucleotide probes is described and a protocol for the hybridization assays is provided.

2. Materials

2.1. Synthesis of Pyridinium (4,4′-Dimethoxytrityloxy) Acetate (1)

1. The following common solvents are required: pyridine (dried by distillation from powdered CaH_2 and stored over 4 Å molecular sieves), toluene, methanol (MeOH; ≥99.8% pure), and dichloromethane (CH_2Cl_2, ≥99% pure) (*see* **Note 1**).
2. The following salts are required: anhydrous sodium sulfate (Na_2SO_4) and sodium bicarbonate ($NaHCO_3$) (*see* **Note 2**).
3. Water purified by reversed osmosis and ion-exchange chromatography (Milli-Q Gradient, Millipore Corporation, Billerica, MA) (*see* **Note 3**).
4. Glycolic acid (Sigma-Aldrich Chemie GmbH, Steinheim, Germany).
5. 4,4′-Dimethoxytrityl chloride (DMTrCl; 97%, Aldrich) (*see* **Note 4**).
6. Sintered glass column (3 × 20 cm^2, porosity 3) for chromatography.
7. Silica gel: 0.040–0.063-mm Fluka Kieselgel 60 (Fluka Sigma-Aldrich. Logistic GmbH, Schnell Dorf, Germany).
8. TLC plates: silica-coated aluminium plate with fluorescent indicator (Merck silica gel 60 F_{254} [Merck KGaA, Darmstadt, Germany]).

2.2. Synthesis of 15-(4,4′-Dimethoxytrityloxy)-12, 13-Dithiapentadecanoic Acid 6

1. Hexane, 1,4-dioxane, ethyl acetate, diethyl ether and acetonitrile (MeCN; high-performance liquid chromatography (HPLC) grade, stored over 4 Å molecular sieves) solvents, in addition to pyridine, MeOH and CH_2Cl_2 (*see* **Subheading 2.1.**, **item 1**).
2. Salts (*see* **Subheading 2.1.**, **item 2**).
3. Aqueous hydrogen chloride (HCl), a concentrated solution (36–38%), and ampules (Titrisol Merck KGaA, Darmstadt, Germany) giving 1 and 0.1 M solutions (*see* **Note 3**).
4. Bis(2-pyridyl)-disulfide (Aldrich).
5. 2-Mercaptoethanol (Riedel-de Haën, Sigma-Aldrich Laborchemi Kalien GmbH, Seelze, Germany).
6. 4,4′-Dimethoxytrityl chloride (DMTrCl), (*see* **Subheading 2.1.**, **item 5**).
7. Thiolacetic acid (Aldrich) (*see* **Note 5**).
8. Dibenzoylperoxide (Merck) (*see* **Note 6**).
9. 10-Undecenoic acid (Fluka).
10. Methylamine, 40% aqueous solution (Aldrich).
11. Triethylamine (Acros Organics, Geel Belgium).
12. Sintered glass column and silica gel (*see* **Subheading 2.1.**, **items 6** and **7**).
13. TLC plates (*see* **Subheading 2.1.**, **item 8**).

2.3. Derivatization of Microparticles for Oligonucleotide-Coating S1,S2

1. Uniformly sized porous microparticles (diameter 50 µm) made of a copolymer of glycidyl methacrylate (40%) and ethylene dimethylacrylate (60%), SINTEF (Trondheim, Norway). The particles are functionalized with primary amino groups (1 mmol/g) obtained by reacting the particle-bound epoxy groups with diethylene triamine. The average pore size is 30 nm, total pore volume 0.822 mL/g and surface area of 137 m²/g.
2. Pyridine, MeOH, CH_2Cl_2, and diethyl ether solvents (*see* **Subheading 2.1.**, **item 1** and *see* **Subheading 2.2.**, **item 1**).
3. N-Hydroxysuccinimide (Aldrich).
4. N,N′-Diisopropylcarbodiimide (DIC) (Aldrich).
5. Dichloroacetic acid (>99%; Aldrich).
6. 1-Methylimidazole (>99%, redistilled) (Aldrich).
7. Acetic anhydride (97%) (J. T. Baker Chemicals B. V., Deventer, Holland).
8. 4-(Dimethylamino)pyridine (Aldrich) (*see* **Note 7**).

2.4. Preparation of Oligonucleotide-Coated Microparticles

1. ABI 392 Oligonucleotide Synthesizer, with standard chemicals (Applied Biosystems, Foster City, CA).
2. Speed vac evaporator (HETO-Holten A/S, Denmark).
3. 2′-Deoxyribonucleoside CE-phosphoramidite building blocks (Glenn Research, Sterling, VA).
4. Supports dA-, dG-, dC-, and dT-CPG 1000 Å for oligonucleotide synthesis.
5. Amino-modifier C6 dT-CE-phosphoramidite (Glenn Research).

6. Aqueous ammonia >33%. Store at 4°C.
7. Buffer for ion-exchange chromatography: pH 5.6 (with phosphoric acid); solution A: 0.05 M KH_2PO_4 in 50% aqueous formamide; solution B: solution A + 0.6 M aqueous $(NH_4)SO_4$ (*see* **Note 8**).
8. SynChropak AX-300 ion-exchange HPLC-column; 250×4.6 mm^2, 6.5 µm.
9. TSKgel G 2000 S HPLC-gel column; 300×7.5 mm^2, 10 µm^2.
10. Dimethyl sulfoxide (Riedel-de Haën).
11. Triethylamine (Acros).
12. N-Succinimidyl 3-(2-pyridyldithio)propionate (SPDP) (Pierce, Rockford, IL).
13. ThermoHypersil ODs RP-HPLC column; 250×4.6 mm^2, 5 µm.
14. NH_4OAc buffers: solution A: 50 mM aqueous NH_4Oac; solution B: 50 mM NH_4OAc in 65% aqueous MeCN.
15. Dithiotreitol (Aldrich).
16. Maleimide (Fluka).

2.5. Preparation of Fluorescently Tagged Oligonucleotides

1. {2,2′,2″,2‴−{[4′−(4‴−isothiocyanatophenyl)−2,2′:6′,2″-terpyridine-6′,6″-diyl]bis(methylenenitrilo)}tetrakis(acetato)}europium(III) (PerkinElmer Life and Analytical Sciences, Wallac, OY).
2. $NaHCO_3$/Na_2CO_3-buffer, 50 mM, pH 9.8–10.0.
3. NAP™ 5 column, Amersham Biosciences (GE Healthcare Europe GmbH, Branch Finland).
4. ThermoHypersil ODS RP-HPLC column (*see* **Subheading 2.4., item 11**).
5. Triethylammonium acetate (TEAA) buffer: pH 7.0; solution A: 100 mM TEAA, solution B: 100 mM TEAA in 65% aqueous MeCN.

2.6. Hybridization Assays

1. Hybridization buffer: 50 mM aqueous Tris-HCl buffer, pH 7.5, containing 0.5 M NaCl and 0.01% Tween-20.
2. Victor^2 luminometer (Perkinelmer Life and Analytical Sciences, Wallacoy, Finland).

3. Methods

3.1. Synthesis of Pyridinium (4,4′-Dimethoxytrityloxy)Acetate 1

1. Dissolve 0.30 g (3.94 mmol) of glycolic acid in 5 mL of dry pyridine and evaporate to dryness. Repeat twice with dry pyridine. Dissolve the oily residue finally to 20 mL of dry pyridine. Add 1.35 g (3.98 mmol) of DMTrCl. Stopper the flask and agitate with magnetic stirrer overnight at room temperature, **Scheme 1**.
2. Evaporate the solution to dryness. Dissolve the residue in 40 mL of toluene and evaporate to dryness. Repeat this twice (*see* **Note 9**).
3. Dissolve the residue in 40 mL of CH_2Cl_2 and wash the mixture three times with 50 mL of saturated aqueous sodium bicarbonate ($NaHCO_3$).
4. Dry the organic phase over anhydrous Na_2SO_4 and filter off the drying agent. Evaporate the solution to dryness.

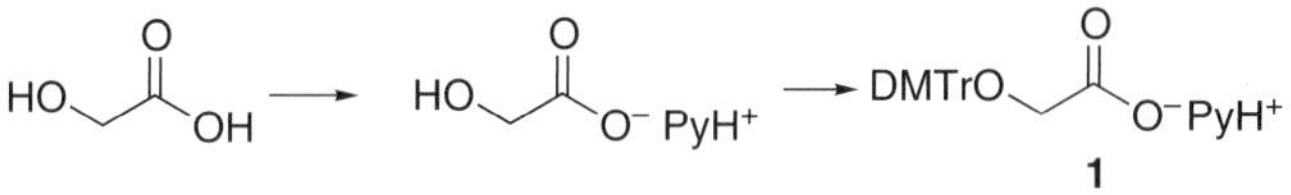

Scheme 1. Preparation of Pyridinium (4,4′-Dimethoxytrityloxy)Acetate **1**.

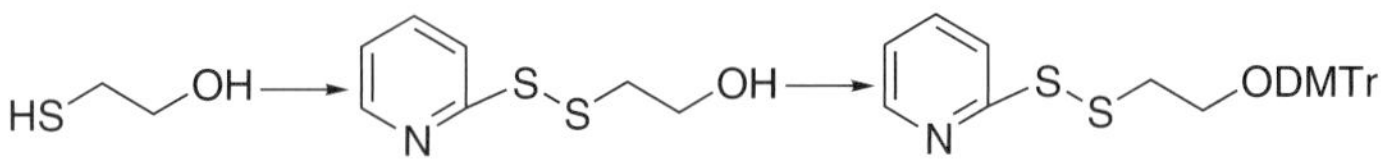

Scheme 2. Preparation of 2-(4,4′-Dimethoxytrityloxy)ethyl 2-Pyridyl disulfide **3**.

5. Purify the residue by silica gel chromatography. Pack a 3×20-cm^2 glass column with silica gel suspended in CH_2Cl_2 containing 1% (v/v) pyridine. Apply the sample onto the column and elute using a stepwise gradient: first with a 1:5:94 (v/v/v) mixtures of pyridine, MeOH and CH_2Cl_2, then with a 1:10:89 mixture of the same solvents and finally with a 1:20:79 mixture.

6. Monitor the fractions by TLC. Combine the fractions containing the product (R_f = 0.44 when a 1:5:94 mixture of pyridine, MeOH and CH_2Cl_2 used as an eluent) and evaporate to dryness. ^{1}H NMR ($CDCl_3$): 6.82–7.44 (13H, DMTr), 3.75 (6H, 2 × OMe), 3.44 (2H, DMTrOCH$_2$). The spectrum also exhibited the signals referring to pyridinium ion.

3.2. Synthesis of 15-(4,4′-Dimethoxytrityloxy)-12, 13-Dithiapentadecanoic Acid 6

3.2.1. Synthesis of 2-(4,4′-Dimethoxytrityloxy)ethyl 2-Pyridyl Disulfide (3)

1. Dissolve 250 mg (1.1. mmol) of bis(2-pyridyl) disulfide in 10 mL of MeOH containing 1% (v/v) pyridine, **Scheme 2**.
2. Add 72-μL (1.mmol) of 2-mercaptoethanol drop wise to the magnetically stirred solution.
3. Stopper the flask and stir the solution for 48 h.
4. Evaporate the mixture to dryness and coevaporate three times with 5 mL of dry pyridine.
5. Dissolve the residue **2** in 10 mL of dry pyridine. Add 375 mg (1.1 mmol) of DMTrCl. Stopper the flask and stir overnight.
6. Stop the reaction by adding 0.5 mL of MeOH and 5 mL of 5% aqueous $NaHCO_3$.
7. Evaporate the solution to oil in vacuum.
8. Dissolve the oil in 40 mL of CH_2Cl_2 and wash it twice with 20 mL of water. Dry the organic phase over anhydrous Na_2SO_4 and filter off the drying agent.
9. Evaporate the solution to dryness.
10. Purify the residue by silica gel chromatography. Pack a 3×20-cm^2 glass column with silica gel suspended in a 3:1 (v/v) mixture CH_2Cl_2 and hexane containing

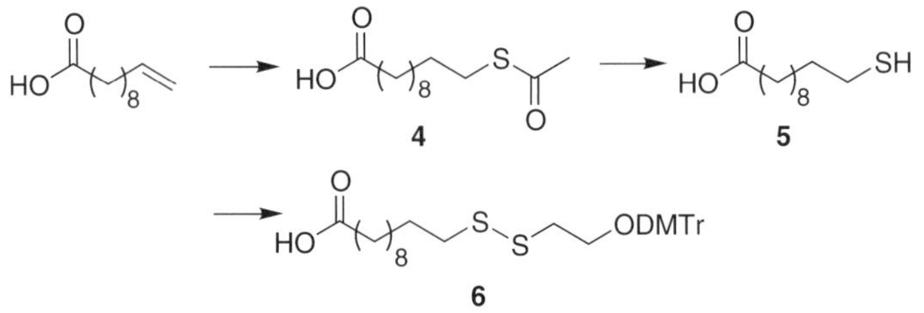

Scheme 3. Preparation of 15-(4,4′-dimethoxytrityloxy)-12,13-dithiapentadecanoic acid.

0.1% pyridine. Apply the sample onto the column and elute using a stepwise gradient from the 3:1 mixture of CH_2Cl_2 and hexane to pure CH_2Cl_2.

11. Monitor the fractions by TLC. Combine the fractions containing the product ($R_f =$ 0.62 when a 49:1 mixture of CH_2Cl_2 and MeOH used as an eluent) and evaporate to dryness. ^1H NMR (CDCl$_3$): 6.78–7.43 (13H, DMTr), 3.77 (6H, 2 × OMe), 3.38 (2H, DMTrOCH$_2$), 2.94 (2H, t, PyrSSCH$_2$). The spectrum also exhibited the signals referring to pyridinium ion.

3.2.2. Synthesis of 15-(4,4′-Dimethoxytrityloxy)-12, 13-Dithiapentadecanoic Acid (6)

1. Dissolve 1 g of thiolacetic acid (13 mmol) in 1 mL of dioxane, **Scheme 3**.
2. Add a few crystals (1–2 mg) of dibenzoylperoxide (*see* **Note 6**).
3. Dissolve 1.21 g (6.6 mmol) of 10-undecenoic acid in 20 mL of dioxane and add the mixture to the thiolacetic acid solution (**steps 1–2**). Stopper the flask and stir overnight at room temperature.
4. Add 20 mL of water and acidify the solution with aqueous HCl.
5. Extract twice with 30 mL of ethyl acetate. Dry over anhydrous Na_2SO_4 and filter off the drying agent.
6. Crystallize the product **4** as a white solid from ethyl acetate.
7. Take 0.5 g of the 11-acetylthioundecanoic acid **4** prepared and dissolves it to 20 mL of MeOH and add 5 mL of 40% aqueous methylamine. This results in instant deacetylation to 11-mercaptoundecanoic acid **5**.
8. Evaporate to dryness.
9. Dissolve the residue **5** and make up to 30 mL of water and adjust the pH to 2.0 with aqueous HCl.
10. Extract three times with 30 mL of diethyl ether. Dry the organic phase over anhydrous Na_2SO_4 and filter off the drying agent.
11. Evaporate to dryness. Take a 0.070-g (0.32 mmol) sample from the residue and dissolve it in 0.9 mL of a 1:9 (v/v) mixture of pyridine and MeOH.
12. Dissolve 0.2 g (0.4 mmol) of 2-(4,4′-dimethoxytrityloxy)ethyl 2-pyridyl disulfide (**3**; *see* **Subheading 3.2.1.**) in 5 mL of a 1:1 mixture of MeOH and MeCN. Add this to the mixture prepared in **step 11**.
13. Stopper the flask and stir the solution overnight at room temperature.
14. Evaporate to dryness.
15. Dissolve the residue **6** in 5 mL of CH_2Cl_2 containing 5% (v/v) of Et_3N.

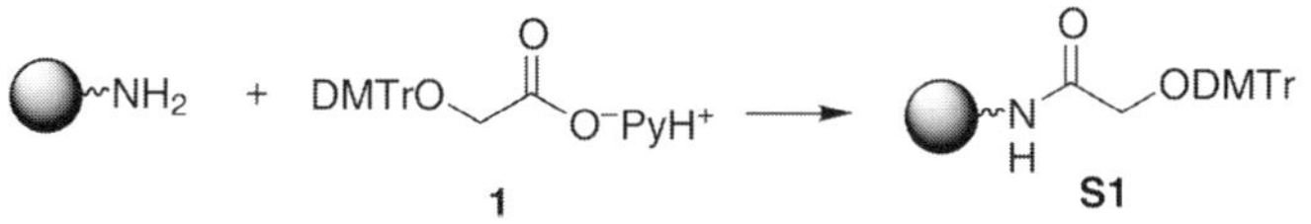

Scheme 4. Preparation of microparticles for *in situ* synthesis of oligonucleotide probes **S1**.

16. Purify the crude product by silica gel chromatography. Pack a 3×20-cm^2 glass column with silica gel suspended in CH_2Cl_2 containing 5% (v/v) of Et_3N. Apply the sample onto the column and elute using a stepwise gradient: first with a 5:95 (v/v) mixture of Et_3N and CH_2Cl_2 and then with the same mixture containing additionally 5% (v/v) of MeOH. ^{1}H NMR (CDCl$_3$): 6.78–7.46 (13H, DMTr), 3.79 (6H, $2 \times$ OMe), 3.35 (2H, OCH$_2$CH$_2$), 2.85 (2H, CH$_2$CH$_2$S), 2.56 (2H, CH$_2$COO), 2.29 (SSCH$_2$), 1.60 (2H, SSCH$_2$CH$_2$), 1.25–1.40 (14H, $7 \times$ CH$_2$). The spectrum also exhibited the signals referring to triethylammonium ion.

3.3. Derivatization of Microparticles for Oligonucleotide-Coating

3.3.1. Microparticles for In Situ Synthesis of Oligonucleotide Probes S1

1. Weight 50 mg of microparticles (50 μmol of amino functions) to a round glass bottle and add 1 mL of dry pyridine, **Scheme 4**.
2. Dissolve 25 mg (217 μmol) N-hydroxysuccinimide in 1 mL of dry pyridine and add the solution to the suspension of microparticles.
3. Add 68 μL (434 μmol) of DIC to the mixture.
4. Dissolve pyridinium-(4,4′-dimethoxytrityloxy)acetate **1** (3.2 mmol, *see* **Subheading 3.1.**) in 3 mL of dry pyridine. Take 200 μL of this solution (contains 217 μmol of **1**) and add it to the mixture prepared in **step 3** (*see* **Note 10**).
5. Agitate the suspension (not with a magnetic stirrer) for 3–5 min (*see* **Note 11**).
6. Filter the microparticles and wash them on the filter with pyridine and MeOH. Dry the microparticles on the filter in vacuum for 2 h (*see* **Note 12**).
7. Measure the 4,4′-dimethoxytrityl-loading of the microparticles by the following procedure. Weight accurately 2–3 mg of microparticles into a 10-mL volumetric flask. Add 10 mL of 3% dichloroacetic acid in CH_2Cl_2. Stir a moment and let the particles settle at the bottom. Measure the absorbance of the 4,4′-dimethoxytrityl-cation released on a UV-VIS (Perkin-Elmer Lambda 2, Global Medical Instrumentation Inc., Ramsey, MN) spectrophotometer at 503.7 nm. Calculate the loading by the following equation.

$$\text{Loading } (\mu\text{mol/g}) = \{[\text{absorbance } (503.7 \text{ nm}) \times V \text{ (mL)}]/76\} \times [1000/\text{mass of particles (mg)}]$$

8. Cap the unreacted amino groups on the microparticles by a 15 min treatment with 7 mL of a 1:5:1 (v/v) mixture of 1-methylimidazole, dry pyridine, and acetic anhydride. Repeat the same treatment twice. Filter the microparticles and wash with pyridine, MeOH and finally with diethyl ether. Dry in vacuum for 2 h. Store the microparticles at 4°C.

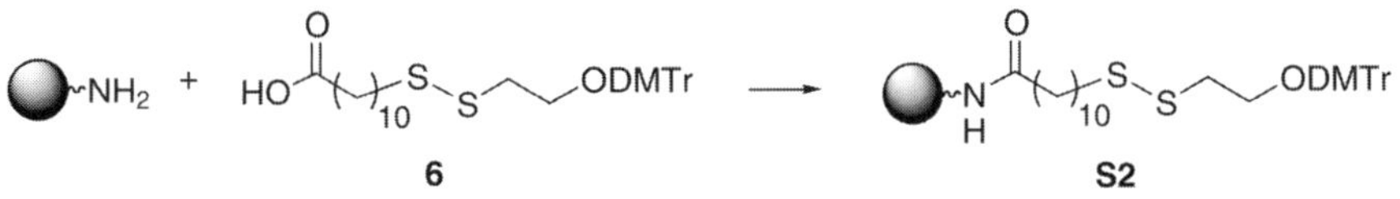

Scheme 5. Preparation of disulfide activated microparticles **S2** for postsynthetic immobilization of mercapto-functionalized oligonucleotides.

3.3.2. Microparticles for Postsynthetic Immobilization of Oligonucleotides S2

1. Weight 50 mg of microparticles (50 μmol of amino functions) to a round glass bottle and add 1 mL of dry pyridine, **Scheme 5**.
2. Dissolve 25 mg (217 μmol) of *N*-hydroxysuccinimide in 0.5 mL of dry pyridine and add the solution to the suspension of microparticles.
3. Add 68 μL (434 μmol) of DIC to the mixture.
4. Dissolve 50 mg of 15-(4,4′-dimethoxytrityloxy)-12,13-dithiapentadecanoic acid (**6**; 85 μmol, **Subheading 3.2.**) in 0.5 mL of dry pyridine and add it to the mixture prepared.
5. Add a catalytic amount of 4-(dimethylamino)pyridine.
6. Agitate the suspension (not with a magnetic stirrer) overnight (*see* **Note 11**).
7. Filter the microparticles and wash them on the filter with pyridine and MeOH. Dry the microparticles on the filter in vacuum for 2 h (*see* **Note 12**).
8. Measure the 4,4′-dimethoxytrityl-loading of the microparticles as described in **step 7**, *see* **Subheading 3.3.1**. Loadings higher than 2 μmol/g are sufficient (*see* **Note 13**).
9. Cap the microparticles as described (*see* **Subheading 3.3.1.**, **step 8**) Store the microparticles at 4°C.

3.4. Preparation of Oligonucleotide-Coated Microparticles

3.4.1. In Situ *Synthesis of Oligonucleotide Probes (7)*

1. Prepare a 0.1 mol/L solution of each 2′-deoxynucleoside CE-phosphoramidite in anhydrous MeCN, **Scheme 6**.
2. Weight the appropriately derivatized microparticles (**S1**, *see* **Subheading 3.3.1.**) for the synthesis column. Use 0.2 μmol scale (*see* **Note 14**).
3. Start the automated solid-phase oligonucleotide synthesis and assemble the desired oligonucleotide chain using standard protocol and 2′-deoxyribonucleoside CE-phosphoramidites. Use the DMTr-OFF mode and monitor the 4,4′-dimethoxytrityl-loadings of the microparticles. Calculate the loading as described (*see* **Subheading 3.3.1.**, **step 7**).
4. After completion of the assembly, remove the column and dry in vacuum.
5. Remove the base moiety protections from the immobilized oligonucleotides by allowing the microparticles to stand in aqueous ammonia (33%) for 48 h at room temperature (*see* **Note 15**). Dry the microparticles in vacuum.

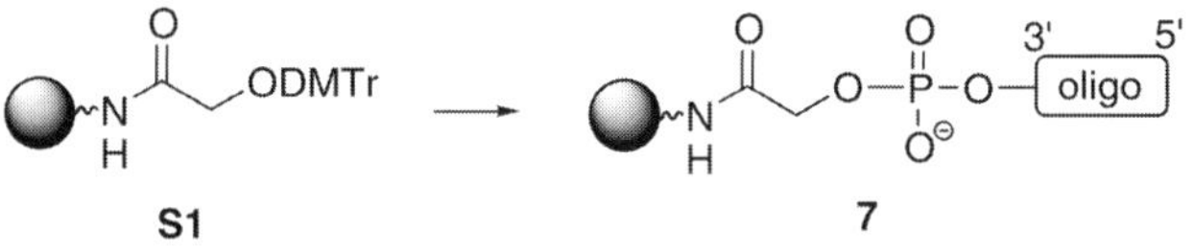

Scheme 6. Oligonucleotide synthesis on microparticles **S1**.

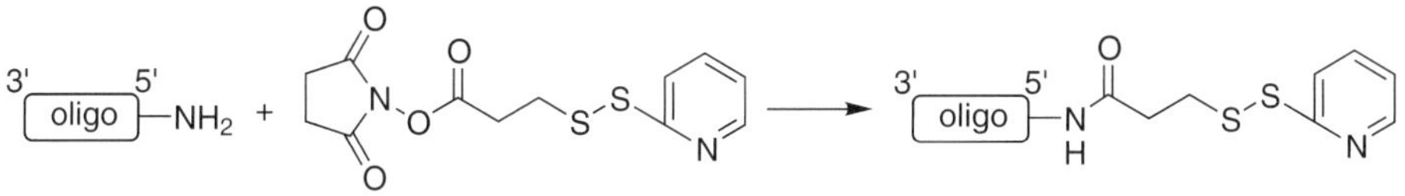

Scheme 7. Conversion of amino-functionalized oligonucleotides to disulfide activated conjugates.

3.4.2. Synthesis of Oligonucleotides for Postsynthetic Immobilization

1. Prepare a 0.1-*M* solution of each nucleoside phosphoramidite in anhydrous MeCN **Scheme 7**.
2. Weight the desired support (either dT-, dA-, dG- or dC-CPG) for the synthesis column. Use 1 µmol scale.
3. Start the automated solid-phase oligonucleotide synthesis and assemble the desired oligonucleotide chain using standard protocol and 2′-deoxyribonucleoside CE-phosphoramidites. Use amino-modifier C6 dT-CE-phosphoramidite as the last building block. Use the DMTr-OFF mode and monitor the 4,4′-dimethoxytrityl-loadings.
4. After completion of the assembly, remove the column and dry in vacuum.
5. Remove the protecting groups from the base moieties and the amino-modifier by incubation in aqueous ammonia (33%) for 16 h at 55°C in silanized glass bottle (*see* **Note 16**). This treatment also releases the oligonucleotides from the support. Evaporate to dryness using a Speed vac evaporator.
6. Add 1.0 mL of water and centrifuge 5 min at 9190*g*.
7. Isolate the oligonucleotide by ion-exchange HPLC. HPLC conditions: SynChropak AX-300 ion-exchange HPLC-column; 250×4.6 mm^2, 6.5 µm. Solution A = 0.05 *M* KH$_2$PO$_4$ in 50% aqueous formamide, solution B, eluent A + 0.6 *M* aqueous (NH$_4$)SO$_4$. Elution is performed using a linear gradient from 0 to 35% solution B in solution A in 35 min (for an ~12 nucleotide long oligonucleotide, R_t about 20 min). For long oligonucleotides, a gradient from 0 to 60% solution B in solution A in 40 min should be used (R_t of a 30-mer ~30 min).
8. Desalt the oligonucleotides on a TSKgel G 2000 S HPLC-gel column; 300×7.5 mm^2, 10 µm, eluting with water at a rate 1 mL/min.
9. Evaporate the desalted solution to dryness with Speed vac evaporator.
10. Dissolve the oligonucleotide in water and quantify by measuring the UV absorbance at 260 nm. Store the solution frozen (*see* **Note 17**).

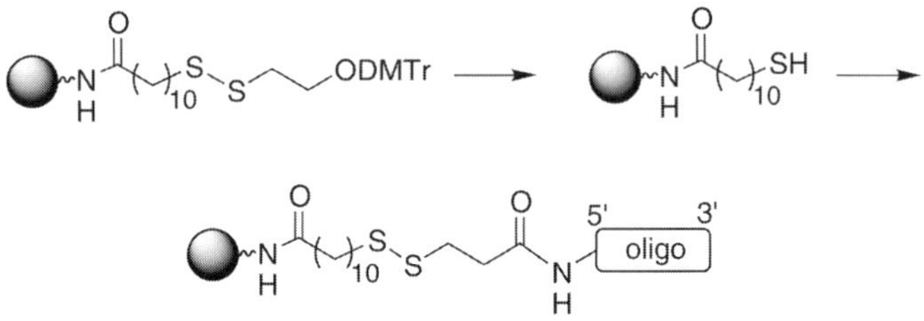

Scheme 8. Postsynthetic immobilization of disulfide activated oligonucleotides to microparticles.

11. Dissolve 100 nmol of the amino-functionalized oligonucleotide in 10 µL of water in an Eppendorf tube. Add 75 µL MeCN and 1 µL Et$_3$N. Add 10 µmol of SPDP in 10 µL of dimethyl sulfoxide (*see* **Note 18**).

12. Rotate the Eppendorf tube overnight with rotamix at 15 rpm.

13. Purify the oligonucleotide conjugate obtained by RP-HPLC. HPLC conditions: ThermoHypersil ODS RP-HPLC column; 250 × 4.6 mm^2, 5 µm, flow rate = 1 mL/min. Solution A = 50 mmol/L aqueous NH$_4$OAc, solution B = 50 mmol/L NH$_4$OAc in 65% aqueous MeCN. Elution is performed applying first a linear gradient from 0 to 25% of solution B in solution A in 30 min and then a similar gradient from 25 to 80% in 30 to 40 min (for an ~12 nt long oligonucleotide, R_t ~30 min).

3.4.3. Postsynthetic Immobilization of Oligonucleotides to Microparticles

1. Weight 10 mg of microparticles derivatized as described (*see* **Subheading 3.3.2. [S2]**) and add 15 mg (0.1 mmol) of dithiotreitol in 1 mL of MeOH. Add 1 µL of Et$_3$N **Scheme 8**.

2. Agitate the suspension overnight (not with a magnetic stirrer).

3. Filter the microparticles and wash them with MeOH and diethyl ether. Dry in vacuum for 2 h.

4. Weight accurately 1–2 mg of particles into an Eppendorf tube. Dissolve the disulfide activated oligonucleotide prepared as described (*see* **Subheading 3.4.2.**) in 200 µL of water (~10 nmol) and measure the absorbance at 260 nm. Add the solution onto the particles in the Eppendorf tube and shake overnight with rotamix.

5. Measure the absorbance at 260 nm after the incubation and calculate the amount of the oligonucleotide attached to the particles. $(n_{initial} - n_{finish})/m$ (particles) = loading (µmol/g).

6. Wash the particles with water and pyridine.

7. Cap the unreacted mercapto groups on the particles by 2 h treatment with 1 mL of 0.5 *M* maleimide in a 1:1 (v/v) mixture of EtOH and pyridine.

8. Wash the particles with pyridine, MeOH, and diethyl ether. Dry in vacuum for 2 h (*see* **Note 19**).

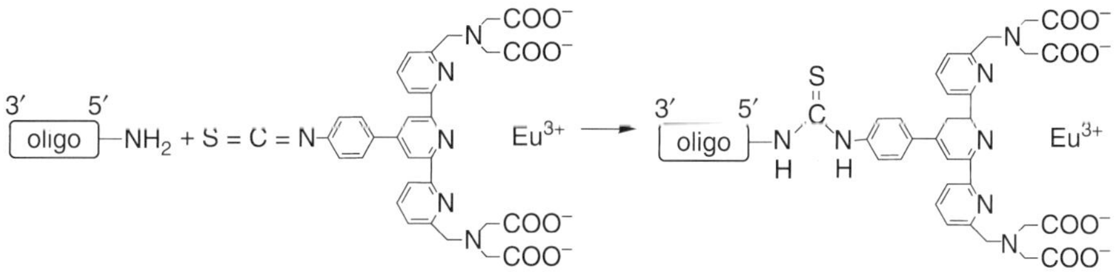

Scheme 9. Fluorescent tagging of oligonucleotides.

3.5. Preparation of Fluorescently Tagged of Oligonucleotides

1. Dissolve approx 1–2 mg of the europium(III) chelate in 20 µL of $NaHCO_3/Na_2CO_3$-buffer, pH 9.8–10.0 **Scheme 9**.
2. Measure the concentration on a UV-VIS spectrometer as follows. Make a 1:1000 dilution of the chelate-solution in water. Measure the absorbance at 260 and 320 nm. $\varepsilon(260\ nm) = 14\ 800$ L/mol/s and $\varepsilon(330\ nm) = 27{,}000$ L/mol/cm.
3. Add appropriate volume of the chelate stock-solution (*see* **Subheading 3.4.3., step 1**) to dry oligonucleotide (50 µg) so that the chelate is present in 12-fold excess.
4. Incubate overnight in dark.
5. Remove the excess of the chelate on a NAP™ 5 column eluting with water.
6. Purify the oligonucleotide fraction by RP-HPLC. HPLC conditions: ThermoHypersil ODS RP-HPLC column; $250 \times 4.6\ mm^2$, 5 µm, flow rate 1 mL/min. Eluent A, 100 m*M* TEAA, eluent B, 100 m*M* TEAA in 65% aqueous MeCN. Elution is performed using a linear gradient from 0 to 40% B in A in 40 min (for an ~17 nucleotide long oligonucleotide).
7. Desalt the oligonucleotide on a TSKgel G 2000 S HPLC-gel column (Tosoh Biosciences GmbH, Stuttgart, Germany); $300 \times 7.5\ mm^2$, 10 µm, eluting with water 1 mL/min.
8. Evaporate the desalted solution to dryness with Speed vac evaporator.
9. Dissolve the oligonucleotide in water and quantify by measuring the UV absorbance at 260 nm. Store the solution frozen. The concentration of the fluorescently tagged oligonucleotide conjugate is calculated by the following equation:

$$\text{Conjugate concentration } (M) = \frac{\text{Absorbance of the conjugate at 260 nm}}{\varepsilon\ (\text{oligo, 260 nm}) + \varepsilon\ (\text{chelate, 260 nm})}.$$

Here, $\varepsilon(\text{chelate, 260 nm})$ is 14,800 L/mol/cm (*see* **item 2**) and $\varepsilon(\text{oligo, 260 nm})$ is calculated from the known molar absorptivities of the base moieties. For this, a program at http://www.gensetoligos.com/Calculation/calculation.html is recommended.

The amount of the conjugate (mol) = Conjugate concentration (mol/L) × volume of the solution (L).

10. Determine the degree of labeling by DELFIA®: Take a known volume (2–5 µL) of the oligonucleotide solution and add 0.1 *M* aqueous HCl (total volume 20 µL).

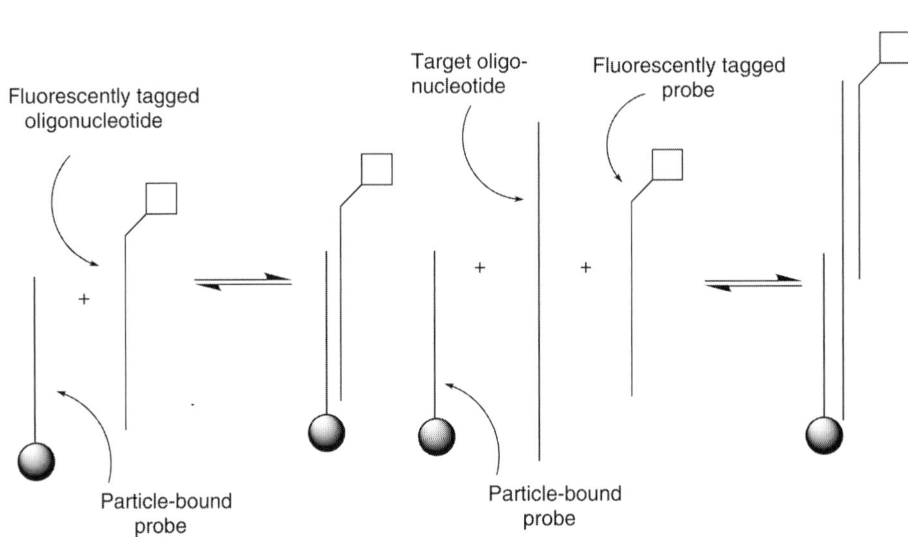

Scheme 10. Hybridization assays.

Incubate for 1 h. Pipet 198 μL of DELFIA® enhancement solution to wells of a 96-plate. Add 2-μL of the oligonucleotide solution to the well. Add to another well 2-μL of the Eu-calibrator (100 nM) solution. Wait for 5 min and measure the fluorescence emission intensities.

11. Calculate the amount of Eu^{3+} ions released from the fluorescently tagged oligonucleotide by comparing the signal to the signal of the calibrator solution. Eu^{3+}-concentration in the well of the oligonucleotide (M) = (Signal obtained from oligonucleotide well/Signal obtained from the Eu-calibrator well on using concentration 1 nM) × 10^{-9}. Concentration of Eu^{3+} in 20 μL oligonucleotide solution = 100 × concentration in the well. The amount (mol) of Eu^{3+} in oligonucleotide solution = concentration in 20 μL × 20 × 10^{-6}.

   Labeling degree = Amount of Eu^{3+} (mol)/Amount of oligonucleotide (mol, *see* **step 9**).

3.6. Hybridization Assays

1. Add 1 mL of hybridization buffer to an Eppendorf tube containing the oligonucleotide-coated microparticles, prepared either by *see* **Subheadings 3.4.1.** or **3.4.3.**, shake the mixture by vortex and centrifuge the microparticles to the bottom at 1020g. Remove the solution and repeat the treatment (*see* **Note 20**) **Scheme 10**.

2. Add a desired volume of hybridization buffer to the tube, and make a dilution to another tube. Use the dilution to count the average number of microparticles in a given volume (preferably 2.5–5 μL) with the aid of a microscope.

3. Pipet a known volume of the microparticle-suspension (and, hence, a known number of particles) to a PCR tube (200–600 µL by volume). Add the fluorescently tagged oligonucleotide probe (*see* **Subheading 3.5.**, from 1×10^9 to 3×10^{11} molecules) in 2.5–5 µL of the hybridization buffer. Add so much of the hybridization buffer that the density of particles reaches the value 5 microparticles in 1 µL of solution. On using a sandwich-type assay, add the target oligonucleotide. The amount of the target oligonucleotide should be equal to that of the fluorescently tagged probe.

4. Mix the solution overnight with rotamix at 15 rpm.

5. Wash the particles with 195 and 200 µL of the hybridization buffer. Collect the removed solutions to a bigger Eppendorf tube.

6. Add 20 µL of 0.11 mol/L aqueous HCl onto the particles. Add 33-µL of 1.0 mol/L aqueous HCl and 66 µL of water to the tube containing the collected hybridization buffers.

7. Shake the tubes in rotamix for 1 h (*see* **Note 21**).

8. Pipet 198 µL of DELFIA® enhancement solution to the wells of a 96-plate. Add a 2-µL aliquot from each tube to a well containing the enhancement solution. Add to one well 2-µL of Eu-calibrator (100 nmol/L) solution. Wait for 5 min.

9. Measure the plate on a Victor² luminometer using manufacture's protocol for Eu^{3+} ion detection: excitation 340 nm, emission 615 nm, delay time 400 µs and decay time 400 µs.

10. Calculate the number of Eu^{3+} ions bound to the microparticles and the number remained in solution upon the oligonucleotide hybridization by comparing the respective signals to the signal of the calibration sample.

Eu^{3+}-concentration in the well of oligonucleotide (mol/L) = [(Signal obtained from the oligonucleotide well)/(Signal obtained from the Eu-calibrator well using a concentration of 1 nmol/L)] $\times$ 10^{-9}. Concentration of Eu^{3+} in 20 µL of oligonucleotide solution = 100 $\times$ concentration in the well. The amount (mol) of Eu^{3+} bound to the particles = concentration in 20 µL $\times$ 20 $\times$ 10^{-6}.

Determination of Eu^{3+} remained in solution: Eu^{3+}-concentration in the well of the oligonucleotide (mol/L) = [(signal obtained from the oligonucleotide well)/(signal obtained from the Eu-calibrator well on using a concentration of 1 nmol/L) $\times$ 10^{-9}. Concentration of Eu^{3+} in 500 µL of oligonucleotide solution = 100 $\times$ concentration in a well. The amount (mol) of Eu^{3+} remained in the solution = concentration in 500 µL $\times$ 500 $\times$ 10^{-6}.

Hybridization efficiency = amount (mol) of oligonucleotide bound to the particles/amount (mol) of oligonucleotide used for the hybridization assay (*see* **Note 22**).

4. Notes

1. Distillation from CaH_2 must be carried out with extreme care. Do not distill to dryness. Destroy the unreacted CaH_2 by adding slowly methanol in the pyridine left undistilled on CaH_2. Dry the molecular sieves at 200°C and let them cool to room

temperature (protected from moisture) before to use. Perform all operations involving organic solvents and reagents in a well-ventilated fume hood, and wear gloves and protective glasses.

2. Dry Na_2SO_4 at 200°C and let it cool to room temperature (protected from moisture) before to use.
3. All aqueous solutions should be prepared in water that has a resistitivity of 18.2 MΩ.
4. Use fresh 4,4′-dimethoxytrityl chloride. An aged reagent may well be partly hydrolyzed.
5. Thiolacetic acid and its derivatives are evil-smelling. Use a well-ventilated fume hood and be very cautious.
6. Dibenzoylperoxide may explode by shock, friction fire, or other sources of ignition.
7. 4-(Dimethylamino)pyridine should be recrystallized from ethyl acetate before use.
8. Dissolve the salts first into water before adding the formamide.
9. Repeated evaporation of the residue from toluene removes pyridine, which otherwise may complicate purification.
10. All the reagents should be mixed fast and (4,4′-dimethoxytrityloxy)acetate should be added last.
11. (1) Mechanical stirring harms the particles seriously. (2) The reaction time affects to the loading; elongation of the reaction time leads to increased loading. A 3- to 5-min treatment gives a suitable loading for oligonucleotide synthesis and hybridization experiments.
12. The microparticles may be dried on the filter when covered by Parafilm. Tiny holes should be done into the parafilm to allow drying.
13. A loading between 2 and 20 µmol/g is optimal.
14. Although 0.2 µmol scale is otherwise used, the amount of particle-bound hydroxyl functions on which the oligonucleotide chains are assembled should, however, be less than 0.2 µmol.
15. The particles do not withstand base moiety deprotection in aqueous ammonia at elevated temperatures.
16. Silanize the vessels according to the manual of the synthesizer.
17. The same procedure, i.e., **Notes 1–10**, is used for preparation of target oligonucleotides needed in a sandwich-type hybridization assay. However, coupling of the amino-modifier (*see* **Note 3** above) is then omitted.
18. Add more MeCN if SPDP is not dissolved.
19. This postsynthetic procedure allows immobilization of the oligonucleotides via the 5′-terminus, while the *in situ* assembly gives 3′-immobilized probes. Accordingly, the two methods are complementary to each other.
20. The microparticles should not be centrifuged faster than at 1020*g*.
21. The release of Eu(III) ions from the TERPY-chelate takes about 1 h in 0.1 mol/L aqueous HCl.
22. The fluorescence emission intensity may also be measured directly from a single particle by microfluorometer, such as the prototype equipment prepared by PerkinElmer Life and Analytical Sciences, Wallac OY *(17)*.

References

1. O'Donnell-Maloney, M. J., Smith, C. L., and Cantor, C. R. (1996) The development of microfabricated arrays for DNA sequencing and analysis. *Tibtechnol.* **14,** 401–407.
2. Niemeyer, C. M. and Blohm, D. (1999) DNA Microarrays. *Angew. Chem. Int. Ed.* **38,** 2865–2869.
3. Epstein, C. B. and Butow, R. A. (2000) Microarray technology–enhanced versatility, persistent challenge. *Curr. Opin. Biotechnol.* **11,** 36–41.
4. Jung, A. (2002) DNA chip technology. *Anal. Bioanal. Chem.* **372,** 41–42.
5. Gerhold, D. L., Jensen, R. V., and Gullans, S. R. (2002) Better therapeutics through microarrays. *Nature Genet. Suppl.* **32,** 547–552.
6. Lockhart, D. J., Dong, H., Byrne, M. C., et al. (1996) Expression monitoring by hybridization to high-density oligonucleotide arrays. *Nature Biotechnol.* **14,** 1675–1680.
7. Czarnik, A. (1998) Illuminating the SNP genomic code. *Modern Drug Discov.* **1,** 49–55.
8. Southern, E., Mir, K., and Shchepinov, M. (1999) Molecular interactions on microarrays. *Nat. Genet.* **21,** 5–9.
9. Fahy, E., Davis, G. R., DiMichele, L. J., and Ghosh, S. S. (1993) Design and synthesis of polyacrylamide-based oligonucleotide supports for use in nucleic acid diagnostics. *Nucleic Acids Res.* **21,** 1819–1826.
10. Hakala, H. and Lönnberg, H. (1997) Time-resolved fluorescence detection of oligonucleotide hybridizations on a single microparticle: covalent immobilization of oligonucleotides and quantification of a model system. *Bioconjugate Chem.* **8,** 232–237.
11. Hakala, H., Heinonen, P., Iitiä, A., and Lönnberg, H. (1997) Detection of olionucleotide hybridization on a single microparticle by time-resolved fluorometry: hybridization assays on polymer particles obtained by direct solid phase assembly of the oligonucleotide probes. *Bioconjugate Chem.* **8,** 378–384.
12. Hakala, H., Mäki, E., and Lönnberg, H. (1998) Detection of oligonucleotide hybridization on a single microparticle by time-resolved fluorometry: quantification and optimization of a sandwich type assay. *Bioconjugate Chem.* **9,** 316–321.
13. Ketomäki, K., Hakala, H., and Lönnberg, H. (2002) Mixed-phase hybridization of short oligodeoxyribonucleotides on microscopic polymer particles: Effect of one-base mismatches on duplex stability. *Bioconjugate Chem.* **13,** 542–547.
14. Ketomäki, K., Hakala, H., Kuronen, O., and Lönnberg, H. (2003) Hybridizations properties of support-bound oligonucleotides: the effect of the site of immobilization on the duplex stability and selectivity of duplex formation. *Bioconjugate Chem.* **14,** 811–816.
15. Fulton, R. J., McDade, R. L., Smith, P. L., Kienker, L. J., and Kettman, J. R., Jr. (1997) Advanced multiplexed analysis with the FlowMetrix™ system. *Clin. Chem.* **43,** 1749–1756.

16. Hakala, H., Virta, P., Salo, H., and Lönnberg, H. (1998) Simultaneous detection of several oligonucleotides by time-resolved fluorometry: the use of a mixture of categorized microparticles in a sandwich type mixed-phase hybridization assay. *Nucleic Acids Res.* **26,** 5581–5588.
17. Lövgren, T., Heinonen, P., Lehtinen, P., et al. (1997) Sensitive bioaffinity assays with individual microparticles and time-resolved fluorometry. *Clin. Chem.* **43,** 1937–1943.

21

Measurement of the Sugar-Binding Specificity of Lectins Using Multiplexed Bead-Based Suspension Arrays

Kazuo Yamamoto, Fumiko Yasukawa, and Seiichiro Ito

Summary

A multiplexed glyco-bead array method for determining the sugar-binding specificities of plant lectins was described by a bead-based flow cytometric assay. Glycopeptides with N- and O-glycans were immobilized on multiplexed beads, and the specificities of several kinds of sugar chains were measured in a single reaction. This strategy is easy, rapid, reproducible, and suitable for small samples, and allows the reliable elucidation of sugar-binding properties of lectins under identical conditions.

Key Words: Flow cytometry; fluorescent microsphere; lectin; multiplexed analysis; sugar specificity; glycopeptide.

1. Introduction

Sugar moieties on the cell surface play one of the most important roles in these cellular recognition including development, embryogenesis, immune recognition, infection, and so on. Intracellular function of carbohydrates attached to proteins is also reported in the folding, degradation, sorting, and transport of glycoproteins within the secretary system, and intracellular signal transduction. Glycosylation, which is one of posttranslational modifications, provides further functions beyond the one-to-one correlation of genetic sequence; therefore, it is necessary to develop a large-scale analysis of glycome, especially elucidation of detailed interaction between carbohydrates and lectins.

Carbohydrate-binding specificity of lectins was studied by conventional hapten-inhibition of haemagglutination using various sugars and sugar derivatives as inhibitors. Other approaches for examining specificities of lectins have been described so far, including equilibrium dialysis (*1*), affinity chromatography on immobilized lectin columns (*2,3*), titration calorimetry using a microcalorimeter (*4*),

From: *Methods in Molecular Biology, vol. 381: Microarrays: Second Edition: Volume 1: Synthesis Methods*
Edited by: J. B. Rampal © Humana Press Inc., Totowa, NJ

and current surface plasmon resonance *(5)*. However, the intra- and extracelular ligands of lectins are sometimes limited in amounts and it makes difficult to elucidate the detailed specificities at the molecular level. Suspension arrays of microspheres analyzed using flow cytometry offer a sensitive, accurate, reproducible approach to a few dozen analyses in solution *(6)*. Luminex microspheres are dye polystyrene microspheres with two spectrally distinct fluorochromes. Using ratios of these two fluorochromes, a spectral array is created encompassing 100 different microsphere sets **(Fig. 1)**. The reactants (antibodies, oligonucleotides, peptides, substrates, antigens, ligands, and so on) are coupled to the surface of uniquely fluorescent beads. A third fluorochrome coupled to a reporter molecule quantifies the extent of binding occurred on the each microsphere surface. Three kinds of fluorescence intensity of thousands of microspheres are measured per second by flow cytometer having a reporter and a classification laser. The currently available microsphere arrays offer a 100-element array and it is almost enough for the analysis of interaction between carbohydrates and lectins.

In this chapter, arrays of microspheres that bear *N*- and *O*-linked glycopeptides derived from several glycoproteins were prepared and demonstrated their binding to plant lectins to identify the widerange of specificity and affinity of these lectins **(Table 1)** *(7)*. This method provides specific, sensitive, simple, reproducible, multiplexed, and rapid assays compared with conventional systems.

2. Materials

2.1. Instrumentation

1. Bath sonicator: for example Bronson (Camberyl, UK), ED-40 (*see* **Note 1**).
2. Microcentrifuge that can spin 1.5-mL microcentrifuge tubes (*see* **Note 2**).
3. Vacuum manifold for a 96-well filter plate (Millipore, Bedford, MA) (*see* **Note 3**).
4. Plate shaker (*see* **Note 4**).
5. Vortex mixer.
6. Luminex 100 cytometer (Luminex Corp., Austin, TX). The luminex 100 cytometer contains two solid state lasers; a reporter laser (532 nm) that excites fluorescent molecules bound to the microsphere surface and a classification laser (635 nm) that excites the fluorochrome on the microsphere.
7. Rotator.
8. Data analysis and graphing software program (e.g., Microsoft Excel, MasterPlex, or Luminex 100 IS [Hitachi Software Engineering America]).

2.2. Coupling of Glycopeptides to Microspheres

1. Glycopeptides for coupling (*see* **Note 5**).
2. Microspheres (polystyrene beads, 5.5 μm in diameter) with a carboxylated surface and different ratios of red and orange fluorescence (Luminex Corp.). Fluorescent microspheres are light sensitive and should be protected from light during storage and usage (*see* **Notes 6** and **7**).

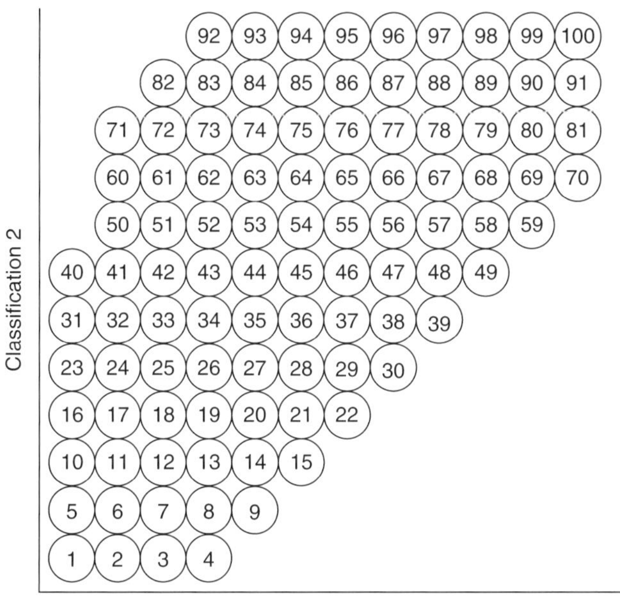

Fig. 1. About 100 different microsphere sets with specific spectral addresses.

3.　Coupling buffer I: 50 mM 2-(N-morpholino)-ethanesulfonic acid, pH 4.5.
4.　EDC: 1-ethyl-3-(3-dimethylaminopropyl)carbodiimide hydrochloride (20 mg/mL in water) (Pierce Biotechnology, Rockford, IL). Make fresh as required.
5.　Wash buffer: 0.15 M Tris-HCl, pH 8.0, containing 0.02% (v/v) Tween-20.
6.　Activation buffer: 0.1 M phosphate buffer, pH 6.2.
7.　Coupling buffer II: 10 mM phosphate buffer, pH 7.4, containing 0.15 M NaCl.
8.　Sulfo-NHS: N-hydroxysulfosuccinimide (50 mg/mL in activation buffer) (Pierce Biotechnology). Prepare fresh as required.
9.　Binding buffer: 10 mM phosphate buffer, pH 7.4, containing 0.15 M NaCl, and 10 mg/mL bovine serum albumin.
10.　PBST: 10 mM phosphate buffer, pH 7.4, containing 0.15 M NaCl, and 0.05% (v/v) Tween-20.
11.　Biotin-labeled lectins (Seikagaku Kogyo, Tokyo, Japan).
12.　R-phycoerythrin (PE)-conjugated streptavidin (Molecular Probes Eugene, CO) (*see* **Note 8**).
13.　96-Well PVDF filter plate: pore size 1.2 µm (Millipore).

3. Methods

3.1. Coupling of Glycopeptides to Microspheres (see Notes 9 and 10)

3.1.1. One-Step Coupling of Glycopeptides to Carboxylated Microspheres

1.　Dispense 2.5×10^6 of carboxylated microspheres into a 1.5-mL microcentrifuge tube.
2.　Centrifuge the microspheres at 18,000g for 1 min.

Table 1
Shows Reactivities of Several Lectins With 13 *N*- and *O*-Linked Glycopeptides From Bovine Thyroglobulin, Porcine Thyroglobulin, Ovalbumin, Bovine Submaxillary Mucin, and Porcine Submaxillary Mucin

Sample	CONT	BTG- A1C(+)	BTG- A1C(−)	PTG UA-I	PTG UA-II	PTG UA-III	OA GP-I	OA GP-II	OA GP-III	OA GP-IV	OA GP-V	OA GP-VI	BSMA1	PSMA2
RCA-120	313	3758	20,675	1819	2646	1610	7907	10,659	2960	3245	3990	494	347	6172
WGA	139	609	2135	86	172	196	682	4329	1886	2808	829	432	292	1187
PHA-E4	289	2260	10,986	375	760	739	1260	2234	2266	3066	1809	460	192	1800
UEA-1	763	621	1276	784	592	693	817	989	791	889	811	862	124	376
DBA	620	457	1588	570	379	521	575	985	770	830	591	768	124	320
LCA	142	901	6474	170	293	239	420	590	539	431	692	170	67	817
LTA	199	225	748	212	172	204	236	387	266	309	253	298	51	115
SBA	364	201	848	277	180	239	279	426	404	409	329	425	114	549
MAN	286	65	695	85	63	82	79	144	173	169	120	161	41	47
ABA	468	711	1483	364	403	373	609	695	543	600	585	322	689	1637
DSA	98	427	5418	420	717	486	2087	3252	2741	2707	1223	441	32	522
GNA	35	144	225	26	25	29	68	121	153	636	937	963	42	62
PHA-L4	62	67	339	93	75	94	102	184	158	158	116	141	20	42
CSA	29	17	55	15	16	19	25	41	46	44	35	18	14	21
PNA	575	391	1224	526	342	467	501	784	664	716	520	629	86	428
NPA	51	170	309	52	45	55	85	132	161	429	672	731	40	82
SSA	1021	777	1802	915	701	832	962	1107	899	956	878	860	288	2828

See **ref.** *7.*
Values are mean fluorescence intensity.

3. Remove the supernatant carefully, without disturbing the pellet.
4. Add 100 µL of coupling buffer I. Vortex and sonicate.
5. Add 50 nmole of glycopeptides in 100 µL of 50 mM 2-(N-morpholino)-ethanesulfonic acid (*see* **Note 11**).
6. Before use, add 1 mL of water to 20 mg of EDC powder (*see* **Note 12**).
7. Vortex immediately after adding 10-µL aliquots of the fresh EDC solution to the microspheres. Vortex immediately.
8. Incubate for 30 min at room temperature with occasional sonication in the dark.
9. Repeat **steps 6–8** three times with fresh EDC.
10. Centrifuge the microspheres at 18,000g for 1 min.
11. Remove the supernatant carefully, without disturbing the pellet.
12. Add 0.2 mL of wash buffer and vortex gently.
13. Centrifuge the microspheres at 18,000g for 1 min.
14. Repeat **steps 11–13**.
15. Remove the supernatant carefully, without disturbing the pellet.
16. Resuspend the microspheres in 100 µL of wash buffer and store in the dark at 4°C (*see* **Note 13**).
17. Count microspheres.

3.1.2. Two-Step Coupling of Glycopeptides to Carboxylated Microspheres (see *Note 14*)

3.1.2.1. ACTIVATION OF MICROSPHERES

1. Dispense 2.5×10^6 microspheres into a 1.5-mL microcentrifuge tube.
2. Centrifuge the microspheres at 18,000g for 1 min.
3. Remove the supernatant carefully, without disturbing the pellet.
4. Wash twice with 200 µL of activation buffer.
5. Resuspend the microspheres in 80 µL of activation buffer.
6. Before use, make a 50 mg/mL sulfo-NHS solution in activation buffer.
7. Add 10 µL of the sulfo-NHS solution to the microsphere suspension and vortex gently.
8. Make a 100 mg/mL EDC solution in activation buffer.
9. Add 10 µL of the EDC solution to the microsphere suspension. Vortex gently.
10. Incubate the suspension for 1 h at room temperature with occasional sonication in the dark.
11. Centrifuge the microspheres at 18,000g for 1 min (*see* **Note 15**).
12. Remove the supernatant carefully, without disturbing the pellet.
13. Add 200 µL of coupling buffer II to the activated microspheres and vortex.
14. Centrifuge the microspheres at 18,000g for 1 min.
15. Remove the supernatant carefully, without disturbing the pellet.
16. Resuspend the microspheres in 100 µL of coupling buffer II and go immediately to the following coupling steps.

3.1.2.2. COUPLING, BLOCKING, AND STORAGE

1. Add 50 nmole of glycopeptides (in 100 µL of water) to the activated microspheres and vortex gently.

 2. Incubate the suspension for 2 h at room temperature with occasional sonication in the dark.
 3. Centrifuge the microspheres at 18,000g for 1 min.
 4. Remove the supernatant carefully, without disturbing the pellet.
 5. Add 0.2 mL of wash buffer and vortex gently.
 6. Centrifuge the microspheres at 18,000g for 1 min.
 7. Repeat **steps 4–6**.
 8. Remove the supernatant carefully, without disturbing the pellet.
 9. Resuspend the microspheres in 100 µL of wash buffer and store in the dark at 4°C (*see* **Note 13**).
10. Count microspheres.

3.2. Multiplexed Bead–Lectin Binding Assay

 1. Prepare a microsphere mixture by diluting the coupled microsphere stocks at a concentration of 10^5/mL.
 2. Dispense 20 µL of microsphere mixture into wells of 96-well filter plate.
 3. Add 100 µL of biotin-labeled lectin solution (10 µg/mL) in binding buffer to each sample well (*see* **Note 16**).
 4. Add 100 µL of binding buffer to a background well.
 5. Mix the reactions gently by pipetting.
 6. Incubate for 1 h at room temperature in the dark on a plate shaker with shaking at 550 rpm.
 7. Aspirate the supernatant by vacuum manifold (*see* **Note 17**).
 8. Wash each well three times with 100 µL of PBST and aspirate by vacuum manifold (*see* **Note 18**).
 9. Dilute the PE-conjugated streptavidin to 2.5 µg/mL in binding buffer.
10. Add 100 µL of the diluted PE-streptavidin to each well.
11. Mix the reactions gently by pipetting.
12. Incubate for 30 min at room temperature in the dark on a plate shaker with shaking at 550 rpm.
13. Aspirate the supernatant by vacuum manifold.
14. Wash each well twice with 100 µL of PBST and aspirate by vacuum manifold.
15. Add 100 µL of PBST each well and mix the reactions gently by pipetting.
16. Analyze 50–75 µL on the Luminex 100 analyzer according to the system manual (flow rate 60 µL/min, bead count 100 each).

3.3. Flow Cytometric Analysis

3.3.1. Calibration, Settings, and Data Collection of Flow Cytometer

 1. Switch on the instrument and warmup for 30 min.
 2. Dispense classification calibration microspheres and reporter calibration microspheres from storage. Vortex the calibration microspheres into each labeled sample tube.
 3. Wrap the sample tubes or plates in aluminum foil to protect the microspheres from light. Allow the samples to warm to room temperature.

4. Flush the cell lines using 1.2 mL of 70% isopropanol or 70% ethanol for the alcohol flush. The alcohol flush removes any air bubbles that may be present.
5. Wash the cell lines using sheath fluid.
6. Load the classification calibration microsphere tube into the sample port and start calibration according to manufacturer's protocol.
7. Wash the cell lines using sheath fluid.
8. Load the reporter calibration microsphere tube into the sample port and start calibration according to manufacturer's protocol (*see* **Note 19**).
9. Make sure the instrument has been warmed up and calibrated.
10. Set the condition for data collection according to manufacturer's protocol (*see* **Note 20**).
11. Load the sample tube into the sample port and start data collection (*see* **Note 21**).
12. Progress of data collection usually can be seen on the histogram or the bar graph.

4. Notes

1. Microspheres should be distributed before use. Gentle vortexing or sonication is the effective method of separating aggregated microspheres.
2. In the coupling protocols, microspheres are separated from reaction mixtures or wash buffer by centrifugation. Microcentrifuge that can pellet microspheres at 18,000g in 1 min.
3. Instead of centrifugation, a vacuum separation method is useful for binding and washing steps. Binding and washing are performed in microtiter filter-bottom 96-well plates and resuspension is accomplished by repetitive pipeting.
4. Agitation with a plate shaker during incubation steps prevents microspheres from settling.
5. Preparation of glycopeptides is usually performed by pronase digestion of glycoproteins followed by gel filtration and ion-exchange chromatography. Free amino groups are essential for coupling.
6. Coupling efficiency of glycopeptides to microspheres is measured by released oligosaccharides from glycopeptide-coupled microspheres, which have been treated with endoglycosidases or N-glycanases. PE-labeled lectins, which can bind to the glycopeptides, are also used to monitor the coupling efficiency.
7. To avoid photobleaching, wrap microsphere containers in foil. Once photobleached, the beads are no longer usable.
8. Alexa532 and Cy3 are also used as reporter fluorochrome.
9. For preparing control microspheres, triethanolamine hydrochloride is used instead of glycopeptides to block charged carboxyl groups.
10. When oligosaccharides are used as ligands, biotin-hydrazide (Dojindo Laboratories, Kumamoto, Japan) is sometimes introduced at reducing termini by reductive amination *(8)*. Biotin-labeled oligosaccharides are then incubated with avidin-coupled microspheres (Lumavidin microsphres; Luminex Corp.).
11. Free amino groups are essential for coupling. If glycopeptides were previously solubilized in an amine-containing buffer (e.g., Tris-HCl, sodium azide, and so on), then followed by desalting is required by gel filtration on a Sephadex G-10

column with water. Coupling efficiency of glycopeptides depends on the concentration of glycopeptides. To optimal coupling concentration for a given glycopeptides is determined by coupling at various concentrations within useable range.

12. Use aliquots immediately and discard containers after use. Make a fresh 20 mg/mL EDC solution before each addition.

13. Microspheres should be protected from light during storage and usage. Freezing conditions and organic solvents should be avoided.

14. EDC coupling agent helps link any carboxyl group with an amino group. Glycopeptides have the amino groups that react with the target carboxyl group on the microspheres. If the glycopeptides have the sialic acid's carboxyl group, the use of two step coupling method is recommended to avoid unwanted reaction.

15. During activation, centrifuged microspheres form a sheet-like layer on the side of the microfuge tube. This formation is normal.

16. It is recommended that samples are filtered before to mix with microspheres.

17. Less than 5 psi is suitable when vacuum aspiration is performed.

18. **Steps 8, 13,** and **14** can be omitted for a no-wash binding assay. This procedure is suitable when the binding reaction is performed in a tube.

19. Calibration microspheres are very concentrated; wash the cell lines with three wash cycles using sheath fluid after calibration.

20. Flow rate, sample volume, and bead concentration should be used according to manufacturer's recommendation.

21. It is recommended that each bead set selected acquires 100 events before sample acquisition is complete. For adequate confidence, 100 collected events per microsphere set in each reaction tube are usually sufficient. If you choose total events instead of events per bead, some bead sets may not have enough number of events.

References

1. Crowley, J. F., Goldstein, I. J., Arnarp, J., and Lonngren, J. (1984) Carbohydrate binding studies on the lectin from Datura stramonium seeds. *Arch. Biochem. Biophys.* **231,** 524–533.

2. Yamamoto, K., Tsuji, T., and Osawa, T. (2002) Affinity chromatography of oligosaccharides and glycopeptides with immobilized lectins, in *Protein Protocols Handbook, 2nd ed.,* (Walker, J. M., ed.), Humana, Totowa, NJ, pp. 917–931.

3. Kasai, K., Oda, Y., Nishikawa, M., and Ishii, S. (1986) Frontal affinity chromatography: theory for its application to studies on specific interactions of biomolecules. *J. Chromatogr.* **376,** 11–32.

4. Gupta, D., Cho, M., Cummings, R. D., and Brewer, C. F. (1996) Thermodynamics of carbohydrate binding to galectin-1 from chinese hamster ovary cells and two mutants. *Biochemistry* **35,** 15,236–15,243.

5. Shinohara, Y., Hasegawa, Y., Kaku, H., and Shibuya, N. (1997) Elucidation of the mechanism enhancing the avidity of lectin with oligosaccharides on the solid phase surface. *Glycobiology* **7,** 1201–1208.

6. Nolan, J. P. and Sklar, L. A. (2002) Suspension array technology: evolution of the flat-array paradigm. *Trends Biotechnol.* **20,** 9–12.

7. Yamamoto, K., Ito, S., Yasukawa, F., Konami, Y., and Matsumoto, N. (2005) Measurement of the carbohydrate-binding specificity of lectins by a multiplexed bead-based flow cytometoric assay. *Anal. Biochem.* **336,** 28–38.

8. Shinohara, Y., Sota, H., Kim, F., et al. (1995) Use of a biosensor based on surface plasmon resonance and biotinyl glycans for analysis of sugar binding specificities of lectins. *J. Biochem.* **117,** 1076–1082.

22

Nanotechnology

Moving From Microarrays Toward Nanoarrays

Hua Chen and Jun Li

Summary

Microarrays are important tools for high-throughput analysis of biomolecules. The use of microarrays for parallel screening of nucleic acid and protein profiles has become an industry standard. A few limitations of microarrays are the requirement for relatively large sample volumes and elongated incubation time, as well as the limit of detection. In addition, traditional microarrays make use of bulky instrumentation for the detection, and sample amplification and labeling are quite laborious, which increase analysis cost and delays the time for obtaining results. These problems limit microarray techniques from point-of-care and field applications. One strategy for overcoming these problems is to develop nanoarrays, particularly electronics-based nanoarrays. With further miniaturization, higher sensitivity, and simplified sample preparation, nanoarrays could potentially be employed for biomolecular analysis in personal healthcare and monitoring of trace pathogens. In this chapter, it is intended to introduce the concept and advantage of nanotechnology and then describe current methods and protocols for novel nanoarrays in three aspects: (1) label-free nucleic acids analysis using nanoarrays, (2) nanoarrays for protein detection by conventional optical fluorescence microscopy as well as by novel label-free methods such as atomic force microscopy, and (3) nanoarray for enzymatic-based assay. These nanoarrays will have significant applications in drug discovery, medical diagnosis, genetic testing, environmental monitoring, and food safety inspection.

Key Words: Atomic force microscopy (AFM); carbon nanotube (CNT); CCD camera; chemical vapor deposition (CVD); CNT nanoelectrode array; electrochemical; enzymatic assay; multiwalled carbon nanotube (MWCNT); nanoarray; nanochip array; nanoelectrode; nanolithography; nanotechnology; nucleic acid analysis; plasma-enhanced chemical vapor deposition (PECVD); protein detection; scanning probe microscopy; self-assembly; single-walled carbon nanotube.

From: *Methods in Molecular Biology, vol. 381: Microarrays: Second Edition: Volume 1: Synthesis Methods*
Edited by: J. B. Rampal © Humana Press Inc., Totowa, NJ

1. Introduction

Microarrays are important tools for high-throughput analysis of biomolecules. The use of microarrays for parallel screening of nucleic acid and protein profiles has recently become an industry standard for drug discovery and biomarker identification. But it requires relatively large sample volumes and elongated incubation time because of the large spot size. This also limits the sensitivity of detection. In addition, traditional microarrays utilize bulky instruments for detection and require extensive efforts in sample amplification and labeling, which increases the cost and time for getting the results. These drawbacks are the biggest impediment to move the nucleic acid- and protein-based detection to point-of-care and field applications. Nanoarrays brought by recent development in nanotechnologies have potential to overcome these limitations. Two types of nanoarrays are discussed in this chapter, i.e., carbon nanotube (CNT) nanoelectrode array and nanochip array *(1–10)*. CNT nanoelectrode arrays are produced by reducing the electrode size to <100 nm using carbon nanotubes. This type of array has been demonstrated for ultrasensitive nucleic acid analysis by label-free electrochemical (EC) detection *(1–3)*. Nanochip arrays, on the other hand, are fabricated by reducing the spot size of the traditional microarray to the nanometer scale using nanolithography techniques such as dip pen nanolithography *(4–8)*. Nanochip arrays have been demonstrated for ultrasensitive protein detection in small volume by conventional optical fluorescence microscopy and also using novel label-free methods such as atomic force microscopy, eliminating the need for specialized optical readout instrumentation *(9,10)*. Both types of arrays have also been applied to enzymatic assays *(11,12)*. These characteristics suggest that nanoarrays will be advantageous over microarrays for many applications in biomedicine and biotechnology.

1.1. What is Nanotechnology?

The concept of nanoscience arose in a lecture by the physicist Feynman in 1959 *(13)*, and the remarkable possibilities of nanotechnology in molecular manufacturing were discussed by Drexler in 1986 *(14)*. After Iijima discovered the cylindrical structure of carbon nanotube in 1991 *(15)*, nanoscience and nanotechnology have undergone great advances recently.

Nanotechnology is the creation of useful materials, devices, and systems through the manipulation of matter on this miniscule scale. The emerging field of nanotechnology involves scientists from many different disciplines, including physicists, chemists, engineers, and biologists. Nanotechnology is in a broad sense the science and engineering concerned with the design, synthesis, characterization, and application of materials and devices that have a functional organization in at least one dimension on the nanometer scale, ranging from a

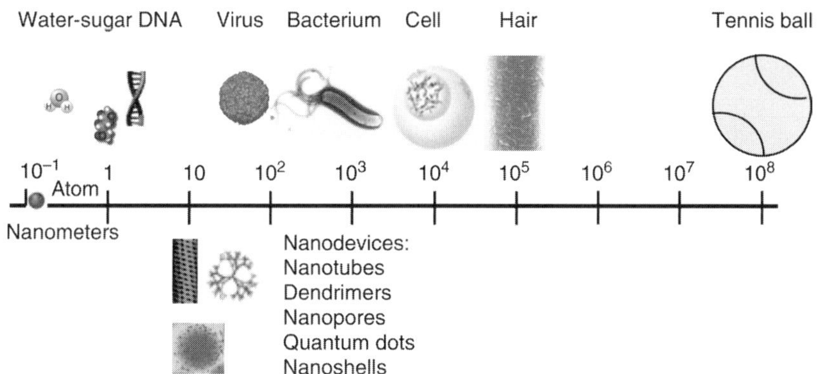

Fig. 1. Size of nanomaterials, molecules and live organisms.

few to about 100 nm. A nanometer is a billionth of a meter as illustrated in **Fig. 1**. It is difficult to imagine anything so small, but think of something only 1/80,000 the width of a human hair. Ten hydrogen atoms could be laid side-by-side in a single nanometer. A nanometer is roughly the size of a molecule itself (e.g., the width of duplex DNA is about 2.5 nm, whereas, the diameter of a sodium atom is about 0.2 nm) *(16–18)*. There are many interesting nanodevices being developed for both diagnostics and therapeutic applications. For example, quantum dots have been demonstrated for cancer diagnostics, CNT and nanowire-based field-effect-transistor for biosensors, and dendrimers for drug delivery *(1,19–22)*. The development of nanoarrays is focused in this chapter.

1.2. Why Nanoarray?

Nanomaterials offer significant advantages over conventional methods in biodetection regarding sensitivity, specificity, speed, portability, throughput, and cost because of their (1) small size (1–100 nm) and correspondingly large surface-to-volume ratio; (2) chemically tailorable physical properties, which directly relate to size, composition, and shape; (3) unusual target-binding properties; and (4) overall structural robustness *(20)*. **Figure 2** summarizes the three types of arrays. Conventional microarray spot sizes are typically 200 μm in diameter (*see* **Fig. 2A**). In nanoelectrode array platform, individually addressed ultramicrospot are about 20×20 μm², consist of tens or hundreds of nanoelectrodes. Each nanoelectrode has a diameter of 50–100 nm, which are laid in insulating materials such as SiO_2 with only the end being exposed. The average spacing between neighboring nanoelectrodes is over 1500 nm to avoid the overlap of the hemispherical diffusion layer around each individual

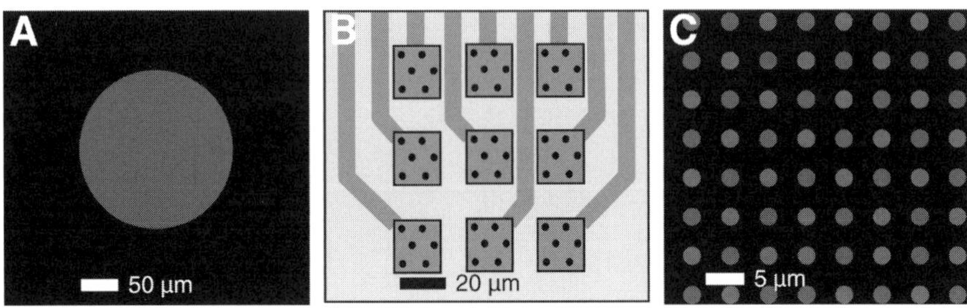

Fig. 2. Three types of array (**A**) microarray, (**B**) CNT nanoelectrode array, and (**C**) nanochip array.

one (**Fig. 2B**). In nanochip arrays spot sizes are typically 0.25–1 μm^2 with pitch >2 μm (*see* **Fig. 2C**).

1.3. CNT Nanoelectrode Array

CNT technology is ideal as a nanoelectrode array platform. A CNT is a cylindrical form of carbon, configurationally equivalent to a two-dimensional graphene sheet rolled into a tube (*23,24*). A CNT might consist of a single wall, known as a single-walled carbon nanotube, or multiple concentric cylindrical grapheme walls, known as multiwalled carbon nanotube (MWCNT). The outside diameter of a MWCNT varies from a few nanometers to about 200 nm and the length varies from a few microns to hundreds of microns. The physical dimension of MWCNTs is ideal for fabricating nanoelectrodes. MWCNT shows highly conductive metallic properties. The open end is an ideal electrode similar to graphite edge plane, while the sidewall is very inert similar to graphite basal plane. The difference in electron transfer rate at these two sites differs by five to six orders of magnitude (*25*). This makes MWCNT an ideal nanoelectrode, which can pick up the signal at the tip and transfer it to the measuring unit connected at the other end with minimum interference by the surrounding environment. With proper EC etching or acid treatment, the open end of MWCNTs contains –COOH and –OH groups, which can be used for attaching biomolecules through the formation of amide bonds (*26*).

The sensitivity of an electrode is mainly determined by its *signal-to-noise ratio*. The noise is the background current mainly from the capacitive charging/discharging current at the electrode/electrolyte interface and thus proportional to the surface area (A) of the inlaid electrode, i.e.,

$$i_n \propto C_d^{\,0} A \tag{1}$$

where $C_d^{\,0}$ is the specific capacitance at the interface. In most common voltammetry measurements, the magnitude of the peak current of the redox signal is the sum of two terms: a linear diffusion as described in the Cottrell equation, and a nonlinear radial diffusion *(27)*

$$i_{1,peak} = nFAC_0 * \sqrt{\frac{D_0}{\pi t}} + nFAC_0 * \left(\frac{D_0}{r}\right) \tag{2}$$

where n is the number of electrons involved in the reaction with one electroactive species, F is the Faraday constant, C_0 is the concentration of electroactive species, D_0 is the diffusion coefficient, t is time, and r is the electrode radius. As the radius is decreased below 25 μm, the second term increases and dominates the signal, $i_{1,peak}$, which is typically referred to as ultramicroelectrodes (UME) *(28)*. In this regime, the signal-to-noise ratio follows the equation

$$i_{1peak}/i_n \propto nFC_0D_0/r \tag{3}$$

Clearly, if the electrode size is reduced from 200 μm to 200 nm, *the signal-to-noise will be improved by 1000 times*. At the same time, the electrode also *responds 1000 times faster*, similar to the fact that smaller transistor features afford faster computer speed.

It is clear that the performance of an electrode with respect to the temporal and spatial resolution scales inversely with the electrode radius. Therefore, the sensitivity can be dramatically improved by reducing the size of the electrodes to nanoscale. Even a single redox molecule can be detected using a Pt-Ir electrode with 15 nm diameter *(29)*. With the diameter approaching the size of the target molecules, nanoelectrodes can interrogate biomolecules much more efficiently than conventional electrodes. For practical applications, hundreds of nanoelectrodes are desired to be attached to each microelectrode as a local nanoelectrode array in order to increase the statistical reliability. From our previous work *(2)*, each individual electrode will behave as an independent nanoelectrode as long as the average spacing between the neighbors is approx >1 μm. The total signal summed over all nanoelectrodes on the same local microspot will increase the signal amplitude so that it will not be interfered by instrumental and environmental noise. The EC behavior of such low-density nanoelectrode arrays or ensembles is similar to a single nanoelectrode, which is fundamentally different from the high-density nanoelectrode ensemble fabricated with a templated method by Menon *(30)*. The diffusion layer of each nanoelectrode in a high-density array/ensemble will overlap giving similar behavior as a solid macroelectrode.

The nanoarray format also has a second advantage in improving the efficiency of target capture and kinetics of the target/probe hybridization. Peterson has demonstrated that 100% of probes can be hybridized following Langmuir-like

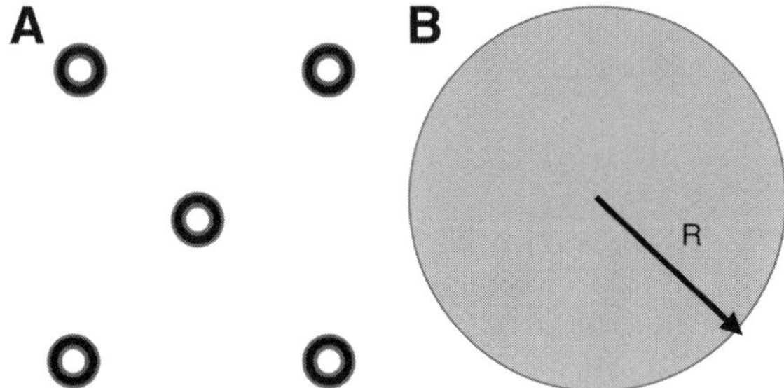

Fig. 3. Schematic of the relative amount of the probes at the edge (in red) vs those at the high density area (in black). **(A)** At the end of an inlaid MWCNT nanoelectrode array and **(B)** at a microspot of a conventional DNA microarray.

kinetics at a low probe density (2×10^{12} probes/cm^2) and only about 10% probes are hybridized with a high probe density (12×10^{12} probes/cm^2) with much slower kinetics *(31,32)*. Better mismatch discrimination is obtained by reducing the probe density. This observation can be as a result of the steric hindrance from the neighboring molecules. In nanoelectrode array platform, the efficiency and kinetics for the hybridization of target molecules faces the same problem, particularly if long single-stranded DNA or mRNA are the target molecules *(33,34)*. Because of the small size and geometry of each nanoelectrode, a large portion of capture probes is well-exposed at the edge and can be easily accessed by the target molecules. The hollow structure of MWCNTs forms an additional inner circular edge. Approximately, as high as 27% of the probes are in the 2 nm wide edge at the end of an inlaid MWCNT nanoelectrode with the outer radius of $R_{out} = 25$ nm and inner radius of $R_{in} = 10$ nm **(Fig. 3)**. In contrast, the percentage of probes at the edge is negligible in conventional DNA microarrays. Therefore, the nanoelectrode platform would naturally be more efficient for hybridization and mismatch discrimination.

For electronic nano-chip technology, a key advantage is that a large local electrical field can be applied right at the tip of nanoelectrodes on each microspot. Because DNA and RNA molecules are negatively charged, the electric field might enhance or retard hybridization and denature surface immobilized DNA duplexes. Discrimination between fully matched and mismatched hybrids can be achieved by programming each microsites with a specific electrode potential *(35–37)*. The electrostatic force generated by the local electric field coupled with global temperature control provides much better stringency control over the whole chip. The electrode potential can be further programmed into biphasic pulses, which is synchronized with the fluid flow to accelerate the process for the right target to hybridize onto the electrode.

1.4. Nanochip Array

Nanochip arrays are obtained by printing nanometer size spots on surface using dip pen nanolithography technology *(4–8)*. The surface area of a single nanoarray spot is about several thousand times less than that of conventional microarray, with estimated volumes of approx 30 attoliters/spot. Therefore, the amount of analyte in one microarray spot (100 µm in diameter) is adequate for fabrication of many thousands of nanoarray spots. This suggests that nanoarrays are compatible with picoliter volumes of analyte, which is consistent with the volumes of a single cell *(9–10)*. The small volume requirement of nanoarrays enables sensitive detection for less analyte such as protein extracts generated from just a few cells of interest harvested by laser capture microdissection from a cancer biopsy.

According to Einstein's square root law of Brownian motion *(38)*, $< |r|^2 > = 6\,D\,t$, the distance of molecular diffusion, r, will be proportional to the square root of the time, t, and therefore, a modest reduction in volume will yield a significant decrease in diffusion time. A reduction in array size statistically ensures that every analyte molecule will have an opportunity to sample the entire capture surface in a reasonable amount of time. A traditional microarray utilizing tens or hundreds of microliters of sample spread out across several centimeters of capture domains results in the effective wasting of the majority of the sample because it will never be in close proximity to a relevant capture domain unless additional energy is added to enhance the movement of the analyte (e.g., agitation, mixing, or electromotive force). In typical static 2-h incubation, each microarray spot theoretically samples not more than a 1.7-mm radius half-sphere. In contrast, a multiplexed nanoarray in a tiny microfluidic chamber measuring 100×100 µm^2 could expose each capture domain to every analyte molecule in a matter of minutes. The small size of array with small volume of sample enables fast hybridization and efficient target capture *(6,9,10)*.

Because of the small size, nanoarrays are more accessible optically than microarrays as the entire nanoarray field can be read in a single optical microscopy image with conventional optical fluorescence microscopy or by novel label-free methods such as atomic force microscopy, eliminating the need for specialized optical readout instrumentation *(6,9,10)*.

2. CNT Nanoelectrode Array for Nucleic Acid Analysis

CNTs belong to a family of materials consisting of seamless graphitic cylinders with extremely high aspect ratios *(39–45)*. The typical diameter varies from about one nanometer to hundreds of nanometers and the length spans from tens of nanometers to hundreds of microns or even centimeters.

As a result of the intriguing nanometer-scale structures and unique properties, CNTs have quickly attracted intensive attention in the past few years in many fields such as nanoelectronic devices *(46–51)*, composite materials *(52)*, field-emission devices *(53,54)*, atomic force microscope probes *(55–58)*, and hydrogen/lithium ion storage *(59,60)*. Many studies reported potential for ultrahigh sensitivity sensors as well *(51,61–64)*. The extremely high surface-to-volume ratio of CNTs is ideal for efficient adsorption. The one-dimensional quantum wire nature makes their electronic properties very sensitive to gas or chemical adsorption. Both of these factors are essential for high-sensitivity sensors.

In the past few years, CNT sensors have been demonstrated in many applications involving gas molecules, liquid-phase chemicals, to biomolecules with improved performance compared with conventional sensors utilizing micro- or macromaterials and thin films *(65)*. This section summarizes the recent progress in the fabrication of CNT nanoelectrode for rapid, ultrasensitive, labelless nucleic acid analysis using nanotechnologies.

2.1. The Fabrication of CNT Nanoelectrode Array

CNTs are critical components of the sensing devices, which are integrated either directly or indirectly during the fabrication routes. A variety of methods, ranging from advanced micro- or nanolithography techniques to handmade processes, have been demonstrated by various researchers to build functional devices. Generally, CNTs could be the sensing elements whose properties are changed on occurrence of the biological events or by the transducers that transfer the changed signal to the measuring units. The sensing device could use a single CNT or an ensemble of such materials.

An innovative bottom-up approach based on vertically aligned MWCNTs was developed to fabricate CNT nanoelectrode arrays *(2,65–67)*. This technique is compatible with semiconductor processes and can be employed to mass produce nanoelectrode array chips. The processing flow is illustrated in **Fig. 4** and the details are described next in six steps:

1. A microelectrode array with individual addressing lines is fabricated on a Si chip using standard ultraviolet (UV)-lithography patterning techniques.
2. A thin layer (~10–30 nm) of catalyst metal such as Ni is deposited at the microelectrode site to define the location for MWCNT growth.
3. Arrays of vertical aligned MWCNTs are grown at the catalyst covered sites with plasma-enhanced chemical vapor deposition (PECVD) *(68–70)*.
4. A SiO_2 layer is deposited by tetraethoxysilane chemical vapor deposition (CVD) to encapsulate all MWCNTs as well as the metal lines.
5. A chemical mechanical polishing or etching step is then applied to remove excess SiO_2 and expose only the very ends of MWCNTs.

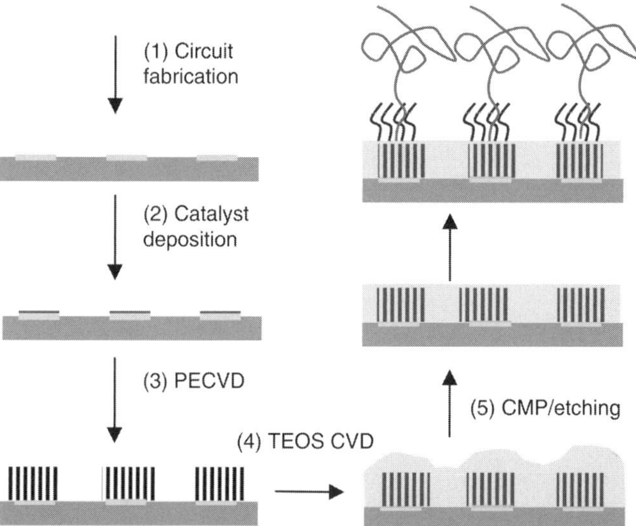

Fig. 4. Schematic of the processing flow for fabricating MWCNT nanoelectrode array Nano-chips for DNA/RNA analysis. MWCNT nanoelectrode arrays are connected to individually addressed microelectrodes on a Si chip in an array-in-array format.

6. Such nanoelectrode arrays are then functionalized with specific oligonucleotide probes by forming amide bonds. Further hybridization and washing steps are carried out using the similar protocols for conventional DNA microarrays.

As a first step for demonstrating the feasibility of multiplex electronic Nanochip, we designed a 3 × 3 array and successfully fabricated well-defined MWCNT nanoelectrode arrays using the procedure described in **Fig. 4**. **Figure 5A,B** shows scanning probe microscopy images of a 3 × 3 array of individually addressed electrodes on a Si (100) wafer covered with 500 nm thermal oxide. The electrodes and contact lines are 200 nm thick Cr patterned with UV-lithography. Each electrode can be varied from 2 × 2 to 200 × 200 μm^2, consisting of a vertically aligned MWCNT array grown by PECVD from 10–30 nm thick Ni catalyst films. **Figure 5C,D** shows MWNT arrays grown on 2 μm and 200 nm diameter Ni spots defined by UV- and e-beam lithography, respectively. The spacing and spot size can be precisely controlled. The diameter of the MWCNTs is uniform over the whole chip and can be controlled between 30 and 100 nm by PECVD conditions. The number of CNTs at each spot can be varied by changing the thickness of the Ni film. Single nanotubes can be grown at each catalyst spot if their size is reduced to <100 nm. The tetraethoxysilane CVD process produces a conformal SiO_2 film, which encapsulates each nanotube and the substrate surface, resulting in a mechanically stable and well-insulated matrix so that aggressive polishing or etching processes can be applied.

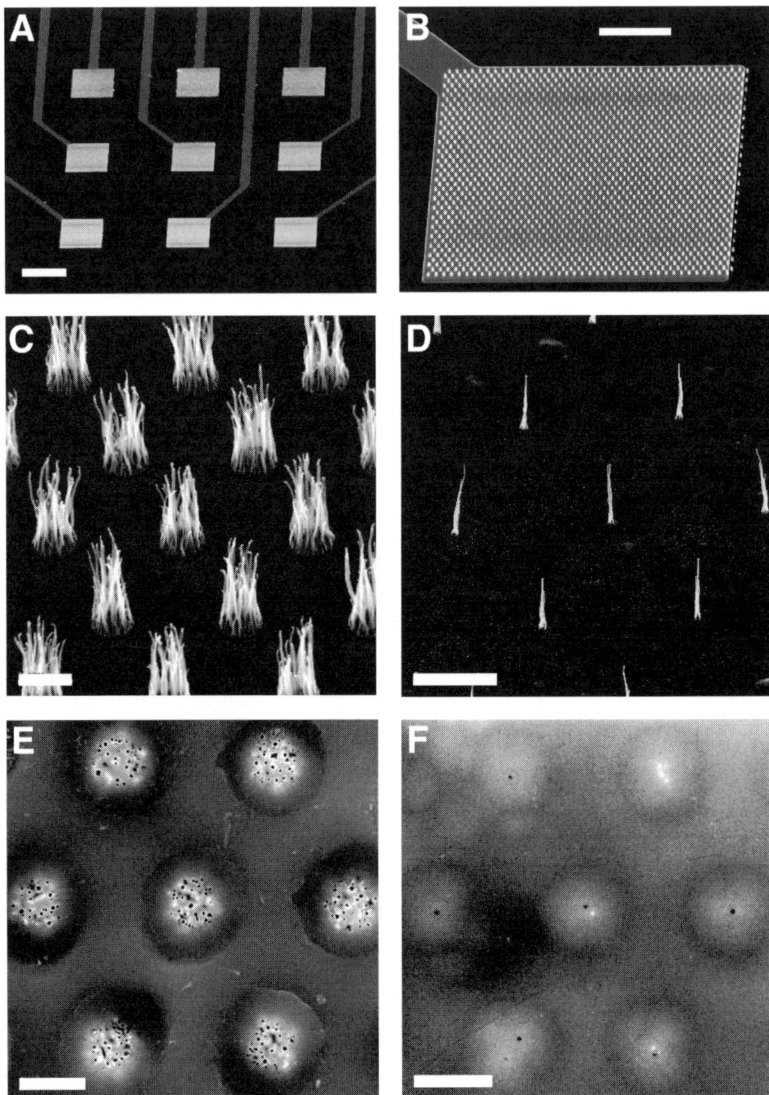

Fig. 5. Scanning probe microscopy images of **(A)** a 3 × 3 electrode array, **(B)** an array of MWNT bundles on one of the electrode pads, **(C)** and **(D)** array of MWNTs at UV-lithography and e-beam patterned Ni spots, respectively, **(E)** and **(F)** the surface of polished MWNT array electrodes grown on 2 μm and 200 nm spots, respectively. **(A)–(D)** are 45° perspective views and **(E–F)** are top views. The scale bars are 200, 50, 2, 5, 2, and 2 μm, respectively.

Figure 4E,F shows the top-view of MWCNT nanoelectrode arrays inlaid in SiO_2 after polishing from the UV and e-beam patterned samples, respectively. Clearly, CNTs retain their integrity and are separated from each other. The CNTs in as polished samples normally protrude over the SiO_2 matrix by about 30–50 nm because of their high mechanical resilience. An EC etching method is employed to shorten CNTs and level the tips approximately in the surface plane of the SiO_2 matrix. The MWCNTs grown by our PECVD technique present a bamboo-like structure with a series of closed shells along the tube, which seal off the hollow channel, leaving only the very end accessible by electrolytes.

2.2. Functionalization of Oligonucleotide Probes

A common feature for biological sensing is the requirement of immobilization of biomolecules with specific functionalities to the sensing device. These biomolecules serve as probes to specifically either bind particular species in the testing sample or catalyze the reaction of the analyte. Such an event produces a change in chemical or physical properties that can be converted into a measurable signal by the transducer. The specific recognition of the target molecules is the essential feature for biological sensing. The common probe and target (analyte) recognition mechanisms include: (1) antibody/antigen interactions, (2) nucleic acid hybridizations, (3) enzymatic reactions, and (4) cellular interactions. Depending on the device and its sensing mechanism, different functionalization methods have to be adapted.

The oxides at the CNT ends can mostly be converted to carboxylic groups using EC etching. A ferrocene (Fc) derivative, Fc $(CH_2)_2NH_2$, can be selectively functionalized at the tube ends through amide bonds facilitated by the coupling reagents dicyclohexylcarbodiimide (DCC) and N-hydroxysuccinimide (NHS) *(71)*. Similar carbodiimide chemistry can be applied to functionalize primary amine terminated oligonucleotides using water-soluble coupling reagents 1-ethyl-3-(3-dimethyl aminopropyl)carbodiimide hydrochloride and N-hydroxysulfo-succinimide as shown in **Fig. 6**. A probe [Cy3] 5′-CTI-IATTTCICAIITCCT-3′ [AmC7-Q] and a target [Cy5] 5′-AGGACCTGC-GAAATCCAGGGGGGGGGGGGG-3′ are used, which are related to the wild-type gene of *BRCA1* Arg1443 stop *(72)*. Fluorescence images of Cy3 and Cy5 labels are taken with a laser scanner after each step of functionalization, hybridization, and washing to confirm the attachment. The Cy3 and Cy5 images after hybridization and washing obtained from a functionalized spot of 3 mm in diameter on a piece of MWCNT nanoelectrode array chip with a size of ~1 × 1 cm². Clearly, the hybridization only occurs at the spot with probe

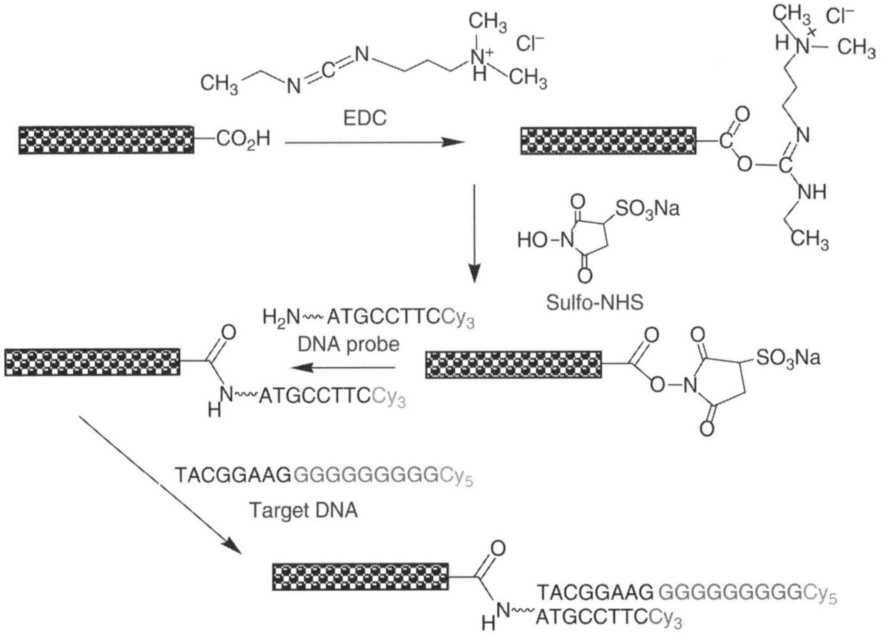

Fig. 6. The schematic of functionalization of oligonucleotide probe to the end of a CNT through the formation of amide bonds.

molecules, indicating that the target binds to the surface only through specific hybridization with the probe. Nonspecific binding was removed by washing procedures.

2.3. Nanoelectrode Array-Based EC Detection

It has been reported that the guanine bases in DNA can be oxidized at about 1.05 V *(73)*. This oxidation reaction can be used for DNA detection *(74)*. EC detection of an oligonucleotide target was first demonstrated as shown in **Fig. 7B**. A 10 bp polyG tag was attached to the sequence fully matched with the probe, which serves as the signal moiety. The guanine bases in the probe are replaced with inosine to ensure no guanine contribution from the probe. To further amplify the guanine oxidation signal, a mediator Ru $(bpy)_3^{2+}$ was introduced, which can transfer electrons from the guanine bases dangling near the electrode surface as schematically illustrated in **Fig. 7**. Combining the advantages of nanoelectrode array, guanine oxidation, mediator amplification, and AC voltammetry altogether, extremely high sensitivity of EC detection has been achieved. The net AC voltammetry data is presented in **Fig. 8A**. If scaling down with a MWCNT array on a 20×20 μm^2 micro-contact, we estimated that $<10^6$ *oligonucleotide target* attached to this spot

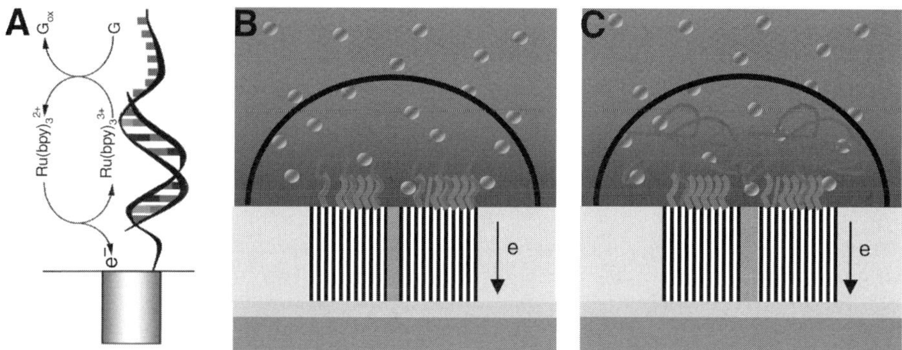

Fig. 7. The schematic of (**A**) the mechanism of $Ru(bpy)_3^{2+}$ mediator amplified guanine oxidation, (**B**) the detection of oligonucleotide targets, and (**C**) PCR amplicon targets.

could be easily detected. It needs to be noted that this number is the maximum estimation, whereas the real number of target DNA that were detected could be orders of magnitude smaller.

Direct EC detection of *PCR amplicons* using CNT nanoelectrode arrays has also been demonstrated *(3)*. In this case, the abundant inherent guanine bases are directly used as the signal moieties without employing a signal tag labeling process as fluorescent techniques do. The same *BRCA1* gene probe was employed. But the target was changed to about 300 bases normal allele within *BRCA1* gene containing 5′-AGGACCTGCGAAATCCAG-3′, which is complementary to the specific oligonucleotide probe. Genomic DNA from a healthy donor was used in PCR amplification. Considering the physical size (radius of gyration of ~6 nm for 300 base ssDNA), it is expected that not more than ~70 targets can hybridized with the probes on each MWNT of ~100 nm diameter. The nonspecific binding is removed by stringent washing with 2X SSC/0.1% SDS, 1X SSC, and 0.1% SSC at 40°C for 15 min, respectively. The major part of the target molecules likely dangles near but might not be necessarily be in direct contact with the electrode surface (**Fig. 8C**). Ru $(bpy)_3^{2+}$ mediators can efficiently transport electrons from the guanine bases to MWCNT nanoelectrodes to provide an amplified guanine oxidation signal as long as target DNA molecules are within the three-dimensional diffusion layer (typically a hemisphere with a radius of ~300 nm, i.e., ~6 R_{ave}, where R_{ave} is the average nanoelectrode radius). Because the approx 300 base long PCR amplicon statistically contains about 75 inherent guanine bases, about the same level of net EC signal was obtained (as shown in **Fig. 8B**) compared with the 10-mer polyG tagged oligonucleotide target even though the number of PCR amplicon target are much less. The sensitivity for detecting *<1000 target molecules* can be achieved at a 20×20 µm² microcontact. In contrast, target samples that underwent the

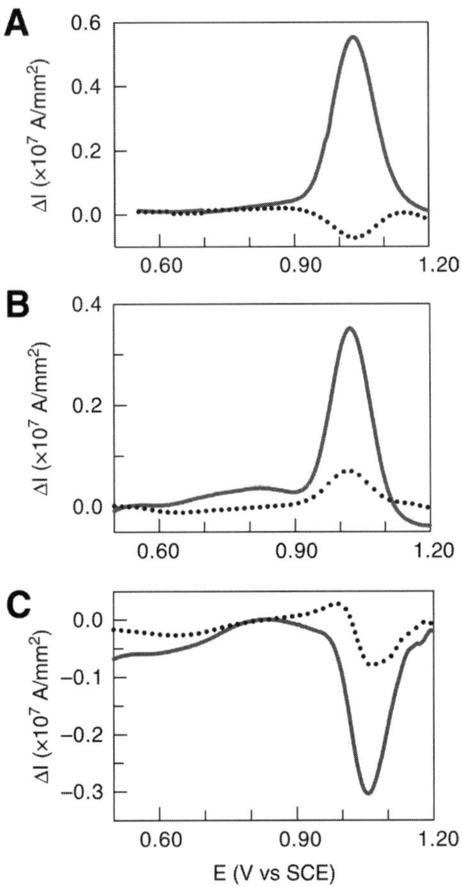

Fig. 8. The net AC voltammetry signal in 5 mM Ru(bpy)$_3^{2+}$ and 0.20 M NaOAc (pH 5.2) with an AC sinusoidal wave of 10 Hz and 25 mV amplitude on top of a staircase DC ramp after incubation in **(A)** specific polyG tagged oligonucleotide target, **(B)** specific PCR amplicon, and **(C)** random PCR amplicon. The horizontal dashed lines indicate the maximum background noise.

same incubation condition, except were unrelated about 400 bases PCR amplicon, give net negative EC signal (as shown in **Fig. 8C**). The normalized net EC signal from a clean MWCNT nanoelectrode array without probe oligonucleotide has shown a reproducible background noise as marked by the dashed lines in **Fig. 8**. The hybridization of specific targets (whether oligonucleotides **[Fig. 8A]** or PCR amplicons **[Fig. 8B]**) generates a distinct positive signal over the noise, while the absence of hybridization gives a distinct negative signal **(Fig. 8C)**. Such high reliability has not been reported by using other nanotechnology-based DNA sensor yet.

One big advantage of the CNT nanoelectrode array platform is that it can be integrated with electrical field stringency control and microfluidics to improve the accuracy and hybridization speed. Controlled electric fields are known to be able to regulate the transport and concentration of biomolecules *(75)*, and control the hybridization and denaturization of DNA molecules *(35,76)*. Such electric methods can be readily applied to MWCNT nanoelectrode array to speed up the hybridization. Ideally specific electric field can be applied at each individual microspot to gain specific stringency control for each probe. This makes the screening of single-nucleotide polymorphism (SNP) much more precisely *(77,78)*, particularly in highly multiplexed analysis, which is critical for nucleic acid analysis.

3. Nanochip Array for Protein Detection

The advantages of ultra miniaturization and multiplexing enable one to use smaller sample volumes and achieve greater sensitivity and quicker analysis owing to the small size of entire array and the individual features that comprise the array. There is significant development in use of nanolithography for patterning surfaces with proteins on the submicrometer length scale to create protein nanoarrays. Each 1 μ spot in these array-based assays covers less than 1/1000th of surface area of a conventional microarray spot while still maintaining enough antibodies to provide a useful dynamic range *(9–10,15)*. Nanoarray-based protein assay exceeds the limit of detection of conventional enzyme-linked immunosorbent assay-based immunoassays by >1000-fold *(10)*. This could lead to the discovery of other useful diagnostic biomarkers that currently escape detection. This section summarize the practical methods and protocols for protein nanoarrays, especially for the surface modification and antibody immobilization that are the key factors in determining the overall success of a nanoarray.

3.1. Nanoarrayer

The basic principle of the NanoArrayer is mechanically mediated direct deposition of materials on surfaces with high spatial precision, surface sensing capabilities and environmental control. Briefly, a microfabricated deposition tool is loaded by immersing the distal end in a drop of the protein solution. The surface to be patterned is positioned on a piezoelectric inchworm-driven XY stage with 20 nm resolution over 25 mm of travel (Burleigh exfo, New York). Manipulating the local humidity at the interface between the deposition tool and surface with gentle bursts of wet or dry air allows precise control of the molecular transfer rate. All stage movement and patterning parameters are controlled with a custom software package called NanoWare™ (Phoenix, AZ) *(5,6,9,10)*.

3.2. Surface Modification

The use of gold-based substrates for monitoring protein interactions has a relatively long history. Instead of using polycrystalline gold surfaces, thin gold films (10–200 nm) can easily be deposited on solid supports similar to glass or silicon wafers by evaporation or sputtering in an ultrahigh vacuum *(79)*. Because the adhesion of gold on glass or silicon is poor, precoating of a thin layer of Cr or Ti (1–10 nm) is generally required ahead of gold deposition.

Since the work of Allara *(80,81)* and Whitesides *(82,83)* in the mid-1980s, the spontaneous adsorption of organosulfur compounds onto gold has been a widely used method for the preparation of self-assembled monolayers (SAMs). If the alkyl chain is of sufficient length, the resulting molecular architecture is very stable and orients itself nearly perpendicular to the surface *(84,85)*. The close proximity of the adsorbed molecules causes van der Waals forces to play an important role in intermolecular stabilization. The thickness of these monolayers is typically in the nanometer range in the vertical direction and as a result, they can be termed nanoassemblies. Unlike glass- or polymer-based arrays, SAMs on gold also have the added advantage of the surface properties being efficiently characterized by optical, mechanical, and EC analytical tools.

A major limitation to the use of SAMs as an immobilization matrix is their low temperature stability resulting from the weak bonding (interactions) between sulfur and gold. Although this is not a concern in the case of protein arrays (where experiments are conducted at low temperatures), such nanoarrays are found to be unsuitable for DNA applications where highly stringent washing and temperature conditions are employed.

Efficient immobilization of biomolecules is a key factor in determining the overall success of a nanoarray or, for that matter, any high-throughput screening tool. If the immobilized probes are not correctly oriented on the microarray surface or are denatured, it can dramatically affect the downstream biomolecular interaction events. **Figure 9** summarizes three types of SAMs for antibody immobilization. Protein A/G surface is very useful for properly orienting antibodies and boosting specific net target binding, because protein A/G is the mixture of protein A and Protein G that specifically bind immunoglobulin (Ig)G Fc and leave IgG F(ab)$_2$ open for antigen capture as shown in **Fig 9A**. Protein A and protein G have different binding affinity to IgG Fc of different species. The combination of protein A and protein G work well for all types of IgG (antibody). Salicylhydroxamic acid (SHA) binds IgG functionalized with 1,3-phenyldiboronic acid (PDBA) while exhibiting very low nonspecific binding to target protein without a PDBA functional group as shown in **Fig. 9B**. Once formed, SHA–PDBA complex is stable to a wide variety of conditions. PDBA-IgG conjugates can be released from these SHA-affinity surfaces in a

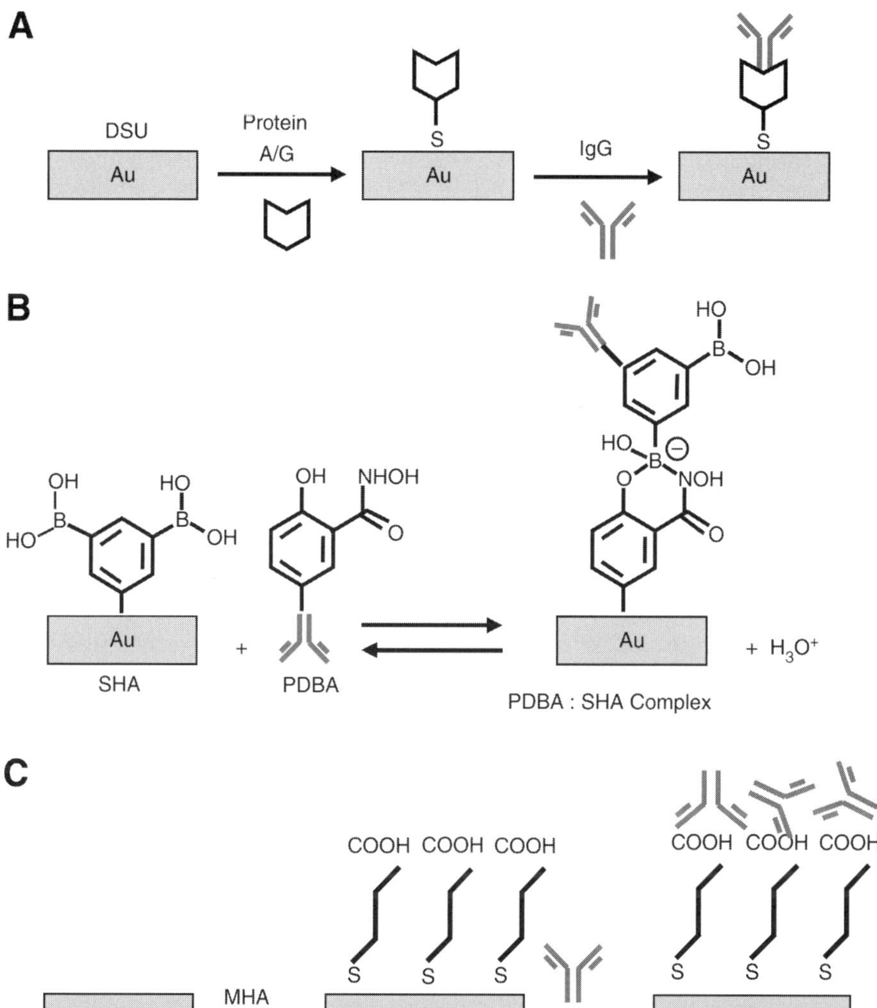

Fig. 9. Schematic of surface modification and functionalization of (**A**) Protein A/G chip, (**B**) SHA-SAM chip, (**C**) MHA chip.

controlled manner that allow separation of IgG captured protein for the further analysis. The 16-mercaptohexadecanoic acid (MHA) is deprotonated at pH 7.4, and therefore the nanofeatures are negatively charged as shown in **Fig. 9C**. Antibody will adhere to negatively charged surface features through electrostatic interactions, thus retaining its biological activity toward its targets. MHA-based chips have robust two-dimensional structure and less steps for antibody immobilization. This antibody immobilization is simple, but not directional

(5,6,9,10,86–89). The following descriptions summarize the three common surface modification and functionalization protocols.

3.2.1. Protein A/G Chips

1. Sputtered gold chips are immersed in alkanethiolate solutions (0.5 mM dithiobis-succinimidyl undecanoate (DSU) *[89]* in 1,4-dioxane) to allow SAM formation overnight.
2. Wash the chips in dioxane and dry the chips with 100% ethanol.
3. The chips are blown dry with argon and allowed to react with 0.1 mg/mL protein A/G (Pierce) in phosphate buffered saline (PBS) for 30 min at 25°C.
4. Unreacted succinimide groups are blocked with 10 mM Tris-HCl, pH 7.5, 1 M glycine for 30 min at 25°C.
5. Without washing, chips are transferred into StabilGuard (SurModics, Eden Prairie, MN) protein stabilizing and blocking agent and incubated for 1 h at 25°C.
6. Chips are then rinsed briefly with a few drops of double-distilled water, dried with argon, and stored in desiccators at 4°C before use.

3.2.2. SHA SAM Chips

1. Fresh gold-coated silicon wafers are incubated in 5 mM SHA alkanethiolate SAM reagents and ethanol as directed by the manufacturer's protocol (Versalinx™, Prolinx, Bothell, WA) *(87)*.
2. Chips are sonicated and washed successively in ethanol and double-distilled water before use.

3.2.3. MHA Chips

1. Pattern MHA into dot features on a gold thin film.
2. To minimize nonspecific binding of proteins on the inactive portions of the array, the areas surrounding the MHA patterned features are passivated with PEG-alkythiol (11-mercaptoundecyl-tri[ethylene glycol]) by placing a droplet of a 1 mM ethanolic solution of the surfactant on the patterned area for 2 h.
3. Rinse the chips with ethanol and, then, Nanopure water.

3.3. Antibody Immobilization and Hybridization

Proper antibody immobilization and background blocking are critical for the highly specific and sensitive detection of target proteins. Here summarizes the common antibody immobilization and hybridization protocol for these three chips.

1. The capture antibody are immobilized by arrayed deposition of IgG onto a protein A/G chip, PDBA-labeled IgG on SHA SAM chip, or immersing MHA chip in a solution containing IgG at 0.2–1 mg/mL in 10 mM PBS for 1 h.
2. The chips are stored overnight in a humid environment for 1 h at 25°C or overnight at 4°C to facilitate binding of antibody to activated surface.

3. Wash chips with 500 µL of PBS + 0.1% Tween-80 (PBST) for 5 min and double-distilled water for 5 min to remove salts and detergent.
4. Add prehybridization buffer containing 5 mg/mL bovine IgG (Sigma St. Louis, MO) in PBST to block remaining binding sites for 30 min at 25°C.
5. Incubate antigen in PBST + 4% IgG-free, protease-free BSA with agitation for 30–45 min at 25°C.
6. For atomic force microscopy (AFM) imaging, stop here with three successive washes for 3 min each in 500 µL of PBST. Otherwise, continue next steps for optical image acquisition with CCD camera.
7. The chips are then immersed in 500 µL of PBST + 1.5 µg/mL Cy3-labeled detection antibody for 30–45 min at 25°C.
8. Alternatively, biotinylated detection antibody is added to the reaction at 0.25 µg/mL and incubated with agitation for 30–45 min at 25°C.
9. The chips are then transferred into 500 µL of Cy3-streptavidin (Jackson ImmunoResearch Laboratories, West Grove, PA) at 1.5 µg/mL in PBST and incubated with agitation for 30–45 min at 25°C.
10. Three successive washes are carried out for 3 min each in 500 µL of PBST before image acquisition.

3.4. Detection

Nanoarrays that utilized fluorescent reporting molecules are visualized on optical microscope equipped with a Cy3 filter and a Texas Red filter from Chroma Technology (Rockingham, VT). Images are captured using a scientific grade CCD camera (Cohu, Inc., San Diego, CA). Immediately before visualization, chips are briefly rinsed with a few drops of double-distilled water and inverted onto a drop of *n*-propyl gallate/glycerol antifade solution (5% *n*-propylgallate, 70% glycerol, 25% 0.5 M Tris-HCl, pH 9.0, adjusted to a final pH 7.4). Fluorescent images were analyzed for spot size, intensity, and coefficient of variance with the Array Pro Analyzer software package (Media Cybernetics, Carlsbad, CA).

AFM images are recorded in tapping mode with a Dimension 3100 AFM from Digital Instruments (Santa Barbara, CA). The chips are washed 5 min in PBS + 1% Tween-80, 5 min in double-distilled water, and dried before final AFM imaging. To amplify the signal associated with antigen binding to the array, gold nanoparticle probes, which are heavily functionalized with detection antibodies to the antigen, are reacted with the nanoarray by soaking the array in a solution containing detection antibody coated gold nanoparticles (20 nM, 10 nM in 10 mM PBS) for 1 h.

4. Nanoarray for Enzymatic Assays

Enzymatic assays are extremely useful for clinical diagnostics, environmental monitoring, and drug discovery. The development of nanoarray for enzymatic assay significantly reduces the amount of biological and chemical reagents and samples and shortens the throughput time and improves the detection sensitivity.

Both CNT nanoelectrode arrays and nanochip arrays have been applied for these assays *(11,12)*. CNT nanoelectrode array-based EC transducers offer substantial improvements in the performance of amperometric enzyme electrodes. The greatly enhanced EC reactivity for numerous of hydrogen peroxide and NADH at CNT-modified electrodes make these nanomaterials extremely attractive for numerous oxidase- and dehydrogenase-based amperometric biosensors. There are over 200 dehydrogenases and 100 oxidases. Many of these enzymes catalyze specifically the reactions of clinically important analytes (e.g., glucose, lactate, cholesterol, amino acids, urate, pyruvate, glutamate, alcohol, hydroxybutyrate) to generate the electrochemically detectable products NADH and hydrogen peroxide. Similar sensitivity and stability improvements have been illustrated for EC biosensors based on other enzymes, including tyrosinase, peroxidase, organophosphorous hydrolase, or alkaline phosphatase *(11)*.

Nanochip array-based enzymatic assays for alcohol dehydrogenase, pyruvate kinase, and enolase were compared with conventional assays monitored in 96-well microtiter plates. All miniaturized reactions could be performed in maximum volumes of 6.3–8 nL and were read out with a conventional fluorescence microsystem equipped a scientific grade CCD camera. Enzymatic-based assays using nanoarrays are a promising low-volume, low-reagents, and low sample-consuming alternative to current methodology *(12)*.

5. Conclusion

This chapter summarized recent progress in the development of CNT nanoelectrode array for nucleic acid analysis and nanochip array for protein detection with improved miniaturization, higher sensitivity and rapid analysis. The CNT nanoelectrode array for nucleic acid analysis platform can be adapted for protein detection and enzymatic assay. Nanochip array for protein detection platform can also be adapted for nucleic acid analysis and enzymatic assay. Single copy detection within minutes for these novel devices is expected by integrating microfluidics and electric field control of probe–target hybridization. CNT nanoelectrode array-based EC sensors are more amenable to miniaturization. In addition, the continuous response of an electrode system allows for computer control, which makes the EC detection simpler, more reliable and less expensive than other analytical techniques.

Acknowledgments

We acknowledge Dr. M. Meyyappan, Alan Cassell, Wendy Fan, and Harry Partridge for encouragement and technical discussions during preparation of the manuscript.

References

1. Li, J., Ng, H. T., and Chen, H. (2005) Carbon nanotube and nanowires for biological sensing, in *Protein Nanotechnology, Protocols, Instrumentation and Applications,* (Vo-Dinh, T., ed.), Humana, Totowa, NJ, pp. 191–223.
2. Li, J., Ng, H. T., Cassell, A., et al. (2003) Carbon nanotube nanoelectrode array for ultrasensitive DNA detection. *Nano Lett.* **3,** 597–602.
3. Koehne, J., Chen, H., Li, J., et al. (2003) Ultrasensitive label-free electronic method for DNA analysis using carbon nanotube nanoelectrode array. *Nanotechnology* **14,** 1239–1245.
4. Piner, R. D., Zhu, J., Xu, F., Hong, S., and Mirkin, C. A. (1999) "Dip-Pen" nanolithography. *Science* **283,** 661.
5. Lee, K. -B., Park, S. J., Mirkin, C. A., Smith, J. C., and Mrksich, M. (2002) Protein nanoarrays generated by dip-pen nanolithography. *Science* **295,** 1702.
6. Demers, L. M., Ginger, D. S., Park, S. J., Li, Z., Chung, S. W., and Mirkin, C. (2002) Direct patterning of modified oligonucleotides on metals and insulators by dip-pen nanolithography. *Science* **296,** 1836.
7. Ginger, D. S., Zhang, H., and Mirkin, C. A. (2004) The evolution of Dip-Pen nanolithography. *Angew. Chem., Int. Ed.* **43,** 30.
8. Bruckbauer, A., Zhou, D., Kang, D. J., Korchev, Y. E., Abell, C., and Klenerman, D. (2004) An addressable antibody nanoarray produced on a nanostructured surface. *J. Am. Chem. Soc.* **126,** 6508–6509.
9. Lynch, M., Mosher, C., Huff, J., Nettikadan, S., Johnson, J., and Henderson, E. (2004) Functional protein nanoarrays for biomarker profiling. *Proteomics* **4,** 1695–1702.
10. Lee, K. B., Kim, E. -Y., Mirkin, C. A., and Wolinsky, S. M. (2004). The use of nanoarrays for highly sensitive and selective detection of human immunodeficiency Virus Type 1 in Plasma. *Nano Lett.* **4,** 1869–1872.
11. Wang, J. (2005) Carbon-nanotube based electrochemical biosensors: a review. *Electroanalysis* **17,** 7–14.
12. Dietrich, H. R., Knoll, J., van den Doel, L. R., et al. (2004) Nanoarrays: a method for performing enzymatic assays. *Anal. Chem.* **76,** 4112–4117.
13. Feynman, R. P. (1960) There's plenty of room at the bottom: an invitation to enter a new field of physics. Engineering Sci: http://www.zyvex.com/nanotech/feynman.html (accessed Feb 2, 2005).
14. Drexler, K. E. (1986) *Engines of Creation*. Anchor Press/Doubleday, New York.
15. Iijima, S. (1991) Helical microtubules of graphitic carbon. *Nature* **354,** 56–58.
16. Ferrari, M. (2005) Cancer nanotechnology: opportunities and challenges. *Nat. Rev. Cancer* **5,** 161–171.
17. Silva, G. A. (2004) Introduction to nanotechnology and its applications to medicine. *Surg. Neurol.* **61,** 216–220.
18. Stix, G. (2001) Little big science. *Scientific American* **285,** 32–37.
19. Fortina, P., Kricka, L. J., Surrey, S., and Grodzinski, P. (2005) Nanobiotechnology: the promise and reality of new approaches to molecular recognition. *Trends Biotechnol.* **23,** 168–173.

20. Rosi, N. L. and Mirkin, C. A. (2005) Nanostructures in biodiagnostics. *Chem. Rev.* **105,** 1547–1562.
21. Emerich, D. F. (2005) Nanomedicine—prospective therapeutic and diagnostic applications. *Expert. Opin. Biol. Ther.* **5,** 1–5.
22. Silva, G. A. (2005) Nanotechnology approaches for the regeneration and neuro-protection of the central nervous system. *Surg. Neurol.* **63,** 301–306.
23. Dresselhaus, M. S., Dresselhaus, G., and Eklund, P. C. (eds.) (1996) *Science of Fullerenes and Carbon Nanotubes.* Academic Press, New York.
24. Meyyappan, M. (ed.) (2004) *Carbon Nanotubes: Science and Applications.* CRC Press, Boca Raton, FL.
25. McCreery, R. L. (1991) Carbon electrodes: structural effects on electron transfer kinetics, in *Electroanalytical Chemistry*, vol. 1 (Bard, A. J., ed.), Marcel Dekker, New York, pp. 221–374.
26. Koehne, J., Li, J., Cassell, A. M., et al. (2004) The fabrication and electrochemical characterization of carbon nanotube nanoelectrode arrays. *J. Matr. Chem.* **14,** 676–684.
27. Bard, A. J. and Faulkner, L. R. (eds.) (2001) *Electrochemical Methods: Fundamentals and Applications, 2nd ed.* Wiley, New York.
28. Wightman, R. M. (1981) Microvoltammetric electrodes. *Anal. Chem.* **53,** 1125A–1134A.
29. Fan, F. R. F. and Bard, A. J. (1995) Electrochemical detection of single molecules. *Science* **267,** 871.
30. Menon, V. P. and Martin, C. R. (1995) Fabrication and evaluation of nanoelectrode ensemble. *Anal. Chem.* **67,** 1920.
31. Peterson, A. W., Heaton, R. J., and Georgiadis, R. M. (2001) The effect of surface probe density on DNA hybridization. *Nucleic Acids Res.* **29,** 5163–5168.
32. Peterson, A. W., Wolf, L. K., and Georgiadis, R. M. (2002) Hybridization of mis-matched or partially matched DNA at surfaces. *J. Am. Chem. Soc.* **124,** 14,601–14,607.
33. Dorris, D. R., Nguyen, A., Gieser, L., et al. (2003) Oligodeoxyribonucleotide probe accessibility on a three-dimensional DNA microarray surface and the effect of hybridization time on the accuracy of expression ratios. *BMC Biotechnol.* **3,** 1472–1483.
34. Yao, D., Kim, J., Yu, F., Nielsen, P. E., Sinner, E. -K., and Knoll, W. (2005) Surface density dependence of PCR amplicon hybridization on PNA/DNA probe layers. *Biophys. J.* **88,** 2745–2751.
35. Heaton, R. J., Peterson, A. W., and Georgiadis, R. M. (2001) Electrostatic surface plasmon resonance: direct electric field-induced hybridization and denaturation in monolayer nucleic acid films and label-free diacrimination of base mis-matches. *Proc. Natl. Acad. Sci. USA* **98,** 3701–3704.
36. Sosnowski, R. G., Tu, E., Butler, W. F., O'Connell, J. P., and Heller, M. J. (1997) Rapid determination of single base mismatch mutations in DNA hybrids by direct electric field control. *Proc. Natl. Acad. Sci. USA* **94,** 1119.
37. Popovich, N. D. and Thorp, H. H. (2002) New strategies for electrochemical nucleic acid detection. *Interface* **11,** 30.

38. Einstein, A. (1956) On the movement of small particles suspended in a stationary liquid demanded by the molecular kinetic theory of heat, in *Investigations on the Theory of the Brownian Movement* (Furth, R., and Cowper, A. D. eds.), Dover, New York, pp. 1–18.

39. Dresselhaus, M. S., Dresselhaus, G., and Eklund, P. C. (eds.) (1996) *Science of Fullerenes and Carbon Nanotubes*. Academic Press, New York.

40. Ebbesen, T. W. (ed.) (1996) *Carbon Nanotubes: Preparation and Properties.* CRC Press, Boca Raton, FL.

41. Saito, R., Dresselhaus, M. S., and Dresselhaus, G. (eds.) (1998) *Physical Properties of Carbon Nanotubes*. World Scientific, New York.

42. Tománek, D. and Enbody, R. (eds.) (2000) *Science and Application of Nanotubes.* Kluwer Academic, New York.

43. Collins, P. G., Arnold, M. S., and Avouris, P. (2001) Engineering carbon nanotubes and nanotube circuits using electrical breakdown. *Science* **292,** 706–709.

44. Tans, S. J., Verschueren, A. R. M., and Dekker, C. (1998) Room-temperature transistor based on a single carbon nanotube. *Natur*e **393,** 49–52.

45. Fuhrer, M. S., Nygard, J., Shih, L., et al. (2000) Crossed nanotube junctions. *Scienc*e **288,** 494–497.

46. Zhou, C. W., Kong, J., Yenilmez, E., and Dai, H. (2000) Modulated chemical doping of individual carbon nanotubes. *Science* **290,** 1552–1555.

47. Rueckes, T., Kim, K., Joselevich, E., Tseng, G. Y., Cheung, C. L., and Lieber, C. M. (2000) Carbon nanotube-based nonvolatile random access memory for molecular computing. *Science* **289,** 94–97.

48. Derycke, V., Martel, R., Appenzeller, J., and Avouris, Ph. (2001) Carbon Nanotube inter- and intramolecular logic gates. *Nano Lett.* **1,** 453–456.

49. Bachtold, A., Hadley, P., Nakanishi, T., and Dekker, C. (2001) Logic circuits with carbon nanotube transistors. *Science* **294,** 1317–1320.

50. Liu, X. L., Lee, C., Zhou, C. W., and Han, J. (2001) Carbon nanotube field-effect inverters. *Appl. Phys. Lett.* **79,** 3329–3331.

51. Rosenblatt, S., Yaish, Y., Park, J., Gore, J., Sazonova, V., and McEuen, P. L. (2002) High performance electrolyte gated carbon nanotube transistors. *Nano Lett.* **2,** 869–872.

52. Vigolo, B., Penicaud, A., Coulon, C., et al. (2000) Macroscopic fibers and ribbons of oriented carbon nanotubes. *Science* **290,** 1331–1334.

53. de Heer, W. A., Chatelain, A., and Ugarte, D. (1995) A carbon nanotube field-emission electron source. *Science* **270,** 1179–1180.

54. Rinzler, A. G., Hafner, J. H., Nikolaev, P., et al. (1995) Unraveling nanotubes: field emission from an atomic wire. *Science* **269,** 1550–1553.

55. Dai, H., Hafner, J. H., Rinzler, A. G., Colbert, D. T., and Smalley, R. E. (1996) Nanotubes as nanoprobes in scanning probe microscopy. *Nature* **384,** 147–150.

56. Wong, S., Joselevich, E., Woolley, A., Cheung, C., and Lieber, C. M. (1998) Covalently functionalized nanotubes as nanometer-sized probes in chemistry and biology. *Nature* **394,** 52–55.

57. Li, J., Cassell, A., and Dai, H. (1999) Carbon nanotubes as AFM tips: measuring DNA molecules at the liquid/solid interfaces. *Surf. Interface Anal.* **28,** 8–11.

58. Nguyen, C. V., Chao, K. J., Stevens, R. M. D., et al. (2001) Growth of carbon nanotubes by thermal and plasma chemical vapour deposition processes and applications in microscopy. *Nanotechnology* **12,** 363–367.

59. Liu, C. F., Fan, Y. Y., Liu, M., Cong, H. T., Chen, H. M., and Dresselhaus, M. S. (1999) Hydrogen storage in single-walled carbon nanotubes at room temperature. *Science* **286,** 1127–1129.

60. Che, G., Lakshmi, B. B., Fisher, E. R., and Martin, C. R. (1998) Carbon nanotubule membranes for electrochemical energy storage and production. *Nature* **393,** 346–349.

61. Kong, J., Franklin, N. R., Zhou, C. W., et al. (2000) Nanotube molecular wires as chemical sensors. *Science* **287,** 622–625.

62. Collins, P. G., Bradley, K., Ishigami, M., and Zettl, A. (2000) Extreme oxygen sensitivity of electronic properties of carbon nanotubes. *Science* **287,** 1801–1804.

63. Sumanasekera, G. U., Adu, C. K. W., Fang, S., and Eklund, P. C. (2000) Effects of gas adsorption and collisions on electrical transport in single-walled carbon nanotubes. *Phys. Rev. Lett.* **85,** 1096–1099.

64. Ng, H. T., Fang, A., Li, J., and Li, S. F. Y. (2001) Flexible carbon nanotube membrane sensory system: a generic platform. *J. Nanosci. Nanotech.* **1,** 375–379.

65. Li, J. and Ng, H. T. (2003) Carbon nanotube sensors, in *Encyclopedia of Nanoscience and Nanotechnology, vol. 1* (Nalwa, H. S., ed.), American Scientific Publishers, Santa Barbar, CA, pp. 591–601.

66. Li, J., Ye, Q., Cassell, A., et al. (2003) Bottom-up approach for carbon nanotube interconnects. *Appl. Phys. Lett.* **82,** 2491–2493.

67. Ren, Z. F., Huang, Z. P., Xu, J. W., et al. (1998) Synthesis of large arrays of well-aligned carbon nanotubes on glass. *Science* **282,** 1105–1107.

68. Delzeit, L., McAninch, I., Cruden, B. A., et al. (2002) Growth of multiwall carbon nanotubes in an inductively coupled plasma reactor. *J. Appl. Phys.* **91,** 6027–6033.

69. Cassell, A. M., Ye, Q., Cruden, B. A., et al. (2004) Combinatorial chips for optimizing the growth and integration of carbon nanofibre based devices. *Nanotechnology* **15,** 9–15.

70. Cruden, B. A., Cassell, A. M., Ye, Q., and Meyyappan, M. (2003) Reactor design considerations in the hot filament/direct current plasma synthesis of carbon nanofibers. *J. Appl. Phys.* **94,** 4070–4078.

71. Staros, J. V. (1982) N-hydroxysulfosuccinimide active esters—bis(N-hydroxysulfosuccinimide) esters of 2 dicarboxylic-acids are hydrophilic, membrane-impermeant, protein cross-linkers. *Biochemistry* **21,** 3950–3955.

72. Miki, Y., Swensen, J., Shattuck-Eidens, D., et al. (1994) A strong candidate for the breast and ovarian cancer susceptibility gene BRCA1. *Science* **266,** 66–71.

73. Staros, J. V. (1982) N-hydroxysulfosuccinimide active esters—bis(N-hydroxy-sulfosuccinimide) esters of 2 dicarboxylic-acids are hydrophilic, membrane-impermeant, protein cross-linkers. *Biochemistry* **21,** 3950–3955.

74. Popovich, N. D. and Thorp, H. H. (2002) New strategies for electrochemical nucleic acid detection. *Interface* **11,** 30.

75. Cheng, J., Sheldon, E. L., Wu, L., et al. (1998) Preparation and hybridization analysis of DNA/RNA from E. coli on microfabricated bioelectronic chips. *Nat. Biotechnol.* **16,** 541–546.

76. Edman, C. F., Raymond, D. E., Wu, D. J., et al. (1997) Electric field directed nucleic acid hybridization on microchips. *Nucleic Acids Res.* **25,** 4907–4914.

77. Sosnowski, R. G., Tu, E., Butler, W. F., O'Connell, J. P., and Heller, M. J. (1997) Rapid determination of single base mismatch mutations in DNA hybrids by direct electric field control. *Proc. Natl. Acad. Sci. USA* **94,** 1119.

78. Gilles, P. N., Wu, D. J., Foster, C. B., Dillon, P. J., and Chanock, S. J. (1999) Single nucleotide polymorphoric discrimination by an electronic dot blot assay on semiconductor microchip. *Nat. Biotechnol.* **17,** 365–370.

79. Bertilsson, L. and Liedberg, B. (1993) Infrared study of thiol monolayer assemblies on gold: preparation, characterization and functionalization of mixed monolayers. *Langmuir* **9,** 141–149.

80. Nuzzo, R. G. and Allara, D. L. (1983) Adsorption of bifunctional organic disulfides on gold surfaces. *J. Am. Chem. Soc.* **105,** 4481–4483.

81. Porter, M. D., Bright, T. B., Allara, D. L., and Chidsey, C. D. (1987) Spontaneously organized molecular assemblies. 4. Structural characterization of n-Alkyl Thiol monolayers on gold by optical ellipsometry, infrared spectroscopy, and electrochemistry. *J. Am. Chem. Soc.* **109,** 3559–3568.

82. Whitesides, G. M. and Bain C. D. (1988) Molecular-level control over surface order in self-assembled monolayer films of thiols on gold. *Science* **240,** 62–63.

83. Strong, L. and Whitesides, G. M. (1988) The structures of self-assembled monolayer films of organosulfur compounds adsorbed on gold single crystals: electron diffraction studies. *Langmuir* **4,** 546–558.

84. Dubois, L. H., Zegarski, B. R., and Nuzzo, R. G. (1987) Fundamental studies of the interactions of adsorbates on organic surfaces. *Proc. Natl. Acad. Sci. USA* **84,** 4739–4742.

85. Bain, C. D. and Whitesides, G. M. (1988) Correlation between wettability and structure in monolayers in alkanethiols adsorbed on gold. *J. Am. Chem. Soc.* **110,** 3665–3666.

86. Springer, A. L., Gall, A. S., Hughes, K. A., Kaiser, R. J., Li, G. S., and Lund, K. P. (2003) Salicylhydroxamic acid functionalized affinity membranes for specific immobilization of proteins and oligonucleotides. *J. Biomol. Tech.* **14,** 183–190.

87. Schaeferling, M., Stefan, S. S., Hubert, P. H., et al. (2002) Application of self-assembly techniques in the design of biocompatible protein microarray surfaces. *Electrophoresis* **23,** 3097–3105.

88. Kenseth, J. R., Harnisch, J. R., Vivian, W., Jones, V. W., and Porter, M. D. (2001) Investigation of approaches for the fabrication of protein patterns by scanning probe lithography. *Langmuir* **17,** 4105–4112.
89. Nakano, K., Taira, H., Maeda, M., and Takagi, M. (1993) New bifunctional dialkyl disulfide reagent for the fabrication of a gold surface with the bioaffinity ligand. *Anal. Sci.* **9,** 133–136.

Index